# Climate Change and Ecosystems

The global population is projected to increase by 3.3 billion from 6.7 billion in 2008 to 10 billion in 2100. As a result, soil degradation and desertification are growing due to the increasing demand for food, feed, fiber, and fuel on finite soil resources. The problem of global food insecurity may be further worsened by the threat of global warming. Climate change is showing its impacts in terms of increasing temperatures, variable rainfall, and an increase in climate-related extremes such as floods, droughts, cyclones, sea-level rise, salinity, and soil erosion. The agriculture sector is the most sensitive to climate change because the climate of a region/country determines the nature and characteristics of vegetation and crops. Increase in the mean seasonal temperature and decrease in effective precipitation can reduce the duration of many crops, may lead to outbreaks of pests and diseases, and hence reduce final yield ultimately affecting the food security of the country. Despite the positive impact of $CO_2$ fertilization, the net productivity may decrease because of an increase in respiration rate, drought stress, and nutrient deficiency. For example, for every 75 ppm increase in $CO_2$ concentration, rice yields will increase by 0.5 t/ha, but the yield will decrease by 0.6 t/ha for every 1°C increase in temperature. The global agricultural productivity is expected to decrease from 3% to 16% by 2080. The estimated decrease in agricultural productivity in the developing countries is 10%–25% in the 2080s, where average air temperature is already near or above crop tolerance levels. This book is intended to serve as a stimulating collection that will contribute to debate and reflection on the sustainable future of agriculture and food production in the face of global change.

**Features:**

- This book brings together a multidimensional group of international scholars exploring the ethical dimensions of climate change and ecosystem.
- New strategies have been pointed out in this book for better sustainable development.
- This book has been designed to provide a good overview of major challenges facing policymakers, researchers, and ultimately humankind in dealing with climate change.
- This book summarizes the diverse features of vulnerability, adaptation, and amelioration of climate change in respect to plants, crops, soil, and microbes for the sustainability of the agricultural sector, and, ultimately, food security for the future.
- This book provides a state-of-the-art description of the physiological, biochemical, and molecular status of the understanding of abiotic stress in plants.

# Footprints of Climate Variability on Plant Diversity

*Series Editor:*
Shah Fahad

Improvement of Plant Production in the Era of Climate Change
*Shah Fahad, Muhammad Adnan, and Shah Saud*

Climate Change and Ecosystems: Challenges to Sustainable Development
*Shah Fahad, Muhammad Adnan, Shah Saud, and Lixiao Nie*

Climate Change and Plants: Biodiversity, Growth and Interactions
*Shah Fahad, Osman Sönmez, Shah Saud, Depeng Wang, Chao Wu, Muhammad Adnan, and Veysel Turan*

Developing Climate Resilient Crops: Improving Global Food Security and Safety
*Shah Fahad, Osman Sönmez, Shah Saud, Depeng Wang, Chao Wu, Muhammad Adnan, and Veysel Turan*

Sustainable Soil and Land Management and Climate Change
*Shah Fahad, Osman Sönmez, Veysel Turan, Muhammad Adnan, Shah Saud, Chao Wu, and Depeng Wang*

Plant Growth Regulators for Climate-Smart Agriculture
*Shah Fahad, Osman Sönmez, Veysel Turan, Muhammad Adnan, Shah Saud, Chao Wu, and Depeng Wang*

Engineering Tolerance in Crop Plants Against Abiotic Stress
*Shah Fahad, Osman Sönmez, Shah Saud, Depeng Wang, Chao Wu, Muhammad Adnan, Muhammad Arif, and Amanullah*

# Climate Change and Ecosystems
## Challenges to Sustainable Development

Edited by
Shah Fahad
Hainan University, China
University of Haripur, Pakistan
Muhammad Adnan
University of Swabi, Pakistan
Shah Saud
Northeast Agricultural University, China
Lixiao Nie
Hainan University, China

CRC Press
Taylor & Francis Group
Boca Raton London New York

CRC Press is an imprint of the
Taylor & Francis Group, an **informa** business

First edition published 2023
by CRC Press
6000 Broken Sound Parkway NW, Suite 300, Boca Raton, FL 33487-2742

and by CRC Press
2 Park Square, Milton Park, Abingdon, Oxon, OX14 4RN

*CRC Press is an imprint of Taylor & Francis Group, LLC*



*Library of Congress Cataloging-in-Publication Data*
Names: Fahad, Shah (Assistant professor of agriculture), editor. |
Adnan, Muhammad (Lecturer in agriculture), editor. | Saud, Shah, editor. | Nie, Lixiao, editor.
Title: Climate change and ecosystems : challenges to sustainable development / edited by Shah Fahad, University of Haripur, Pakistan, Muhammad Adnan, University of Haripur, Pakistan, Shah Saud, Northeast Agricultural University, China, Lixiao Nie, Hainan University, China.
Description: First edition. | Boca Raton : CRC Press, 2022. |
Series: Footprints of climate variability on plant diversity |
Includes bibliographical references and index.
Identifiers: LCCN 2022001603 (print) | LCCN 2022001604 (ebook) |
ISBN 9781032260686 (hardback) | ISBN 9781032260693 (paperback) | ISBN 9781003286400 (ebook)
Subjects: LCSH: Crops and climate. | Climatic changes. | Sustainable development. |
Food supply–Environmental aspects.
Classification: LCC S600.7.C54 C625 2022 (print) | LCC S600.7.C54 (ebook) |
DDC 338.1/4–dc23/eng/20220126
LC record available at https://lccn.loc.gov/2022001603
LC ebook record available at https://lccn.loc.gov/2022001604

ISBN: 9781032260686 (hbk)
ISBN: 9781032260693 (pbk)
ISBN: 9781003286400 (ebk)

DOI: 10.1201/9781003286400

Typeset in Times
by codeMantra

# Contents

# *Acknowledgments*

Words are bound and knowledge is limited to praise ALLAH, the instant and sustaining source of all mercy and kindness, and the sustainer of the world. My greatest and ultimate gratitude is due to ALLAH (Subhanahu wa Taqadus). I thank ALLAH with all my humility for everything that I can think of. His generous blessing and exaltation succeeded my thoughts and thrived my ambition to have the cherish fruit of my modest efforts in the form of this piece of literature from the blooming spring of blossoming knowledge. May ALLAH forgive my failings and weaknesses, strengthen and enliven my faith in HIM, and endow me with knowledge and wisdom. All praises and respects are for **Holy Prophet Muhammad** *Salle Allah Alleh Wassalam*, the greatest educator, the everlasting source of guidance, and knowledge for humanity. He taught the principles of morality and eternal values, and enabled us to recognize our Creator. I have a deep sense of obligation to my parents, brothers, sisters, and son. Their unconditional love, care, and confidence in my abilities helped me achieve this milestone in my life. For this and much more, I am forever in their debt. It is to them that I dedicate this book. In this arduous time, I also appreciate the patience and serenity of my wife, who brought joy to my life in so many different ways. It is indeed on account of her affections and prayers that I was able to achieve something in my life.

**Shah Fahad**

# Editors

**Dr. Shah Fahad** is a Professor in the Hainan Key Laboratory for Sustainable Utilization of Tropical Bioresource, College of Tropical Crops, Hainan University, Haikou, China. He obtained his Ph.D. in Agronomy from Huazhong Agriculture University, China, in 2015. After doing his postdoctoral research in Agronomy at the Huazhong Agriculture University (2015–2017), he accepted the position of Assistant Professor at the University of Haripur. He has published over 332 peer-reviewed papers (impact factor 1050.18) with more than 300 research and 32 review articles, on important aspects of climate change, plant physiology and breeding, plant nutrition, plant stress responses and tolerance mechanisms, and exogenous chemical priming-induced abiotic stress tolerance. He has also contributed 58 book chapters to various book editions published by Springer, Wiley-Blackwell, and Elsevier. He has edited 17 book volumes, including this one, published by CRC Press, Springer, and Intech Open. He won the Young Rice International Scientist Award, Distinguished Scholar Award, Young Investigator Award, Dr. Shah Jahan Agriculture Award (PAS) in 2014, 2015, 2019, and 2021, respectively. He won 15 projects from international and national donor agencies. Dr. Shah Fahad has been figured twice among the top 2% scientists in a global list compiled by the Stanford University, USA. Dr. Shah Fahad is also included in the distinguished list of 2021 highly cited researchers announced by the Clarivate/Web of Sciences. He has been working on a wide range of topics, including climate change, greenhouse emission gasses, abiotic stresses tolerance, roles of phytohormones and their interactions in abiotic stress responses, heavy metals, and regulation of nutrient transport processes.

**Dr. Muhammad Adnan** is a lecturer in the Department of Agriculture at the University of Swabi, Pakistan. He has completed his Ph.D. (Soil Fertility and Microbiology) in the Department of Soil and Environmental Sciences at the University of Agriculture, Peshawar, Pakistan, and the Department of Plant, Soil and Microbial Sciences, Michigan State University, USA. He has also received his M.Sc. and B.Sc. (Hons.) from the Department of Soil and Environmental Sciences at the University of Agriculture, Peshawar, Pakistan.

**Dr. Shah Saud** is currently working as an Assistant Professor in the College of Life Science, Linyi University, Linyi, China. He received his Ph.D. in Turf Grasses (Horticulture) from Northeast Agricultural University, Harbin, China. He is currently working as a Post Doctorate researcher in the Department of Horticulture, Northeast Agricultural University, and Harbin, China. Dr. Shah Saud has published over 160 research publications in peer-reviewed journals. He has edited 5 books and written 35 book chapters on important aspects of plant physiology, plant stress responses, and environmental problems in relation to agricultural plants. According to Scopus®, Dr. Shah Saud's publications have received roughly 3000 citations with an h-index of 34.

**Dr. Lixiao Nie** is a Professor and Ph.D. supervisor in Crop Cultivation and Farming System at College of Tropical Crops, Hainan University, Haikou, China. He received a Ph.D. degree in Crop Cultivation and Farming System from Huazhong Agricultural University, China in 2008. In 2005–2007, He conducted his Ph.D. thesis research at the Philippines-based International Rice Research Institute under the supervision of Drs. Shaobing Peng and Bas Bouman. He then joined Huazhong Agricultural University and worked as a Crop Physiologist from 2008 to 2019. In 2019, he moved to Hainan University. His research mainly focuses on crop simplified and green planting and cultivation. He teaches courses related to plant physiology and crop cultivation, and has published more than 90 papers/chapters in international journals and books. His h-index and RG score are 28 and 35.83, respectively. He served as Editorial Board Member for international journals such as *Field Crops Research* (2018–now) and *Scientific Reports* (2015–now), and Deputy Editor in chief for a Chinese national journal (2020–now). He received "leading talents" and "one thousand talents project" from Hainan Province, China in 2020. In recent years, he has presided over four NSFC projects and two national key R&D projects.

# Contributors

**Mohamed Shebl Abd Elfattah**
Faculty of Agriculture, Department of Plant Protection
Suez Canal University
Ismailia, Egypt

**Abdullah**
Department of Plant Sciences
Quaid-i-Azam University
Islamabad, Pakistan

**Muhammad Adnan**
Department of Agriculture
University of Swabi
Swabi, Khyber Pakhtunkhwa, Pakistan

**Ashfaq Ahmad**
US-Pakistan Centre for Advanced Studies in Agriculture and Food Security
University of Agriculture
Faisalabad, Pakistan

**Mushtaq Ahmad**
Department of Plant Sciences
Quaid-i-Azam University
Islamabad, Pakistan

**Zeeshan Ahmad**
Department of Plant Sciences
Quaid-i-Azam University
Islamabad, Pakistan

**Rida Akram**
Department of Environmental Sciences
COMSATS University Islamabad
Vehari Campus, Islamabad, Pakistan

**Hasnain Alam**
Department of Biological Sciences
International Islamic University
Islamabad, Pakistan
and
Department of Biology, College of Sciences
United Arab Emirates University
Al Ain, United Arab Emirates

**Haifa Abdulaziz S. Alhaithloul**
Department of Biology, College of Science
Jouf University
Sakaka, Kingdom of Saudi Arabia

**Kawsar Ali**
Department of Agriculture
Abdul Wali Khan University
Mardan, Pakistan

**Musaddiq Ali**
Department of Environmental Sciences
COMSATS University Islamabad
Vehari Campus, Islamabad, Pakistan

**Shafaqat Ali**
Department of Environmental Sciences and Engineering
Government College University Faisalabad
Faisalabad, Pakistan
and
Department of Biological Sciences and Technology
China Medical University (CMU)
Taichung City, Taiwan

**Abdel Rahman Altawaha**
Department of Biological Sciences
Al Hussein Bin Talal University
Ma'an, Jordan

**Abdel Razzaq Al-Tawaha**
Faculty of Agriculture, Department of Crop Science
Universiti Putra Malaysia
Serdang, Selangor, Malaysia

**Khizer Amanet**
Department of Environmental Sciences
COMSATS University Islamabad
Vehari Campus, Islamabad, Pakistan

**Amanullah**
Department of Agronomy
The University of Agriculture
Peshawar, Pakistan

**Muhammad Arif**
Department of Agronomy
The University of Agriculture
Peshawar, Pakistan

**Hossam Eldean A. Awad**
Faculty of Science,
Department of Botany and Microbiology
Cairo University
Giza, Egypt

**Gohar Ayub**
Department of Horticulture
The University of Agriculture
Peshawar, Pakistan

**Hamida Bibi**
Department of Soil and Environmental Sciences
The University of Agriculture
Peshawar, Pakistan

**Amr A. El Kelish**
Faculty of Science, Department of Botany
Suez Canal University
Ismailia, Egypt

**Ayman El Sabagh**
Faculty of Agriculture, Department of Agronomy
Kafrelsheikh University
Kafr El Sheikh Governorate, Egypt

**Shah Fahad**
Hainan Key Laboratory for Sustainable Utilization of Tropical Bioresource, College of Tropical Crops
Hainan University
Hainan, China
and
Department of Agronomy
The University of Haripur
Haripur, Khyber Pakhtunkhwa, Pakistan

**Hafiz Umar Farid**
Department of Agricultural Engineering
Bahauddin Zakariya University
Multan, Pakistan

**Maham Fatima**
Department of Environmental Sciences
COMSATS University Islamabad
Vehari Campus, Islamabad, Pakistan

**Bibi Haleema**
Agricultural Research Institute
Tarnab, Peshawar, Pakistan

**Asmaa A. Halema**
Faculty of Agriculture, Department of Genetics
Cairo University
Giza, Egypt

**Ahmed Ragab Henawy**
Faculty of Agriculture, Department of Agriculture Microbiology
Cairo University
Giza, Egypt

**Xunbing Huang**
Institute of Plant protection
Chinese of Academy of Agricultural Sciences
Beijing, China

**Sajjad Hussain**
Department of Environmental Sciences
COMSATS University Islamabad
Vehari Campus, Islamabad, Pakistan

**Marian Hýbl**
Faculty of AgriSciences, Department of Zoology, Fishering, Hydrobiology and Apidology
Mendel University in Brno
Brno, Czech Republic

**Muhammad Ilyas**
Department of Botany
University of Swabi
Swabi, Khyber Pakhtunkhwa, Pakistan

**Imran**
Department of Agronomy
The University of Agriculture
Peshawar, Pakistan

**Javed Iqbal**
Department of Environmental Sciences
COMSATS University Islamabad
Vehari Campus, Islamabad, Pakistan

**Fazal Jalal**
College of Plant Science and Technology
Huazhong Agricultural University
Wuhan, China
and
Department of Agriculture
Abdul Wali Khan University
Mardan, Pakistan

**Talha Jan**
Department of Agronomy
The University of Agriculture
Peshawar, Pakistan

**Shah Khalid**
Department of Agronomy
The University of Agriculture
Peshawar, Pakistan

**Abdul Saboor Khan**
National Key Laboratory of Crop Genetic Improvement
Huazhong Agricultural University
Wuhan, China

**Bushra Khan**
Department of Environmental Sciences
University of Peshawar
Peshawar, Pakistan

**Hamdan Ali Khan**
Department of Agronomy
The University of Agriculture
Peshawar, Pakistan

**Nazish Huma Khan**
Department of Environmental Sciences
University of Swabi
Swabi, Khyber Pakhtunkhwa, Pakistan

**Shujaul Mulk Khan**
Department of Plant Sciences
Quaid-i-Azam University
Islamabad, Pakistan

**Mohsin Khurshid**
Department of Microbiology
Government College University Faisalabad
Faisalabad, Pakistan

**Denisa Kovářová**
Faculty of AgriSciences, Department of Crop Science, Breeding and Plant Medicine
Mendel University in Brno
Brno, Czech Republic

**Ekaterina Kozuharova**
Faculty of Pharmacy, Department of Pharmacognosy
Medical University of Sofia
Sofia, Bulgaria

**Tariq Mahmood**
Department of Plant Sciences
Quaid-i-Azam University
Islamabad, Pakistan

**Sana Malik**
Department of Bioinformatics & Biotechnology
Government College University Faisalabad
Faisalabad, Pakistan

**Irfan Manzoor**
Department of Bioinformatics & Biotechnology
Government College University Faisalabad
Faisalabad, Pakistan

**Muhammad Aamer Mehmood**
Department of Bioinformatics & Biotechnology
Government College University Faisalabad
Faisalabad, Pakistan

**Ishaq Ahmad Mian**
Department of Soil Environmental Science
The University of Agriculture
Peshawar, Pakistan

**Petr Mráz**
Faculty of Agriculture, Department of Genetics and Special Plant Production
University of South Bohemia in České Budějovice
České Budějovice, Czech Republic

**Muhammad Mubeen**
Department of Environmental Sciences
COMSATS University Islamabad
Vehari Campus, Islamabad, Pakistan

**Sher Muhammad**
Department of Bioinformatics & Biotechnology
Government College University Faisalabad
Faisalabad, Pakistan

**Sahar Mumtaz**
Division of Science and Technology, Department of Botany
University of Education
Lahore, Pakistan

**Fazal Munsif**
Department of Agronomy
The University of Agriculture
Peshawar, Pakistan

**Nadia**
Department of Agronomy
The University of Agriculture
Peshawar, Pakistan

**Wajid Nasim**
Department of Environmental Sciences
COMSATS University Islamabad
Vehari Campus, Islamabad, Pakistan

**Moona Nazish**
Department of Plant Sciences
Quaid-i-Azam University
Islamabad, Pakistan

**Brajendra Parmar**
ICAR-Indian Institute of Rice Research, Soil Science
Hyderabad, India

**Ali M. Qaisi**
Department of Pharmaceutical Sciences, School of Pharmacy
University of Jordan
Amman, Jordan

**Xinghu Qin**
University of St Andrews
St Andrews, United Kingdom

**Rafiullah**
Department of Agriculture
University of Swabi
Swabi, Khyber Pakhtunkhwa, Pakistan

**Fahd Rasul**
Department of Agronomy
University of Agriculture
Faisalabad, Pakistan

**Abdur Rauf**
Department of Chemistry
University of Swabi
Swabi, Khyber Pakhtunkhwa, Pakistan

**Muhammad Riaz**
Department of Environmental Sciences and Engineering
Government College University Faisalabad
Faisalabad, Pakistan

**Tooba Saeed**
National Centre of Excellence in Physical Chemistry
University of Peshawar
Peshawar, Pakistan

**Saifullah**
Department of Plant Sciences
Quaid-i-Azam University
Islamabad, Pakistan

**Muhammad Sajid**
Department of Horticulture
The University of Agriculture
Peshawar, Pakistan

**Muhammad Hamzah Saleem**
College of Plant Science and Technology
Huazhong Agricultural University
Wuhan, China

**Shah Saud**
College of Life Science
Linyi University
Linyi, Shandong, China

**Azizullah Shah**
Pakistan Agriculture Research Council
Islamabad, Pakistan

**Hamayun Shaheen**
Department of Botany
University of Azad Jammu and Kashmir
Muzaffarabad, Pakistan

**Muhammad Shahid**
Department of Bioinformatics & Biotechnology
Government College University Faisalabad
Faisalabad, Pakistan

**Munazzam Jawad Shahid**
Department of Environmental Sciences and Engineering
Government College University Faisalabad
Faisalabad, Pakistan

**Mona H. Soliman**
Faculty of Science, Department of Biology
Taibah University
Al-Sharm, Yanbu El-Bahr, Kingdom of Saudi Arabia
and
Faculty of Science, Department of Botany and Microbiology
Cairo University
Giza, Egypt

**Shazia Sultana**
Department of Plant Sciences
Quaid-i-Azam University
Islamabad, Pakistan

**Abdel Rahman Mohammad Tawaha**
Department of Biological Sciences
Al Hussein Bin Talal University
Ma'an, Jordan

**Laila Trioui**
Faculty of Sciences Ain Chock, Laboratory of Microbiology, Pharmacology, Biotechnology and Environment
University Hassan II
Casablanca, Morocco

**Zahoor ul Haq**
Department of Plant Sciences
Quaid-i-Azam University
Islamabad, Pakistan

**Anila Ulfat**
Department of Botany
University of Poonch
Rawlakot, Pakistan

**Amjad-ur-Rahman**
Department of Botany
University of Swabi
Swabi, Khyber Pakhtunkhwa, Pakistan

**Hristo Valchev**
Faculty of Pharmacy, Department of Pharmacognosy
Medical University of Sofia
Sofia, Bulgaria

**Fazli Wahid**
Department of Agriculture
University of Swabi
Swabi, Khyber Pakhtunkhwa, Pakistan

**Huihui Wu**
Institute of Plant Protection
Chinese of Academy of Agricultural Sciences
Beijing, China

**Muhammad Zafar**
Department of Plant Sciences
Quaid-i-Azam University
Islamabad, Pakistan

**Sajjad Zaheer**
Department of Agronomy
The University of Agriculture
Peshawar, Pakistan

**Muhammad Zamin**
Department of Agriculture
University of Swabi
Swabi, Khyber Pakhtunkhwa, Pakistan

**Weizhong Zhang**
Institute of Plant Protection
Chinese of Academy of Agricultural Sciences
Beijing, China

**Zehua Zhang**
Institute of Plant Protection
Chinese of Academy of Agricultural Sciences
Beijing, China

# 1

# *Microbial Flora of Marble Waste-Polluted Environment in the Phylogenetic Perspectives*

**Zeeshan Ahmad and Shujaul Mulk Khan**
*Quaid-i-Azam University*

**CONTENTS**



## 1.1 Introduction

Microbial flora is one of the most diverse and abundant forms of life on the earth (Dawson et al. 2017; Lu et al. 2013; Reshi et al. 2015). Microbes are present in every type of natural habitat, i.e., water, soil, air, hot springs, rocks, deep in the ocean and up to high in the atmosphere (Brock 2012; Sterflinger et al. 2012). According to a study, 10 million microorganisms were present in 1 mm of ocean water and 40 million in 1 g of soil (Richardson and Jones 1958). Microbes make up 10% of human dry weight. Microbes are the first living organisms on the earth's surface. Majority of them are beneficial. They keep natural ecosystems clean by removing toxins from soil, water, and degraded organic remnants of animals and plants. Microbes are crucial in the nutrient recycling of an ecosystem as they act as decomposers. They are also used as an indicator of landmines and build radioactive metals inert. Microorganisms help in the digestion process and prevent humans from harmful effects. In short, one can say that life could not be possible without microbes. Biogeographic patterns of an ecosystem provide evidences of

DOI: 10.1201/9781003286400-1

occurrence, distribution pattern, and variation in communities' structure of microorganisms (Feinstein and Blackwood 2012; He et al. 2010). Microbe's study is mostly concerned with the distribution structure of one biome, but little attention has been given to controlling and mediating transfers along landscape variables including aquatic and terrestrial ecosystems that resulting various biogeographic microbial diversity (Adnan et al. 2016; Kallmeyer et al. 2012; Whitman et al. 1998).

The soil is harboring diverse flora of microbes due to temporally and spatially heterogeneous type of nature in terms of microorganism habitat (Quince et al. 2008; Whitman et al. 1998). More than 33,000 archaeal and bacterial taxon would be identified in a single sample of soil through phylochip (Berendsen et al. 2012; Overmann and Lepleux 2016; Wolf 2014). These microbial floras play important roles in a number of functions of different ecosystems. It helps in sustaining plant growth, carbon storage, recycling of nutrients, soil structure maintenance, anthropogenic pressure, and many other services to humans (Young and Crawford 2004; Young et al. 2008). The decomposable products of plants exhibited a wide and more variable ratio of phosphorus, nitrogen, and carbon for microorganisms (Sistla et al. 2012; Sistla and Schimel 2012). There is a wide range of microbial diversity in varying soil, moisture, pH, and nitrogen variability (Griffiths et al. 2012; Griffiths et al. 2011; Hawkes et al. 2011; Landesman and Dighton 2010; Rosling et al. 2011). They vary in different plant communities (Kowalchuk et al. 2002) due to human disturbance, i.e., pollution and alteration of natural ecosystems (Cevallos-Cevallos et al. 2009; Flores et al. 2011; McIntyre et al. 2008).

Water is one of the major habitat types, covering approximately 70% of the earth's surface. It provides the largest habitat space for uncountable numbers of living organisms, especially, microbes. It generally experiences highly fluctuating sort of environmental conditions. Such types of conditions not only differ from one aquatic micro-habitat to other but also vary in macro environments. Different habitat types could be recognized within the same ecosystem characterized by more than one microbial community. This variability contributes to well-organized microbial diversity in the water ecosystem. Microbes of water ecosystem not only thrive on the surface of water but also flourish in the abyssal depths of oceans in specific niches from the black smoker of hot thermal vent at sea floor to the blue water of coral reefs (Azam et al. 1983; Das et al. 2006; Fierer et al. 2007). Aquatic ecosystems are regarded as key habitat types hosting diverse groups of microorganisms as compared to other environmental types (Sibille et al. 1998). In such a habitat, microbial mats and biofilms develop that are believed to be responsible for primary production in the system (Lee et al. 2012; Perri et al. 2012). Drier habitats have relatively less visible signs of life and such primary productivity (Cary et al. 2010).

### 1.1.1 Isolation and Characterization of Microbes

The isolation and identification of microbes is one of the most important steps in studying microbes. There are numerous practical solicitations for identification of unknown microbes. The method used in the diagnostic laboratories is mostly based on phenotypic characteristics of microbes. Preliminary procedures include simple tests, such as isolation in pure forms, staining reactions, morphological features, culture characteristics, metabolism, biochemical tests, and growth on different types of culture media.

Molecular techniques are the major reliable tools for identification of microorganisms. It was one of the most effective and fastest technologies for identification of microbes in various types of habitats. Organisms are identified from some unique part of their RNA or DNA providing definite information regarding its diversity. Molecular techniques have many advantages as compared to traditional methods which were not much appropriate for identification of microbial diversity. Fungi and bacteria are generally identified through 18S rRNA and 16S rRNA sequencing techniques. The rRNA (rDNA) is the most conserved region in all types of microbe cells due to the hyper-variable region. It is extensively used to determine phylogeny (evolutionary relationship), a taxonomy that estimates the rate of divergence in species. Primers are designed to bind with the conserved region to amplify the variable region. These types of sequences are extensively used for identification of a large number of species. The sequences from tens of thousands of isolates are available on NCBI (National Center for Biotechnology Information) weblink. It also provides the search algorithms to compare new sequences with those already available on the database.

### 1.1.2 Marble Waste-Polluted System

Marble is a type of metamorphic rock generally composed of calcite or dolomite (Török et al. 2011). Most of the marbles are formed at boundaries of the convergent plate where the maximum area of earth crust is bare to regional metamorphism. Some of the marble stones are produced/formed by contact metamorphism when hot magma passionate/heated adjacent dolomite/limestone. It usually contains chemical compounds such as calcium oxide (CaO), ferric oxide ($Fe_2O_3$,) phosphorus pentoxide ($P_2O_5$), aluminum oxide ($Al_2O_3$), magnesium oxide (MgO), silicon dioxide ($SiO_2$), titanium dioxide ($TiO_2$), and sodium oxide ($Na_2O$) (Knoche et al. 1995). It has distinctive properties as these are used for architecture, sculpture, agriculture, cosmetics, paper, paint, and pharmaceutical purposes (André et al. 2014; Karaşahin and Terzi 2007). It plays a significant role in the socio-economic situation of people. But it also wastes 70% of minerals resources throughout processing, mining, and polishing dispensation (Gazi et al. 2012). The marble waste produced during the quarrying process is dumped in roads, pasture lands/agriculture fields, empty pits, and river beds, which ultimately leads to environmental pollution (Akbulut and Gürer 2007; Aukour and Al-Qinna 2008; Mendoza et al. 2014). Mining and processing of marbles result in environmental deterioration in the form of soil, air, and water pollution. It directly or indirectly disturbs physical/chemical nature of the environment and the composition, diversity, and abundance of plants and animals (Figures 1.1 and 1.2) (Yu et al. 2005).

Water, one of the most important natural resource, is used in processing marble stone. During marble processing, i.e., cutting, cleaning, washing, and polishing, a great quantity of water is consumed which results in a huge quantity of wastewater. 50–150 $m^3$ of water per day is required for an average size factory. Majority of the marble industries are situated near the river sides as a result of which waste from these units goes directly into rivers and streams and pollutes them. These pollutants affect not only aquatic flora and fauna but agricultural fields and crop production as well.

There is a rapid change in the structure and function of natural ecosystems due to industrialization. Some of such problems are very slow, but few others are quite fast (Deepranjan et al. 2021; Haider et al. 2021; Huang Li et al.2021; Ikram et al. 2021; Jabborova et al. 2021; Khadim et al. 2021a, b; Manzer et al. 2021; Muzammal et al. 2021; Abdul et al. 2021a, b; Ashfaq et al. 2021; Amjad et al. 2021; Atif et al. 2021; Athar et al. 2021; Adnan et al. 2018a, b; Adnan et al. 2019; Akram et al. 2018a, b; Aziz et al. 2017a, b; Chang et al. 2021; Chen et al. 2021;Emre et al. 2021;Habib et al. 2017; Hafiz et al. 2016; Hafiz et al. 2019; Ghulam et al. 2021; Guofu et al. 2021; Hafeez et al. 2021; Khan et al. 2021; Kamaran et al. 2017;

**FIGURE 1.1** A glimpse of marble waste-polluted ecosystem at Khyber Pakhtunkhwa, Pakistan.

**FIGURE 1.2** River water becomes white/polluted due to direct discharge of marble wastewater effluents.

Muhmmad et al. 2019; Safi et al. 2021; Sajjad et al. 2019; Saud et al. 2013; Saud et al. 2014; Saud et al. 2017; Saud et al. 2016; Shah et al. 2013; Saud et al. 2020; Qamar et al. 2017; Hamza et al. 2021; Irfan et al. 2021; Wajid et al. 2017; Yang et al. 2017; Zahida et al. 2017; Depeng et al. 2018; Hussain et al. 2020; Hafiz et al. 2020a, b; Shafi et al. 2020; Wahid et al. 2020; Subhan et al. 2020; Zafar-ul-Hye et al. 2020a, b; Zafar-ul-Hye et al. 2021; Adnan et al. 2020; Ilyas et al. 2020; Saleem et al. 2020a, b, c; Rehman et al. 2020; Farhat et al. 2020; Wu et al. 2020; Mubeen et al. 2020; Farhana et al. 2020; Jan et al. 2019; Wu et al. 2019; Ahmad et al. 2019; Baseer et al. 2019; Hafiz et al. 2018; Tariq et al. 2018). Such alteration of the environment may be stern for different organisms that cannot adapt themselves to such environmental changes (Davis and Shaw 2001; Fahad and Bano 2012; Fahad et al. 2017; Fahad et al. 2013; Fahad et al. 2014a, b; Fahad et al. 2016a, b, c, d; Fahad et al. 2015a, b; Fahad et al. 2018a, b; Fahad et al. 2019a, b; Fahad et al. 2020; Fahad et al. 2021a, b, c, d, e, f; Hesham and Fahad 2020. Iqra et al. 2020; Akbar et al. 2020; Mahar et al. 2020; Noor et al. 2020; Bayram et al. 2020; Amanullah, Fahad S 2018a, b; Amanullah and Fahad S 2017; Amanullah et al. 2020; Amanullah et al. 2021; Rashid et al. 2020; Arif et al. 2020; Amir et al. 2020; Saman et al. 2020; Muhammad Tahir et al. 2020; Md Jakirand Allah 2020; Mahmood et al. 2021; Farah et al. 2020; Sadam et al. 2020; Unsar et al. 2020; Fazli et al. 2020; Md. Enamul et al. 2020; Gopakumar et al. 2020; Zia-ur-Rehman 2020; Ayman et al. 2020; Al-Wabel et al. 2020a, b; Senol 2020; Amjad et al. 2020; Ibrar et al. 2020; Sajid et al. 2020; Muhammad et al. 2021; Sidra et al. 2021; Zahir et al. 2021; Sahrish et al. 2022).

### 1.1.3 Introduction to Study Area

The Buner district is located in the Khyber Pakhtunkhwa province with an area of 1865 $km^2$ (720 square miles). This small mountainous district can further be divided into seven subdivisions. Eelam and Mora ranges of mountains separate it from the Swat district. A scenic route to Swat, i.e., Ambela Pass, loomed easily from the Mardan district via Rustam at a distance of 50 km. Yousafzai, Mander, and Syed are the major tribes of the district. Whole population of the area are Muslims (Sunni sect of Islam) and wear Shalwar, Qamees, and Chaddar. Majority of the people are living in joint family system with male-dominant society where females have a negligible role in decision-making. Their foods include fried fish, meat dishes, rice, different kinds of vegetables, sweets, beaten curd, wheat and corn bread, etc. They grandly celebrate both Eid-ul-fitar and Eid-ul-azha. Children look forward to receiving eid money (edi) from their elders. They usually gather in large numbers with their friends and family to celebrate Eid with different types of entertainment. The Buner district is full of natural resources. Majority of its people are farmers depending upon agriculture and livestock. Few rich people are engaged in the marble business. There are more than 600 marble factories, and 50–100 are in construction stage. It leads to a gigantic environmental contamination badly affecting both fauna and flora of the area.

### 1.1.4 Justification

The increased proportion of human population and industrialization is an immense challenge to protect the natural environment. Lack of adequate measures to treat effluents from marble industry have significant impact on the surrounding environment/ecosystem and hence, harm the ecosystem. Local dwellers depend on the system. Pollution affects people, animals, and plants directly in the form of diseases and indirectly in the form of food safety and or habitat destruction. Marble industries have a deleterious effect on the environment. In such situations, it turns imperative for young students and researchers to study environmental issues including pollutants, microbial strains, bioremediation procedure, and heavy metals accumulation (Khan et al. 2013; Guisan and Zimmermann 2000; Økland and Eilertsen 1993).

Therefore, this chapter aims to isolate, characterize, and identify microbial flora of marble waste-polluted ecosystems via biochemical tests, morphological, microscopic characteristics, and PCR-based sequencing techniques. It evaluates phylogenetic relationships among various species of microfungi collected from marble waste-polluted ecosystems under varying conditions.

## 1.2 Materials and Methods

### 1.2.1 Collection of Samples

The present research work was carried out on the variety and extent of pollutants produced by marble industries located in the Buner district of Khyber Pakhtunkhwa, Pakistan. A total of 15 marble factories at a distance of 3–4 km were unsystematically designated for the sampling of microbial biota in light of the marble wastewater possessing environment/ecosystem (Figure 1.3). Three samples from each factory, i.e., two samples of marble wastewater effluent and one control sample from the control environment (a total of 45 samples from 15 factories), were collected in sterilized falcon tubes. All the collected samples were stored in the refrigerator at 4°C before processing for isolation of fungi and bacteria. Four samples out of 45 were randomly selected for DNA-based analyses, i.e., three each water sample and one control sample (samples of factory numbers 1, 8, 14, and 7).

### 1.2.2 Isolation, Morphological and Microscopic Identification of Fungal Flora

Fungal populations in the samples were determined via serial dilution following Harley and Prescott methods (Prescott et al. 1993). Malt Extract Agar Media was prepared and autoclaved for fungal isolation (Ottow and Glathe 1968). 20 mL of agar media was poured into Petri dishes to prepare agar plates and labeled them properly in the laminar flow hood. 100 μL solutions were taken from each sample through a pipette and spread on agar plates. The Petri dishes were wrapped by cling foil (Wiegand et al. 2008) and were kept at 28°C for 48–72 hours. The fungal colonies were formed after the incubation process. Isolation of pure fungal strains was done using Hyphal Tip Isolation Techniques. The shape, structure, and type of each colony in terms of colony appearance and spore color were analyzed through the observation method. The Lactophenol Cotton Blue staining procedure was carried out for further morphological identification. All the prepared slides were observed under $10^x$, $40^x$, and $100^x$ lenses of microscope. Immersion oil was used for clear observation of microscopic characteristics under $100^x$ lens. Color reactions of fungi tissues were noted with the help of Melzer's reagents, lactic acid, and 5%–10% KOH. The hyphae length and width of fungi were measured in micrometer (μm) using Piximeter software.

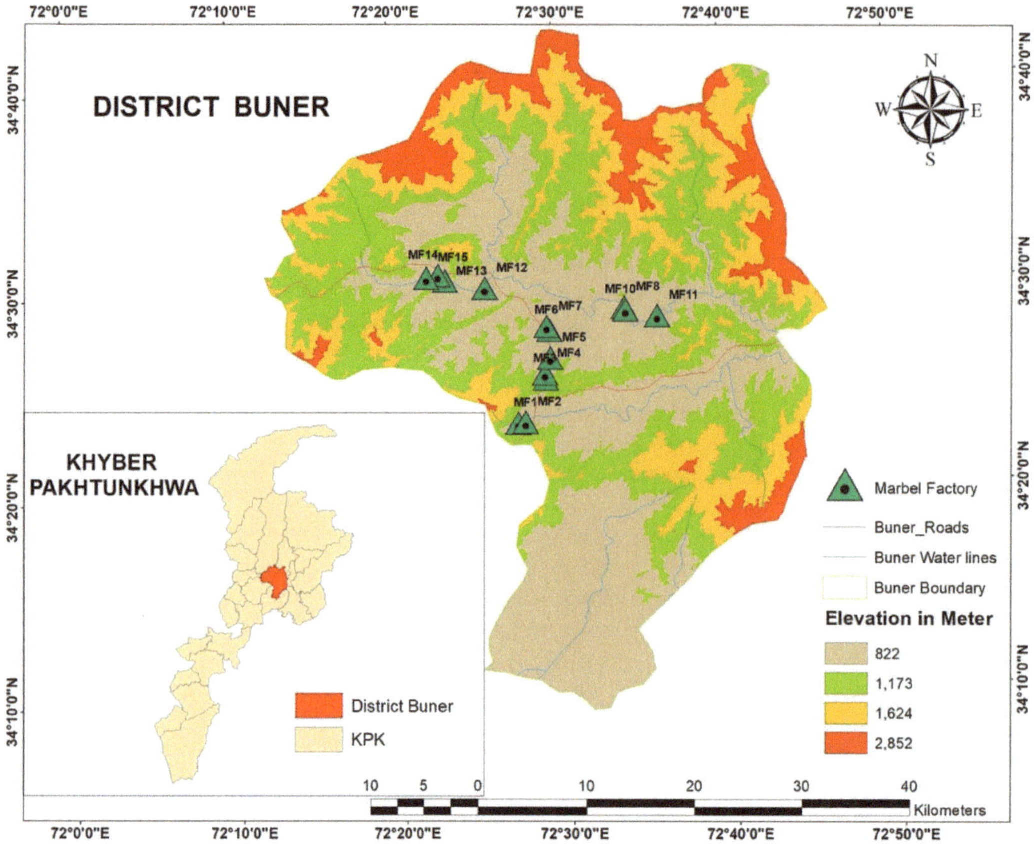


FIGURE 1.3 Location of the studied marble factories in the Buner district of Khyber Pakhtunkhwa, Pakistan.

#### *1.2.2.1 Molecular and Phylogenetic Analyses*

DNA of fungal samples was extracted by using a standard protocol of Dentinger et al. (2010). Gel electrophoresis was done, UV trans-illuminator was used for gel visualization, and pictures were taken on gel documentation system. Nuclear ribosomal Internal Transcribed Spacer (ITS) was amplified by means of ITS4-R and ITS1-F universal primers (Dentinger et al. 2010; White et al. 1990). ExoSAP-IT® was used to purify PCR products. The Bigdye® terminator version 3.1 cycles sequencing kit was combined with primers in 10 µL reactions to perform dye-terminated unidirectional sequencing. Ethanol precipitation was used to clean sequencing reaction subsequent after the manufacturer injunctions, and then again suspended it in 30 µL distilled water and run on ABI 3730 DNA sequencer in the Beijing Genomics Institute, Shenzhen, China (Dentinger et al. 2010). ITS sequences were matched/compared via the BLAST service of the National Center for Biotechnology Information to confirm identification. BioEdit software was used to edit and prepared the sequence and aligned with help of other sequences present in the GenBank through Muscle Alignment Tool. Phylogenetic tree of fungal species sequences was made separately. The maximum likelihood (ML) analysis was conceded through molecular evolutionary genetic analysis (Tamura et al. 2011).

### 1.2.3 Isolation, Morphological and Microscopic Identification of Bacterial Flora

A series of consecutive dilutions were used in order to decrease the dense culture of bacteria to useable concentrations. Nutrient agar media was prepared and autoclaved for bacterial growth (Wiegand et al. 2008; Hadi et al. 2014). In laminar flow hood, the first 20 mL of nutrient agar media was poured in

Petri dishes to prepare LB agar plates for bacterial growth and labeled properly. 100 µL solutions were taken from each dilution through a pipette and spread on agar plates. Petri plates were covered by cling foil incubated in an incubator in upside-down position at 37°C for 24 hours (Wiegand et al. 2008). The streaking technique was used to obtain pure culture or strains from the resulting colonies (Kaeberlein et al. 2002).

Morphology of the bacterial isolates/colonies was observed through naked eyes for recording color, growth, and form pattern. Gram staining techniques were used for identification of Gram-positive and Gram-negative bacterial strains (Gregersen 1978; Nugent et al. 1991). Shape, color, and pattern of the bacterial cells were observed through Fluorescence Microscope under $10^x$, $40^x$, and $100^x$ lens. Biochemical tests, i.e., Citrate, oxidase, Catalase, indole, Methyl Red (MR), and Voges–Proskauer (VP), were carried out for further identification of bacterial strains. Both fungi and bacteria cultures were persevered in 20% Glycerol at −20°C for future research work.

## 1.3 Results

A total of 26 microbial strains were isolated from the marble waste-polluted ecosystem of the Buner district, Khyber Pakhtunkhwa, Pakistan. It revealed 6 fungal and 20 bacterial strains. The detailed descriptions of its physicochemical and molecular analysis are as follows.

### 1.3.1 Characterization and Phylogenetic Studies of Fungal Flora

Morphologically most of these strains comprehended aseptate hyphae and black, brown, green, white to dark green colors. Anatomically these strains range from cylindrical to round, hyaline in lactophenol blue, thick to thin walled, smooth to ornamented surface with sharp scale, and fusoid to ellipsoid in shape (Table 1.1; Figure 1.4).

**TABLE 1.1**

Morphological and Anatomical Characteristics of Fungal Colonies Grown on Malt Extract Agar Plates and Stained after 48–72 hours of Incubation Collected from Marble Wastewater-Polluted Ecosystem at Buner, Pakistan

| S. No. | Scientific Name | Spore Color | Appearance of Colonies | Anatomical Characters/Microscopic Characters | Shape of Hyphae |
|---|---|---|---|---|---|
| 1 | *Aspergillus sydowii* | Green | Foam like | Spherical, black in color, ornamented, thick-walled, 6.4–7.6 µm in length and 6.1–7.3 µm in width. | Aseptate |
| 2 | *Aspergillus brasiliensis* | Black & white spores black | Foam & spread | Hyaline in lactophenol, thick-walled, spherical and retienlate in shape, 6.5–8.3 µm in length and 6.3–6.9 µm in width, smooth surface. | Aseptate |
| 3 | *Fusarium petroliphilum* | Black | Pointed | Cylindrical, Hyaline in lactophenol, thin walled, 37–77 µm in length and 4–7 µm in width, smooth surface. | Aseptate |
| 4 | *Alternaria alternata* | Black | Pointed | Clavate, sometimes spherical, ellipsoid, brown color, thick-walled, 21–39 µm in length and 11–20 µm in width, smooth surface, just like telio spores. | Septate |
| 5 | *Aureobasidum leucosper* | White/ brown | Foam/ snow-like | Hyaline in lactophenol, thin walled, spherical in shape, surface ornamentation present, 6.1–4.6 µm in length and 4.6–5.3 µm in width. | Aseptate |
| 6 | *Curvularia aeria* | Pure black | Feather like/ spread | Olive brown color, thick-walled, ellipsoid, clavate, 30–34 µm in length and 12–15 µm in width, smooth surface, just like telio spores. | Septate |

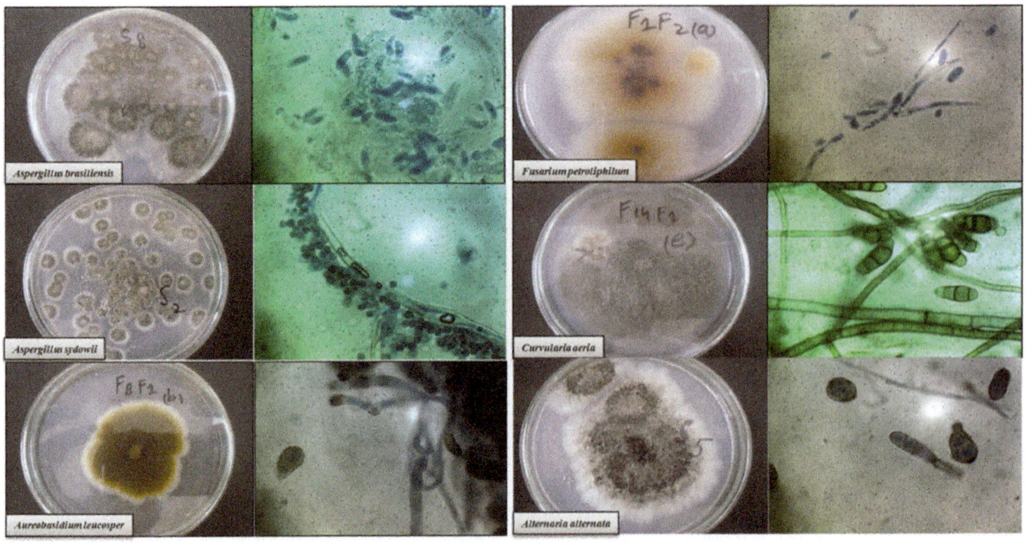

**FIGURE 1.4** The microscopic and morphological characterization of fungal isolates collected from the marble wastewater ecosystem of the Buner district of Pakistan.

### *1.3.1.1 Phylogenetic Analysis of the Selected Microfungi*

Based on the aforementioned results, i.e., isolation, morphological, anatomical/microscopic characterization, some of the microfungi were selected for further molecular analyses. Its detailed descriptions are as follows.

The BLAST analysis was carried out to identify the most similar sequence. All the consensus of ITS1F and ITS4 were a blast in gene bank search. ZAF 04 and ZAF 05 strains showed 99% similarity with *Aspergillus brasiliensis* (KX011592), having 95% query coverage. ZAF 02 and ZAF 03 resemble 99% with *Aspergillus sydowii* (KU687806) along with 97% query coverage. ZAF 07 strain showed 100% resemblance with *Aureobasidium leucospermi* JN712487 having 99% query coverage. ZAF 06 resembles 99% with *Fusarium petrophilium* (KP132225), acquiring 93% query coverage. ZAF 08 resembled 99% in conjunction with 99% query coverage to *Alternaria alternata* (KY609180). ZAF 09 comprehends 99% similarity with *Curvularia aeria* (KT933642.1) along with 98% query coverage. The final complete dataset of 54 sequences resulted in 1150 characters including gaps; of these, 486 characters were removed on both 3' and 5' ends. The final dataset consists of 664 characters, out of which 227 were conserved sequences, 395 were variable, 297 were parsimony informative sites, and 98 were singleton sites. The final phylogram is represented by three clades with *Lyomyces organensis* as root of the tree (out-group). The clade one is further divided into three subclades. The subclade A with 100% ML bootstrap value contained ZAF 04 and ZAF 05 strains having close affinities with accessions KX011592, JQ316521, FJ195349, KC796389, KT378129, KM491891, FJ629321, FJ195348, KX098315, AM295181, and FJ717684 of *Aspergillus brasiliensis*. In subclade B our strains ZAF 02 and ZAF 03 have close leaning with *Aspergillus sydowii* having accession numbers KU687806, LN898728, KJ413376, and KU687806 at 93% ML bootstrap value. The consensus of these sequences comprehends polymorphism on one position which indicated minimum divergence. The subclade C with 100% bootstrap value contained ZAF 07 along with *Aureobasidium leucospermi* JN712487 and JN712488. Clade 2 represents ZAF 06 in conjunction with *Fusarium petroliphilum* having accession KP132225, KC254043, LC184213, LC184243, and LC184196 at 100% ML bootstrap value. Clade 3 was further divided into two subclades: A and B. Our strain ZAF 09 has close affinities with accession KT933642, KP131919, KT933641, KT933639, and KU856631 of *Curvularia aeria* at 100% ML bootstrap value. The subclade B contained ZAF 08 strain with 74% bootstrap value with accession KY609180, KU182490, KX377683, KP739874, KP739875, KY075667, KX926578, KJ002064, AY154682, and KU179665 of *Alternaria alternata*. It shared 100% genetic character with *Alternaria alternata* (KY609180) and presented minimum divergence (Figure 1.5).

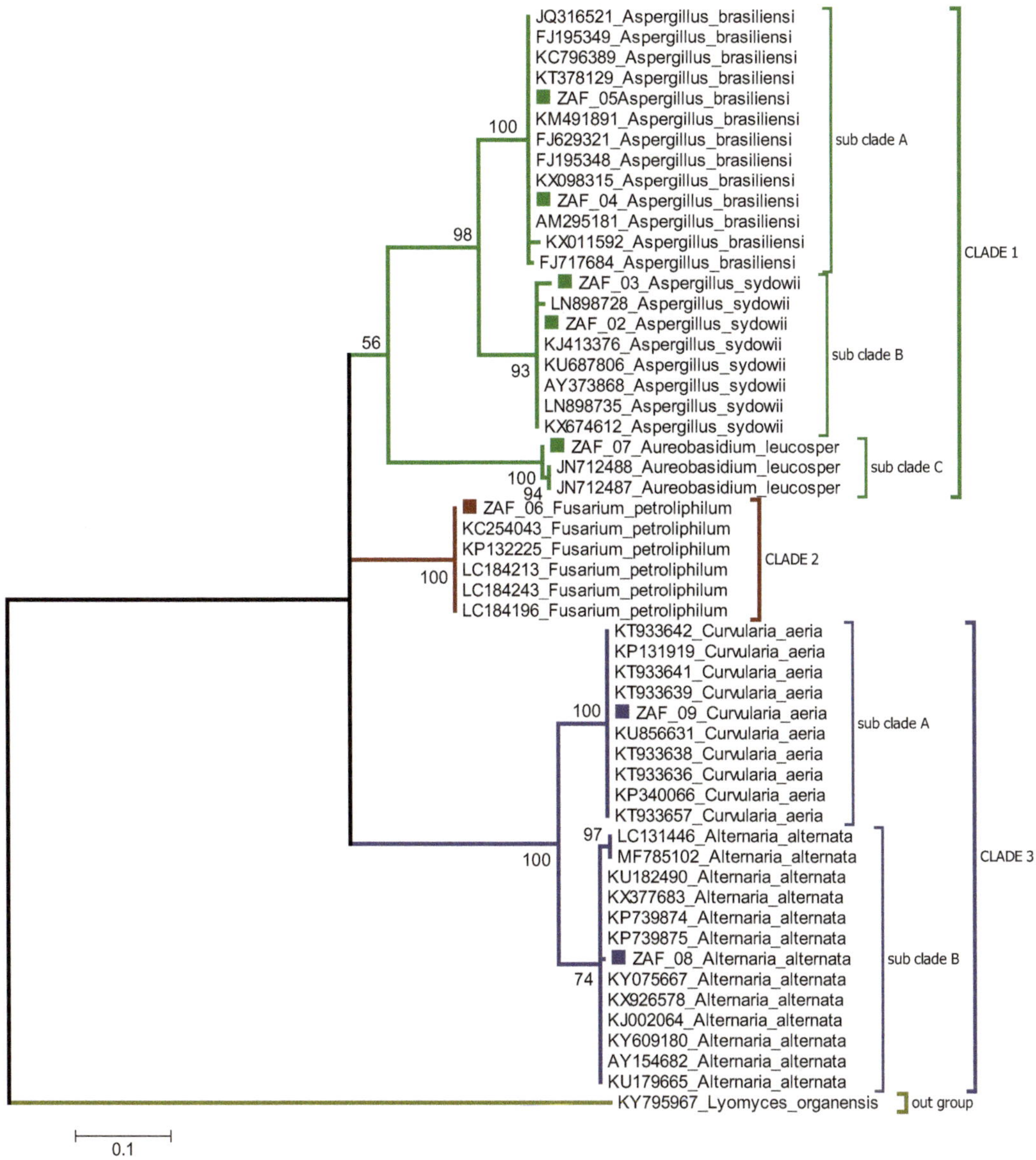


**FIGURE 1.5** Molecular phylogenetic relationship of the microfungi along with their allies inferred via ML technique based upon Kimura two-parameter model by means of nr ITS-rDNA data. It involved 54 nucleotide sequences and a total of 298 positions in the final data. Species collected from marble wastewater-polluted ecosystems were labeled with (■) box.

### 1.3.2 Characterization of Bacterial Strains

A total of 20 bacterial colonies/strains were recorded from the marble waste-polluted ecosystem. Its morphology in terms of colonies color, shape, and margins was studied that varied from transparent to light yellow, with sharp edges to sticky nature and round to spread in appearance. Based on morphology, the following types of morphological characteristics were recorded for various bacteria.

#### *1.3.2.1 Gram Staining Microscopic Characterization*

Microscopic characterization is one of the most important techniques in identification of unknown bacteria. Gram staining helps identify whether a strain is Gram-positive or Gram-negative. It was observed

**TABLE 1.2**

Morphological and Microscopic Characterization of Each Bacterial Colonies Growing on Nutrient Agar Plate after Incubation of 24 Hours

| S. No. | Bacterial Strains | Colony Morphology | Microscopic Characterization | Gram Staining |
|---|---|---|---|---|
| 1 | C1 | Transparent with sharp color edges, spread | Rod shaped | Negative |
| 2 | C2 | White & round | Spiral shaped | Negative |
| 3 | C3 | Transparent, regular, slimy | Rod shaped | Negative |
| 4 | C4 | White, regular, slimy | Rod shaped | Negative |
| 5 | F1A | Light yellow, round spread colonies | Rod shaped | Negative |
| 6 | F1B | Thick white spread colonies | Spiral shaped | Negative |
| 7 | F1C | White & round | Rod shaped | Positive |
| 8 | F8A | Transparent, round, sharp edges and white in color | Spiral shaped | Negative |
| 9 | F8B | Transparent & spread colonies | Spiral shaped | Negative |
| 10 | F8C | Light yellow, cellular | Spiral shaped | Negative |
| 11 | F8D | Transparent & slimy | Rod shaped | Negative |
| 12 | F8E | White growth like branches | Spiral shaped | Negative |
| 13 | F8F | White, colonies spread | Spiral shaped | Negative |
| 14 | F14A | Yellow round, regular elevated, sticky | Spiral shaped | Positive |
| 15 | F14B | White & slim, spread | Rod shaped | Positive |
| 16 | F14C | Transparent, round, flower like edge | Rod shaped | Negative |
| 17 | F14F | Light yellow & regular shape | Rod shaped | Positive |
| 18 | F14G | Spread colonies, round white in color | Spiral shaped | Negative |
| 19 | F14H | Small, round colonies, white in color | Rod shaped | Negative |
| 20 | F14IJ | Round, transparent colonies | Rod shaped | Negative |

C=bacterial colonies obtained from the control sample, while its numbers 1, 2, 3, and 4 demonstrated various colony numbers. F1, F8, and F14=bacterial colonies/strains acquired from marble factory numbers 1, 8, and 14, while A, B, C, D, E, F, etc. comprehended colony numbers obtained from each factory effluent.

that C1, C3, C4, F1A, F8D, F14C, F14H, and F14IJ strains were Gram-negative and rod-shaped. The strains F1C, F14B, and F14F were Gram-positive and rod-shaped. Furthermore, the Gram-negative spiral-shaped revealed C2, F1B, F8A, F8B, F8C, F8E, F14G, and F8F bacterial strains, while the strain F14A was observed as Gram-positive and spiral-shaped (Table 1.2 and Figure 1.6).

#### *1.3.2.2 Biochemical Characteristics*

Biochemical tests are performed to identify of various types of bacterial strains based on variation in the biochemical activities. Its physiology may be different from one species to another species. The MR, VP, indole, Catalase, oxidase, and Simmons citrate biochemical tests were carried out, and their detailed description is as follows.

##### *1.3.2.2.1 Methyl Red Test*

The MR biochemical test is used to identify bacterial strains that are producing stable acids via mixed-acid fermentation of glucose. It is carried out to determine which pathway of fermentation is used for glucose utilization. In mixed-acid fermentation, glucose is fermented and several organic acids are produced like lactic, succinic, formic, and acetic acids. Its stable production overawed by the phosphate buffer, which resulted in pH less than 4.4. Addition of MR (pH indicator) an aliquot of culture broth (having pH below 4.4) resulted in red color. Whenever MR turns yellow color, the culture doesn't utilize the mixed-acid fermentation pathway and a result its pH above 6. Few drops of MR indicator were added to each test tube culture after 24 hours of incubation. It was observed that the culture strains C2, C3, F1A, F1B, F8A, F8C, F8D, F8E, F8F, F14 A, F14C, F14 H, and F14 IJ were changed

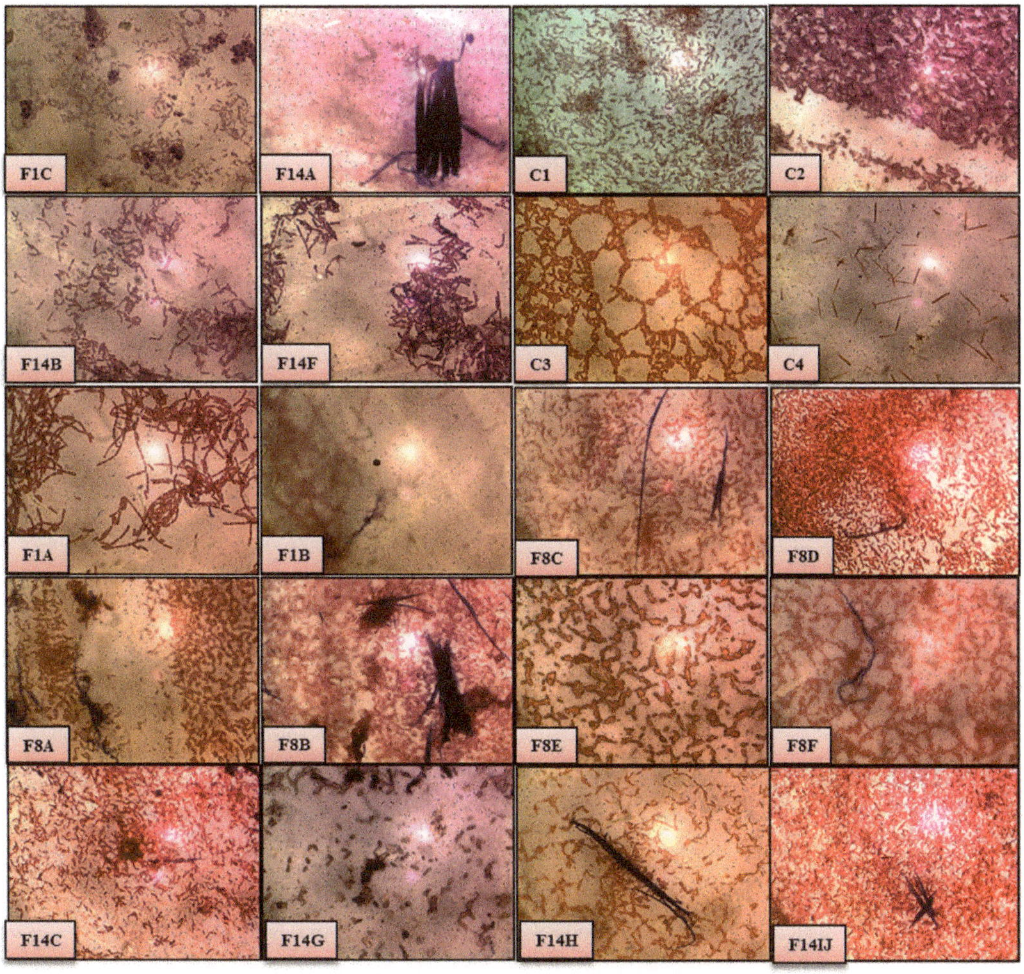


**FIGURE 1.6** Gram staining microscopic characterization of the recorded bacterial strains collected from marble wastewater-polluted ecosystem.

to red color showing MR-positive result (acid production positive result), while C1, C4, F1C, F8B, F14B, F14H, and F14G strains comprehended yellow color showing MR-negative result (Table 1.3 and Figure 1.7).

### *1.3.2.2.2 Indole Test*

Several types of bacteria produce indole from the amino acid tryptophan using tryptophanase enzyme. The indole biochemical test is used to determine the ability of a bacterium to convert the amino acid tryptophanase into indole. The formation of indole can be determined by Kovacs reagents. Indole reacts with aldehyde in reagents and forms a red color ring at the top which was taken as positive. Few drops of Kovacs reagent were added into each bacterial culture after incubation of 24 hours. The strains C3, C4, F1A, F1B, F8C, F8D, F8E, F8F, F14A, F14B, and F14IJ were indole positive because when Kovacs reagent was added, it formed indole red rings, while the strains C1, C2, F1C, F8A, F8B, F14C, F14F, F14G, and F14H were Indole negative (Figure 1.8).

### *1.3.2.2.3 Voges–Proskauer Test*

Some types of bacteria can be differentiated on the basis of acetone formation. For this purpose, VP test is carried out by adding alpha-naphthol and potassium hydroxide to VP broth. It turns brownish-red

**TABLE 1.3**
Biochemical Test Result of the Isolated Bacterial Strains

| S. No. | Strains | MR | Indole | VP | Catalase | Oxidase | Simmons Citrate |
|---|---|---|---|---|---|---|---|
| 1 | C1 | – | – | + | + | – | – |
| | C2 | + | – | + | + | – | + |
| | C3 | + | + | + | + | – | – |
| | C4 | – | + | + | + | – | + |
| 2 | F1 A | + | + | – | + | – | + |
| | F1 B | + | + | – | – | – | – |
| | F1 C | – | – | + | + | – | + |
| 3 | F8 A | + | – | – | + | – | – |
| | F8 B | – | – | – | + | – | – |
| | F8 C | + | + | – | + | – | – |
| | F8 D | + | + | + | + | – | + |
| | F8 E | + | + | + | + | – | + |
| | F8 F | + | + | + | + | – | + |
| 4 | F14 A | + | + | + | + | – | + |
| | F14 B | – | + | + | + | – | + |
| | F14 C | + | – | + | – | – | + |
| | F14 F | – | – | – | + | – | + |
| | F14 G | – | – | – | + | – | + |
| | F14 H | + | – | + | + | – | + |
| | F14 IJ | + | + | + | + | – | + |

C = bacterial colonies obtained from the control sample, while its numbers 1, 2, 3, and 4 represented colony numbers. F1, F8, and F14 = bacterial colonies/strains acquired from marble factory numbers 1, 8, and 14, while A, B, C, D, E, F, G, H, etc. comprehended colony number obtained from each factory effluent.

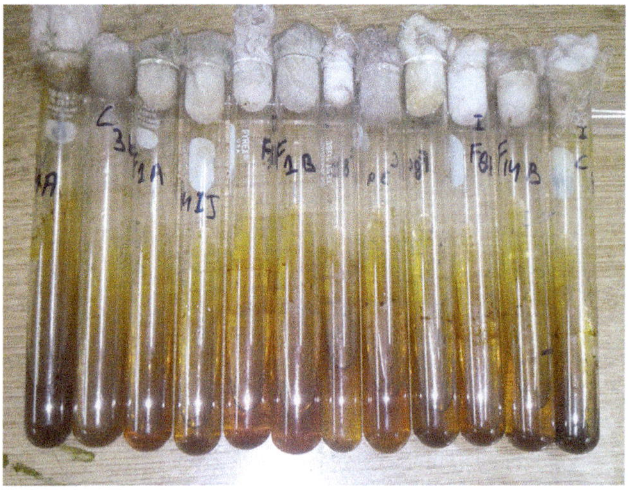

**FIGURE 1.7** MR biochemical test for the identification of bacterial strains.

to pink color when the culture is positive for acetone and brownish-green to yellow when the culture is negative for acetone. Two to three drops of alpha-naphthol and potassium hydroxide were added to VP media tubes after the incubation. C1, C2, C3, C4, F1C, F8D, F8E, F8F, F14A, F14B, F14C, F14H, and F14IJ samples were VP positive, whereas, F1A, F1B, F8A, F8B, F8C, F14F, and F14G bacterial strains were VP negative (Figure 1.9).

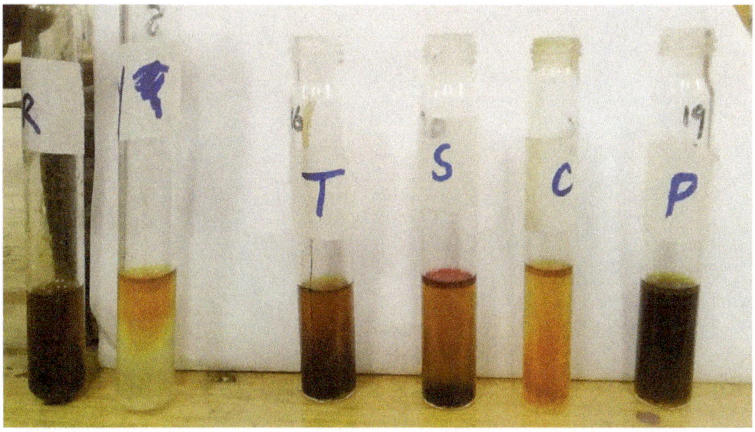


**FIGURE 1.8** Indole test for differentiation of bacteria via Kovacs reagents.

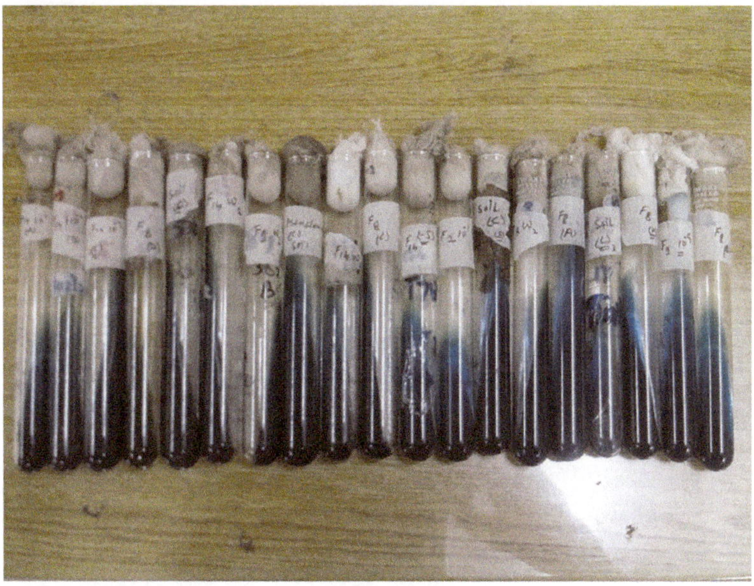

**FIGURE 1.9** Voges–Proskauer test for differentiation of bacteria based on acetone formation.

### *1.3.2.2.4 Catalase Test*

Some bacteria possess a protective enzyme catalase. It is capable of destroying the dangerous chemical hydrogen peroxide. Hydrogen peroxide is decomposed into water and oxygen by catalase enzyme. When oxygen gas freed as a result, bubbles or effervescence is produced. The C1, C2, C3, C4, FIC, F8A, F8B, F14A, F14B, and F14F bacterial culture were giving positive results to catalase biochemical test. When a small amount of colony is smeared, it gives copious bubbles by adding hydrogen peroxide and indicated the presence of catalase enzyme. At the same time bacterial culture F1B and F14C showed a negative catalase result (Figure 1.10).

### *1.3.2.3.5 Oxidase Test*

Various types of bacteria produced oxidase enzymes (cytochrome oxidase). The oxidase biochemical test was used to differentiate among the families of numerous bacteria, especially those that utilize oxygen as a final electron acceptor in aerobic respiration (oxidase-positive). All of the bacterial strains that do not comprehend Wurster blue-colored smear were considered as oxidase-negative (Figure 1.11).

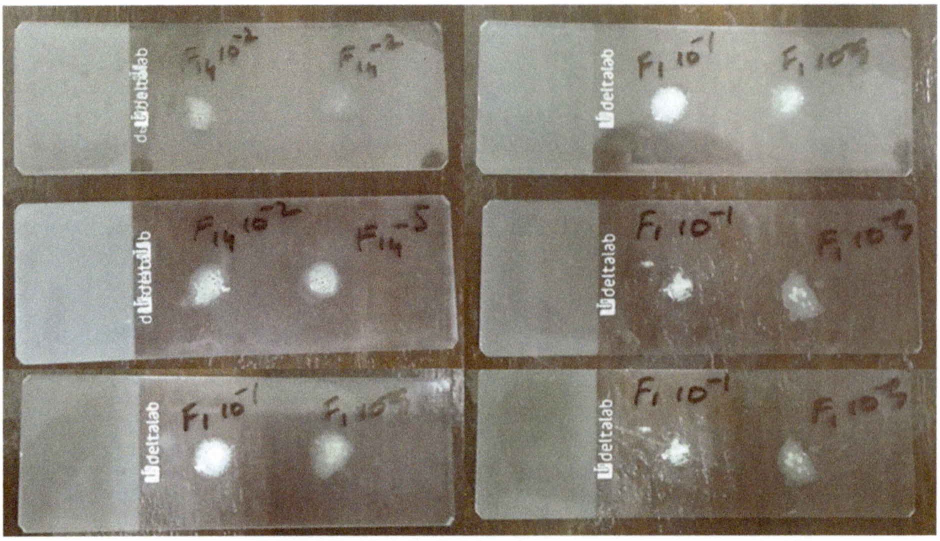

**FIGURE 1.10** Result of Catalase biochemical test of bacterial cultures.

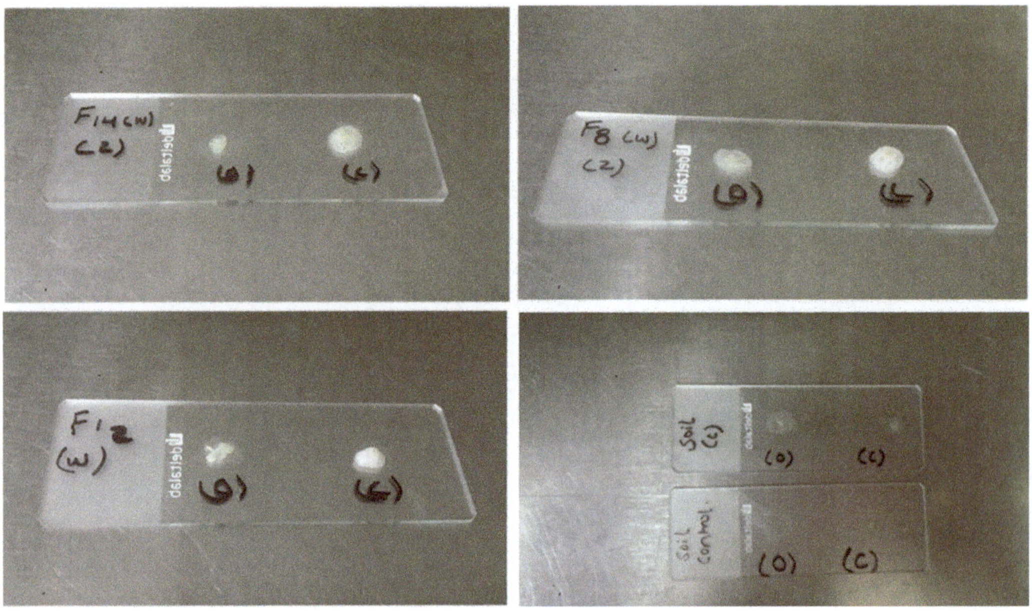

**FIGURE 1.11** Oxidase biochemical test of bacterial strains comprehended negative oxidase.

### *1.3.2.2.6 Simmons Citrate Test*

The Simmons citrate agar test is carried out to check citrate utilization as a carbon source of an organism. It contains sodium citrate as a source of carbon, ammonium dihydrogen phosphate as a nitrogen source, bromothymol blue as pH indicator, and other nutrients. Bacterial strains C2, C4, F1A, F1C, F8C, F8D, F8E, F8F, F14A, F14B, F14C, and F14F were observed as citrate positive because these cultures utilized the ammonium dihydrogen phosphate and sodium citrate and showed blue color, while the strains C1, C3, F1B, F8A, and F8C were citrate negative (Figure 1.12).

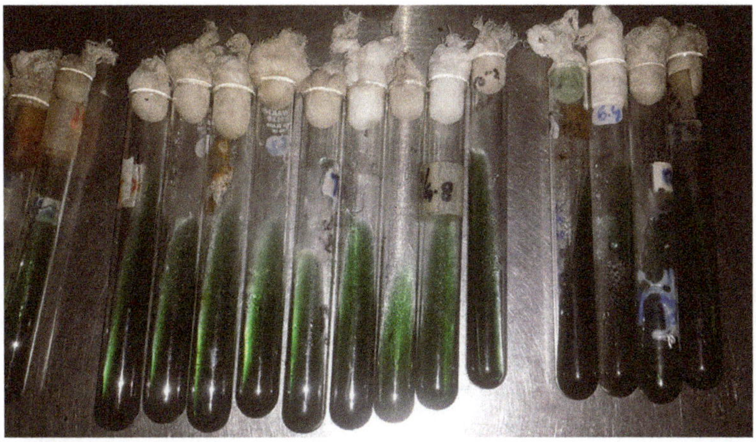

**FIGURE 1.12** The Simmons citrate agar test of bacterial culture for citrate utilization as a carbon source.

## 1.4 Discussion

The traditional identification of microorganisms in terms of phenotypic features was not considered as an accurate method for the identification of microbes as compared to genotypic techniques. Comparison of the microfungi based on 18S rRNA gene sequence is a nascent preferred genetic tool. Currently, such techniques are widely used to estimate the phylogeny of various organisms. Nowadays, it has become a key tool for identification of unknown fungi up to species and variety levels (Shahid et al. 2014). It included function of 18S rRNA gene which has not been changed intensely and indicating random sequence changes more significant measure of evolution/time. The rRNA molecular examination is a significant method/tool not only to evaluate diversity but also to identify new strains of different species.

In this chapter, molecular approaches were found out using consensuses of ITS1F and ITS4 primers for each species. Molecular identification and phylogeny of isolated microfungi revealed that most of the species belong to genus *Aspergillus,* i.e., *Aspergillus sydowii* and *Aspergillus brasiliensis.* Other species included *Aureobasidium leucospermi*, *Fusarium petroliphilum*, *Curvularia aeria*, and *Alternaria alternata.* It comprehended close harmony with the work of Roussos et al. (1995), where they reported a total of 272 strains of filamentous fungi from soil, leaves of coffee plant, and coffee cherries grown in Mexico. Most of the isolated microorganism included belong to *Aspergillus*, *Fusarium*, *Trichoderma*, *Penicillium*, and *Humicola* genera having the ability to degrade 100% of the caffeine in liquid media. Moreover, Maiti et al. (2013) sequenced the ITS region of two fungi isolated from aerobic compost environment.

DNA extraction was carried out following the glass fiber filtration and enzymatic digestion method (Dentinger et al. 2010). ITS1-F and ITS4-R primers were used for amplification of ITS region. ExoSAP-IT® was used to purify PCR products. The Bigdye Terminator version 3.1 cycle sequence kit along with ITS1-F and ITS4-R primers in 10 μL was used to perform dye-terminated unidirectional sequencing. Guo et al. (2003) conceded molecular identification of the endophytic fungi with the help of rDNA sequencing analysis. 5.8S gene and ITS1 and ITS2 regions of nrDNA were amplified and sequenced. Further identification was done with the help of phylogenetic analysis of ITS regions (Dentinger et al. 2010; White et al. 1990). Similarly, Manter and Vivanco (2007) used ITS primers for characterization of fungal diversity using qPCR techniques. De Carolis et al. (2012) used matrix-assisted laser desorption ionization time of flight mass spectrometry for identification of *Aspergillus*, *Mucorales*, and *Fusarium* species.

The DNA sequencing resulted *Aspergillus sydowii* (ZAF 02 and ZAF 03), *Aspergillus brasilensi* (ZAF 04 and ZAF 05), *Fusarium petrophilium* (ZAF 06), *Aureobasidium leucospermi* (ZAF 07), *Alternaria alternata* (ZAF 08), and *Curvularia aeria* (ZAF 09), respectively. *Aspergillus sydowii* is the predominant saprophytic fungus mostly found in soil and is occasionally pathogenic to humans (causing aspergillosis, keratomycosis, and onychomycosis), and also cause contamination of food. In 1990, it was found in seawater and shown to be the cause of aspergillosis. Matkar et al. (2013) worked on the production of cellulose from *Aspergillus sydowii*, a new isolated strain obtained from the terrestrial region and also reported from marine habitats. They identified isolate using 18S rDNA sequencing. The current study also revealed the isolation, characterization, and phylogeny of *Curvularia aeria* from marble wastewater. *Curvularia* spp. was the most common pathogenic fungi on Gramineae plants and was frequently found in other cellulosic substrates (Nakada et al. 1994). Furthermore, *Aureobasidium leucospermi* was reported from marble wastewater, which is mostly ubiquitous black in color and found in different types of environments (water, soil, air, and limestone) (Andrews et al. 2002). Smithson et al. (2013) worked on halophilic fungi (*Aureobasidium* spp.) that have a significant effect on clear-out slats formed on the surface of sandstone in Medamoud, Egypt. *Aureobasidium* spp. has a significant importance in the field of biotechnology in terms of making various types of enzymes as well as using them in the biological control of plant disease (Ferreira-Pinto et al. 2006; Zhang et al. 2012). Di Francesco et al. (2015) worked on the production of volatile organic compounds from *Aureobasidium* spp. and demonstrated a significant role against five fruit postharvest pathogens. *Aureobasidium pullulans* decreased the incidence of blue and gray mold of apples by 67% and 89%, respectively (Ippolito et al. 2000). *Aspergillus brasiliensis* causes black mold ailment of certain vegetables and fruits. It was ubiquitous in nature (Varga et al. 2007). *Aspergillus brasiliensis* was extensively exploited by fermentation industries for the production of organic acids like citric acid and certain types of enzymes (Pel et al. 2007). It has the ability to degrade xenobiotics through different oxidative, demethylation, and hydroxylation reactions, playing a potential role in bioremediations. Kapoor et al. (1999) worked on the removal of heavy metals like cadmium, copper, nickel, and lead using *Aspergillus brasiliensis* and concluded a significant role of *Aspergillus* spp. in heavy metal removal. Wen et al. (2005) co-cultured *Aspergillus phoenicis* with *Trichoderma reesei* by means of dairy manure as a substrate that produced a high-level of β-glucosidase.

A total of 20 bacterial isolates were stained and examined under the $100^x$ lens of a microscope. Most of the observed bacteria were Gram-negative and rod-shaped including C1, C3, C4 (of control sample), F1A, F8D, F14C, F14H, and F14IJ strains. The Gram-positive and rod-shaped comprised F1C, F14B, and F14F strains, whereas the Gram-negative and spiral-shaped included C2, F1B, F8A, F8B, F8C, F8E, F14G, and F8F strains. The F14A strain was observed as Gram-positive and spiral-shaped. Similarly, Abrusci et al. (2005) worked on characterization and isolation of fungi and bacteria from cinematographic. They reported 14 different colonies of bacteria: 5 species of *Staphylococcus*, such as *S. hominis*, *S. haemolyticus*, *S. lugdunensis*, *S. lentus*, and *S. epidermidis*, and 5 *Bacillus* species such as *B. subtilis*, *B. pichinotyi*, *B. pumilus*, *B. megaterurella*, and *B. amyloliquefaciens* were identified together with *Kocuria kristinae*, *Pasteurella haemolytica*, and *Sphingomonas paucimobilis*. Similar to our study, initial characterizations were carried out via morphological and biochemical methods that indicated most bacteria were Gram-positive. Furthermore, Leung et al. (2000) recorded a total of 19 heavy metal resistant and non-resistant bacteria from activated sludge. Like the current study, it also revealed both Gram-positive and Gram-negative bacteria. Abrusci et al. (2005) isolated rod-shaped bacteria that revealed catalase- and oxidase-positive and six strains of Gram-positive cocci that were catalase-positive. In contrast, current project comprehended that rod-shaped bacteria were mostly catalase- and oxidase-negative. The Gram-negative rod-shaped isolated (S. paucimobilis) was varied spread in nature owed its physiological and metabolic adaptability (Azeredo and Oliveira 2000). Habitat and morphological characteristics of the bacterial isolates were of the significance's in explaining the prospective reasons of contamination of the films. Finding of this research showed isolation and characterization of microorganism (fungi and bacteria) from the wastewater ecosystem. Similarly, Bauer et al. (2002) investigated the bacteria and fungi in aerosols, generated via two different types of wastewater treatment plants. They concluded wastewater as a potential hauler of pathogenic microbes that may pose a health risk when it is aerosolized. Gray (1989) examined microorganism communities in wastewater treatment plants and found various types of fungal and bacterial communities in all types of anaerobic and aerobic treatment plants. Leung et al. (2000) recorded a total of 19 heavy metal resistant and non-resistant bacteria from activated sludge treating both metal-contaminated municipal wastewater and industrial effluents. Like the current project, Cai et al. (2003) worked on isolation and characterization of an atrazine-degrading *Arthrobacter* from industrial wastewater in China. Similarly, Arutchelvan et al. (2005) conducted research on isolation and identification of high-strength phenol-degrading bacteria from phenol industrial wastewater. Based on biochemical tests, they identified *Bacillus brevis* and *Pseudomonas cepacia*. These species showed an efficient role in phenol degradation. Congeevaram et al. (2007) worked on biosorption of nickel and chromium through bacterial and fungal heavy metals resistant isolates. They isolated fungal and bacterial cultures resistant to heavy metals from the soil of the electroplating industry. They counted the species of *Aspergillus* and *Micrococcus* from industrial wastewater. Various species of *Aspergillus, Saprophyticus, Phanerochaete, Bacillus*, and *Pseudomonas* have been reported as efficient nickel and chromium reducers. Ahluwalia and Goyal (2007) worked on microbes and plants derived biomass for the removal of heavy metals from contaminated water. Chemical composition of biomass isolates heavy metal ions through metal forming complexes from the solution that removes the inevitability to uphold potential growth supporting conditions. They concluded that species of *Aspergillus*, *Penicillium*, *Rhizopus*, *Chlorella*, *Oscillatoria*, *Bacillus*, and *Streptomyces* had higher heavy metal adsorption capacities for lead, cadmium, zinc, copper, nickel, and chromium, respectively.

## 1.5 Conclusion

It was concluded that marble wastewater mostly revealed six different types of microfungal strains in the Buner district, Pakistan. Their molecular identification and phylogeny identified *Aspergillus sydowii* (ZAF 02 and ZAF 03), *Aspergillus brasilensi* (ZAF 04 and ZAF 05), *Fusarium petrophilium* (ZAF 06), *Aureobasidium leucospermi* (ZAF 07), *Alternaria alternata* (ZAF 08), and *Curvularia aeria* (ZAF 09) species. Morphologically, many of these strains exhibited aseptate hyphae and different types of colors. Anatomically, these strains range from cylindrical to round, hyaline in lactophenol blue, thick to thin walled, smooth to ornamented surface with sharp scale, and fusoid to ellipsoid in shape. A total of 20 bacterial strains were isolated and characterized. Morphologically, bacterial colonies ranged from transparent to light yellow in colors, sharp edges to sticky nature, and spread to round shapes in appearance. It was observed that C1, C3, C4 (of control sample), F1A, F8D, F14C, F14H, and F14IJ strains were Gram-negative and rod-shaped. The F1C, F14B, and F14F strains were Gram-positive and rod-shaped. C2, F1B, F8A, F8B, F8C, F8E, F14G, and F8F were Gram-negative and spiral-shaped bacterial strains, and F14A was observed as Gram-positive and spiral-shaped strain. Biochemically, most of the cultures were MR, VP, Catalase, Simmons citrate, and indole positive. All the strains exhibit no reaction to the oxidase enzyme. Identification and confirmation of fungal species exhibit certain disease-causing strains as well which can be further recommended for detailed studies. Confirmation of bacterial strains through molecular techniques is also recommended.

## Acknowledgments

We are thankful to Dr. Ishtiaq Ali, Department of Microbiology, Quaid-i-Azam University Islamabad and Prof. Dr. Abdul Nasir Khalid, Department of Botany, University of the Punjab Lahore for facilitating the research work in one way or the other.

## REFERENCES

Abdul S, Muhammad AA, Shabir H, Hesham A. El E, Sajjad H, Niaz A, Abdul G, Sayyed RZ, Fahad S, Subhan D, Rahul D (2021a) Zinc nutrition and arbuscular mycorrhizal symbiosis effects on maize (Zea mays L.) growth and productivity. J Saudi Soc Agric Sci. https://doi.org/10.1016/j.sjbs.2021.06.096.

Abdul S, Muhammad AA, Subhan D, Niaz A, Fahad S, Rahul D, Mohammad JA, Omaima N, Muhammad Habib ur R, Bernard RG (2021b) Effect of arbuscular mycorrhizal fungi on the physiological functioning of maizeunder zinc-deficient soils. Sci Rep 11:18468.

Abrusci C, Martín-González A, Del Amo A, Catalina F, Collado J, Platas G (2005) Isolation and identification of bacteria and fungi from cinematographic films. Int Biodeter Biodegr 56:58–68.

Adnan M, Fahad S, Khan IA, Saeed M, Ihsan MZ, Saud S, Riaz M, Wang D, Wu C (2019) Integration of poultry manure and phosphate solubilizing bacteria improved availability of Ca bound P in calcareous soils. 3 Biotech 9(10):368.

Adnan M, Fahad S, Muhammad Z, Shahen S, Ishaq AM, Subhan D, Zafar-ul-Hye M, Martin LB, Raja MMN, Beena S, Saud S, Imran A, Zhen Y, Martin B, Jiri H, Rahul D (2020) Coupling phosphate-solubilizing bacteria with phosphorus supplements improve maize phosphorus acquisition and growth under lime induced salinity stress. Plants 9(900). https://doi.org/10.3390/plants9070900.

Adnan M, Shah Z, Sharif M, Rahman H (2018b) Liming induces carbon dioxide ($CO_2$) emission in PSB inoculated alkaline soil supplemented with different phosphorus sources. Environ Sci Pollut Res 25(10):9501–9509.

Adnan M, Shah Z, Ullah H, Khan B, Arshad M, Mian, IA, khan, GA, Alam, M, Basir, A, Rahman, IU, Ali M, khan WU (2016) Yield response of wheat to nitrogen and potassium fertilization. Pure Appl Bio 5(4):868–875.

Adnan M, Zahir S, Fahad S, Arif M, Mukhtar A, Imtiaz AK, Ishaq AM, Abdul B, Hidayat U, Muhammad A, Inayat-Ur R, Saud S, Muhammad ZI, Yousaf J, Amanullah, Hafiz MH, Wajid N (2018a) Phosphate-solubilizing bacteria nullify the antagonistic effect of soil calcification on bioavailability of phosphorus in alkaline soils. Sci Rep 8:4339. https://doi.org/10.1038/s41598-018-22653-7.

Ahluwalia SS, Goyal D (2007) Microbial and plant derived biomass for removal of heavy metals from wastewater. Bioresour Technol 98:2243–2257.

Ahmad S, Kamran M, Ding R, Meng X, Wang H, Ahmad I, Fahad S, Han Q (2019) Exogenous melatonin confers drought stress by promoting plant growth, photosynthetic capacity and antioxidant defense system of maize seedlings. PeerJ 7:e7793. http://doi.org/10.7717/peerj.7793.

Akbar H, Timothy JK, Jagadish T, M. Golam M, Apurbo KC, Muhammad F, Rajan B, Fahad S, Hasanuzzaman M (2020) Agricultural land degradation: Processes and problems undermining future food security. In: Fahad S, Hasanuzzaman M, Alam M, Ullah H, Saeed M, Khan AK, Adnan M (eds) Environment, climate, plant and vegetation growth. Springer Publ Ltd, Springer Nature Switzerland AG. Part of Springer Nature, pp 17–62. https://doi.org/10.1007/978-3-030-49732-3.

Akbulut H, Gürer C (2007) Use of aggregates produced from marble quarry waste in asphalt pavements. Build Environ 42:1921–1930.

Akram R, Turan V, Hammad HM, Ahmad S, Hussain S, Hasnain A, Maqbool MM, Rehmani MIA, Rasool A, Masood N, Mahmood F, Mubeen M, Sultana SR, Fahad S, Amanet K, Saleem M, Abbas Y, Akhtar HM, Waseem F, Murtaza R, Amin A, Zahoor SA, ul Din MS, Nasim W (2018a) Fate of organic and inorganic pollutants in paddy soils. In: Hashmi MZ, Varma A (eds) Environmental pollution of paddy soils, soil biology. Springer International Publishing AG, Switzerland, pp 197–214.

Akram R, Turan V, Wahid A, Ijaz M, Shahid MA, Kaleem S, Hafeez A, Maqbool MM, Chaudhary HJ, Munis, MFH, Mubeen M, Sadiq N, Murtaza R, Kazmi DH, Ali S, Khan N, Sultana SR, Fahad S, Amin A, Nasim W (2018b) Paddy land pollutants and their role in climate change. In: Hashmi MZ, Varma A (eds) Environmental pollution of paddy soils, soil biology. Springer International Publishing AG, Switzerland, pp 113–124.

Al-Wabel MI, Abdelazeem S, Munir A, Khalid E, Adel RAU (2020b) Extent of climate change in Saudi Arabia and its impacts on agriculture: A case study from Qassim Region. In: Fahad S, Hasanuzzaman M, Alam M, Ullah H, Saeed M, Khan AK, Adnan M (eds) Environment, climate, plant and vegetation growth. Springer Publ Ltd, Springer Nature Switzerland AG. Part of Springer Nature. pp 635–658. https://doi.org/10.1007/978-3-030-49732-3.

Al-Wabel MI, Ahmad M, Usman AR, Akanji M, Rafique MI (2020a) Advances in pyrolytic technologies with improved carbon capture and storage to combat climate change. In: Fahad S, Hasanuzzaman M, Alam M, Ullah H, Saeed M, Khan AK, Adnan M (eds) Environment, climate, plant and vegetation growth. Springer Publ Ltd, Springer Nature Switzerland AG. Part of Springer Nature. pp 535–576. https://doi.org/10.1007/978-3-030-49732-3.

Amanullah, Fahad S (eds) (2017) Rice - technology and production. IntechOpen, Croatia. http://dx.doi.org/10.5772/64480.

Amanullah, Fahad S (eds) (2018a) Corn - Production and human Health in changing climate. IntechOpen, United Kingdom. http://dx.doi.org/10.5772/intechopen.74074.

Amanullah, Fahad S (eds) (2018b) Nitrogen in agriculture – updates. IntechOpen, Croatia. http://dx.doi.org/10.5772/65846.

Amanullah, Muhammad I, Haider N, Shah K, Manzoor A, Asim M, Saif U, Izhar A, Fahad S, Adnan M et al. (2021) Integrated foliar nutrients application improve wheat (Triticum Aestivum L.) productivity under calcareous soils in drylands. Commun Soil Sci Plant Anal. https://doi.org/10.1080/00103624.2021.1956521.

Amanullah, Shah K, Imran, Hamdan AK, Muhammad A, Abdel RA, Muhammad A, Fahad S, Azizullah S, Brajendra P (2020) Effects of climate change on irrigation water quality. In: Fahad S, Hasanuzzaman M, Alam M, Ullah H, Saeed M, Khan AK, Adnan M (eds) Environment, climate, plant and vegetation growth. Springer Publ Ltd, Springer Nature Switzerland AG. Part of Springer Nature. pp 123–132. https://doi.org/10.1007/978-3-030-49732-3.

Amir M, Muhammad A, Allah B, Sevgi Ç, Haroon ZK, Muhammad A, Emre A (2020) Bio fortification under climate change: The fight between quality and quantity. In: Fahad S, Hasanuzzaman M, Alam M, Ullah H, Saeed M, Khan AK, Adnan M (eds) Environment, climate, plant and vegetation growth. Springer Publ Ltd, Springer Nature Switzerland AG. Part of Springer Nature. pp 173–228. https://doi.org/10.1007/978-3-030-49732-3.

Amjad I, Muhammad H, Farooq S, Anwar H (2020) Role of plant bioactives in sustainable agriculture. In: Fahad S, Hasanuzzaman M, Alam M, Ullah H, Saeed M, Khan AK, Adnan M (eds) Environment, climate, plant and vegetation growth. Springer Publ Ltd, Springer Nature Switzerland AG. Part of Springer Nature. pp 591–606. https://doi.org/10.1007/978-3-030-49732-3.

Amjad SF, Mansoora N, Din IU, Khalid IR, Jatoi GH, Murtaza G, Yaseen S, Naz M, Danish S, Fahad S, et al (2021) Application of zinc fertilizer and mycorrhizal inoculation on physio-biochemical parameters of wheat grown under water-stressed environment. Sustainability 13:11007. https://doi.org/10. 3390/su131911007.

André A, de Brito J, Rosa A, Pedro D (2014) Durability performance of concrete incorporating coarse aggregates from marble industry waste. J Cleaner Prod 65:389–396.

Andrews JH, Spear RN, Nordheim EV (2002) Population biology of Aureobasidium pullulans on apple leaf surfaces. Can J Microbiol 48:500–513.

Arif M, Talha J, Muhammad R, Fahad S, Muhammad A, Amanullah, Kawsar A, Ishaq AM, Bushra K, Fahd R (2020) Biochar; a remedy for climate change. In: Fahad S, Hasanuzzaman M, Alam M, Ullah H, Saeed M, Khan AK, Adnan M (eds) Environment, climate, plant and vegetation growth. Springer Publ Ltd, Springer Nature Switzerland AG. Part of Springer Nature. pp 151–172. https://doi.org/10.1007/978-3-030-49732-3.

Arutchelvan V, Kanakasabai V, Nagarajan S, Muralikrishnan V (2005) Isolation and identification of novel high strength phenol degrading bacterial strains from phenol-formaldehyde resin manufacturing industrial wastewater. J Hazard Mater 127:238–243.

Ashfaq AR, Uzma Y, Niaz A, Muhammad AA, Fahad S, Haider S, Tayebeh Z, Subhan D, Süleyman T, Hesham AElE, Pramila T, Jamal MA, Sulaiman AA, Rahul D (2021) Toxicity of Cadmium and nickel in the context of applied activated carbon biochar for improvement in soil fertility. Saudi Soc Agric Sci. https://doi.org/10.1016/j.sjbs.2021.09.035.

Athar M, Masood IA, Sana S, Ahmed M, Xiukang W, Sajid F, Sher AK, Habib A, Faran M, Zafar H, Farhana G, Fahad S (2021) Bio-diesel production of sunflower through sulphur management in a semi-arid subtropical environment. Environ Sci Pollut Res. https://doi.org/10.1007/s11356-021-16688-z.

Atif B, Hesham A, Fahad S (2021) Biochar coupling with phosphorus fertilization modifies antioxidant activity, osmolyte accumulation and reactive oxygen species synthesis in the leaves and xylem sap of rice cultivars under high-temperature stress. Physiol Mol Biol Plants. https://doi.org/10.1007/s12298-021-01062-7.

Aukour FJ, Al-Qinna MI (2008) Marble production and environmental constrains: Case study from Zarqa Governorate, Jordan. Jordan J Earth Environ Sci 1:11–21.

Ayman EL Sabagh, Hossain A, Barutçular C, Iqbal MA, Islam MS, Fahad S, Sytar O, Çig F, Meena RS, Erman M (2020) Consequences of salinity stress on the quality of crops and its mitigation strategies for sustainable crop production: An outlook of arid and semi- arid regions. In: Fahad S, Hasanuzzaman M, Alam M, Ullah H, Saeed M, Khan AK, Adnan M (eds) Environment, climate, plant and vegetation growth. Springer Publ Ltd, Springer Nature Switzerland AG. Part of Springer Nature. pp 503–534. https://doi.org/10.1007/978-3-030-49732-3.

Azam F, Fenchel T, Field JG, Gray J, Meyer-Reil L, Thingstad F (1983) The ecological role of water-column microbes in the sea. Mar Ecol Progr Ser:257–263.

Azeredo J, Oliveira R (2000) The role of exopolymers in the attachment of Sphingomonas paucimobilis. Biofouling 16:59–67.

Aziz K, Daniel KYT, Fazal M, Muhammad ZA, Farooq S, Fan W, Fahad S, Ruiyang Z (2017a) Nitrogen nutrition in cotton and control strategies for greenhouse gas emissions: A review. Environ Sci Pollut Res 24:23471–23487. https://doi.org/10.1007/s11356-017-0131-y.

Aziz K, Daniel KYT, Muhammad ZA, Honghai L, Shahbaz AT, Mir A, Fahad S (2017b) Nitrogen fertility and abiotic stresses management in cotton crop: A review. Environ Sci Pollut Res 24:14551–14566. https://doi.org/10.1007/s11356-017-8920-x.

Baseer M, Adnan M, Fazal M, Fahad S, Muhammad S, Fazli W, Muhammad A, Jr. Amanullah, Depeng W, Saud S, Muhammad N, Muhammad Z, Fazli S, Beena S, Mian AR, Ishaq AM (2019) Substituting urea by organic wastes for improving maize yield in alkaline soil. J Plant Nutr. https://doi.org/10.1080/01904167.2019.1659344.

Bauer H, Fuerhacker M, Zibuschka F, Schmid H, Puxbaum H (2002) Bacteria and fungi in aerosols generated by two different types of wastewater treatment plants. Water Res 36:3965–3970.

Bayram AY, Seher Ö, Nazlican A (2020) Climate change forecasting and modeling for the year of 2050. In: Fahad S, Hasanuzzaman M, Alam M, Ullah H, Saeed M, Khan AK, Adnan M (eds) Environment, climate, plant and vegetation growth. Springer Publ Ltd, Springer Nature Switzerland AG. Part of Springer Nature. pp 109–122. https://doi.org/10.1007/978-3-030-49732-3.

Berendsen RL, Pieterse CM, Bakker PA (2012) The rhizosphere microbiome and plant health. Trends Plant Sci 17:478–486.

Brock TD (2012) Thermophilic microorganisms and life at high temperatures. Springer Science & Business Media New York.

Cai B, Han Y, Liu B, Ren Y, Jiang S (2003) Isolation and characterization of an atrazine-degrading bacterium from industrial wastewater in China. Lett App Microbiol 36:272–276.

Cary SC, McDonald IR, Barrett JE, Cowan DA (2010) On the rocks: The microbiology of Antarctic dry valley soils. Nat Rev Microbiol 8:129–138.

Cevallos-Cevallos JM, Reyes-De-Corcuera JI, Etxeberria E, Danyluk MD, Rodrick GE (2009) Metabolomic analysis in food science: A review. Trends Food Sci Technol 20:557–566.

Chang W, Qiujuan J, Evgenios A, Haitao L, Gezi L, Jingjing Z, Fahad S, Ying J (2021) Hormetic effects of zinc on growth and antioxidant defense system of wheat plants. Sci Total Environ. https://doi.org/10.1016/j.scitotenv.2021.150992.

Chen Y, Guo Z, Dong L, Fu Z, Zheng Q, Zhang G, Qin L, Sun X, Shi Z, Fahad S, Xie F, Saud S. (2021) Turf performance and physiological responses of native Poa species to summer stress in Northeast China. PeerJ 9:e12252. http://doi.org/10.7717/peerj.12252.

Congeevaram S, Dhanarani S, Park J, Dexilin M, Thamaraiselvi K (2007) Biosorption of chromium and nickel by heavy metal resistant fungal and bacterial isolates. J Hazard Mater 146:270–277.

Das S, Lyla P, Khan SA (2006) Marine microbial diversity and ecology: Importance and future perspectives. Curr Sci 1325–1335.

Davis MB, Shaw RG (2001) Range shifts and adaptive responses to quaternary climate change. Science 292:673–679.

Dawson W, Hör J, Egert M, Van Kleunen M, Pester M (2017) A small number of low-abundance bacteria dominate plant species-specific responses during Rhizosphere colonization. Front Microbiol 8.

De Carolis E, et al (2012) Species identification of Aspergillus, Fusarium and Mucorales with direct surface analysis by matrix-assisted laser desorption ionization time-of-flight mass spectrometry. Clin Microbiol Infect 18:475–484.

Dentinger B, Margaritescu S, Moncalvo J (2010) Rapid and reliable high-throughput methods of DNA extraction for use in barcoding and molecular systematics of mushrooms. Mol Ecol Res 10:628–633.

Depeng W, Fahad S, Saud S, Muhammad K, Aziz K, Mohammad NK, Hafiz MH, Wajid N (2018) Morphological acclimation to agronomic manipulation in leaf dispersion and orientation to promote "Ideotype" breeding: Evidence from 3D visual modeling of "super" rice (*Oryza sativa* L.). Plant Physiol Biochem 135:499–510. https://doi.org/10.1016/j.plaphy.2018.11.010.

Deepranjan S, Ardith S, Siva OD, Sonam S, Shikha, Manoj P, Amitava R, Sayyed RZ., Abdul G, Mohammad JA, Subhan D, Fahad S, Rahul D (2021) Optimizing nutrient use efficiency, productivity, energetics, and economics of red cabbage following mineral fertilization and biopriming with compatible rhizosphere microbes. Sci Rep 11:15680.

Di Francesco A, Ugolini L, Lazzeri L, Mari M (2015) Production of volatile organic compounds by Aureobasidium pullulans as a potential mechanism of action against postharvest fruit pathogens. Biolog Con 81:8–14.

Emre B, Ömer SU, Martín LB, Andre D, Fahad S, Rahul D, Muhammad Z-ul-H, Ghulam SH, Subhan D (2021) Studying soil erosion by evaluating changes in physico-chemical properties of soils under different land-use types. J Saudi Soc Agric Sci 20:190–197.

Fahad S, Abdul B, Adnan M (eds) (2018b) Global wheat production. IntechOpen, United Kingdom. http://dx.doi.org/10.5772/intechopen.72559.

Fahad S, Adnan M, Hassan S, Saud S, Hussain S, Wu C, Wang D, Hakeem KR, Alharby HF, Turan V, Khan MA, Huang J (2019b) Rice responses and tolerance to high temperature. In: Hasanuzzaman M, Fujita M, Nahar K, Biswas JK (eds) Advances in rice research for abiotic stress tolerance. Woodhead Publ Ltd, Abington Hall Abington, Cambridge, UK, pp 201–224.

Fahad S, Bajwa AA, Nazir U, Anjum SA, Farooq A, Zohaib A, Sadia S, NasimW, Adkins S, Saud S, Ihsan MZ, Alharby H, Wu C, Wang D, Huang J (2017) Crop production under drought and heat stress: Plant responses and management options. Front Plant Sci 8:1147. https://doi.org/10.3389/fpls.2017.01147.

Fahad S, Bano A (2012) Effect of salicylic acid on physiological and biochemical characterization of maize grown in saline area. Pak J Bot 44:1433–1438.

Fahad S, Chen Y, Saud S, Wang K, Xiong D, Chen C, Wu C, Shah F, Nie L, Huang J (2013) Ultraviolet radiation effect on photosynthetic pigments, biochemical attributes, antioxidant enzyme activity and hormonal contents of wheat. J Food Agric Environ 11(3&4):1635–1641.

Fahad S, Hasanuzzaman M, Alam M, Ullah H, Saeed M, Ali Khan I, Adnan M (eds) (2020) Environment, climate, plant and vegetation growth. Springer Nature Switzerland AG. https://doi.org/10.1007/978-3-030-49732-3.

Fahad S, Hussain S, Bano A, Saud S, Hassan S, Shan D, Khan FA, Khan F, Chen Y, Wu C, Tabassum MA, Chun MX, Afzal M, Jan A, Jan MT, Huang J (2014a) Potential role of phytohormones and plant growth-promoting rhizobacteria in abiotic stresses: Consequences for changing environment. Environ Sci Pollut Res 22(7):4907–4921. https://doi.org/10.1007/s11356-014-3754-2.

Fahad S, Hussain S, Matloob A, Khan FA, Khaliq A, Saud S, Hassan S, Shan D, Khan F, Ullah N, Faiq M, Khan MR, Tareen AK, Khan A, Ullah A, Ullah N, Huang J (2014b) Phytohormones and plant responses to salinity stress: A review. Plant Growth Regul 75(2):391–404. https://doi.org/10.1007/s10725-014-0013-y.

Fahad S, Hussain S, Saud S, Hassan S, Chauhan BS, Khan F, et al (2016a) Responses of rapid viscoanalyzer profile and other rice grain qualities to exogenously applied plant growth regulators under high day and high night temperatures. PLoS One 11(7):e0159590. https://doi.org/10.1371/journal.pone.0159590.

Fahad S, Hussain S, Saud S, Hassan S, Ihsan Z, Shah AN, Wu C, Yousaf M, Nasim W, Alharby H, Alghabari F, Huang J (2016c) Exogenously applied plant growth regulators enhance the morphophysiological growth and yield of rice under high temperature. Front Plant Sci 7:1250. https://doi.org/10.3389/fpls.2016.01250.

Fahad S, Hussain S, Saud S, Hassan S, Tanveer M, Ihsan MZ, Shah AN, Ullah A, Nasrullah KF, Ullah S, Alharby H, Wu C, Huang J (2016d) A combined application of biochar and phosphorus alleviates heat-induced adversities on physiological, agronomical and quality attributes of rice. Plant Physiol Biochem 103:191–198.

Fahad S, Hussain S, Saud S, Khan F, Hassan S, Nasim W, Arif M, Wang F, Huang J (2016b) Exogenously applied plant growth regulators affect heat-stressed rice pollens. J Agron Crop Sci 202:139–150.

Fahad S, Hussain S, Saud S, Tanveer M, Bajwa AA, Hassan S, Shah AN, Ullah A, Wu C, Khan FA, Shah F, Ullah S, Chen Y, Huang J (2015a) A biochar application protects rice pollen from high-temperature stress. Plant Physiol Biochem 96:281–287.

Fahad S, Muhammad ZI, Abdul K, Ihsanullah D, Saud S, Saleh A, Wajid N, Muhammad A, Imtiaz AK, Chao W, Depeng W, Jianliang H (2018a) Consequences of high temperature under changing climate optima for rice pollen characteristics-concepts and perspectives, Archives Agron Soil Sci. https://doi.org/10.1080/03650340.2018.1443213.

Fahad S, Nie L, Chen Y, Wu C, Xiong D, Saud S, Hongyan L, Cui K, Huang J (2015b) Crop plant hormones and environmental stress. Sustain Agric Rev 15:371–400.

Fahad S, Rehman A, Shahzad B, Tanveer M, Saud S, Kamran M, Ihtisham M., Khan SU, Turan V, Rahman MHU (2019a) Rice responses and tolerance to metal/metalloid toxicity. In: Hasanuzzaman M, Fujita M, Nahar K, Biswas JK (eds) Advances in rice research for abiotic stress tolerance. Woodhead Publ Ltd, Abington Hall Abington, Cambridge, UK, pp 299–312.

Fahad S, Saud S, Yajun C, Chao W, Depeng W (eds) (2021f) Abiotic stress in plants. IntechOpen, United Kingdom. http://dx.doi.org/10.5772/intechopen.91549.

Fahad S, Sönmez O, Saud S, Wang D, Wu C, Adnan M, Arif M, Amanullah (eds) (2021e) Engineering tolerance in crop plants against abiotic stress, 1st edn. ed, Footprints of climate variability on plant diversity. CRC Press, Boca Raton, FL.

Fahad S, Sönmez O, Saud S, Wang D, Wu C, Adnan M, Turan V (eds) (2021b) Climate change and plants: biodiversity, growth and interactions, 1st edn. ed, Footprints of climate variability on plant diversity. CRC Press, Boca Raton, FL.

Fahad S, Sönmez O, Saud S, Wang D, Wu C, Adnan M, Turan V (eds) (2021c) Developing climate resilient crops: improving global food security and safety, 1st edn. ed, Footprints of climate variability on plant diversity. CRC Press, Boca Raton, FL.

Fahad S, Sönmez O, Saud S, Wang D, Wu C, Adnan M, Turan V (eds) (2021a) Plant growth regulators for climate-smart agriculture, 1st edn. ed, Footprints of climate variability on plant diversity. CRC Press, Boca Raton, FL.

Fahad S, Sönmez O, Turan V, Adnan M, Saud S, Wu C, Wang D (eds) (2021d) Sustainable soil and land management and climate change, 1st edn. ed, Footprints of climate variability on plant diversity. CRC Press, Boca Raton, FL.

Farah R, Muhammad R, Muhammad SA, Tahira Y, Muhammad AA, Maryam A, Shafaqat A, Rashid M, Muhammad R, Qaiser H, Afia Z, Muhammad AA, Muhammad A, Fahad S (2020) Alternative and non-conventional soil and crop management strategies for increasing water use efficiency. In: Fahad S, Hasanuzzaman M, Alam M, Ullah H, Saeed M, Khan AK, Adnan M (eds) Environment, climate, plant and vegetation growth. Springer Publ Ltd, Springer Nature Switzerland AG. Part of Springer Nature. pp 323–338. https://doi.org/10.1007/978-3-030-49732-3.

Farhana G, Ishfaq A, Muhammad A, Dawood J, Fahad S, Xiuling L, Depeng W, Muhammad F, Muhammad F, Syed AS (2020) Use of crop growth model to simulate the impact of climate change on yield of various wheat cultivars under different agro-environmental conditions in Khyber Pakhtunkhwa, Pakistan. Arabian J Geosci 13:112 https://doi.org/10.1007/s12517-020-5118-1.

Farhat A, Hafiz MH, Wajid I, Aitazaz AF, Hafiz FB, Zahida Z, Fahad S, Wajid F, Artemi C (2020) A review of soil carbon dynamics resulting from agricultural practices. J Environ Manage 268:110319.

Fazli W, Muhmmad S, Amjad A, Fahad S, Muhammad A, Muhammad N, Ishaq AM, Imtiaz AK, Mukhtar A, Muhammad S, Muhammad I, Rafi U, Haroon I, Muhammad A (2020) Plant-microbes interactions and functions in changing climate. In: Fahad S, Hasanuzzaman M, Alam M, Ullah H, Saeed M, Khan AK, Adnan M (eds) Environment, climate, plant and vegetation growth. Springer Publ Ltd, Springer Nature Switzerland AG. Part of Springer Nature. pp 397–420. https://doi.org/10.1007/978-3-030-49732-3.

Feinstein LM, Blackwood CB (2012) Taxa–area relationship and neutral dynamics influence the diversity of fungal communities on senesced tree leaves. Environ Microbiol 14:1488–1499.

Ferreira-Pinto M, Moura-Guedes M, Barreiro M, Pais I, Santos M, Silva M (2006) Aureobasidium pullulans as a biocontrol agent of blue mold in" Rocha"pear. Comm Agr Appl Biol Sci 71:973–978.

Fierer N, Bradford MA, Jackson RB (2007) Toward an ecological classification of soil bacteria. Ecology 88:1354–1364.

Flores GE, Bates ST, Knights D, Lauber CL, Stombaugh J, Knight R, Fierer N (2011) Microbial biogeography of public restroom surfaces. PLoS One 6:e28132.

Gazi A, Skevis G, Founti M (2012) Energy efficiency and environmental assessment of a typical marble quarry and processing plant. J Clean Prod 32:10–21.

Ghulam M, Muhammad AA, Donald LS, Sajid M, Muhammad FQ, Niaz A, Ateeq ur R, Shakeel A, Sajjad H, Muhammad A, Summia M, Aqib HAK, Fahad S, Rahul D, Mazhar I, Timothy DS (2021) Formalin fumigation and steaming of various composts differentially influence the nutrient release, growth and yield of muskmelon (*Cucumis melo* L.). Sci Rep 11:21057.

Gopakumar L, Bernard NO, Donato V (2020) Soil microarthropods and nutrient cycling. In: Fahad S, Hasanuzzaman M, Alam M, Ullah H, Saeed M, Khan AK, Adnan M (eds) Environment, climate, plant and vegetation growth. Springer Publ Ltd, Springer Nature Switzerland AG. Part of Springer Nature. pp 453–472. https://doi.org/10.1007/978-3-030-49732-3.

Gray NF (1989) Biology of wastewater treatment. Oxford University Press, Oxford.

Gregersen T (1978) Rapid method for distinction of Gram-negative from Gram-positive bacteria. Eur J Appl Microbiol Biotechnol 5:123–127.

Griffiths MJ, van Hille RP, Harrison ST (2012) Lipid productivity, settling potential and fatty acid profile of 11 microalgal species grown under nitrogen replete and limited conditions. J App Phycol 24:989–1001.

Griffiths RI, Thomson BC, James P, Bell T, Bailey M, Whiteley AS (2011) The bacterial biogeography of British soils. Enviorn Microbiol 13:1642–1654.

Guisan A, Zimmermann NE (2000) Predictive habitat distribution models in ecology. Ecol Model 135:147–186.

Guo LD, Huang GR, Wang Y, He WH, Zheng WH, Hyde KD (2003) Molecular identification of white morphotype strains of endophytic fungi from Pinus tabulaeformis. Mycol Res 107:680–688.

Guofu L, Zhenjian B, Fahad S, Guowen C, Zhixin X, Hao G, Dandan L, Yulong L, Bing L, Guoxu J, Saud S (2021) Compositional and structural changes in soil microbial communities in response to straw mulching and plant revegetation in an abandoned artificial pasture in Northeast China. Global Ecol Conserv 31:e01871.

Habib ur R, Ashfaq A, Aftab W, Manzoor H, Fahd R, Wajid I, Md. Aminul I, Vakhtang S, Muhammad A, Asmat U, Abdul W, Syeda RS, Shah S, Shahbaz K, Fahad S, Manzoor H, Saddam H, Wajid N (2017) Application of CSM-CROPGRO-Cotton model for cultivars and optimum planting dates: Evaluation in changing semi-arid climate. Field Crops Res. http://dx.doi.org/10.1016/j.fcr.2017.07.007.

Hadi F, Rahman AU, Ibrar M, Dastagir G, Arif M, Naveed K, Adnan M (2014) Weed diversity with special reference to their ethnomedicinal uses in wheat and maize at Rech valley, Hindokush Range, Chitral, Pakistan. Pak J Weed Sci Res 20(3):335–346.

Hafeez M, Farman U, Muhammad MK, Xiaowei L, Zhijun Z, Sakhawat S, Muhammad I, Mohammed AA, Mandela F-G, Nicolas D, Muzammal R, Fahad S, Yaobin L (2021) Metabolic-based insecticide resistance mechanism and ecofriendly approaches for controlling of beet armyworm Spodoptera exigua:a review. Environ Sci Pollut Res. https://doi.org/10.1007/s11356-021-16974-w.

Hafiz MH, Abdul K, Farhat A, Wajid F, Fahad S, Muhammad A, Ghulam MS, Wajid N, Muhammad M, Hafiz FB (2020b) Comparative effects of organic and inorganic fertilizers on soil organic carbon and wheat productivity under arid region. Commun Soil Sci Plant Anal. https://doi.org/10.1080/00103624.2020.1763385.

Hafiz MH, Farhat A, Ashfaq A, Hafiz FB, Wajid F, Carol Jo W, Fahad S, Gerrit H (2020a) Predicting kernel growth of maize under controlled water and nitrogen applications. Int J Plant Prod. https://doi.org/10.1007/s42106-020-00110-8.

Hafiz MH, Farhat A, Shafqat S, Fahad S, Artemi C, Wajid F, Chaves CB, Wajid N, Muhammad M, Hafiz FB (2018) Offsetting land degradation through nitrogen and water management during maize cultivation under arid conditions. Land Degrad Dev:1–10. https://doi.org/10.1002/ldr.2933.

Hafiz MH, Muhammad A, Farhat A, Hafiz FB, Saeed AQ, Muhammad M, Fahad S, Muhammad A (2019) Environmental factors affecting the frequency of road traffic accidents: A case study of sub-urban area of Pakistan. Environ Sci Pollut Res. https://doi.org/10. 1007/s11356-019-04752-8.

Hafiz MH, Wajid F, Farhat A, Fahad S, Shafqat S, Wajid N, Hafiz FB (2016) Maize plant nitrogen uptake dynamics at limited irrigation water and nitrogen. Environ Sci Pollut Res 24(3):2549–2557. https://doi.org/10. 1007/s11356-016-8031-0.

Haider SA, Lalarukh I, Amjad SF, Mansoora N, Naz M, Naeem M, Bukhari SA, Shahbaz M, Ali SA, Marfo TD, Subhan D, Rahul D, Fahad S (2021) Drought stress alleviation by potassium-nitrate-containing chitosan/montmorillonite microparticles confers changes in Spinacia oleracea L. Sustainability 13:9903. https://doi.org/10.3390/su13179903.

Hamza SM, Xiukang W, Sajjad A, Sadia Z, Muhammad N, Adnan M, Fahad S et al. (2021) Interactive effects of gibberellic acid and NPK on morpho-physio-biochemical traits and organic acid exudation pattern in coriander (Coriandrum sativum L.) grown in soil artificially spiked with boron. Plant Physiol Biochem 167:884–900.

Hawkes CV, Kivlin SN, Rocca JD, Huguet V, Thomsen MA, Suttle KB (2011) Fungal community responses to precipitation. Global Change Biol 17:1637–1645.

He Z et al. (2010) GeoChip 3.0 as a high-throughput tool for analyzing microbial community composition, structure and functional activity. ISME J 4:1167–1179.

Hesham FA, Fahad S (2020) Melatonin application enhances biochar efficiency for drought tolerance in maize varieties:Modifications in physio-biochemical machinery. Agron J 112(4):1–22.

Huang Li Y, Li Xiao X, Zhang Yun B, Fahad S, Wang F (2021) dep1 improves rice grain yield and nitrogen use efficiency simultaneously by enhancing nitrogen and dry matter translocation. J Integrative Agric. https://doi.org/10.1016/S2095-3119(21)63795-4.

Hussain MA, Fahad S, Rahat S, Muhammad FJ, Muhammad M, Qasid A, Ali A, Husain A, Nooral A, Babatope SA, Changbao S, Liya G, Ibrar A, Zhanmei J, Juncai H (2020) Multifunctional role of brassinosteroid and its analogues in plants. Plant Growth Regul. https://doi.org/10.1007/s10725-020-00647-8.

Ibrar K, Aneela R, Khola Z, Urooba N, Sana B, Rabia S, Ishtiaq H, Mujaddad Ur Rehman, Salvatore M (2020) Microbes and environment: Global warming reverting the frozen zombies. In: Fahad S, Hasanuzzaman M, Alam M, Ullah H, Saeed M, Khan AK, Adnan M (eds) Environment, climate, plant and vegetation growth. Springer Publ Ltd, Springer Nature Switzerland AG. Part of Springer Nature. pp 607–634. https://doi.org/10.1007/978-3-030-49732-3.

Ikram U, Khadim D, Muhammad T, Muhammad S, Fahad S (2021) Gibberellic acid and urease inhibitor optimize nitrogen uptake and yield of maize at varying nitrogen levels under changing climate. Environ Sci Pollut Res. https://doi.org/10.1007/s11356-021-16049-w.

Ilyas M, Mohammad N, Nadeem K, Ali H, Aamir HK, Kashif H, Fahad S, Aziz K, Abid U (2020) Drought tolerance strategies in plants: A mechanistic approach. J Plant Growth Regul. https://doi.org/10.1007/s00344-020-10174-5.

Ippolito A, El Ghaouth A, Wilson CL, Wisniewski M (2000) Control of postharvest decay of apple fruit by Aureobasidium pullulans and induction of defense responses. Postharvest Biol Technol 19:265–272.

Iqra M, Amna B, Shakeel I, Fatima K, Sehrish L, Hamza A, Fahad S (2020) Carbon cycle in response to global warming. In: Fahad S, Hasanuzzaman M, Alam M, Ullah H, Saeed M, Khan AK, Adnan M (eds) Environment, climate, plant and vegetation growth. Springer Publ Ltd, Springer Nature Switzerland AG. Part of Springer Nature. pp 1–16. https://doi.org/10.1007/978-3-030-49732-3.

Irfan M, Muhammad M, Muhammad JK, Khadim MD, Dost M, Ishaq AM, Waqas A, Fahad S, Saud S, et al (2021) Heavy metals immobilization and improvement in maize (Zea mays L.) growth amended with biochar and compost. Sci Rep 11:18416.

Jabborova D, Sulaymanov K, Sayyed RZ, Alotaibi SH, Enakiev Y, Azimov A, Jabbarov Z, Ansari MJ, Fahad S, Danish S, et al (2021) Mineral fertilizers improves the quality of turmeric and soil. Sustainability 13:9437. https://doi.org/10.3390/su13169437.

Jan M, Anwar-ul-Haq M, Adnan NS, Muhammad Y, Javaid I, Xiuling L, Depeng W, Fahad S (2019) Modulation in growth, gas exchange, and antioxidant activities of salt-stressed rice (Oryza sativa L.) genotypes by zinc fertilization. Arab J Geosci 12:775. https://doi.org/10.1007/s12517-019-4939-2.

Kallmeyer J, Pockalny R, Adhikari RR, Smith DC, D'Hondt S (2012) Global distribution of microbial abundance and biomass in subseafloor sediment. Proc Natl Acad Sci USA 109:16213–16216.

Kapoor A, Viraraghavan T, Cullimore DR (1999) Removal of heavy metals using the fungus Aspergillus niger. Bioresour Technol 70:95–104.

Kamarn M, Wenwen C, Irshad A, Xiangping M, Xudong Z, Wennan S, Junzhi C, Shakeel A, Fahad S, Qingfang H, Tiening L (2017) Effect of paclobutrazol, a potential growth regulator on stalk mechanical strength, lignin accumulation and its relation with lodging resistance of maize. Plant Growth Regul 84:317–332. https://doi.org/10.1007/s10725-017-0342-8.

Karaşahin M, Terzi S (2007) Evaluation of marble waste dust in the mixture of asphaltic concrete. Constr Build Mater 21:616–620.

Khadim D, Fahad S, Jahangir MMR, Iqbal M, Syed SA, Shah AK, Ishaq AM, Rahul D, et al (2021a) Biochar and urease inhibitor mitigate $NH_3$ and $N_2O$ emissions and improve wheat yield in a urea fertilized alkaline soil. Sci Rep 11:17413.

Khadim D, Saif-ur-R, Fahad S, Syed SA, Shah AK, et al (2021b) Influence of variable biochar concentration on yield-scaled nitrous oxide emissions, Wheat yield and nitrogen use efficiency. Sci Rep 11:16774.

Khan A, Sharif M, Ali AS, Shah NM, Mian IA, Wahid F, Jan B, Adnan M, Nawaz S, Ali N (2013) Potential of AM fungi in phytoremediation of heavy metals and effect on yield of wheat crop. Am J Plant Sci 5:1578–1586.

Khan MMH, Niaz A, Umber G, Muqarrab A, Muhammad AA, Muhammad I, Shabir H, Shah F, Vibhor A, Shams H. A-H, Reham A, Syed MBA, Nadiyah MA, Ali TKZ, Subhan D, Rahul D (2021) Synchronization of Boron application methods and rates is environmentally friendly approach to improve quality attributes of Mangifera indica L. On sustainable basis. Saudi J Bio Sci. https://doi.org/10.1016/j.sjbs.2021.10.036.

Kaeberlein T, Lewis K, Epstein SS (2002) Isolating "uncultivable" microorganisms in pure culture in a simulated natural environment. Science 296:1127–1129.

Knoche R, Dingwell D, Webb S (1995) Melt densities for leucogranites and granitic pegmatites: Partial molar volumes for SiO2, $Al_2O_3$, Na2O, K2O, Li2O, Rb2O, Cs2O, MgO, CaO, SrO, BaO, B2O3, P2O5, F2O– 1, TiO2, Nb2O5, Ta2O5, and WO3. Geochimica et cosmochimica acta 59:4645–4652.

Kowalchuk GA, Buma DS, de Boer W, Klinkhamer PG, van Veen JA (2002) Effects of above-ground plant species composition and diversity on the diversity of soil-borne microorganisms. Antonie Van Leeuwenhoek 81:509.

Landesman WJ, Dighton J (2010) Response of soil microbial communities and the production of plant-available nitrogen to a two-year rainfall manipulation in the New Jersey Pinelands. Soil Biol Biochem 42:1751–1758.

Lee CK, Barbier BA, Bottos EM, McDonald IR, Cary SC (2012) The inter-valley soil comparative survey: The ecology of dry valley edaphic microbial communities. ISME J 6:1046–1057.

Leung WC, Wong M-F, Chua H, Lo W, Yu P, Leung C (2000) Removal and recovery of heavy metals by bacteria isolated from activated sludge treating industrial effluents and municipal wastewater. WaterSci Tech 41:233–240.

Lu F, Kang X, Jiang C, Lou B, Jiang M, Way MO (2013) Isolation and characterization of bacteria from mid-gut of the rice water weevil (Coleoptera: Curculionidae). Environ Entomol 42:874–881.

Maiti S, Ray D, Mitra D, Mukhopadhyay A (2013) Isolation and characterisation of starch/polyvinyl alcohol degrading fungi from aerobic compost environment. Int Biodeter Biodegr 82:9–12.

Mahar A, Amjad A, Altaf HL, Fazli W, Ronghua L, Muhammad A, Fahad S, Muhammad A, Rafiullah, Imtiaz AK, Zengqiang Z (2020) Promising technologies for Cd-contaminated soils: Drawbacks and possibilities. In: Fahad S, Hasanuzzaman M, Alam M, Ullah H, Saeed M, Khan AK, Adnan M (eds) Environment, climate, plant and vegetation growth. Springer Publ Ltd, Springer Nature Switzerland AG. Part of Springer Nature. pp 63–92. https://doi.org/10.1007/978-3-030-49732-3.

Mahmood Ul H, Tassaduq R, Chandni I, Adnan A, Muhammad A, Muhammad MA, Muhammad H-ur-R, Mehmood AN, Alam S, Fahad S (2021) Linking plants functioning to adaptive responses under heat stress conditions: A mechanistic review. J Plant Growth Regul. https://doi.org/10.1007/s00344-021-10493-1.

Manter DK, Vivanco JM (2007) Use of the ITS primers, ITS1F and ITS4, to characterize fungal abundance and diversity in mixed-template samples by qPCR and length heterogeneity analysis. J Microbiol Meth 71:7–14

Manzer HS, Saud A, Soumya M, Abdullah A. Al-A, Qasi DA, Bander MA. Al-M, Hayssam MA, Hazem MK, Fahad S, Vishnu DR, Om PN (2021) Molybdenum and hydrogen sulfide synergistically mitigate arsenic toxicity by modulating defense system, nitrogen and cysteine assimilation in faba bean (*Vicia faba* L.) seedlings. Environ Pollut 290:117953.

Matkar K, Chapla D, Divecha J, Nighojkar A, Madamwar D (2013) Production of cellulase by a newly isolated strain of Aspergillus sydowii and its optimization under submerged fermentation. Int Biodeter Biodegr 78:24–33.

McIntyre NE, Knowles-Yánez K, Hope D (2008) Urban Ecol as an interdisciplinary field: Differences in the use of "urban" between the social and natural sciences. Urban Ecol:49–65.

Md Jakir H, Allah B (2020) Development and applications of transplastomic plants; a way towards eco-friendly agriculture. In: Fahad S, Hasanuzzaman M, Alam M, Ullah H, Saeed M, Khan AK, Adnan M (eds) Environment, climate, plant and vegetation growth. Springer Publ Ltd, Springer Nature Switzerland AG. Part of Springer Nature. pp 285–322. https://doi.org/10.1007/978-3-030-49732-3.

Md. Enamul H, Shoeb AZM, Mallik AH, Fahad S, Kamruzzaman MM, Akib J, Nayyer S, K. M. Mehedi A, Swati AS, Md Yeamin A, Most SS (2020) Measuring vulnerability to environmental hazards: Qualitative to quantitative. in: Fahad S, Hasanuzzaman M, Alam M, Ullah H, Saeed M, Khan AK, Adnan M (eds) Environment, climate, plant and vegetation growth. Springer Publ Ltd, Springer Nature Switzerland AG. Part of Springer Nature. pp 421–452. https://doi.org/10.1007/978-3-030-49732-3.

Mendoza J-MF, Feced M, Feijoo G, Josa A, Gabarrell X, Rieradevall J (2014) Life cycle inventory analysis of granite production from cradle to gate. Int J Life Cycle Assess 19:153–165.

Mubeen M, Ashfaq A, Hafiz MH, Muhammad A, Hafiz UF, Mazhar S, Muhammad Sami ul Din, Asad A, Amjed A, Fahad S, Wajid N (2020) Evaluating the climate change impact on water use efficiency of cotton-wheat in semi-arid conditions using DSSAT model. J Water Climate Change. https://doi.org/10.2166/wcc.2019.179/622035/jwc2019179.pdf.

Muhammad N, Muqarrab A, Khurram S, Fiaz A, Fahim N, Muhammad A, Shazia A, Omaima N, Sulaiman AA, Fahad S, Subhan D, Rahul D (2021) Kaolin and Jasmonic acid improved cotton productivity under water stress conditions. J Saudi Soc Agric Sci 28:6606–6614. https://doi.org/10.1016/j.sjbs.2021.07.043.

Muhammad Z, Abdul MK, Abdul MS, Kenneth BM, Muhammad S, Shahen S, Ibadullah J, Fahad S (2019) Performance of Aeluropus lagopoides (mangrove grass) ecotypes, a potential turfgrass, under high saline conditions. Environ Sci Pollut Res. https://doi.org/10.1007/s11356-019-04838-3.

Muhammad Tahir ul Q, Amna F, Amna B, Barira Z, Xitong Z, Ling-Ling C (2020) Effectiveness of conventional crop improvement strategies vs. omics. In: Fahad S, Hasanuzzaman M, Alam M, Ullah H, Saeed M, Khan AK, Adnan M (eds) Environment, climate, plant and vegetation growth. Springer Publ Ltd, Springer Nature Switzerland AG. Part of Springer Nature. pp 253–284. https://doi.org/10.1007/978-3-030-49732-3.

Muzammal R, Fahad S, Guanghui D, Xia C, Yang Y, Kailei T, Lijun L, Fei-Hu L, Gang D (2021) Evaluation of hemp (Cannabis sativa L.) as an industrial crop: A review. Environ Sci Pollut Res. https://doi.org/10.1007/s11356-021-16264-5.

Nakada M, Tanaka C, Tsunewaki K, Tsuda M (1994) RFLP analysis for species separation in the genera Bipolaris and Curvularia. Mycoscience 35:271–278.

Noor M, Naveed ur R, Ajmal J, Fahad S, Muhammad A, Fazli W, Saud S, Hassan S (2020) Climate change and costal plant lives. In: Fahad S, Hasanuzzaman M, Alam M, Ullah H, Saeed M, Khan AK, Adnan M (eds) Environment, climate, plant and vegetation growth. Springer Publ Ltd, Springer Nature Switzerland AG. Part of Springer Nature. pp 93–108. https://doi.org/10.1007/978-3-030-49732-3.

Nugent RP, Krohn MA, Hillier SL (1991) Reliability of diagnosing bacterial vaginosis is improved by a standardized method of gram stain interpretation. J Clin Microbiol 29:297–301.

Økland RH, Eilertsen O (1993) Vegetation-environment relationships of boreal coniferous forests in the Solhomfjell area, Gjerstad, S. Norway. Sommerfeltia 16:1–259

Ottow J, Glathe H (1968) Rose bengal-malt extract-agar, a simple medium for the simultaneous isolation and enumeration of fungi and actinomycetes from soil. App Microbiol 16:170.

Overmann J, Lepleux C (2016) Marine bacteria and archaea: Diversity, adaptations, and culturability. In: Stal LJ, Cretoiu MS (eds) The marine microbiome. Springer, pp 21–55.

Pel HJ, et al (2007) Genome sequencing and analysis of the versatile cell factory Aspergillus niger CBS 513.88. Nat Biotechnol 25:221–231.

Perri E, Manzo E, Tucker ME (2012) Multi-scale study of the role of the biofilm in the formation of minerals and fabrics in calcareous tufa. Sediment Geol 263:16–29.

Prescott L, Harley J, Klein D (1993) Microbiology. William C Brown, Dubuque, pp 112–122.

Qamar uz Z, Zubair A, Muhammad Y, Muhammad ZI, Abdul K, Fahad S, Safder B, Ramzani PMA, Muhammad N (2017) Zinc biofortification in rice: Leveraging agriculture to moderate hidden hunger in developing countries. Arch Agron Soil Sci 64:147–161. https://doi.org/10.1080/03650340.2017.1338343.

Quince C, Curtis TP, Sloan WT (2008) The rational exploration of microbial diversity.ISME J 2:997–1006.

Rashid M, Qaiser H, Khalid SK, Mohammad I. Al-Wabel, Zhang A, Muhammad A, Shahzada SI, Rukhsanda A, Ghulam AS, Shahzada MM, Sarosh A, Muhammad FQ (2020) Prospects of biochar in alkaline soils to mitigate climate change. In: Fahad S, Hasanuzzaman M, Alam M, Ullah H, Saeed M, Khan AK, Adnan M (eds) Environment, climate, plant and vegetation growth. Springer Publ Ltd, Springer Nature Switzerland AG. Part of Springer Nature. pp 133–150. https://doi.org/10.1007/978-3-030-49732-3.

Rehman M, Fahad S, Saleem MH, Hafeez M, Muhammad Habib ur R, Liu F, Deng G (2020) Red light optimized physiological traits and enhanced the growth of ramie (Boehmeria nivea L.). Photosynthetica 58(4):922–931.

Reshi ZA, Shah MA, Rasool N (2015) Microbial diversity in Kashmir Himalaya: Present status and future prospects. ENVIS Bull Himalayan Ecol 23:15.

Richardson RL, Jones M (1958) A bacteriologic census of human saliva. J Dental Res 37:697–709.

Rosling A, et al (2011) Archaeorhizomycetes: Unearthing an ancient class of ubiquitous soil fungi. Science 333:876–879.

Roussos S, et al (1995) Biotechnological management of coffee pulp—isolation, screening, characterization, selection of caffeine-degrading fungi and natural microflora present in coffee pulp and husk. App Microbiol Biotechnol 42:756–762

Sadam M, Muhammad Tahir ul Q, Ghulam M, Muhammad SK, Faiz AJ (2020) Role of biotechnology in climate resilient agriculture. In: Fahad S, Hasanuzzaman M, Alam M, Ullah H, Saeed M, Khan AK, Adnan M (eds) Environment, climate, plant and vegetation growth. Springer Publ Ltd, Springer Nature Switzerland AG. Part of Springer Nature. pp 339–366. https://doi.org/10.1007/978-3-030-49732-3.

Safi UK, Ullah F, Mehmood S, Fahad S, Ahmad Rahi A, Althobaiti F, et al (2021) Antimicrobial, antioxidant and cytotoxic properties of Chenopodium glaucum L.. PLoS One 16(10):e0255502. https://doi.org/10.1371/journal.pone.0255502.

Sahrish N, Shakeel A, Ghulam A, Zartash F, Sajjad H, Mukhtar A, Muhammad AK, Ahmad K, Fahad S, Wajid N, Sezai E, Carol Jo W, Gerrit H (2022) Modeling the impact of climate warming on potato phenology. Eur J Agron 132:126404.

Sajid H, Jie H, Jing H, Shakeel A, Satyabrata N, Sumera A, Awais S, Chunquan Z, Lianfeng Z, Xiaochuang C, Qianyu J, Junhua Z (2020) Rice production under climate change: Adaptations and mitigating strategies. In: Fahad S, Hasanuzzaman M, Alam M, Ullah H, Saeed M, Khan AK, Adnan M (eds) Environment, Climate, Plant and Vegetation Growth. Springer Publ Ltd, Springer Nature Switzerland AG. Part of Springer Nature. pp 659–686. https://doi.org/10.1007/978-3-030-49732-3.

Sajjad H, Muhammad M, Ashfaq A, Waseem A, Hafiz MH, Mazhar A, Nasir M, Asad A, Hafiz UF, Syeda RS, Fahad S, Depeng W, Wajid N (2019) Using GIS tools to detect the land use/land cover changes during forty years in Lodhran district of Pakistan. Environ Sci Pollut Res. https://doi.org/10.1007/s11356-019-06072-3.

Saleem MH, Fahad S, Adnan M, Mohsin A, Muhammad SR, Muhammad K, Qurban A, Inas AH, Parashuram B, Mubassir A, Reem MH (2020a) Foliar application of gibberellic acid endorsed phytoextraction of copper and alleviates oxidative stress in jute (Corchorus capsularis L.) plant grown in highly copper-contaminated soil of China. Environ Sci Pollut Res. https://doi.org/10.1007/s11356-020-09764-3.

Saleem MH, Fahad S, Shahid UK, Mairaj D, Abid U, Ayman ELS, Akbar H, Analía L, Lijun L (2020c) Copper-induced oxidative stress, initiation of antioxidants and phytoremediation potential of flax (Linum usitatissimum L.) seedlings grown under the mixing of two different soils of China. Environ Sci Poll Res. https://doi.org/10.1007/s11356-019-07264-7.

Saleem MH, Rehman M, Fahad S, Tung SA, Iqbal N, Hassan A, Ayub A, Wahid MA, Shaukat S, Liu L, Deng G (2020b) Leaf gas exchange, oxidative stress, and physiological attributes of rapeseed (Brassica napus L.) grown under different light-emitting diodes. Photosynthetica 58(3):836–845.

Saman S, Amna B, Bani A, Muhammad Tahir ul Qamar, Rana MA, Muhammad SK (2020) QTL mapping for abiotic stresses in cereals. In: Fahad S, Hasanuzzaman M, Alam M, Ullah H, Saeed M, Khan AK, Adnan M (eds) Environment, climate, plant and vegetation growth. Springer Publ Ltd, Springer Nature Switzerland AG. Part of Springer Nature. pp 229–252. https://doi.org/10.1007/978-3-030-49732-3.

Saud S, Chen Y, Fahad S, Hussain S, Na L, Xin L, Alhussien SA (2016) Silicate application increases the photosynthesis and its associated metabolic activities in Kentucky bluegrass under drought stress and post-drought recovery. Environ Sci Pollut Res 23(17):17647–17655. https://doi.org/10.1007/s11356-016-6957-x.

Saud S, Chen Y, Long B, Fahad S, Sadiq A (2013) The different impact on the growth of cool season turf grass under the various conditions on salinity and drought stress. Int J Agric Sci Res 3:77–84.

Saud S, Fahad S, Cui G, Chen Y, Anwar S (2020) Determining nitrogen isotopes discrimination under drought stress on enzymatic activities, nitrogen isotope abundance and water contents of Kentucky bluegrass. Sci Rep 10:6415. https://doi.org/10.1038/s41598-020-63548-w.

Saud S, Fahad S, Yajun C, Ihsan MZ, Hammad HM, Nasim W, Amanullah Jr, Arif M, Alharby H (2017) Effects of nitrogen supply on water stress and recovery mechanisms in Kentucky bluegrass plants. Front Plant Sci 8:983. https://doi.org/10.3389/fpls.2017.00983.

Saud S, Li X, Chen Y, Zhang L, Fahad S, Hussain S, Sadiq A, Chen Y (2014) Silicon application increases drought tolerance of Kentucky bluegrass by improving plant water relations and morph physiological functions. SciWorld J 2014:1–10. https://doi.org/10.1155/2014/368694.

Senol C (2020) The effects of climate change on human behaviors. In: Fahad S, Hasanuzzaman M, Alam M, Ullah H, Saeed M, Khan AK, Adnan M (eds) Environment, climate, plant and vegetation growth. Springer Publ Ltd, Springer Nature Switzerland AG. Part of Springer Nature. pp 577–590. https://doi.org/10.1007/978-3-030-49732-3

Shafi MI, Adnan M, Fahad S, Fazli W, Ahsan K, Zhen Y, Subhan D, Zafar-ul-Hye M, Martin B, Rahul D (2020) Application of single superphosphate with humic acid improves the growth, yield and phosphorus uptake of wheat (Triticum aestivum L.) in calcareous soil. Agron 10:1224. https://doi.org/10.3390/agronomy10091224.

Shah F, Lixiao N, Kehui C, Tariq S, Wei W, Chang C, Liyang Z, Farhan A, Fahad S, Huang J (2013) Rice grain yield and component responses to near 2°C of warming. Field Crop Res 157:98–110.

Shahid M, et al (2014) Phylogenetic diversity analysis of Trichoderma species based on internal transcribed spacer (ITS) marker. Afr J Biotechnol 13.

Sibille I, Sime-Ngando T, Mathieu L, Block J (1998) Protozoan bacterivory and Escherichia coli survival in drinking water distribution systems. Appl Enviorn Microbiol 64:197–202.

Sidra K, Javed I, Subhan D, Allah B, Syed IUSB, Fatma B, Khaled DA, Fahad S, Omaima N, Ali TKZ, Rahul D (2021) Physio-chemical characterization of indigenous agricultural waste materials for the development of potting media. J Saudi Soc Agric Sci. https://doi.org/10.1016/j.sjbs.2021.08.058.

Sistla SA, Asao S, Schimel JP (2012) Detecting microbial N-limitation in tussock tundra soil: Implications for Arctic soil organic carbon cycling. Soil Biol Biochem 55:78–84.

Sistla SA, Schimel JP (2012) Stoichiometric flexibility as a regulator of carbon and nutrient cycling in terrestrial ecosystems under change. New Phytol 196:68–78.

Smithson P, Addison K, Atkinson K (2013) Fundamentals of the physical environment. Routledge, London.

Sterflinger K, Tesei D, Zakharova K (2012) Fungi in hot and cold deserts with particular reference to microcolonial fungi. Fungal Ecol 5:453–462.

Subhan D, Zafar-ul-Hye M, Fahad S, Saud S, Martin B, Tereza H, Rahul D (2020) Drought stress alleviation by ACC deaminase producing *Achromobacter xylosoxidans* and *Enterobacter cloacae*, with and without timber waste biochar in maize. Sustainability 12(6286). https://doi.org/10.3390/su12156286.

Tamura K, Peterson D, Peterson N, Stecher G, Nei M, Kumar S (2011) MEGA5: Molecular evolutionary genetics analysis using maximum likelihood, evolutionary distance, and maximum parsimony methods. Mol Biol Evol 28:2731–2739.

Tariq M, Ahmad S, Fahad S, Abbas G, Hussain S, Fatima Z, Nasim W, Mubeen M, ur Rehman MH, Khan MA, Adnan M (2018) The impact of climate warming and crop management on phenology of sunflower-based cropping systems in Punjab, Pakistan. Agric For Meterol 256:270–282.

Török Á, Licha T, Simon K, Siegesmund S (2011) Urban and rural limestone weathering; the contribution of dust to black crust formation. Environ Earth Sci 63:675–693.

Unsar NU, Muhammad R, Syed HMB, Asad S, Mirza AQ, Naeem I, Muhammad H ur R, Fahad S, Shafqat S (2020) Insect pests of cotton crop and management under climate change scenarios. In: Fahad S, Hasanuzzaman M, Alam M, Ullah H, Saeed M, Khan AK, Adnan M (eds) Environment, climate, plant and vegetation growth. Springer Publ Ltd, Springer Nature Switzerland AG. Part of Springer Nature. pp 367–396. https://doi.org/10.1007/978-3-030-49732-3.

Varga J, et al (2007) Aspergillus brasiliensis sp. nov., a biseriate black Aspergillus species with world-wide distribution. Int J Syst Evol Micr 57:1925–1932.

Wahid F, Fahad S, Subhan D, Adnan M, Zhen Y, Saud S, Manzer HS, Martin B, Tereza H, Rahul D (2020) Sustainable management with mycorrhizae and phosphate solubilizing bacteria for enhanced phosphorus uptake in calcareous soils. Agriculture 10(334). https://doi.org/10.3390/agriculture10080334.

Wajid N, Ashfaq A, Asad A, Muhammad T, Muhammad A, Muhammad S, Khawar J, Ghulam MS, Syeda RS, Hafiz MH, Muhammad IAR, Muhammad ZH, Muhammad Habib ur R, Veysel T, Fahad S, Suad S, Aziz K, Shahzad A (2017) Radiation efficiency and nitrogen fertilizer impacts on sunflower crop in contrasting environments of Punjab. Pak Environ Sci Pollut Res 25:1822–1836. https://doi.org/10.1007/s11356-017-0592-z.

Wen Z, Liao W, Chen S (2005) Production of cellulase/β-glucosidase by the mixed fungi culture Trichoderma reesei and Aspergillus phoenicis on dairy manure. Process Biochem 40:3087–3094.

White TJ, Bruns T, Lee S, Taylor J (1990) Amplification and direct sequencing of fungal ribosomal RNA genes for phylogenetics. PCR Protocols: A Guide to Methods and Applications 18:315–322.

Whitman WB, Coleman DC, Wiebe WJ (1998) Prokaryotes: The unseen majority. Proc Natl Acad Sci USA 95:6578–6583.

Wiegand I, Hilpert K, Hancock RE (2008) Agar and broth dilution methods to determine the minimal inhibitory concentration (MIC) of antimicrobial substances. Nat Protoc 3:163–175.

Wolf AB (2014) The effect of soil habitat connectivity on microbial interactions, community structure and diversity: a microcosm-based approach. Utrecht University, Netherlands.

Wu C, Kehui C, She T, Ganghua L, Shaohua W, Fahad S, Lixiao N, Jianliang H, Shaobing P, Yanfeng D (2020) Intensified pollination and fertilization ameliorate heat injury in rice (Oryza sativa L.) during the flowering stage. Field Crops Res 252: 107795.

Wu C, Tang S, Li G, Wang S, Fahad S, Ding Y (2019) Roles of phytohormone changes in the grain yield of rice plants exposed to heat: A review. PeerJ 7:e7792. https://doi.org/10.7717/peerj.7792.

Yang Z, Zhang Z, Zhang T, Fahad S, Cui K, Nie L, Peng S, Huang J (2017) The effect of season-long temperature increases on rice cultivars grown in the central and southern regions of China. Front Plant Sci 8:1908. https://doi.org/10.3389/fpls.2017.01908.

Young IM, Crawford JW (2004) Interactions and self-organization in the soil-microbe complex. Science 304:1634–1637

Young IM, Crawford JW, Nunan N, Otten W, Spiers A (2008) Microbial distribution in soils: Physics and scaling. Adv Agron 100:81–121.

Yu J, Yao S, Chen R, Zhu K, Yu L (2005) A quantitative integrated evaluation of sustainable development of mineral resources of a mining city: A case study of Huangshi, Eastern China. Resour Policy 30:7–19.

Zafar-ul-Hye M, Akbar MN, Iftikhar Y, Abbas M, Zahid A, Fahad S, Datta R, Ali M, Elgorban AM, Ansari MJ et al. (2021) Rhizobacteria inoculation and Caffeic acid alleviated drought stress in lentil plants. Sustainability 13:9603. https://doi.org/10.3390/su13179603.

Zafar-ul-Hye M, Muhammad N, Subhan D, Fahad S, Rahul D, Mazhar A, Ashfaq AR, Martin B, Jiří H, Zahid HT, Muhammad N (2020a) Alleviation of cadmium adverse effects by improving nutrients uptake in bitter gourd through cadmium tolerant rhizobacteria. Environments 7(54). https://doi.org/10.3390/environments7080054.

Zafar-ul-Hye M, Tahzeeb-ul-Hassan M, Muhammad A, Fahad S, Martin B, Tereza D, Rahul D, Subhan D (2020b) Potential role of compost mixed biochar with rhizobacteria in mitigating lead toxicity in spinach. Sci Rep 10:12159. https://doi.org/10.1038/s41598-020-69183-9.

Zahida Z, Hafiz FB, Zulfiqar AS, Ghulam MS, Fahad S, Muhammad RA, Hafiz MH, Wajid N, Muhammad S (2017) Effect of water management and silicon on germination, growth, phosphorus and arsenic uptake in rice. Ecotoxicol Environ Saf 144:11–18.

Zahir SM, Zheng HG, Ala Ud D, Amjad A, Ata Ur R, Kashif J, Shah F, Saud S, Adnan M, Fazli W, Saud A, Manzer HS, Shamsher A, Wajid N, Hafiz MH, Fahad S (2021) Synthesis of silver nanoparticles using Plantago lanceolata extract and assessing their antibacterial and antioxidant activities. Sci Rep 11:20754.

Zhang D, Spadaro D, Valente S, Garibaldi A, Gullino ML (2012) Cloning, characterization, expression and antifungal activity of an alkaline serine protease of Aureobasidium pullulans PL5 involved in the biological control of postharvest pathogens. Int J Food Microbiol 153:453–464.

Zia-ur-Rehman M (2020) Environment, climate change and biodiversity. In: Fahad S, Hasanuzzaman M, Alam M, Ullah H, Saeed M, Khan AK, Adnan M (eds) Environment, climate, plant and vegetation growth. Springer Publ Ltd, Springer Nature Switzerland AG. Part of Springer Nature. pp 473–502. https://doi.org/10.1007/978-3-030-49732-3.

# 2

# *Interaction of Environment and Entomology*

**Nazish Huma Khan and Muhammad Ilyas**
*University of Swabi*

**Tooba Saeed**
*University of Peshawar*

**CONTENTS**



## 2.1 Introduction

The world and its ecosystems are in a continuous state of change (Sunday et al. 2012). In turn, the ecological impacts of Global Environmental Change cause an increasing level of atmospheric carbon dioxide, deposition of fixed nitrogen, species extinction and habitat loss. These changes are increases over short time scales and are considered as important for the human exploitation of the environment (Ceballos et al. 2017; Adnan et al. 2018; Akram et al. 2018a,b; Aziz et al. 2017a; Habib et al. 2017; Hafiz et al. 2016; Hafiz et al. 2019; Basir et al. 2015; Kamaran et al. 2017; Muhmmad et al. 2019; Sajjad et al. 2019; Saud et al. 2013; Saud et al. 2014; Saud et al. 2017; Saud et al. 2016; Shah et al. 2013; Qamar et al. 2017; Wajid et al. 2017; Yang et al. 2017; Zahida et al. 2017; Depeng et al. 2018; Fahad and Bano 2012; Fahad et al. 2017; Fahad et al. 2013; Fahad et al. 2014a,b; Fahad et al. 2016a,b,c,d; Fahad et al. 2015a,b; Fahad et al. 2018a,b; Fahad et al. 2019a,b; Fahad et al. 2020; Fahad et al. 2021a,b,c,d,e,f; Hesham and Fahad 2020; Farah et al. 2020; Farhana et al. 2020; Farhat et al. 2020; Mahmood et al. 2021; Manzer et al. 2021; Md Jakir and Allah 2020; Md. Enamul et al. 2020; Al-Wabel et al. 2020a,b; Mubeen et al. 2020; Muhammad et al. 2021; Muhammad Tahir et al. 2020; Muhammad et al. 2019; Muzammal et al. 2021).

DOI: 10.1201/9781003286400-2

Global Environmental Change not only affects the environment, population, community and organism but also leads to drastic changes in interaction between networks of different species (Tylianakis et al. 2008). Biotic interactions such as parasitism, predation and pollination play a vital role in maintaining the biodiversity, mediating ecosystem responses to Global Environmental Change and stability of the ecosystem services upon which humans depend (Seymour et al. 2003). In this chapter, we focus on insects' interactions with the environment and their effects on the ecosystem.

The natural environment triggers the habitat of species. The specific environment of a species can have a profound effect on its performance, behavior, body size and generation time. Lack of studies on effects of Global Environmental Change on biotic interactions is due to complications in measuring variations in association with variations in biodiversity. However, interactions may be more sensitive to environmental changes because they are sensitive to behavior, psychology and related several species (Davis et al. 2011).

Nowadays, climatic change has a widespread effect on responses of organisms to the thermal environment. Changes in climates affect insects by changing their mean environmental temperature, the time period of growing season and their responses to environmental extremes (Tremmel and Müller 2012; Shah et al. 2016). Changes in environmental conditions such as global warming and climatic changes increased climatic variability and water stress which caused great threats to insects. Most of the insect species have extinct in European countries. Literature reported a 76% decline rate of flying insect biomass in Germany which accounted for 2.8% loss of insect biomass annually (Sanchez-Bayo and Wyckhuys 2019). The decline is attributed to human activities such as intensified agricultural practices, deforestation, urbanization and industrialization which mutually declared 30%–50% encroachment on the ecosystem (Ceballos et al. 2017).

### 2.1.1 Interactions between Environment and Species

Insects and arthropods make up about 90% of all plant and animal species on the earth. They are the energetic species to the ecosystem and biosphere functioning. They play important and efficient roles in the economy of society (Pimentel 1994).

Important factors of the global climate that cause extinction and alter species distribution are (1) climate, (2) nitrogen deposition, (3) enrichment of carbon dioxide, (4) biotic interactions and (5) land use pattern. These factors exert persistent influences on different hostile and mutualistic contacts among organisms (Tylianakis et al. 2008). Environmental changes may lead to variations in the intensity and attacks on herbivorous by enemy and severity of pathogenic impacts on species (Figure 2.1).

Plants are an important part of the environment and they interact with multiple organisms such as microbes and insects. Plants are exposed to many abiotic stresses such as extreme temperatures, drought and nutrient loss due to changes in environmental conditions (Pineda et al. 2010). Environmental changes affect species interactions such as plant–microbe, plant–insect and plant–microbe–insect interactions (Putten et al. 2010; Khan et al. 2016). Essential microbes (e.g., mycorrhizal fungi and rhizobacteria) help the plants to tolerate abiotic stress and to deal with herbivorous insects and pathogens. Therefore, essential microbes play an important role in a changing environment (Pineda et al. 2013).

## 2.2 Insects as Important Component of Environment

Insects are considered powerful and quick adaptive organisms having a short life cycle but a high fecundity rate. They are an important biotic component of the environment and have a role in shaping and maintaining the nature of the environment (Schowalter et al. 2018). The study of insects is called 'entomology', and this field focuses on insects and their relationship to humans, the environment and other organisms. The 'entomology' field is working to enhance insect conditions with technology along with proper strategic plans for future use. Entomologists make a great contribution to fields as diverse as agriculture, chemistry, biology, human/animal health, molecular sciences, criminology and forensics (Borror et al. 1989).

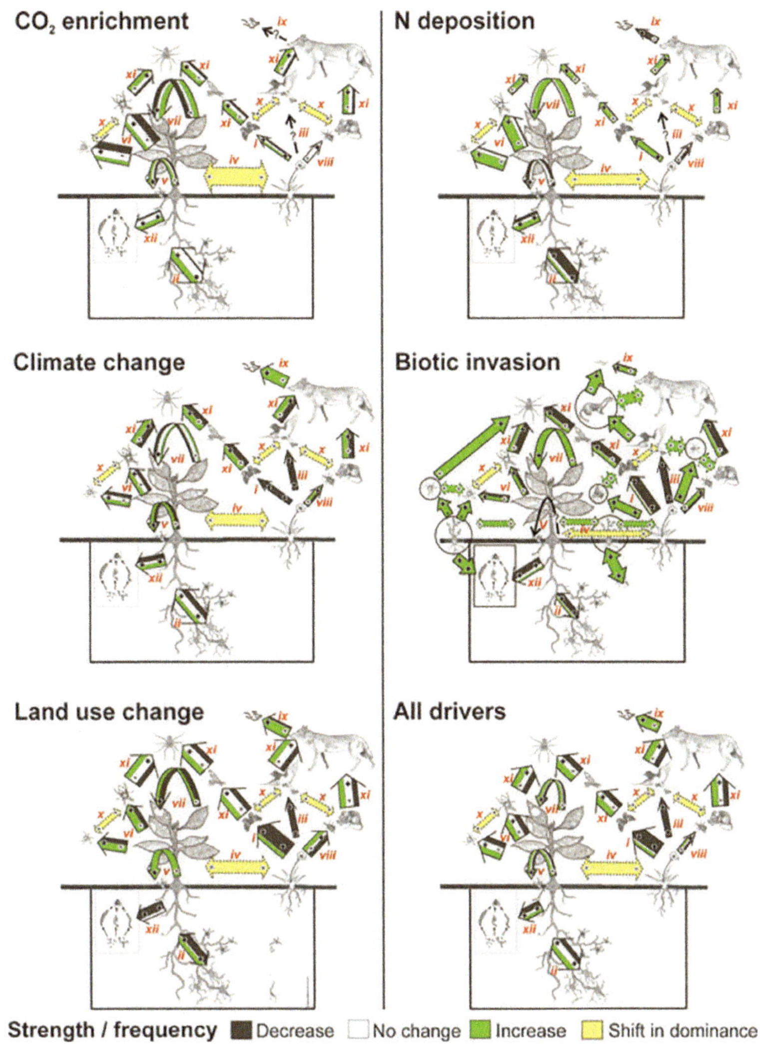


**FIGURE 2.1** . Responses of biotic interactions to environmental change.

Humans seek ways to manage insects to get maximum possible benefits and avoid harm. Entomologists all over the world are trying to find possible solutions and methods to deal with predictable insect problems by taking into account environmental protection (Hunter and Mark 2008). To understand the human–environment interactions, they observe and study the environmental indicators. They try to find better solutions to environmental problems by conserving natural resources and restoring natural habitats. From the insect's study, the entomologist's efforts are appreciated for improving our knowledge regarding the insect's ecology, inheritance and psychology (Potter and Woods 2011).

Insects have been considered important for the food webs of the majority of animal species. They alter plants abundance as herbivores or as a source in plant propagation (Yang and Gratton 2014). On the other hand, insects are considered as humans' chief competitors for food. Humans favored various insect species to develop and modify the environment for their advantage. In many cases, he created a very unnatural environment. Natural regulatory forces are not functioning properly (Yang 2012). Agriculture and monoculture provide host plant resources for many pests that far exceed resources in the natural environment. They are also responsible for transmission of plant, animal and human diseases. Humans concentrated on animal species for the benefit of those who attack and spread the disease. They offer an artificial environment for different pests in homes and industries (Davis et al. 2011).

### 2.2.1 Ecological Importance

Insects are present everywhere and are important for the ecological functioning of natural ecosystems through different activities: from organic matter decomposition to food sources for fish, birds and wildlife (Hunter and Mark, 2008). Moreover, insects perform functions of predators, parasites, saprophages, herbivores and pollinators, and also play important ecological and economic roles in both aquatic and terrestrial ecosystems (Reason 2007).

Therefore, insects' conservation has been considered essential to maintain the environment (Biesmeijer et al. 2006). Environmental disturbances can affect their functions, and insects often respond to these disturbances in a characteristic way. Because of the high diversity of insect species, they are considered important for the functioning of natural ecosystems (Raddi 2004). The presence of insects matters everywhere either in wild lands or urban points by providing vital services of ecosystems such as decomposers, pollinators, plagues enemies and maintenance of food web (Perfecto and Vandermeer 2002).

Being the main regulators of the ecosystem, insects deserve special consideration to maintain the supply of the services. Insects are the biggest competitors for resources and transmitters of human and animal diseases (Schowalter et al. 2018). Insects play an important role in finding solutions for environmental problems. Insects and their products are precious sources of food in many cultures; for example, honey is one of the well-known and essential trade products for centuries (Prather and Laws 2018). Other insects including beetles, caterpillars, grasshoppers, and cicadas are rich sources of protein. As per the record of 113 countries, the consumption rate of edible insect species is about 1500 by 3000 ethnic groups (Wegier et al. 2018, MacEvilly 2000). Some insect species have medical and industrial values such as blow fly and maggots are used for wound healing, whereas silkworms are an important industrial product for commercial silk (Ratcliffe et al. 2011).

## 2.3 Aquatic Insects, Distribution and Salinity Tolerance

Insects exhibit an unequal achievement in terrestrial and aquatic environments. Both terrestrial and aquatic environments support several million and more than 30,000 insect species, respectively (Williams and Williams 1998). It is believed that insects originated first in water (sea), called as 'marine'. They are ranging to several hundred in number. They are believed to be poorly represented in salty water as they are not fitted in saltwater environments. They are counted more in numbers to live in freshwater parts of ecosystems (Williams and Feltmate 1992). While living in saltwater environments, it causes behavioral, psychological, osmotic and genotypic changes in insects to deal with different salinity levels. These changes have been studied in mosquito fish, *Gambusia holbrooki* (Congdon 1994). It is amazing that insects with their evolution (approximately 300 million years) are not more common in saline environments. Majority of aquatic insects were found in coastline environments like mangroves, salt marshes, estuaries and decaying seaweed on the seashore (Cheng 1985).

Studies revealed that the most extreme aquatic insects adapted to the marine environment are *Philanisus plebeious*, which deposits eggs in the ceolomic cavity of starfish. After hatching, the larvae develop in tadpole and leave the host (Winterbourn and Anderson 1980). Another species is *Chironomid pontomvia* which is dependent on tropical coral reefs to complete its life cycle in a depth of 30 m (Bretschko 1982). The gerrid genus Halobates, are completely oceanic comprises of 44 species live on surface of oceans (Australia, Indo-Malaysia and Central Pacific Ocean) (Cheng 1985).

### 2.3.1 Arctic Insects and Their Environment

The Arctic environment possesses less diversity of insect biota. The process of evolution is slower and may be delayed because of climate changes. Several factors influence the insect's life in the Arctic environment such as low temperature, low heat, continuous daylight and variation in climatic conditions (Sinclair et al. 2003). In comparison to insects, the Arctic fauna has the ability to recover. However, it is believed that the Arctic will always tend to be the edge of the psychologically feasible range. Hence, there is a historical record of continuity in the evolution of the tropics where most life forms have been shaped in response to tropical conditions (Downes 1964). Insects depend on water for their food. Those insects living in deserts or terrestrial environments have the capability to resist the loss of water. Therefore, water formation in metabolism is of great importance to insects. Some insects obtain water from the atmosphere, saturated with moisture content. They lay eggs in humid environments. Dry environments damage both egg and embryo (Buxton 1932).

### 2.3.2 Responses of Insects to Low Temperature

Temperature is an important factor in both physical and biological systems (Somme 1995). Temperature variation can affect both marine and terrestrial life and sometimes badly affect the organism due to water–ice transition state (Longhurst 1998). Below-zero temperature occurs in much of the earth's surface. Insects and other ectotherms survive at low temperatures in the situation either to keep their body fluids below the ordinary melting point or survive the ice formation in their tissues (Bale 2002). This practice has been considered an important tool for more than 60 years which has led to a wide understanding of biochemical, physical, molecular and physiological bases of the insect's survival at temperatures below zero (Somme 2000). Insects are prone to quick variations in their body temperatures due to their small body size and ectothermic nature. Therefore, they have the behavioral and physiological capability to respond and stay alive at low temperatures (Teets and Denlinger 2013).

In terrestrial environments, Honeybee and vespine wasps are heterothermic whose body temperature changes with ambient temperature. During foraging state, they act as endothermic where they have to cope with an enormous heat loss due to smaller body size (Stabentheiner et al. 2012). Therefore, thermal tolerance is a fundamental psychological trait which is linked to the insect's distribution on land (Araujo et al. 2013).

Various techniques are applied to examine the cold tolerance of insects and few of these techniques directly measure mortality such as measurement of lower lethal limits and evaluation of survival capacities. Whereas other techniques are adapted to measure the insect's capabilities to the harmful effects of low temperatures such as supercooling point, recovery and chill-coma onset. Chill-coma onset and recovery are considered consistent metrics of thermal plasticity and tolerance in insects (Sinclair et al. 2015). Chill-coma onset has a considerable ecological cost for an individual as the static insects are unable to get nourishment or escape from a predator (Overgaard and MacMillan 2017).

## 2.4 Insects' Role in Ecosystem

In most of the terrestrial ecosystems, insects possess a central position in the biodiversity. But they have not been considered as valuable by neglecting them in the nutrient cycling of biodiversity and ecosystem function (Knapp et al. 2018). Insects play an important role in facilitating the correlation of plants to their ecosystems influencing their activity, composition and plant species (Weisser and Siemann 2004). Insects influence the C:N:P ratios and specific chemical compositions of litter inputs with their direct and indirect effects, microbial activities and ecosystem rates (Figure 2.2). The threat of insects' predation can affect the foraging attitude of herbivores insect and therefore litter inputs and plant communities to the soil are altered. This affects the rates of carbon, nitrogen cycling in the ecosystem.

Insects perform differently, and a few species are discussed below.

### 2.4.1 Shelter-Building Insects

Insects such as caterpillars, beetles, spiders and ants influence plant foliage to build shelters (Lill and Marquis 2007). The shelters are of different types and used by arthropods as part of their life. Shelters vary by insect species and also through ontogenetic stages of species. These structures can also be used by other organisms such as predators and herbivores (Lind et al. 2001). Different insects' larvae build simple external structures on host plants by covering, folding, tying, cutting and rolling plant leaves with silk (Cornelissen et al. 2015), as shown in Figure 2.3.

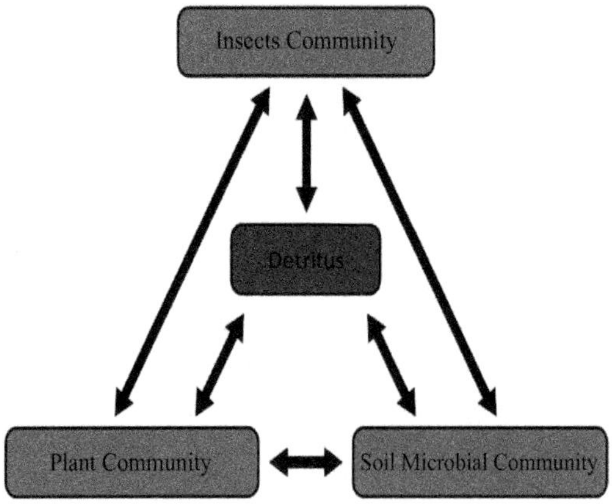


**FIGURE 2.2** A diagrammatic representation for combined and reciprocal interactions of plants, insects, microbes and detrital pool.

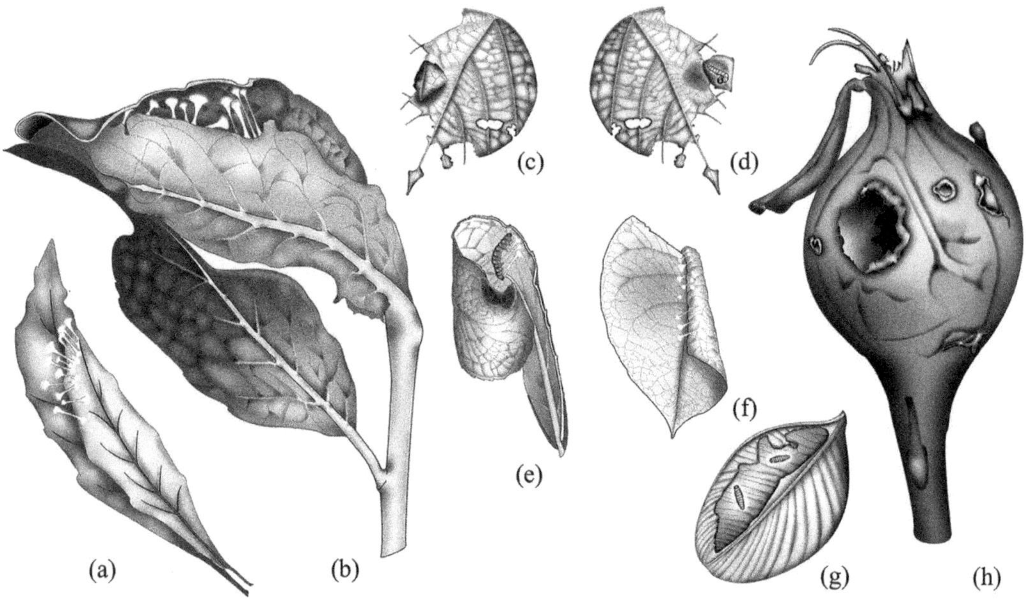


**FIGURE 2.3** Diagrammatic representation of the insect's shelter constructions: (a) closely glued leaves connected by caterpillars with silk, (b) loose leaves and open chamber where the caterpillars growth occur, (c and d) tent (triangle) leaves with exposed caterpillar, (e) roll of leaves where caterpillar develops inside the roll, (f) the roll formed rolling leaf part with silk thread, (g) the visible epidermis and growing larvae and (h) the stem hemia with an opening for insect growth and colonization by secondary users.

These structures are different and clearly distinguishable from other sheltered structures (e.g., mines and galls), where plant manipulation is more widespread and habitually only one structure is used during the growth of insects, confirming nourishment, improvement of microclimatic situations, as well as protection against ordinary competitors (Cornelissen and Stiling 2006). In conjunction, these shelter herbivores (Figure 2.4) can be considered as 'microhabitat manipulators' (Lill and Marquis 2007).

## 2.4.2 Gall Farmers

Insects inducing galls are considered parasites, and their growth on structures and organs of a plant can cause variations in vegetative and reproductive organs of their host plants (Florentine et al. 2005). The

**FIGURE 2.4** (a) A roll of leaf on Miconia species. (b) The developing larvae. (c) The tent leaf shows many caterpillars inside. (d) The larval protection against the ant. (e) Several new leaves induce in the host. (f) A rod-gall of weevil *Collabismus clitellae*. (g) Several galls that build a new shelter. (h) Shelter built of silk.

significant effects of gall insects on the growth and fitness (reduction of flowers, fruits, seeds, leaves, shoots and biomass production) of their host plant have been reported many times (Guimaraes et al. 2014, 2015). Because of their capability to manipulate host plant tissues, gall insects are the most notable and living organisms of interest (Shorthouse et al. 2005). This sophisticated guild is characterized by the induction of cell differentiation structures and organs of their host plants by their food activity (Rohfritsch 2010). These insects are beneficial for microclimatic conditions and larval protection of natural enemies as well as effective plant handling metabolism to improve the nutritious value of tissues (Fernandes et al. 2014).

Galls usually appear on both vegetative and reproductive parts of plants. The average record for gall species has been estimated as 132,930. Gall species can affect the organism's occurrence and distribution in different communities as they provide shelter, food and protection against natural enemies (Fukui 2001). Literature revealed that 7 ants and 16 arthropod species (Araneida, Coleoptera, Hymenoptera, Lepidoptera, Neuroptera and Diplopoda) are produced by cynipid Holcaspis cinerosus Bassett in the galls of stem on live oaks (Wheeler and Longino 1988).

### 2.4.3 Leaf Miners

This category includes small insects inhibited on thick areas of leaves. They are generally able to be seen on the outside of twisting parts of leaf, spots or other characteristics of faded tissues. Mines can occur all over the life of an insect (Cornelissen et al. 2015).

There are about 10,000 insect species described under this category. The structures developed by these insects on their host plants also propose a chance of colonization by insects/arthropods (Kagata and Ohgushi 2004).

### 2.4.4 Leaf Tents

Leaf tents means 'cut and folded' and can be described as the structures in which insect larvae use their mandibles to cut the flap of a leaf in a condition when the leaf is folded and attached to the main leaf with silk (Lind et al. 2001). The insect's larvae inhibit in shelter for a part of its growth and nourish on the leaf under the pointed tent. The tent also provides protection against natural enemies and unsuitable climates. Some insect species (Lepidoptera: Hesperiidae) abandon tents as they grow (Weiss et al. 2004).

### 2.4.5 Leaf Rollers and Leaf Tiers

Leaf rollers and leaf tires can be distinguished by their mode of larval inhabitants. Leaf rollers include species such as beetles and caterpillars that roll the entire leaves with silky threads to use the rollers as shelters. Whereas the larvae of Lepidopteran (particularly, Tortricidae and Gelechiidae) can be easily found on unfolded or partially folded leaves. Some insects nourish inside the rolls and others use them as shelters against natural enemies and abiotic factors (Lill and Marquis 2007). Leaf rolls are considered important defoliators of temperate plants like cottonwood, birch and oaks. Structures developed by insect species are quite different and are used by other arthropods after larval growth and development (Kozlov et al. 2012). These insects provide opportunities to other arthropods for the settlement colonization. Leaf tires mostly attack the terminal buds of plants and bind the leaves with silky threads and create shelter for feeding (Cornelissen et al. 2015).

## 2.5 Effects of Insects on Ecosystem Functioning

Insects are small invertebrates whose activities can affect the ecosystem through their modifications of detrital (dead biomass) in underground systems. Insects are a diverse group of eukaryotes on the earth. Insects do perform different functions such as pollinators, herbivores, parasites, predators, seed disperses and detritivores on terrestrial ecosystems (Weisser and Siemann 2004).

Insects' ecology addresses both their adaptation to the environment and their effects on the environment. Insects have the potential for their dramatic population growth, survival in severe environments and a variety of effects on ecosystems and other species (Schowalter 2011).

Insects can have direct and indirect effects on ecosystem functioning. The reduction of Net Primary Production by herbivores and litter decomposition by detritivores are direct effects. Insects such as cockroaches and termites are key components of soil fauna and soil food web which affect the soil's structure and health (Kehinde et al. 2014). The indirect effects include variations in matter fluxes because of changes in the composition of plant species mediated by insect herbivory. These effects cause changes in nutrient cycling and community productivity (Belovsky and Slade 2000).

Insects also effect the ecosystem functions by interacting with plants. Plants possess important links of aboveground components to belowground of the biosphere and microbes in ecosystems (Gaston 1991). These three groups (plants, microbes and insects) mediate vital ecosystem functions (Figure 2.5).

Insects are small invertebrates and ever-present species of all terrestrial and aquatic (freshwater) environments. Their collective biomass share is small compared to microbes and plants. Therefore, it is believed that they contribute a small share to the ecosystem (Yang 2006). Besides their small size and low cumulative biomass, insects have important effects in nutrient cycling (carbon, nitrogen, phosphorus) by

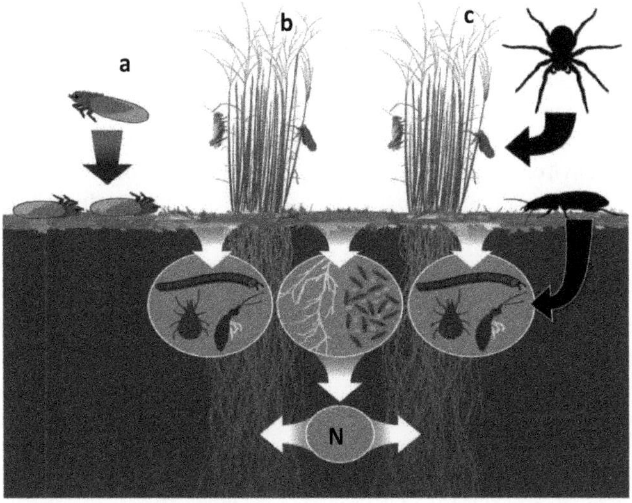


**FIGURE 2.5** (a) Direct inputs of insects to dead biomass. Insects introduce large amounts of biomass (carbon: nitrogen) into underground systems. (b) Insects and arthropods transform dead and live biomass, affecting C, N cycling in the ecosystem. (c) Arthropod's predation of decomposers creates trophic cascades that affect the biomass and decomposition rates.

modulating the quantity and quality of the resources in the food chain with significance to the ecosystem. Such properties show the interaction of insects to plants and microbes at ecosystem levels. These groups (insects, plants and microbes) propose a better way to interact between community level to ecosystem processes (Schmitz et al. 2014; Zahoor et al. 2016).

## 2.5.1 Insects' Role in Carbon Cycle

Insects such as termites perform the carbon cycle during the decomposition process. Termite intestines contain anaerobic micro-sites, resulting in the production of $CO_2$ and methane during plant decomposition. Thus, they can recycle carbon into the atmosphere in the form of methane and $CO_2$ (Schmitz et al. 2014). Methane ($CH_4$) is a major greenhouse gas and its availability can contribute to climate change. Forests are a rich source of organic carbon that accounts for 90% of carbon flux between the earth's surface and the atmosphere (Hartleym and Jones 2004). In forest ecosystems, termites perform an important part in the carbon cycle. Termites are considered a major group of arthropods next to ants. Termites have the largest contribution to carbon flux and have been reported to be responsible for 2% of $CO_2$ and 20% of global $CH_4$ production (Lawton 2000).

## 2.5.2 Insects' Role in Nitrogen Cycle

Insects such as termites also play an important role in the nitrogen cycle. These insects have developed effective mechanisms for conservation of nitrogen during digestion, such as nitrogen fixation organisms have been observed in their intestines (Yang and Gratton 2014). In addition, it has been observed that the soil of termite's habitats is healthier for plants development. Insects are an important part of the scavenger community. Insects' capability of waste metabolism is responsible for their major influences on nitrogen–phosphorus availability to higher plants. For example, Marion Island, located in the sub-Antarctic, mostly a poor plant community has a high primary production. In the results of digestion of plant litter, a moth larva (Pringleophaga marioni) stimulates mineralization of nitrogen 10 times and phosphorus 3 times.

Marion Island hosts a greater populace of house mice that feed on these standing insects at 0.7% per day. The increase of the mice population reduced the number of insect larvae, which indirectly induced an impact on the nutrient cycle and thus caused imbalances between production and decomposition (Weisser and Siemann 2004).

## 2.6 Conclusion

Insects contribute to the environment and ecosystem services in different ways. In few cases, they are useful playing important roles in maintaining the ecological services. In other cases, they mitigate their effects as pests. Insects cannot be ignored as they are a central part of the ecosystem, and some species can provide solutions to environmental issues such as food and fuel production, economy, deforestation and other environmental degradation. This study suggests an opportunity for additional applications of insects' roles. Extreme climatic changes cause the decline of insects, and thus problems of ecological imbalances are increasing. Insects are little things, but they govern the world. Their conservation and safety will not only support the healthy agro-ecosystems but also keep checking the human disease vectors. Therefore, to encourage the sustainability of ecosystem services and ecological balance, the insect's conservation is important.

## REFERENCES

Adnan M, Zahir S, Fahad S, Arif M, Mukhtar A, Imtiaz AK, Ishaq AM, Abdul B, Hidayat U, Muhammad A, Inayat-Ur R, Saud S, Muhammad ZI, Yousaf J, Amanullah Hafiz MH, Wajid N (2018) Phosphate-solubilizing bacteria nullify the antagonistic effect of soil calcification on bioavailability of phosphorus in alkaline soils. Sci Rep 8:4339. https://doi.org/10.1038/s41598-018-22653-7.

Al-Wabel MI, Abdelazeem S, Munir A, Khalid E, Adel RAU (2020b) Extent of climate change in Saudi Arabia and its impacts on agriculture: A case study from Qassim Region. In: Fahad S, Hasanuzzaman M, Alam M, Ullah H, Saeed M, Khan AK, Adnan M (eds) Environment, climate, plant and vegetation growth. Springer Publ Ltd, Springer Nature Switzerland AG. Part of Springer Nature. pp 635–658. https://doi.org/10.1007/978-3-030-49732-3.

Al-Wabel MI, Ahmad M, Usman AR, Akanji M, Rafique MI (2020a) Advances in pyrolytic technologies with improved carbon capture and storage to combat climate change. In: Fahad S, Hasanuzzaman M, Alam M, Ullah H, Saeed M, Khan AK, Adnan M (eds) Environment, climate, plant and vegetation growth. Springer Publ Ltd, Springer Nature Switzerland AG. Part of Springer Nature. pp 535–576. https://doi.org/10.1007/978-3-030-49732-3.

Akram R, Turan V, Hammad HM, Ahmad S, Hussain S, Hasnain A, Maqbool MM, Rehmani MIA, Rasool A, Masood N, Mahmood F, Mubeen M, Sultana SR, Fahad S, Amanet K, Saleem M, Abbas Y, Akhtar HM, Waseem F, Murtaza R, Amin A, Zahoor SA, ul Din MS, Nasim W (2018a) Fate of organic and inorganic pollutants in paddy soils. In: Hashmi MZ, Varma A (eds) Environmental pollution of paddy soils, soil biology. Springer International Publishing AG, Switzerland, pp 197–214.

Akram R, Turan V, Wahid A, Ijaz M, Shahid MA, Kaleem S, Hafeez A, Maqbool MM, Chaudhary HJ, Munis, MFH, Mubeen M, Sadiq N, Murtaza R, Kazmi DH, Ali S, Khan N, Sultana SR, Fahad S, Amin A, Nasim W (2018b) Paddy land pollutants and their role in climate change. In: Hashmi MZ, Varma A (eds) Environmental Pollution of Paddy Soils, Soil Biology. Springer International Publishing AG, Switzerland, pp 113–124.

Araujo MB, Ferri-Yanez F, Bozinovic F, Marquet PA, Valladares F, Chown SL (2013) Heat freezes niche evolution. Ecol Lett 16:1206.

Aziz K, Daniel KYT, Fazal M, Muhammad ZA, Farooq S, Fan W, Fahad S, Ruiyang Z (2017a) Nitrogen nutrition in cotton and control strategies for greenhouse gas emissions: A review. Environ Sci Pollut Res 24:23471–23487. https://doi.org/10.1007/s11356-017-0131-y.

Aziz K, Daniel KYT, Muhammad ZA, Honghai L, Shahbaz AT, Mir A, Fahad S (2017b) Nitrogen fertility and abiotic stresses management in cotton crop: A review. Environ Sci Pollut Res 24:14551–14566. https://doi.org/10.1007/s11356-017-8920-x.

Bale JS (2002) Insects and low temperatures: From molecular biology to distributions and abundance. Philos Trans R Soc Lond Ser B 357:849.

Basir A, Ali R, Alam M, Shah AS, Khilwat A, Adnan M, Ibrahim M, Rehman I, Adnan T (2015) Potential of wheat (Tritium aestivum L.) advanced lines for yield and yield attributes under different planting dates in peshawar valley. Am Eurasian J Agric Environ Sci 15(12):2484–2488.

Belovsky GE, Slade JB (2000) Insect herbivory accelerates nutrient cycling and increases plant production. Proc Natl Acad Sci USA 97:14412–17717.

Biesmeijer JC, Roberts SPM, Reemer M, Ohlemuller R, Edwards M, Peeters T, Schaffers AP, Potts SG, Kleukers R, Thomas CD, Settele J, Kunin WE (2006) Parallel declines in pollinators and insect-pollinated plants in Britain and the Netherlands. Science 313:351–354.

Borror DJ, Triplehorn CA, Johnson NF (1989) Book: An introduction to the study of insects. No. Ed. 6 pp. 875.

Bretschko G (1982) Pontomyia Edwards (Diptera: Chironomidae), a member of the coral reef community at Carrie Bow Bay, Belize The Atlantic barrier reef ecosystem at Carrie Bow Bay, Belize I. Structure and communities. In: Rutzler K, MacIntyre IG (eds) Smithsonian contributions to marine science, Washington, D.C, pp 381–385.

Buxton PA (1932) Terrestrial insects and the humidity of the environment. https://doi.org/10.1111/j.1469-185X.1962.tb01044.x.

Ceballos G, Ehrlich PR, Dirzo R (2017) Biological annihilation via the ongoing sixth mass extinction signaled by vertebrate population losses and declines. Proc Natl Acad Sci 114:6089–6096.

Cheng L (1985) Biology of halobates (Heteroptera: Gerridae). Ann Rev Entomol 30:111–135.

Congdon BC (1994) Salinity-related fitness differences amongst GPI genotypes in the mosquitofish Gambusia holbrooki (Poeciliidea: Teleostei). Biol J Linn Soc 53:343–352.

Cornelissen T, Cintra F, Santos JC (2015) Shelter-building insects and their role as ecosystem engineers. Neotrop Entomol 45:1–12

Cornelissen TG, Stiling P (2006) Does low nutritional quality act as a plant defense? An experimental test of the slow-growth, highmortality hypothesis. Ecol Entomol 31:32–40.

Davis JM, Rosemond AD, Small GE (2011) Increasing donor ecosystem productivity decreases terrestrial consumer reliance on a stream resource subsidy. Oecologia 167:821–834.

Depeng W, Fahad S, Saud S, Muhammad K, Aziz K, Mohammad NK, Hafiz MH, Wajid N (2018) Morphological acclimation to agronomic manipulation in leaf dispersion and orientation to promote "Ideotype" breeding: Evidence from 3D visual modeling of "super" rice (*Oryza sativa* L.). Plant Physiol Biochem. https://doi.org/10.1016/j.plaphy.2018.11.010.

Downes JA (1964) Arctic insects and their environment. Can Entomol 96(1–2):279–307.

Fernandes GW, Coelho M, Santos JC (2014) Neotropical insect galls: Status of knowledge and perspectives. In: Fernandes GW, Santos JC (eds) Neotropical insect galls. Springer, Netherlands, pp 1–14.

Fahad S, Abdul B, Adnan M. (eds) (2018b) Global wheat production. IntechOpen, United Kingdom. http://dx.doi.org/10.5772/intechopen.72559.

Fahad S, Adnan M, Hassan S, Saud S, Hussain S, Wu C, Wang D, Hakeem KR, Alharby HF, Turan V, Khan MA, Huang J (2019b) Rice responses and tolerance to high temperature. In: Hasanuzzaman M, Fujita M, Nahar K, Biswas JK (eds) Advances in rice research for abiotic stress tolerance. Woodhead Publ Ltd, Abington Hall Abington, Cambridge, UK, pp 201–224.

Fahad S, Bajwa AA, Nazir U, Anjum SA, Farooq A, Zohaib A, Sadia S, Nasim W, Adkins S, Saud S, Ihsan MZ, Alharby H, Wu C,Wang D, Huang J (2017) Crop production under drought and heat stress: Plant responses and management options. Front Plant Sci 8:1147. https://doi.org/10.3389/fpls.2017.01147.

Fahad S, Bano A (2012) Effect of salicylic acid on physiological and biochemical characterization of maize grown in saline area. Pak J Bot 44:1433–1438.

Fahad S, Chen Y, Saud S, Wang K, Xiong D, Chen C, Wu C, Shah F, Nie L, Huang J (2013) Ultraviolet radiation effect on photosynthetic pigments, biochemical attributes, antioxidant enzyme activity and hormonal contents of wheat. J Food Agri Environ 11(3&4):1635–1641.

Fahad S, Hasanuzzaman M, Alam M, Ullah H, Saeed M, Ali Khan I, Adnan M (eds) (2020) Environment, climate, plant and vegetation growth. Springer Nature Switzerland AG. https://doi.org/10.1007/978-3-030-49732-3

Fahad S, Hussain S, Bano A, Saud S, Hassan S, Shan D, Khan FA, Khan F, Chen Y, Wu C, Tabassum MA, Chun MX, Afzal M, Jan A, Jan MT, Huang J (2014a) Potential role of phytohormones and plant growth-promoting rhizobacteria in abiotic stresses: Consequences for changing environment. Environ Sci Pollut Res 22(7):4907–4921. https://doi.org/10.1007/s11356-014-3754-2.

Fahad S, Hussain S, Matloob A, Khan FA, Khaliq A, Saud S, Hassan S, Shan D, Khan F, Ullah N, Faiq M, Khan MR, Tareen AK, Khan A, Ullah A, Ullah N, Huang J (2014b) Phytohormones and plant responses to salinity stress: A review. Plant Growth Regul 75(2):391–404. https://doi.org/10. 1007/s10725-014-0013-y.

Fahad S, Hussain S, Saud S, Hassan S, Chauhan BS, Khan F, et al (2016a) Responses of rapid viscoanalyzer profile and other rice grain qualities to exogenously applied plant growth regulators under high day and high night temperatures. PLoS One 11(7):e0159590. https://doi.org/10.1371/journal.pone.0159590.

Fahad S, Hussain S, Saud S, Hassan S, Ihsan Z, Shah AN, Wu C, Yousaf M, Nasim W, Alharby H, Alghabari F, Huang J (2016c) Exogenously applied plant growth regulators enhance the morphophysiological growth and yield of rice under high temperature. Front Plant Sci 7:1250. https://doi.org/10.3389/fpls.2016.01250.

Fahad S, Hussain S, Saud S, Hassan S, Tanveer M, Ihsan MZ, Shah AN, Ullah A, Nasrullah KF, Ullah S, Alharby H, Wu C, Huang J (2016d) A combined application of biochar and phosphorus alleviates heat-induced adversities on physiological, agronomical and quality attributes of rice. Plant Physiol Biochem 103:191–198.

Fahad S, Hussain S, Saud S, Khan F, Hassan S, Nasim W, Arif M, Wang F, Huang J (2016b) Exogenously applied plant growth regulators affect heat-stressed rice pollens. J Agron Crop Sci 202:139–150.

Fahad S, Hussain S, Saud S, Tanveer M, Bajwa AA, Hassan S, Shah AN, Ullah A, Wu C, Khan FA, Shah F, Ullah S, Chen Y, Huang J (2015a) A biochar application protects rice pollen from high-temperature stress. Plant Physiol Biochem 96:281–287.

Fahad S, Muhammad ZI, Abdul K, Ihsanullah D, Saud S, Saleh A, Wajid N, Muhammad A, Imtiaz AK, Chao W, Depeng W, Jianliang H (2018a) Consequences of high temperature under changing climate optima for rice pollen characteristics-concepts and perspectives, Archives Agron Soil Sci. https://doi.org/10.1080/03650340.2018.1443213.

Fahad S, Nie L, Chen Y, Wu C, Xiong D, Saud S, Hongyan L, Cui K, Huang J (2015b) Crop plant hormones and environmental stress. Sustain Agric Rev 15:371–400.

Fahad S, Rehman A, Shahzad B, Tanveer M, Saud S, Kamran M, Ihtisham M., Khan SU, Turan V, Rahman MHU (2019a) Rice responses and tolerance to metal/metalloid toxicity. In: Hasanuzzaman M, Fujita M, Nahar K, Biswas JK (eds) Advances in rice research for abiotic stress tolerance. Woodhead Publ Ltd, Abington Hall Abington, Cambridge, UK, pp 299–312.

Fahad S, Saud S, Yajun C, Chao W, Depeng W (eds) (2021f) Abiotic stress in plants. IntechOpen, United Kingdom. http://dx.doi.org/10.5772/intechopen.91549.

Fahad S, Sönmez O, Saud S, Wang D, Wu C, Adnan M, Arif M, Amanullah (eds) (2021e) Engineering tolerance in crop plants against abiotic stress, 1st edn. ed, Footprints of climate variability on plant diversity. CRC Press, Boca Raton, FL.

Fahad S, Sönmez O, Saud S, Wang D, Wu C, Adnan M, Turan V (eds) (2021a) Plant growth regulators for climate-smart agriculture, 1st edn. ed, Footprints of climate variability on plant diversity. CRC Press, Boca Raton, FL.

Fahad S, Sönmez O, Saud S, Wang D, Wu C, Adnan M, Turan V (eds) (2021b) Climate change and plants: Biodiversity, growth and interactions, 1st edn. ed, Footprints of climate variability on plant diversity. CRC Press, Boca Raton, FL.

Fahad S, Sönmez O, Saud S, Wang D, Wu C, Adnan M, Turan V (eds) (2021c) Developing climate resilient crops: Improving global food security and safety, 1st edn. ed, Footprints of climate variability on plant diversity. CRC Press, Boca Raton, FL.

Fahad S, Sönmez O, Turan V, Adnan M, Saud S, Wu C, Wang D (eds) (2021d) Sustainable soil and land management and climate change, 1st edn. ed, Footprints of climate variability on plant diversity. CRC Press, Boca Raton, FL.

Farah R, Muhammad R, Muhammad SA, Tahira Y, Muhammad AA, Maryam A, Shafaqat A, Rashid M, Muhammad R, Qaiser H, Afia Z, Muhammad AA, Muhammad A, Fahad S (2020) Alternative and non-conventional soil and crop management strategies for increasing water use efficiency. In: Fahad S, Hasanuzzaman M, Alam M, Ullah H, Saeed M, Khan AK, Adnan M (eds) Environment, climate, plant and vegetation growth. Springer Publ Ltd, Springer Nature Switzerland AG. Part of Springer Nature. pp 323–338. https://doi.org/10.1007/978-3-030-49732-3.

Farhana G, Ishfaq A, Muhammad A, Dawood J, Fahad S, Xiuling L, Depeng W, Muhammad F, Muhammad F, Syed AS (2020) Use of crop growth model to simulate the impact of climate change on yield of various wheat cultivars under different agro-environmental conditions in Khyber Pakhtunkhwa, Pakistan. Arabian J Geosci 13:112. https://doi.org/10.1007/s12517-020-5118-1.

Farhat A, Hafiz MH, Wajid I, Aitazaz AF, Hafiz FB, Zahida Z, Fahad S, Wajid F, Artemi C (2020) A review of soil carbon dynamics resulting from agricultural practices. J Environ Manage 268:110319.

Florentine SK, Raman A, Dhileepan K (2005) Effects of gall induction by Epiblema strenuana on gas exchange, nutrients, and energetics in Parthenium hysterophorus. Bio Cont 50:787–801

Fukui A (2001) Indirect interactions mediated by leaf shelter in animal plant communities. Popul Ecol 43:31–40.

Gaston KJ (1991) The magnitude of global insect species richness. Conserv Biol 5:283–296.

Guimaraes ALA, Cruz SMS, Vieira ACM (2014) Structure of floral galls of Byrsonima sericea (Malpighiaceae) induced by Bruggmanniella byrsonimae (Cecidomyiidae, Diptera) and their effects on host plants. Plant Biol 16:467–475

Guimaraes ALA, Neufeld PM, Santiago-Fernandes LDR, Vieira ACM (2015) Structure and development of 'witches' broom' galls in reproductive organs of Byrsonima sericea (Malpighiaceae) and their effects on host plants. Plant Biol 17:493–504

Habib ur R, Ashfaq A, Aftab W, Manzoor H, Fahd R, Wajid I, Md. Aminul I, Vakhtang S, Muhammad A, Asmat U, Abdul W, Syeda RS, Shah S, Shahbaz K, Fahad S, Manzoor H, Saddam H, Wajid N (2017) Application of CSM-CROPGRO-Cotton model for cultivars and optimum planting dates: Evaluation in changing semi-arid climate. Field Crops Res. http://dx.doi.org/10.1016/j.fcr.2017.07.007.

Hafiz MH, Muhammad A, Farhat A, Hafiz FB, Saeed AQ, Muhammad M, Fahad S, Muhammad A (2019) Environmental factors affecting the frequency of road traffic accidents: A case study of sub-urban area of Pakistan. Environ Sci Pollut Res. https://doi.org/10.1007/s11356-019-04752-8.

Hafiz MH, Wajid F, Farhat A, Fahad S, Shafqat S, Wajid N, Hafiz FB (2016) Maize plant nitrogen uptake dynamics at limited irrigation water and nitrogen. Environ Sci Pollut Res 24(3):2549–2557. https://doi.org/10.1007/s11356-016-8031-0.

Hartleym SE, Jones TH (2004) Insect herbivores, nutrient cycling and plant productivity. Ecol Stud 173:27–52.

Hunter MR, Mark DH (2008) Designing for conservation of insects in the built environment. Insect Conserv Diver 1:189–196.

Kagata H, Ohgushi T (2004) Leaf miner as a physical ecosystem engineer: Secondary use of vacant leaf-mines by other arthropods. Ann Entomol Soc Am 97:923–927.

Kehinde T, Amusan B, Ayansola A, Oyelade S, Adu W (2014) Status of insect diversity conservation in Nigeria: A review. Ife J Sci 16 (2):319–327.

Khan I, Hussain H, Shah B, Ullah W, Naeem A, Ali W, Khan N, Adnan M, Junaid K, Shah SRA, Ahmed N, Iqbal M (2016) Evaluation of phytobiocides and different culture media for growth, isolation and control of *Rhizoctoniasolani in vitro*. J Ento Zool Stud 4(2):417–420.

Knapp M, Vernon P, Renault D (2018) Studies on chill coma recovery in the ladybird, Harmonia axyridis: Ontogenetic profile, effect of repeated cold exposures, and capacity to predict winter survival. J Thermal Biol 74:275–280.

Kozlov MV, Lanta V, Zverev VE, Zvereva EL (2012) Delayed local responses of downy birch to damage by leafminers and leafrollers. Oikos 121:428–434.

Lawton JH (2000) Community ecology in a changing world. In: Kinne O (ed) Excellence in ecology. Ecology institute Oldendorf, Germany.

Lill JT, Marquis RJ (2007) Microhabitat manipulation: Ecosystem engineering by shelter-building insects. In: Cuddington K, Byers JE, Hastings A, Wilson WG (eds) Ecosystem engineers: Plants to protists. Academic Press, Elsevier, New York, pp 107–138.

Lind EM, Jones MT, Long J, Weiss M (2001) Ontogenetic changes in leaf shelter construction by larvae of Epargyreus clarus (Hesperiidae), the silver-spotted skipper. J Lepidopterol Soc 54:77–82.

Longhurst A (1998) Ecological geography of the sea, Academic Press, San Diego, p. 27.

MacEvilly C (2000) Bugs in the system. Nutr Bull 25:267–268.

Mahmood Ul H, Tassaduq R, Chandni I, Adnan A, Muhammad A, Muhammad MA, Muhammad H-ur-R, Mehmood AN, Alam S, Fahad S (2021) Linking plants functioning to adaptive responses under heat stress conditions: A mechanistic review. J Plant Growth Regul. https://doi.org/10.1007/s00344-021-10493-1.

Manzer HS, Saud A, Soumya M, Abdullah A. Al-A, Qasi DA, Bander MA. Al-M, Hayssam MA, Hazem MK, Fahad S, Vishnu DR, Om PN (2021) Molybdenum and hydrogen sulfide synergistically mitigate arsenic toxicity by modulating defense system, nitrogen and cysteine assimilation in faba bean (*Vicia faba* L.) seedlings. Environ Pollut 290:117953.

Md. Enamul H, Shoeb AZM, Mallik AH, Fahad S, Kamruzzaman MM, Akib J, Nayyer S, K. M. Mehedi A, Swati AS, Md Yeamin A, Most SS (2020) Measuring vulnerability to environmental hazards: Qualitative to quantitative. In: Fahad S, Hasanuzzaman M, Alam M, Ullah H, Saeed M, Khan AK, Adnan M (eds) Environment, climate, plant and vegetation growth. Springer Publ Ltd, Springer Nature Switzerland AG. Part of Springer Nature. pp 421–452. https://doi.org/10.1007/978-3-030-49732-3.

Md Jakir H, Allah B (2020) Development and applications of transplastomic plants; a way towards eco-friendly agriculture. In: Fahad S, Hasanuzzaman M, Alam M, Ullah H, Saeed M, Khan AK, Adnan M (eds) Environment, climate, plant and vegetation growth. Springer Publ Ltd, Springer Nature Switzerland AG. Part of Springer Nature. pp 285–322. https://doi.org/10.1007/978-3-030-49732-3.

Mubeen M, Ashfaq A, Hafiz MH, Muhammad A, Hafiz UF, Mazhar S, Muhammad Sami ul Din, Asad A, Amjed A, Fahad S, Wajid N (2020) Evaluating the climate change impact on water use efficiency of cotton-wheat in semi-arid conditions using DSSAT model. J Water Climate Change. https://doi.org/10.2166/wcc.2019.179/622035/jwc2019179.pdf.

Muhammad N, Muqarrab A, Khurram S, Fiaz A, Fahim N, Muhammad A, Shazia A, Omaima N, Sulaiman AA, Fahad S, Subhan D, Rahul D (2021) Kaolin and Jasmonic acid improved cotton productivity under water stress conditions. J Saudi Soc Agric Sci 28:6606–6614. https://doi.org/10.1016/j.sjbs.2021.07.043.

Muhammad Tahir ul Q, Amna F, Amna B, Barira Z, Xitong Z, Ling-Ling C (2020) Effectiveness of conventional crop improvement strategies vs. omics. In: Fahad S, Hasanuzzaman M, Alam M, Ullah H, Saeed M, Khan AK, Adnan M (eds) Environment, climate, plant and vegetation growth. Springer Publ Ltd, Springer Nature Switzerland AG. Part of Springer Nature. pp 253–284. https://doi.org/10.1007/978-3-030-49732-3.

Muhammad Z, Abdul MK, Abdul MS, Kenneth BM, Muhammad S, Shahen S, Ibadullah J, Fahad S (2019) Performance of Aeluropus lagopoides (mangrove grass) ecotypes, a potential turfgrass, under high saline conditions. Environ Sci Pollut Res. https://doi.org/10.1007/s11356-019-04838-3.

Muzammal R, Fahad S, Guanghui D, Xia C, Yang Y, Kailei T, Lijun L, Fei-Hu L, Gang D (2021) Evaluation of hemp (Cannabis sativa L.) as an industrial crop: A review. Environ Sci Pollut Res. https://doi.org/10.1007/s11356-021-16264-5.

Overgaard J, MacMillan HA (2017) The integrative physiology of insect chill tolerance. Annu Rev Physiol 79:187–208.

Perfecto I, Vandermeer J (2002) Quality of agroecological matrix in a tropical montane landscape: Ants in coffee plantations in southern Mexico. Conserv Biol 16:174–182.

Pimentel D (1994) Insect population responses to environmental stress and pollutants. Environ Rev 2(1):1–15.

Pineda A, Dickel M, Pieterse CM, Pozo MJ (2013) Plant-Microbe-Insect Interactions. Beneficial microbes in a changing environment: Are they always helping plants to deal with insects. Fun Ecol 27:574–586.

Pineda A, Zheng SJ, Van Loon JJA, Pieterse CMJ, Dicke M (2010) Helping plants to deal with insects: The role of beneficial soil borne microbes. Trends Plant Sci 15:507–514.

Potter KA, Woods HA (2011) No evidence for the evolution of thermal or desiccation tolerance of eggs among populations of Manduca sexta. Fun Ecol 26(1):112–122.

Prather C, Laws A (2018) Insects as a piece of the puzzle to mitigate global problems: An opportunity for ecologists. Basic App Eco 26:71–81.

Putten VD, Macel WH, Visser ME (2010) Predicting species distribution and abundance responses to climate change: Why it is essential to include biotic interactions across trophic levels. Philos Trans R Soc B-Biol Sci 365:2025–2034.

Qamar uz Z, Zubair A, Muhammad Y, Muhammad ZI, Abdul K, Fahad S, Safder B, Ramzani PMA, Muhammad N (2017) Zinc biofortification in rice: Leveraging agriculture to moderate hidden hunger in developing countries. Arch Agron Soil Sci 64:147–161. https://doi.org/10.1080/03650340.2017.1338343.

Raddi AG (2004) Blending biodiversity, bloom and bio-aesthetics: Highway hues. Wastelands News 20(1):20–27.

Ratcliffe NA, Mello CB, Garcia ES, Butt TM, Azambuja P (2011) Insect natural products and processes: New treatments for human disease. Insect Biochem Mol Biol 41:747–769.

Reason P (2007) Education for ecology-science, aesthetics, spirit and ceremony. Manag Learn 38:27–44.

Rohfritsch O (2010) Genesis and development of dipterocecidia. Atti Accad Naz It Entomol 58:55–66.

Sajjad H, Muhammad M, Ashfaq A, Waseem A, Hafiz MH, Mazhar A, Nasir M, Asad A, Hafiz UF, Syeda RS, Fahad S, Depeng W, Wajid N (2019) Using GIS tools to detect the land use/land cover changes during forty years in Lodhran district of Pakistan. Environ Sci Pollut Res. https://doi.org/10.1007/s11356-019-06072-3.

Sanchez-Bayo F, Wyckhuys KAG (2019) Worldwide Worldwide decline of the entomofauna: A review of its drivers. Biol Conserv 232:8–27.

Saud S, Chen Y, Fahad S, Hussain S, Na L, Xin L, Alhussien SA (2016) Silicate application increases the photosynthesis and its associated metabolic activities in Kentucky bluegrass under drought stress and post-drought recovery. Environ Sci Pollut Res 23(17):17647–17655. https://doi.org/10.1007/s11356-016-6957-x.

Saud S, Chen Y, Long B, Fahad S, Sadiq A (2013) The different impact on the growth of cool season turf grass under the various conditions on salinity and drought stress. Int J Agric Sci Res 3:77–84.

Saud S, Fahad S, Yajun C, Ihsan MZ, Hammad HM, Nasim W, Amanullah Jr, Arif M, Alharby H (2017) Effects of nitrogen supply on water stress and recovery mechanisms in Kentucky bluegrass plants. Front Plant Sci 8:983. https://doi.org/10.3389/fpls.2017.00983.

Saud S, Li X, Chen Y, Zhang L, Fahad S, Hussain S, Sadiq A, Chen Y (2014) Silicon application increases drought tolerance of Kentucky bluegrass by improving plant water relations and morph physiological functions. SciWorld J 2014:1–10. https://doi.org/10.1155/2014/368694.

Schmitz OJ, Raymond PA, Estes JA, Kurz WA, Holtgrieve GW, Ritchie ME, Schindler DE, Spivak AC, Wilson RW, Bradford MA, Christensen V, Deegan L, Smetacek V, Vanni MJ, Wilmers CC (2014) Animating the carbon cycle. Ecosystems 17:344–359.

Schowalter TD (2011) Chapter 17: Summary and synthesis. In: Insect ecology: An ecosystem approach, 3rd edn. pp 513–523.

Schowalter TD, Noriegab JA, Tscharntke T (2018) Insect effects on ecosystem Services-Introduction. Basic App Ecol 26:1–7.

Seymour RS, White CR, Gibernau M (2003) Environmental biology: Heat reward for insect pollinators. Nature 426:243–244.

Shah F, Lixiao N, Kehui C, Tariq S, Wei W, Chang C, Liyang Z, Farhan A, Fahad S, Huang J (2013) Rice grain yield and component responses to near 2°C of warming. Field Crop Res 157:98–110

Shah B, Khan IA, Khan A, Din MMU, Junaid K, Adnan M, Shah SRA, Zaman M, Ahmad N, Akbar R, Fayaz W, Rahman IU (2016) Correlation between proximate chemical composition and insect pests of maize cultivars in Peshawar. J Ento Zool Stud 4(1):312–316.

Shorthouse JD, Wool D, Raman A (2005) Gall-inducing insects-nature's most sophisticated herbivores. Basic Appl Ecol 6:407–411

Sinclair BJ, Alvarado LEC, Ferguson LV (2015) An invitation to measure insect cold tolerance: Methods, approaches, and workflow. J Therm Biol 53:180–197.

Sinclair BJ, Vernon P, Klok CJ, Chown SL, (2003) Insects at low temperatures: An ecological perspective. Trends Ecol Evol 18:257–262.

Somme L (1995) Invertebrates in hot and cold arid environments. Springer-Verlag, GmbH & Co. KG, Berlin, Heidelberg.

Somme L (2000) The history of cold hardiness research in terrestrial arthropods. Cryo Lett 21:289–296

Stabentheiner A, Kovaca H, Hetzb SK, Käfera H, Stabentheiner G (2012) Assessing honeybee and wasp thermoregulation and Energetic-New insights by combination of flow-through respirometry with infrared thermography. Thermo Chimica Acta 534:77–86

Sunday JM, Bates AE, Dulvy NK (2012) Thermal tolerance and the global re-distribution of animals. Nat Clim Change 2:686–690.

Teets NM, Denlinger DL (2013) Physiological mechanisms of seasonal and rapid coldhardening in insects. Physiol Entomol 38:105–116.

Tremmel M, Müller C (2012) Insect personality depends on environmental conditions. Behav Ecol 24(2):386–392.

Tylianakis JM, Didham RK, Bascompte J, Wardle DA (2008) Global change and species interactions in terrestrial ecosystems. Ecol Lett 11:1351–1363.

Wajid N, Ashfaq A, Asad A, Muhammad T, Muhammad A, Muhammad S, Khawar J, Ghulam MS, Syeda RS, Hafiz MH, Muhammad IAR, Muhammad ZH, Muhammad Habib ur R, Veysel T, Fahad S, Suad S, Aziz K, Shahzad A (2017) Radiation efficiency and nitrogen fertilizer impacts on sunflower crop in contrasting environments of Punjab. Pak Environ Sci Pollut Res 25:1822–1836. https://doi. org/10.1007/s11356-017-0592-z.

Wegier A, Alavez V, Perez-Lopez J, Calzada L, Cerritos R (2018) Beef or grasshopper hamburgers: The ecological implications of choosing one over the other. Basic App Ecol 26:89–100.

Weiss M, Lind EM, Jones MT, Long JD, Mauplin JL (2004) Uniformity of leaf shelter construction by larvae of Epargyreus clarus (Hesperiidae), the silver-spotted skipper. J Insect Behav 16:465–480.

Weisser WW Siemann E (2004) Insects and ecosystem function. Springer-Verlag, Berlin Heidelberg.

Wheeler J, Longino JT (1988) Arthropods in live oak galls in Texas. Entomol News 991:25–29.

Williams DD, Feltmate BW (1992) Aquatic insects. C.A.B. International, Wallingford, Oxford.

Williams DD, Williams NE (1998) Aquatic insects in an estuarine environment: Densities, distribution and salinity tolerance. Freshwater Biol 39:411–421.

Winterbourn MJ, Anderson NH (1980) The life history of Philanisus plebeius Walker (Trichoptera: Chathamiidae), a caddisfly whose eggs were found in a starfish. Ecol Entomol 5:293–303.

Yang H, Gratton C (2014) Insects as drivers of ecosystem processes. Curr Opin Insect Sci 2:26–32.

Yang L (2012) The ecological consequences of insect outbreaks. In: Barbosa P, Letourneau D, Agrawal A (eds) Insect outbreaks revisited. Wiley-Blackwell, West Sussex, UK.

Yang LH (2006) Interactions between a detrital resource pulse and a detritivore community. Oecologia 147:522–532.

Yang Z, Zhang Z, Zhang T, Fahad S, Cui K, Nie L, Peng S, Huang J (2017) The effect of season-long temperature increases on rice cultivars grown in the central and southern regions of China. Front Plant Sci 8:1908. https://doi.org/10.3389/fpls.2017.01908.

Zahida Z, Hafiz FB, Zulfiqar AS, Ghulam MS, Fahad S, Muhammad RA, Hafiz MH, Wajid N, Muhammad S (2017) Effect of water management and silicon on germination, growth, phosphorus and arsenic uptake in rice. Ecotoxicol Environ Saf 144:11–18.

Zahoor M, Khan N, Ali M, Saeed M, Ullah Z, Adnan M, Ahmad B (2016) Integrated effect of organic waste and NPK fertilizers on nutrients uptake in potato crop and soil fertility. Pure Appl Biol 5(3):601–607.

# 3

# *Climate Change, Insects and Global Food Production*

**Rida Akram, Khizer Amanet, Javed Iqbal, Maham Fatima, Muhammad Mubeen, Sajjad Hussain, Musaddiq Ali, and Wajid Nasim**
*COMSATS University Islamabad*

**Ashfaq Ahmad**
*University of Agriculture*

**Hafiz Umar Farid**
*Bahauddin Zakariya University*

**Ayman El Sabagh**
*Kafrelsheikh University*

**Shah Fahad**
*Hainan University*
*The University of Haripur*

**CONTENTS**



## 3.1 Introduction

Change in climate is producing worldwide disturbance in patterns of rainfall, temperature, and increase in unpredictable, uncertain weather conditions, and it has an important impact on worldwide crop production (Lobell and Field 2007; Ullah et al. 2016; Akram et al. 2019; Fahad and Bano 2012; Fahad et al. 2017; Fahad et al. 2013; Fahad et al. 2014a,b; Fahad et al. 2016a,b,c,d; Fahad et al. 2015a,b; Fahad et al. 2018; Fahad et al. 2019a,b) and will remain to do so into the future

DOI: 10.1201/9781003286400-3

(Challinor et al. 2014). Climatic changings also have direct and indirect influence on the distribution and severity of crop pests, for example "strain or biotype of plant, species, animal or pathogenic agent injurious to plant or plant products" (FAO 2013b), all over the world (Macfadyen et al. 2018), which is further disturbing production of crops. In most of the world, new invasive species are also appearing, channeled by climate change. Without effective management and monitoring systems, these offensive species have the power to become important pests in these new areas (Hellmann et al. 2008). Unfavorable weather can provide some possibility for living species to escape or accept as it is a day-to-day affair, but there is no rescue when the climate becomes unfavorable for any species as it stretches over decades. Climate change is a rule not an exception, and data shows a significant change of climate since the inception of life on planet earth. But man practices like burning of fossil fuel and automobile industry from the last 150 years faster the pace of climate change. Release of greenhouse gases (GHGs) and carbon dioxide ($CO_2$) are major components in recent climate change. Till 1950, atmospheric carbon never crossed the 300 ppm line, but now it is more than 400 ppm. The recent surge in carbon level causes an increase in the average temperature of the earth to 1.1°C (NASA 2018). This trend seems to continue unless a collective effort was not made by the human race. Climate change in the recent past resulted in warming of oceans, shrinking of ice sheets, glaciers retreat, the rise of sea level, ocean acidification, dryness, etc., which disturb human life and frugality directly or indirectly. But the sector that is mostly affected by climate change is agriculture. Soils become drier, and an extra amount of water is required to irrigate them. Plants, especially horticultural crops, are under stress due to dryness and irregular and unique precipitations followed by floods and erosion (Calzadilla et al. 2013). Conversely, increase of temperature due to high amount of atmospheric $CO_2$ would have a negative impact on plant growth (Zhu et al. 2016; Fahad et al. 2016a,b,c,d). Liu et al. (2016) found 1°C rise in global temperature decrease wheat return up to 5%. The other annoying area which makes plant scientists confused is the unexpected behavior of insects and plant pathogens in somewhat warmer weather due to climate change. There might be a shift in the host range due to quicker pathogen evolution and a short incubation period (Phillips et al. 2010).

As some effects of climate change can be beneficial, evidence suggests that overall pest problems are likely to become larger in amplitude and more unpredictable (Gregory et al. 2009). Predictions are further challenged by the fact that climate change can also put its effects on pests, i.e. through the opposing feedback of natural enemies of pests and host crops, a difference in the heat-related priorities of pest crops and their natural enemies can lead to a loss of bringing together between the two and an increased risk of host outbursts. More sideway feedback relate to pest results from changes in the efficacy of pest control plans (e.g. synthetic pesticide, biological control) (Barzman et al. 2015), crop management and changes in land use, which can frequently have a more effect on the pressure of pest than the direct climate change effect (Cock et al. 2013).

## 3.2 Climate Change in Current Scenario

According to a current statement of the Intergovernmental Panel on Climate Change, the worldwide average earth surface temperature till the 20th century has enlarged by around 0.60°C. This value is about 0.15°C more than the previous approximations. The universal mean of the sea and land surface temperatures in May 2003 was the second highest from 1880. Till now, the ten hottest years in the 143–years history of the - global temperature are 1998, 2002 and 2001. It is assessed that the temperature will increase up to 0.3°C–4.8°C (IPCC 2014; Fahad et al. 2016a,b,c,d).

The earth's surface temperature is becoming warmer due to the greenhouse effect. The greatest concern gases of GHGs are $CO_2$, chlorofluorocarbons, methane, water vapor, troposphere ozone and nitrogen oxides. The GHG emission has increased quickly since the industrial revolt in the 1800s and touched their top level in the 2010s (IPCC 2014). Human beings have raised the level of carbon dioxide by 25% and are expecting to double it by 2075.

## 3.3 Role of Insects in Agriculture

Arthropods have occurred for more than 0.40 billion years and subsisted the Permian and Cretaceous mass exterminations. Insects have been enormously fruitful in terms of both species abundance. Insects and global arthropods are seen to be the biggest providers to species abundance (Stork et al. 2015). Insects interpretation for approximately 66% of all known species (Zhang 2011), comprising more than 3/4th of today's overall biodiversity. There are about 1,000,000 species of known insects (Zhang 2011), but only around 7%–10% of insects are systematically defined (Samways 1993). Since many insect classes are not yet known, it is predictable that there could be around 8,000,000 kinds of insects on earth (Samways 2005). Stork et al. (2015) predicted that all insects are in the range of 2.6–7.8 million and an average of 5.5 million species. The aggregate number of insect entities is expected $10^{18}$–$10^{19}$ (Dicke 2017).

For human being proficient crop agriculture, pests have occurred on their crops and insects have been mostly supposed as opponents in the race for existence. The insect–plant link is the foremost biotic interaction (Samways et al. 1993), and almost 50% of insect types are herbivorous, with most herbivorous kinds nourishing on plants in one or a few interrelated plant families (Schoonhoven et al. 2005). Herbivorous insects harm 18% of world agricultural yield, and this is mostly organized by chemical methods (Losey and Vaughan 2006). Despite these harms, less than 0.5% of the total number of the known insect species are measured pests. Aside from anthropocentric perception and societal prejudice, insects are not pests in an ecological or evolutionary context.

Insects are very important for the survival of humanity. Moreover, crops cannot give production without the functions of the ecosystem provided by the insects. Almost 72% of crops of the world depend on insects for pollination. Pollinating insects stabilize or improve the crop production of three-fourths of all crop types (Schwägerl 2016). A variety of insect taxa have been linked with increasing seed sets (Hoehn et al. 2008). There are so many species of insect pollinators such as bees, bumblebees, flies, beetles and butterflies (Garibaldi et al. 2013). All over the world 9.5% of the crop yield is due to pollination services by insects (Gallai et al. 2009).

In agriculture pest control is an evitable fact. Predatory insects comply with important ecosystem functions by controlling pest insects in cultivated areas. It was estimated in 75% of field studies that predators reduce pest populations in arable farmland significantly (Symondson et al. 2002), including ground beetles having predators in arable lands (Woodcock et al. 2014) and effectively reducing population sizes of economically significant agricultural pests such as aphids, slugs, root-feeding flies and phytophagous beetles (Kromp 1999). Insects play a dominant role to improve agricultural soil. The insects increase nitrogen, phosphorous, potassium, calcium or magnesium and total protein contents (De Groot et al. 2002).

## 3.4 Climate Change and Global Food Production

Global warming has now started showing its impact all over the world. From the inception of civilization, climate has become the prime important factor of agriculture productivity (Kumar and Dubey 2017). The sector of agriculture is a very important sector to the climate changes because it determines the aspects and behavior of crops and vegetation (Adnan et al. 2018a,b; Akram et al. 2018a,b; Aziz et al. 2017; Habib et al. 2017; Hafiz et al. 2016; Hafiz et al. 2019; Kamran et al. 2017; Muhammad et al. 2019; Sajjad et al. 2019; Saud et al. 2013; Saud et al. 2014; Saud et al. 2017; Saud et al. 2016; Shah et al. 2013; Qamar et al. 2017; Wajid et al. 2017; Yang et al. 2017; Zahida et al. 2017). Production of the food system is vulnerable to the changing climate because changes in temperature and precipitation lead to outbreaks of disease and pest will ultimately reduce harvest that would affect the food security (IPCC 2014). Figure 3.1 shows the relationship between climate changes, insect, agriculture system and global food production. However, the recent trends in climate change manifest this as a global environmental challenge followed by both natural and anthropogenic activities. Concerning the potential impacts of long term climate change on agriculture has enhanced the research over the past decades and findings to portray that irregular rainfall patterns and increased temperature, affects production and reducing

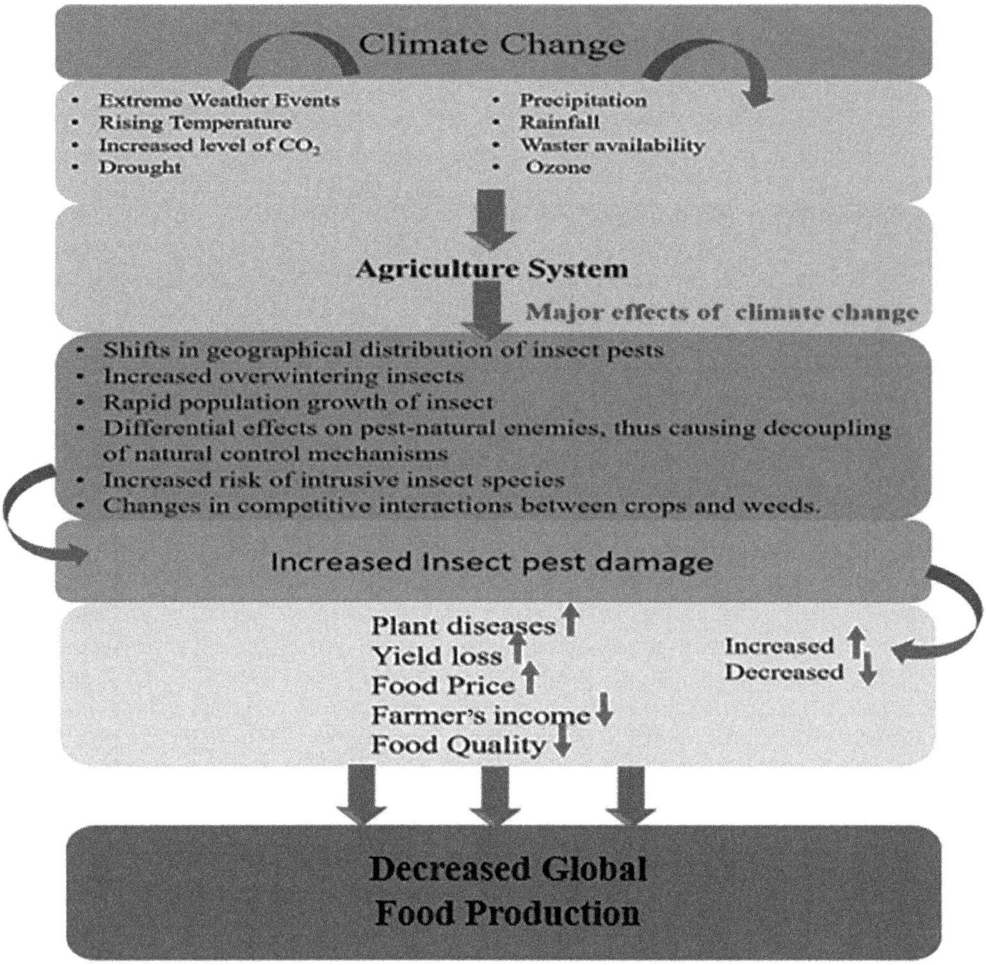


**FIGURE 3.1** Systematic diagram showing the relationship between climate changes, insect, agriculture system and global food production.

yield up to 63%–82% by the end of the 21st century. In the current scenario the weathering parameters are influencing agriculture strongly (67%) compared to other factors like soil and nutrient management (33%) during the season of cropping. It has been confirmed that crop yield falls by 3%–5% for every 1°F increase in the temperature (Kumar and Dubey 2017). These days' mitigation and adaptation strategies have been developed to cope with anticipated climate change impacts which include an adjustment in sowing and plant development stage that is more resilient to climate variability and improvement in agricultural practices. Moreover, the incoming scenario drafts a few speculations regarding issues of potential awareness to interpreting and using the information on smart-climate agriculture which sustain increase productivity, resilience and adaption so that better enhancement in the crop yield can achieve (Table 3.1) (Kumar and Dubey 2017).

## 3.5 Effects of Climate Change on Insects Biodiversity

It is confirmed that insects have body temperature similar to their environment because they are cold blooded. That is why temperature can influence development, insect behavior, reproduction, distribution and survival. It is admitted that the increasing temperature impacts on insects mainly overwhelms the effect of other environmental segments (Bale et al. 2002).

**TABLE 3.1**

Different Types of Forecasting Models

| Forecasting Model | Insect Name | Reference |
|---|---|---|
| Ordinal logistic model | Whitefly, Pyrilla and Fruit fly | Agrawal and Mehta (2007) |
| CLIMAX | *Helicoverpa* sp. | Zalucki and Furlong (2005) |
| SOPRA | *Dysaphis plantaginea* and *Grapholitha lobarzeweskii* | Graf et al. (2002) |
| FLYPAST | *Aphis fabae* | Knight et al. (1992) |
| NAPPFAST | *Scirtothrips dorsalis* | Nietschke (2002) |
| MOTHZV | *Helicoverpa* | West et al. (1985) |
| FRUFLY | *Ceratitis capitate* | Messousi et al. (2007) |

Due to the global increase in temperature, the insect population will boost which results in increased public health pests and insect vector disease. So more severe conditions of climate, i.e. more frequent storms along with increased rainfall, longer and additional droughts and elevated $CO_2$, can occur (Easterling et al. 2000; Stireman et al. 2005). Moreover, due to shorter winter and longer summer seasons, insects will start early breeding (Bale et al. 2002): mainly insects of medical importance are influenced more, i.e. mosquitoes (Epstein 2001).

Raising hotness has exerted impacts on species diversity and distribution. So for a while, in Canada and USA mountain pine beetle catastrophic forest, pest has increased its range to the northward by about 186 miles with 1.9°C increases in temperature. The increasing level of $CO_2$ will enhance carbon–nitrogen ratio in plants (Nasim et al. 2016, 2018), which in turn will affect insect feeding behavior, competition between species and plant compensation responses to insect herbivory and defensive chemical concentration in planets.

## 3.6 Effects of Climate Change on Distribution of Insects

There is a common prediction of poleward shifts in distributional limits in relation to change in climate (Bebber et al. 2013). Svobodova et al. (2014) used species distribution modeling to calculate the impact of changing climate on multiple crop pest species in Europe. Use of different climate change models, a projected shift is observed in the species ranges at the higher places and enhancing in the number of generation of pests in Central Europe by 2055. However, in southern Europe, the generation population was expected to decrease due to temperature increases that would negatively affect the population of the pest. In the same way pest species model in Australia suggests that for the growth and development of some species, many grain production regions will ultimately become less favorable (Hill et al. 2012). Figure 3.2 shows different types of the distribution pattern of insects in response to climate change.

## 3.7 Effects of Climate Change on Population and the Number of Generations of Insects

Changing climate has vulnerable effects on the intensity and incidence of pest generations (Sharma 2014) (Figure 3.3). Climate change can lead to higher densities of residents and the shifting to the temperate regions, which will have an extreme effect on beans and correlated crops (Sharma 2010). Occurrence of an invasive pest in Brazil and North America has been attributed to climate change (Czepak et al. 2013). *Spodoptera litura* (Fab.) has been described emergent as a severe pest under upper levels of $CO_2$ (Kranthi et al. 2009). The woolen aphid, *Ceratovacuna lanigera* Zehntner has appeared as a severe pest in Maharashtra due to climate change (Bade and Ghorpade 2009). Earlier described in West Bengal in 1958, it has now attacked Central and South India. Moreover, sowing dates relating to different temperatures and humidity also affect the pest incidence in legumes. It has been admitted that pigeon pea and chickpea sown in different months showed a differential incidence of insect pests. Pigeon pea sown in

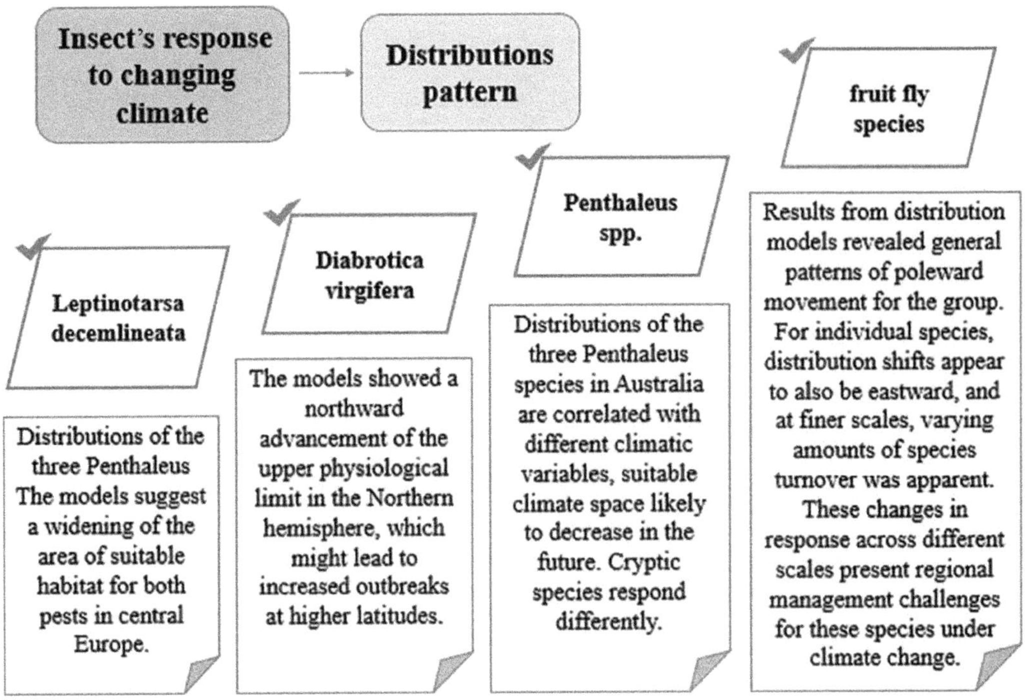


**FIGURE 3.2** Different types of distribution pattern of insects in response to climate change.

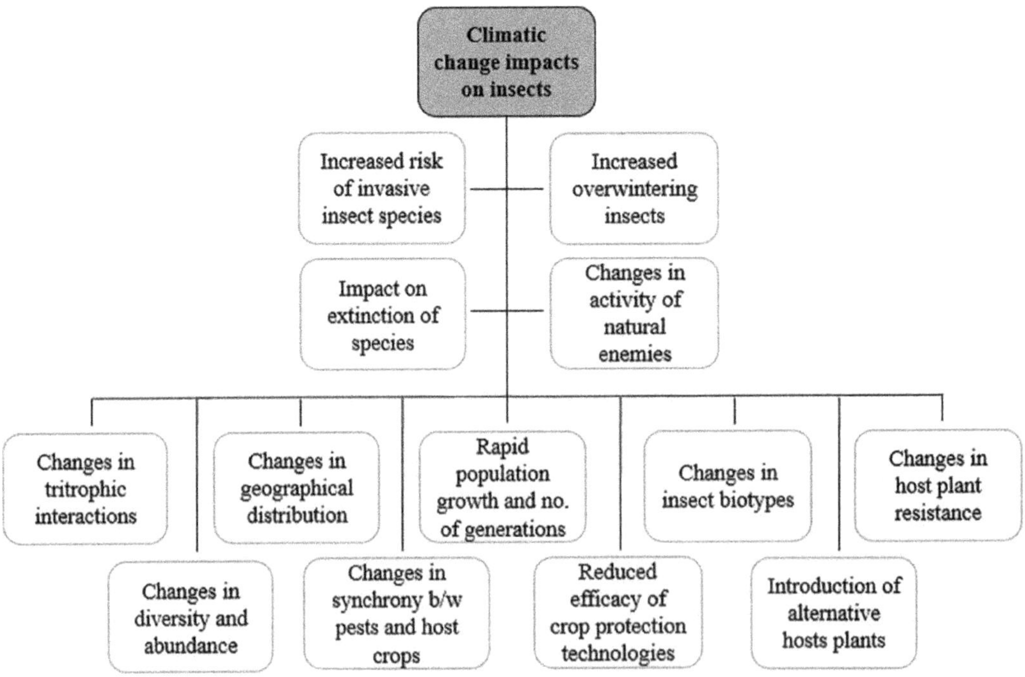


**FIGURE 3.3** Possible impacts of climate change on insects.

September had less incidence of insect pests than the one sown in August (Sharma et al. 2016). Moreover, stress in the plant lowered the defense system of the plant and pest outbreak is resulted (Anderegg and Callaway 2012).

The time from any given stage of the life cycle to the same stage in offspring is called generation. It was severely affected by temperature and the availability of food. The changing waves of temperature may accelerate the rate of reproduction, development and survival of insects. The generation time of insects was shorter in summer but longer in winter. *H. armigera* had three to five generations from March until November in Egypt (Ibrahim 2012). The generation could be predicted at peak when the heat accumulated units recorded 558.18 DDs (EL-Mezayyen and Ragab 2014). The cotton leafworm, *Spodoptera littoralis* (Boisd.), is a polyphaser insect fed on many crops, weeds and vegetables in Egypt causing more damage. The *S. littoralis* was active all over the year and has many groups at the year. due to temperature increase, the generation time of *Spodoptera litura* would reduce to 18%–22% and subsequently increase the number of generations. The development rate of *S. litura* immature stages was enlarged linearly as a function of temperature until approximately 34°C–36°C, after which it became nonlinear. Under a very high temperature of 38°C, there is no development of larval and pupal stage of *S. litura* to the next stage. Females did not lay eggs at 15°C and >35°C (Fand et al. 2015). Male moths of *P. gossypiella* captured in pheromone cons indicated the presence of five to six generations a year. The quantity of male moth varied from one season to another and even during the same season. This might happen due to moth emergence from larval diapause which was affected by crop rotation and weather factors. *P. gossypiella* was three to five generations during the period extended from March to December (Guirguis et al. 1991). "Within one year more number of generations will be completed by the insect and ultimately it will result in more crop damage" (Petzoldt and Seaman 2010) (Figure 3.4).

## 3.8 Conclusion and Recommendations

Lessening the effect of pests is more vital than ever to guarantee worldwide food security, decreased application of pesticides and diminished GHG emanations concentrated per unit of crop production. The pest admonishing frameworks based on climate are vital choice back instruments that offer assistance to agriculturists to assess the chance of pest's attack events under distinctive climatic conditions. The data on climate, crops and insect/pest, is exceptionally critical for the caution frameworks for taking essential activity to anticipate insect pest attacks and reduce financial troubles. So, there is a need to produce modified plants that are adjusted to these unexpected climate conditions. In this situation, the insect pest management approach is a best strategy for agriculturists, expansion laborers, researchers and public/private institutes partners to act in coordination and at scale to reorient pest management approaches and develop an appropriate empowering environment to oversee advancing climate change-induced insect pests dangers and attacks more viably, and so, there's a have to:

- Early discovery of changing insect pests dangers, which needs those expansion experts, are capable and connected to correct data and communications advances and demonstrative frameworks and administrations.
- Conveying farmer-friendly science-based pests administration data specifically into the hands of agriculturists, e.g. utilizing IT-based communication channels, such as portable phones, radio or TV, which are cost-effective and have flexibility.
- Anticipate and outline patterns of potential changes in topographical dissemination, and think about how climate changes will influence improvement, rate and populace flow of insect pests.
- Assess the impact of worldwide warming and climate change impacts on species differing qualities and trimming designs, and their impact on the plenitude of insect pests.
- Understand the modifications in the expression of resistance to insect pests, and recognize steady sources of resistance, and pyramid the resistance qualities in commercial cultivars.
- In the pest management the efficiency of transgenic crops is studied by the effect of global warming.

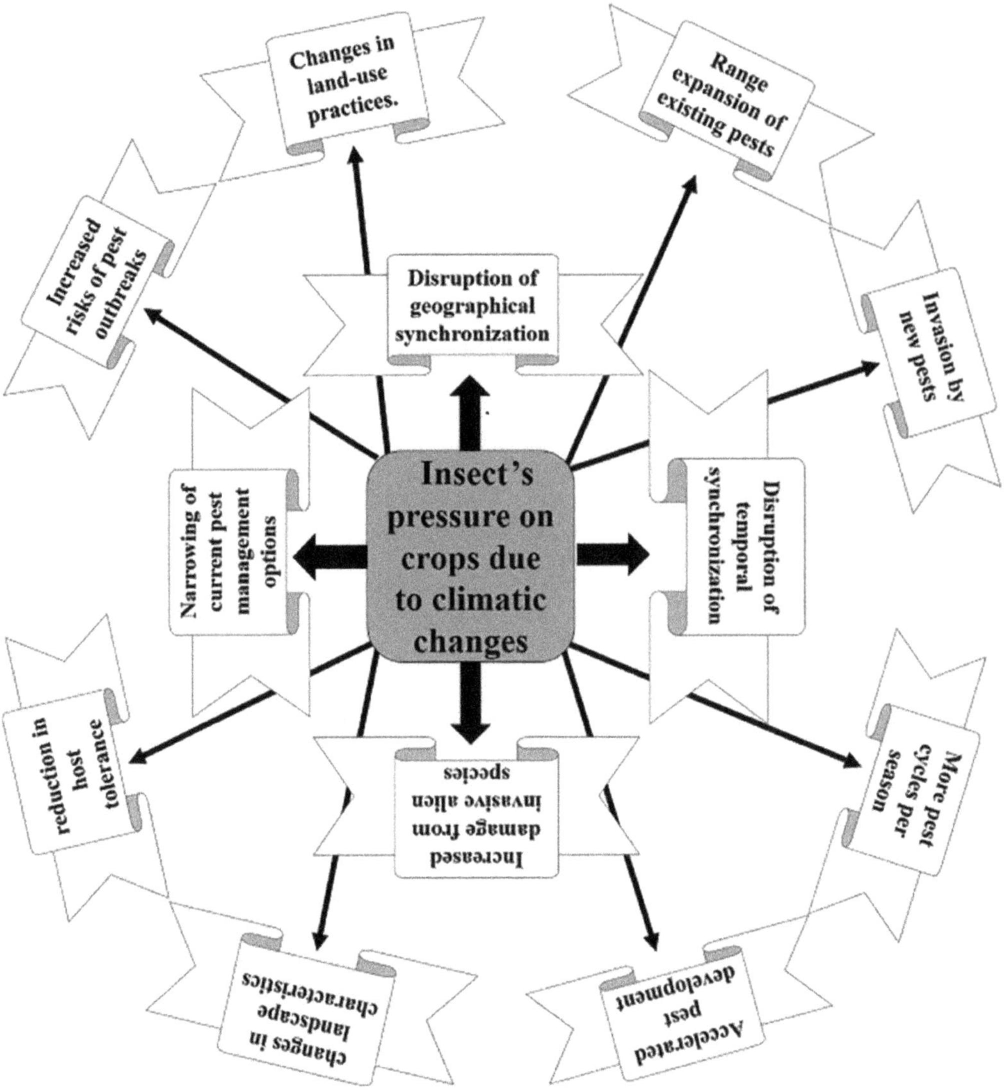


**FIGURE 3.4** Magnify pressure of insects on different food crops due to climate change.

- Assess the adequacy of different pests administration innovations under differing natural conditions, and create suitable procedures for pests administration to relieve the impacts of climate change.
- Climatic and crop prototypes have to be created for arrive utilize criteria, and soil efficiency and the strategies for fitting insecticide/herbicide inputs to climate ought to be created.

## REFERENCES


Adnan M, Shah Z, Sharif M, Rahman H (2018a) Liming induces carbon dioxide ($CO_2$) emission in PSB inoculated alkaline soil supplemented with different phosphorus sources. Environ Sci Pollut Res 25:9501–9509.

Adnan M, Zahir S, Fahad S, Arif M, Mukhtar A, Imtiaz AK, Ishaq AM, Abdul B, Hidayat U, Muhammad A, Inayat-Ur R, Saud S, Muhammad ZI, Yousaf J, Amanullah Hafiz MH, Wajid N (2018b) Phosphate-solubilizing bacteria nullify the antagonistic effect of soil calcification on bioavailability of phosphorus in alkaline soils. Sci Rep 8:4339. https://doi.org/10.1038/s41598-018-22653-7.

Agrawal R, Mehta SC (2007) Weather based forecasting of crop yields, pests and disease IASRI models. J Ind Soc Agril Stati 61:255–263.

Akram R, Fahad S, Hashmi MZ, Wahid A, Adnan M, Mubeen M, Khan N, Rehmani MI, Awais M, Abbas M, Shahzad K (2019) Trends of electronic waste pollution and its impact on the global environment and ecosystem. Environ Sci Pollut Res 1:1–6.

Akram R, Turan V, Hammad HM, Ahmad S, Hussain S, Hasnain A, Maqbool MM, Rehmani MIA, Rasool A, Masood N, Mahmood F (2018b) Fate of organic and inorganic pollutants in paddy soils. In: Zaffar M, Varma HA (eds) Environmental pollution of paddy soils. Springer, Cham, Switzerland, pp 197–214.

Akram R, Turan V, Wahid A, Ijaz M, Shahid MA, Kaleem S, Hafeez A, Maqbool MM, Chaudhary HJ, Munis MFH, Mubeen M (2018a) Paddy land pollutants and their role in climate change. In: Zaffar M, Varma HA (eds) Environmental pollution of paddy soils. Springer, Cham, Switzerland, pp 113–124.

Anderegg WRL, Callaway ES (2012) Infestation and hydraulic consequences of induced carbon starvation. Plant Physiol 159:1866–1874.

Aziz K, Daniel KYT, Fazal M, Muhammad ZA, Farooq S, FanW, Fahad S, Ruiyang Z (2017) Nitrogen nutrition in cotton and control strategies for greenhouse gas emissions: A review. Environ Sci Pollut Res 24:23471–23487. https://doi.org/10.1007/s11356-017-0131-y.

Bade BA, Ghorpade SA (2009) Life fecundity tables of sugarcane woolly aphid, *Ceratovacuna lanigera* Zehntner. J Insect Sci 22:402–405.

Bale JS, Master JG, Hodkinson ID, Awmack C, Bezemer TM, Browm VK, Butterfield J, Buse A, Coulson JC, Frra J (2002) Herbivory in the global climate change reseach:direct effect of rising temperature on insect herbivores. Glob Change Biol 8:1–16.

Barzman M, Lamichhane JR, Booij K, Boonekamp P, Desneux N, Huber L, Kudsk P, Langrell SRH, Ratnadass A, Ricci P, Sarah J-L, Messean A (2015) Research and development priorities in the face of climate change and rapidly evolving pests. In: Lichtfouse E (ed) Sustainable agriculture reviews, vol 17. Springer, Cham pp 1–27.

Bebber DP, Ramotowski MAT, Gurr SJ (2013) Crop pests and pathogens move polewards in a warming world. Nat Clim Change 3:985–988.

Calzadilla A, Rehdanz K, Betts R, Falloon P, Wiltshire A, Tol RS (2013) Climate change impacts on global agriculture. Clim Change 120:357–374.

Challinor AJ, Watson J, Lobell DB, Howden SM, Smith DR, Chhetri N (2014) A meta-analysis of crop yield under climate change and adaptation. Nat Clim Change 4:287–291.

Cock MJW, Biesmeijer JC, Cannon RJC, Gerard PJ, Gillespie D, Jiménez JJ, Lavelle PM, Raina SK (2013) The implications of climate change for positive contributions of invertebrates to world agriculture. CAB Rev Perspect Agric Vet Sci Nutr Nat Resour 8:28.

Czepak C, Albernaz KC, Vivan LM, Guimarães HO, Carvalhais T (2013) First reported occurrence of *Helicoverpa armigera* (Hübner) (Lepidoptera: Noctuidae) in Brazil. Pesq Agropec Trop Goiânia 43:110–113.

De Groot RS, Wilson MA, Boumans RMJ (2002) A typology for the classification, description and valuation of ecosystem functions, goods and services. Ecolog Econ 41:393–408.

Dicke M (2017) Ecosystem services of insects. In: Van Huis A, Tomberlin JK (eds) Insects as food and feed: From production to consumption, Wageningen Academic Publishers, Wageningen, The Netherlands, pp 61–76.

Easterling DR, Meehl GA, Parmesan C, Changnon SA, Karl TR, Mearns LO (2000) Climate extremes: Observations, modelling, and impact. Science 289:2068–2074.

EL-Mezayyen GA, Ragab MG (2014) Predicting the American bollworm, Helicoverpa armigera (Hübner) field generations as influenced by heat unit accumulation. Egypt J Agric Res 92:91–99.

Epstein PR (2001) Climate change and emerging infection diseases. Microbes Infect 3:747–754.

Fahad S, Adnan M, Hassan S, Saud S, Hussain S, Wu C, Wang D, Hakeem KR, Alharby HF, Turan V, Khan MA, Huang J (2019b) Rice responses and tolerance to high temperature. In: Hasanuzzaman M, Fujita M, Nahar K, Biswas JK (eds) Advances in rice research for abiotic stress tolerance. Woodhead Publ Ltd, Abington Hall Abington, Cambridge, UK, pp 201–224.

Fahad S, Bajwa AA, Nazir U, Anjum SA, Farooq A, Zohaib A, Sadia S, NasimW, Adkins S, Saud S, Ihsan MZ, Alharby H, Wu C,Wang D, Huang J (2017) Crop production under drought and heat stress: Plant responses and management options. Front Plant Sci 8:1147. https://doi.org/10.3389/fpls.2017.01147.

Fahad S, Bano A (2012) Effect of salicylic acid on physiological and biochemical characterization of maize grown in saline area. Pak J Bot 44:1433–1438.

Fahad S, Chen Y, Saud S, Wang K, Xiong D, Chen C, Wu C, Shah F, Nie L, Huang J (2013) Ultraviolet radiation effect on photosynthetic pigments, biochemical attributes, antioxidant enzyme activity and hormonal contents of wheat. J Food, Agri Environ 11(3&4):1635–1641.

Fahad S, Hussain S, Bano A, Saud S, Hassan S, Shan D, Khan FA, Khan F, Chen Y, Wu C, Tabassum MA, Chun MX, Afzal M, Jan A, Jan MT, Huang J (2014a) Potential role of phytohormones and plant growth-promoting rhizobacteria in abiotic stresses: Consequences for changing environment. Environ Sci Pollut Res 22(7):4907–4921. https://doi.org/10.1007/s11356-014-3754-2.

Fahad S, Hussain S, Matloob A, Khan FA, Khaliq A, Saud S, Hassan S, Shan D, Khan F, Ullah N, Faiq M, Khan MR, Tareen AK, Khan A, Ullah A, Ullah N, Huang J (2014b) Phytohormones and plant responses to salinity stress: A review. Plant Growth Regul 75(2):391–404. https://doi.org/10. 1007/s10725-014-0013-y.

Fahad S, Hussain S, Saud S, Hassan S, Chauhan BS, Khan F, et al (2016a) Responses of rapid viscoanalyzer profile and other rice grain qualities to exogenously applied plant growth regulators under high day and high night temperatures. PLoS One 11(7):e0159590. https://doi.org/10.1371/journal.pone.0159590.

Fahad S, Hussain S, Saud S, Hassan S, Ihsan Z, Shah AN, Wu C, Yousaf M, Nasim W, Alharby H, Alghabari F, Huang J (2016c) Exogenously applied plant growth regulators enhance the morphophysiological growth and yield of rice under high temperature. Front Plant Sci 7:1250. https://doi.org/10.3389/fpls. 2016.01250.

Fahad S, Hussain S, Saud S, Hassan S, Tanveer M, Ihsan MZ, Shah AN, Ullah A, Nasrullah KF, Ullah S, Alharby H, Wu C, Huang J (2016d) A combined application of biochar and phosphorus alleviates heat-induced adversities on physiological, agronomical and quality attributes of rice. Plant Physiol Biochem 103:191–198.

Fahad S, Hussain S, Saud S, Khan F, Hassan S, Jr A, Nasim W, Arif M, Wang F, Huang J (2016b) Exogenously applied plant growth regulators affect heat-stressed rice pollens. J Agron Crop Sci 202:139–150.

Fahad S, Hussain S, Saud S, Tanveer M, Bajwa AA, Hassan S, Shah AN, Ullah A, Wu C, Khan FA, Shah F, Ullah S, Chen Y, Huang J (2015a) A biochar application protects rice pollen from high-temperature stress. Plant Physiol Biochem 96:281–287.

Fahad S, Muhammad ZI, Abdul K, Ihsanullah D, Saud S, Saleh A, Wajid N, Muhammad A, Imtiaz AK, Chao W, Depeng W, Jianliang H (2018) Consequences of high temperature under changing climate optima for rice pollen characteristics-concepts and perspectives. Archives Agron Soil Sci. https://doi.org/10.10 80/03650340.2018.1443213.

Fahad S, Nie L, Chen Y, Wu C, Xiong D, Saud S, Hongyan L, Cui K, Huang J (2015b) Crop plant hormones and environmental stress. Sustain Agric Rev 15:371–400.

Fahad S, Rehman A, Shahzad B, Tanveer M, Saud S, Kamran M, Ihtisham M., Khan SU, Turan V, Rahman MHU (2019a) Rice responses and tolerance to metal/metalloid toxicity. In: Hasanuzzaman M, Fujita M, Nahar K, Biswas JK (eds) Advances in rice research for abiotic stress tolerance. Woodhead Publ Ltd, Abington Hall Abington, Cambridge, UK, pp 299–312.

Fand BB, Sul NT, Bal SK, Minhas PS (2015) Temperature impacts the development and survival of common cutworm (Spodoptera litura): Simulation and visualization of potential population growth in India under warmer temperatures through life cycle modeling and spatial mapping. PLoS One 10:1–25.

FAO (2013b) International standard for phytosanitary measure. FAO, Rome, Italy.

Gallai N, Salles JM, Settele J, Vaissiere BE (2009) Economic valuation of the vulnerability of world agriculture confronted with pollinator decline. Ecologic Econom 68:810–821.

Garibaldi LA, Steffan-Dewenter I, Winfree R, Aizen MA, et al. (2013) Wild pollinators enhance fruit set of crops regardless of honey bee abundance. Science 339:1608–1611.

Graf B, Hopli HU, Hohn H, Blaise P (2002) SOPRA: A forecasting tool for insect pests in apple orchards. In: *Acta Hort* 584: VI International Symposium on Computer Modelling in Fruit Research in fruit Research and Orchard Management.

Gregory PJ, Johnson SN, Newton AC, Ingram JS (2009) Integrating pests and pathogens into the climate change/food security debate. J Exp Bot 60:2827–2838.

Guirguis MW, El-Feshawi AA, Watson WM, Nassef MA (1991) Occurrence and seasonal abundance of the pink bollworm Pectinophora gossypiella (Saund.) infesting cotton. Egypt J Agric Res 69:63–72.

Habib ur R, Ashfaq A, Aftab W, Manzoor H, Fahd R, Wajid I, Md. Aminul I, Vakhtang S, Muhammad A, Asmat U, Abdul W, Syeda RS, Shah S, Shahbaz K, Fahad S, Manzoor H, Saddam H, Wajid N (2017) Application of CSM-CROPGRO-Cotton model for cultivars and optimum planting dates: Evaluation in changing semi-arid climate. Field Crops Res. http://dx.doi.org/10.1016/j.fcr.2017.07.007.

Hafiz MH, Muhammad A, Farhat A, Hafiz FB, Saeed AQ, Muhammad M, Fahad S, Muhammad A (2019) Environmental factors affecting the frequency of road traffic accidents: A case study of sub-urban area of Pakistan. Environ Sci Pollut Res. https://doi.org/10.1007/s11356-019-04752-8.

Hafiz MH, Wajid F, Farhat A, Fahad S, Shafqat S, Wajid N, Hafiz FB (2016) Maize plant nitrogen uptake dynamics at limited irrigation water and nitrogen. Environ Sci Pollut Res 24(3):2549–2557. https://doi.org/10.1007/s11356-016-8031-0.

Hellmann JJ, Byers JE, Bierwagen BG, Dukes JS (2008) Five potential consequences of climate change for invasive species. Conserv Biol 22:534–543.

Hill MP, Hoffmann AA, McColl SA, Umina PA (2012) Distribution of cryptic blue oat mite species in Australia: Current and future climate conditions. Agric For Entomol 14:127–137.

Hoehn P, Tscharntke T, Tylianakis JM, Steffan-Dewenter I (2008) Functional group diversity of bee pollinators increases crop yield. Proc Royal Soc B: Biol Sci 275:2283–2291.

Ibrahim SS (2012) Impact of certain new cotton genotypes on ecological and biological aspects of cotton leaf and bollworms. PhD thesis, Faculty of Agriculture, Cairo University, pp 178.

IPCC (2014) Climate change 2014: Synthesis Report. In: Core Writing Team, Pachauri RK, Meyer LA (eds) Contribution of Working Groups I, II and III to the Fifth Assessment Report of the Intergovernmental Panel on Climate Change, IPCC, Geneva, Switzerland, pp 151.

Jensen JR (2000) Remote sensing of the environment: An earth resource perspective. Prentice Hall Series of Geographic Information Science. Prentice Hall, Upper Saddle River, NJ, pp 544.

Kamran M, Wenwen C, Irshad A, Xiangping M, Xudong Z, Wennan S, Junzhi C, Shakeel A, Fahad S, Qingfang H, Tiening L (2017) Effect of paclobutrazol, a potential growth regulator on stalk mechanical strength, lignin accumulation and its relation with lodging resistance of maize. Plant Growth Regul 84:317–332. https://doi.org/10.1007/s10725-017-0342-8.

Knight JD, Tatchell GM, Norton G, Harrington R (1992) FLYPAST: An information management system for Rothamsted Aohid Database to aid pest control research and advice. Crop Prot 11:419–426.

Kranthi KR, Kranthi S, Gopalakrishnan N, Asokan R, Mayee CD (2009) Bt resistance-Its management and prospects in the current context of climate change. In: Ramamurthy VV, Gupta GP, Puri SN (eds) Proceedings of National Symposium on IPM Strategies to Combat Emerging Pests in the Current Scenario of Climate Change, January 28–30, 2009, Pasighat, Arunachal Pradesh, pp 237–261.

Kromp B (1999) Carabid beetles in sustainable agriculture: A review on pest control efficacy, cultivation impacts and enhancement. Agric Ecosys Environ 74:187–228.

Kumar R, Dubey V (2017) Climate change and agriculture: Current scenario and future prospective. In: National Seminar on Nutritional Security and Environmental Protection: Present scenario and Future Prospects Society of Biological Sciences and Rural Development, Allahabad (U.P.), India.

Liu B, Asseng S, Müller C, Ewert F, Elliott J, Lobell DB, Martre P, Ruane AC, Wallach D, Jones JW (2016) Similar estimates of temperature impacts on global wheat yield by three independent methods. Nat Clim Change 6:1130.

Lobell DB, Field CB (2007) Global scale climate–crop yield relationships and the impacts of recent warming. Environ Res Lett 2:014002.

Losey JE, Vaughan M (2006) The economic value of ecological services provided by insects. Bioscience 56:311–323.

Macfadyen S, McDonald G, Hill MP (2018) From species distributions to climate change adaptation: Knowledge gaps in managing invertebrate pests in broad-acre grain crops. Agric Ecosyst Environ 253:208–219.

Messousi HE, Hafid H, Lahrouni A, Afif M (2007) Simulation of temperature effect on the population dynamics of the Mediterranean fruit fly, *Ceratitis capitata* (Diptera; Tephritidae). J Agron 6:374–377.

Muhammad Z, Abdul MK, Abdul MS, Kenneth BM, Muhammad S, Shahen S, Ibadullah J, Fahad S (2019) Performance of Aeluropus lagopoides (mangrove grass) ecotypes, a potential turfgrass, under high saline conditions. Environ Sci Pollut Res. https://doi.org/10.1007/s11356-019-04838-3.

NASA (2018) The National Aeronautics and Space Administration

Nasim W, Ahmad A, Amin A, Tariq M, Awais M, Saqib M, Jabran K, Shah GM, Sultana SR, Hammad HM, Rehmani MIA (2018) Radiation efficiency and nitrogen fertilizer impacts on sunflower crop in contrasting environments of Punjab-Pakistan. Environ Sci Pollut Res 25:1822–1836.

Nasim W, Belhouchette H, Tariq M, Fahad S, Hammad HM, Mubeen M, Munis MFH, Chaudhary HJ, Khan I, Mahmood F, Abbas T (2016) Correlation studies on nitrogen for sunflower crop across the agroclimatic variability. Environ Sci Pollut Res 23:3658–3670.

Nietschke BS, Borchert DM, Magarey RD, Matthew AC (2008) Climatological potential for Scirtothrips dorsalis (Thysanoptera: Thripidae) establishment in the United States. Florida Ent 91:79–86.

Petzoldt C, Seaman A (2010) Climate change effects on insects and pathogens. In: Petzoldt C, Seaman, A (eds) Climate change and agriculture: Promoting practical and profitable responses. New York State Agricultural Extension Station, Geneva, NY, pp 6–16.

Phillips BL, Kelehear C, Pizzatto L, Brown GP, Barton D, Shine R (2010) Parasites and pathogens lag behind their host during periods of host range advance. Ecology 91:872–881.

Qamar uz Z, Zubair A, Muhammad Y, Muhammad ZI, Abdul K, Fahad S, Safder B, Ramzani PMA, Muhammad N (2017) Zinc biofortification in rice: Leveraging agriculture to moderate hidden hunger in developing countries. Arch Agron Soil Sci 64:147–161. https://doi.org/10.1080/03650340.2017.1338343.

Sajjad H, Muhammad M, Ashfaq A, Waseem A, Hafiz MH, Mazhar A, Nasir M, Asad A, Hafiz UF, Syeda RS, Fahad S, Depeng W, Wajid N (2019) Using GIS tools to detect the land use/land cover changes during forty years in Lodhran district of Pakistan. Environ Sci Pollut Res. https://doi.org/10.1007/s11356-019-06072-3.

Samways MJ (1993) Insects in biodiversity conservation: Some perspectives and directives. Biodiversity Conserv 2:258–282.

Samways MJ (2005) Insect diversity conservation. Cambridge University Press, New York, pp 25–29.

Saud S, Chen Y, Fahad S, Hussain S, Na L, Xin L, Alhussien SA (2016) Silicate application increases the photosynthesis and its associated metabolic activities in Kentucky bluegrass under drought stress and post-drought recovery. Environ Sci Pollut Res 23(17):17647–17655. https://doi.org/10.1007/s11356-016-6957-x.

Saud S, Chen Y, Long B, Fahad S, Sadiq A (2013) The different impact on the growth of cool season turf grass under the various conditions on salinity and draught stress. Int J Agric Sci Res 3:77–84.

Saud S, Fahad S, Yajun C, Ihsan MZ, Hammad HM, Nasim W, Arif M, Alharby H (2017) Effects of nitrogen supply on water stress and recovery mechanisms in kentucky bluegrass plants. Front Plant Sci 8:983. https://doi.org/10.3389/fpls.2017.00983.

Saud S, Li X, Chen Y, Zhang L, Fahad S, Hussain S, Sadiq A, Chen Y (2014) Silicon application increases drought tolerance of Kentucky bluegrass by improving plant water relations and morph physiological functions. SciWorld J 2014:1–10. https://doi.org/10.1155/2014/368694.

Schoonhoven LM, Van Loon JJA, Dicke M (2005) Insect-plant biology. Oxford University Press, Oxford, UK, pp 400.

Schwägerl C (2016) What's causing the sharp decline in insects, and why it matters. Yale Environment 360, 6 July 2016.

Shah F, Lixiao N, Kehui C, Tariq S, Wei W, Chang C, Liyang Z, Farhan A, Fahad S, Huang J (2013) Rice grain yield and component responses to near 2°C of warming. Field Crop Res 157:98–110.

Sharma HC (2010) Effect of climate change on IPM in grain legumes. In: 5th International Food Legumes Research Conference (IFLRC V), and the 7th European Conference on Grain Legumes (AEP VII), 26–30th April 2010, Anatalaya, Turkey.

Sharma HC (2014) Climate change effects on insects: Implications for crop protection and food security. J Crop Improv 28:229–259.

Sharma HC, Pathania M, War AR, Pavani T, Vashist S (2016) Climate change effects on pest spectrum and incidence in grain legumes. In: Dixit GP, Singh G, Singh NP (eds) Pulses: Challenges and opportunities under challenging climatic scenario. ICAR-IIPR, Kanpur. pp 124–137.

Stireman J, Dyer L, Jazen D, Singer M, Lill J, Marquis R, Reclefs R, Gentry G, Hallwachs W, Coley P (2005) Climatic unpredictability parasitism of caterpillars: Implications of global warming. Proc Nat Acad Sci 102:17384–17387.

Stork NE, McBroom J, Gely C, Hamilton AJ (2015) New approaches narrow global species estimates for beedtles, insects, and terrestrial arthropods. Proc Natl Acad Sci USA 112:7519–7523.

Svobodova E, Trnka M, Dubrovsky M, Semeradova D, Eitzinger J, Stepanek P, Zalud Z (2014) Determination of areas with the most significant shift in persistence of pests in Europe under climate change. Pest Manag Sci 70:708–715.

Symondson WOC, Sunderland KD, Greenstone MH (2002) Can generalist predators be effective biocontrol agents? Ann Rev Entomol 47:561–594.

Ullah H, Khan WU, Alam M, Khalil IH, Adhikari KH, Shahwar D, Jamal Y, Jan I, Adnan M (2016) Assessment of G×E interaction and heritability for simplification of selection in spring wheat genotypes. Can J Plant Sci 96(3):1021–1025.

Wajid N, Ashfaq A, Asad A, Muhammad T, Muhammad A, Muhammad S, Khawar J, Ghulam MS, Syeda RS, Hafiz MH, Muhammad IAR, Muhammad ZH, Muhammad Habib ur R, Veysel T, Fahad S, Suad S, Aziz K, Shahzad A (2017) Radiation efficiency and nitrogen fertilizer impacts on sunflower crop in contrasting environments of Punjab. Pak Environ Sci Pollut Res 25:1822–1836. https://doi.org/10.1007/s11356-017-0592-z.

West JS, Fitt BDL, Townsend JA, Stevens M, Edwards SG, Turner JA, Edmonds J (2015) Impact of climate change on diseases on sustainable crop systems: CLIMDIS. Project Number 539. HCGA.

Woodcock BA, Harrower C, Redhead J, Edwards M, Vanbergen AJ, Heard MS, Roy DB Pywell RF (2014) National patterns of functional diversity and redundancy in predatory ground beetles and bees associated with key UK arable crops. J Appl Ecol 51:142–151.

Yang Z, Zhang Z, Zhang T, Fahad S, Cui K, Nie L, Peng S, Huang J (2017) The effect of season-long temperature increases on rice cultivars grown in the central and southern regions of China. Front Plant Sci 8:1908. https://doi.org/10.3389/fpls.2017.01908.

Zahida Z, Hafiz FB, Zulfiqar AS, Ghulam MS, Fahad S, Muhammad RA, Hafiz MH, Wajid N, Muhammad S (2017) Effect of water management and silicon on germination, growth, phosphorus and arsenic uptake in rice. Ecotoxicol Environ Saf 144:11–18.

Zalucki MP, Furlong MJ (2005) Forecasting Helicoverpa populations in Australia: A comparison of regression based models and a bioclimatic based modeling approach. Insect Sci 12:45–56.

Zhang ZQ (2011) Animal biodiversity: An introduction to higher-level classification and taxonomic richness. Zootaxa 3148:7–12.

Zhu K, Chiariello NR, Tobeck T, Fukami T, Field CB (2016) Nonlinear, interacting responses to climate limit grassland production under global change. Proc Natl Acad Sci:10589–10594.

# 4

# *Extreme High Temperature and Plant Life*

**Muhammad Arif, Talha Jan, Fazal Munsif, Sajjad Zaheer, Amanullah, Fahd Rasul, and Ishaq Ahmad Mian**
*The University of Agriculture*

**Muhammad Riaz**
*Government College University Faisalabad*

**Shah Fahad**
*Hainan University*
*The University of Haripur*

**Shah Saud**
*Linyi University*

**Muhammad Adnan**
*University of Swabi*

**Muhammad Hamzah Saleem**
*Huazhong Agricultural University*

**Bibi Haleema**
*Agricultural Research Institute*

**Fazal Jalal**
*Huazhong Agricultural University*
*Abdul Wali Khan University*

**Muhammad Zamin**
*University of Swabi*

**Kawsar Ali**
*Abdul Wali Khan University*

**Bushra Khan**
*University of Peshawar*

DOI: 10.1201/9781003286400-4

**CONTENTS**



## 4.1 Introduction

Recent estimates and predictions about an increase in ambient temperature have created apprehension among scientists. It has been modeled that global air temperature is increasing at a rate of 0.2°C per decade and by the onset of 2100 A.D. It will be 1.8°C–4.0°C higher than what it currently is. This is in fact a worrying situation because of the tremendous loss that such high temperatures may incur at a global level. This level of rise in temperature can affect all living organisms either directly or by modifying their ambient environment and hence affecting it indirectly (Fahad et al. 2016a,b,c,d; Fahad et al. 2019a,b). Plants among all organisms are at particular risk because of their sessile nature. This sessile nature makes them most vulnerable to the substantial and often lethal impacts of high temperature thus affecting their growth and development processes (Lobell and Field 2007; Ali et al. 2016; Fahad et al. 2016a,b,c,d; Fahad et al. 2019a,b). Heat stress affects plants at every growth stage from germination to storage with particular damage incurred at flowering and reproductive stages which are crucial for getting sustainable yield. Under high-temperature conditions germination is hindered, synchronization of plant morphological stages is disturbed, plant's photosynthetic efficiency is lowered and above all, excess generation of reactive oxygen species (ROS) occurs. All these attributes cause a drastic reduction in plant yield.

### 4.1.1 Effects of High Temperature on Plants

#### *4.1.1.1 Germination*

Seed germination is the first and foremost process adversely affected due to high temperature. Various crop species are documented for adverse impacts of high temperature on their germination mechanisms, and it has been noted that temperature ranges greatly vary for plant species to get adversely affected by high temperature (Casalegno 2011; Kumar et al. 2011; Fahad et al. 2016a,b,c,d; Fahad et al. 2019a,b). Major impacts include reduced germination percentage, reduced plant emergence, emergence of abnormal seedlings, reduced plumule and radicle growth, and poor seedling vigor (Kumar et al. 2011). Inhibition of seed germination is mainly attributed to induction of ABA (Essemine et al. 2010; Khan et al. 2016). In wheat crop, death of embryos and other seed cells was noted at 45°C which reduced the rate of germination of wheat and consequently reduced seedling establishment rate (Cheng et al. 2009).

#### *4.1.1.2 Growth*

After seedling establishment plant needs water for survival. Water availability is greatly affected by high temperature. Loss of water occurs from the cell, reducing cell water content hence reducing cell size and ultimately plant growth is reduced (Ashraf and Hafeez 2004). This reduction in cell water content has drastic effects on almost every plant growth determinant including root and shoot growth, number of tillers, leaf growth, total biomass and final yield. In maize, sugarcane and millet, net assimilation rate and resultantly relative growth rate were noticed to be reduced with induction of high temperature causing a visible reduction in the overall growth of maize and sugarcane (Wahid 2007). In rice crops, plant height, the number of tillers and ultimately total biomass were reduced in response to high temperature (Mitra and Bhatia 2008). Leaves are more susceptible to heat stress as compared to other plant parts. Leaf senescence and abscission, rolling and drying of leaves, necrosis, damage to leaf-tip and margins, visible scorching and sunburns are the major symptoms associated with heat stress-induced leaf deterioration mainly noticed in sugarcane and other plant species (Omae et al. 2012; Rodríguez et al. 2005). Indented leaves (hyponasty) and reduction in green leaf area were also noted in wheat crops due to high temperature (Savin et al. 1997). While in some plants such as common bean, high temperature results in shoot and root growth inhibition (Koini et al. 2009; Naseer et al. 2016), in others, fruit discoloration and damage may occur (Rodríguez et al. 2005). In some plant species, high temperatures cause the formation of unusual elongated stems and reduction occurs in total biomass (Patel and Franklin 2009). In wheat under heat stress, the number of tillers was drastically reduced. At a cellular level, denaturation or aggregation of proteins may occur causing programmed cell death within even seconds which can lead to leaves shedding, flower and fruit abortion, or even death of the entire plant (Rodríguez et al. 2005).

Phenological durations and resultantly plant life periods are reduced due to increased heat stress. Days to anthesis, booting, maturity are reduced in wheat under heat stress reducing the total life period (Yamamoto et al. 2008).

#### *4.1.1.3 Photosynthesis and Gaseous Exchange*

Plant leaf experiences a negative impact with the induction of heat stress. Some of these impacts include reduced leaf area, reduced leaf water potential and pre-mature leaf senescence. All these events have adverse effects on the overall photosynthetic performance of plants (Greer and Weedon 2012; Young et al. 2004; Fahad et al. 2015a; Fahad et al. 2018). Among plant physiological processes, heat stress has a greater influence on photosynthesis than any other process (Crafts-Brandner and Salvucci 2002) with magnitude more pronounced in C3 plants than C4 plants (Yang et al. 2006). Majorly, the amount of photosynthetic pigments and activity of photosystem II is reduced with heat stress (Marchand et al. 2005). Photosynthetic apparatus such as chloroplast and its constituents such as stroma and thylakoid membrane are extremely sensitive to heat stress. Primary sites for heat damage are photochemical reactions in thylakoid lamellae and carbon metabolism of the stroma (Wang et al. 2009). Chloroplasts undergo major alterations under heat stress like altered structural organization of thylakoids and swelling and loss of stacking in grana (Ashraf and Hafeez 2004; Rodríguez et al. 2005).

The ability of plants to sustain leaf gas exchange and $CO_2$ assimilation rates under heat stress is directly correlated with heat tolerance (Kumar et al. 2005; Yang et al. 2006). Heat stress has drastic effects on the leaf water status and leaf stomatal conductance which could lead to the closure of stomata and hence reduction in the intercellular concentration of $CO_2$ (Ashraf and Hafeez 2004; Greer and Weedon 2012; Yang et al. 2006). In sorghum and soybean, photosynthesis and consequent sucrose production were noted to be reduced in the presence of heat stress (Mohammed and Tarpley 2010). Reduction in the photosynthetic rate in rice at 38°C/28°C was found to be around 15% (Hurkman et al. 2009), while the reduction mounted to 60% in *Vitis vinifera* with increasing temperature from 25°C to 45°C most of which was attributed to 15%–30% stomatal closure.

#### 4.1.1.4 Reproductive Development

Reproductive tissues and reproductive stages are most susceptible to heat stress among all plant tissues and growth stages (Lobell et al. 2011). Plant species exhibit great variation in response to heat stress (Sato et al. 2006). Heat stress at the reproductive stage usually results in flower and seed abortion, and may result in plants producing no flowers or producing flowers but no seeds or fruits (Foolad 2005; Maheswari et al. 2012). Main reasons for this are impaired meiosis in both male and female organs, reduction in the number of produced pollen grains and those retained by stigma, reduced pollen germination and impaired pollen tube growth, unfertilized embryo, reduced ovule viability, abnormal style and style positions and hence disturbed fertilization process (Yun-Ying et al. 2008).

Among plant species, grain crops, especially rice are more susceptible to heat stress at a reproductive stage with significant variations among susceptible and tolerant varieties (Hurkman et al. 2009). High-temperature treatment above 33°C at a heading stage results in reduced anther dehiscence and pollen fertility rate. This could lead to less number of pollen grains reaching the stigma resulting in reduced fertilization and subsequent spikelet sterility and production of sterile seed in rice (Ahamed et al. 2010; Hurkman et al. 2009). High temperature especially high night temperatures above 32°C results in decreased pollen germination, leading to increased spikelet sterility in rice (Suwa et al. 2010). High temperature is often associated with excessive production of ethylene which can lead to rice pollen sterility (Hurkman et al. 2009). Key enzymes in sugar–starch metabolism are inhibited weakening sink strength which restricts grain filling and subsequently produces small and sterile rice grains. The same trends have been observed in the wheat crop. In maize crops, heat stress at silking stage results in tremendous yield reduction. The effect was less pronounced at the grain filling stage of maize (Zhang et al. 2013). Due to high temperature, young pods and developing seeds were reduced in soybean due to decreased pollen viability and resulting flowers abortion, resulting in lower seed numbers in soybean (Tan et al. 2011; Tubiello et al. 2007).

#### 4.1.1.5 Yield

High temperature has been noted to affect yields of several cultivated crops including cereals (e.g. wheat, rice, sorghum, maize, barley), oil seed crops (canola, mustard), pulses (e.g. cowpea, chickpea) and many other crop species (Ahamed et al. 2010; Tubiello et al. 2007; Zhang et al. 2013). Crop yield is mainly affected by high temperature through its effect on phenological development processes. In cereals, a 1°C increase in the average temperature decreased the grain yield by 4.1%–10%. The effect is more pronounced in sensitive varieties than tolerant varieties. In rice, major yield attributes affected by heat stress are grain length, width and weight. The main reason behind the reduction in these attributes was heat-induced spikelet sterility. Nitrogen concentration in grain was reported to increase by 44% due to heat stress which lead to low grains weight and resultantly decreased yield by 90% (Suwa et al. 2010). In wheat crop the yield reduction took place due to modification of dough and maturity stages and by the shortening of kernel desiccation period (Saitoh 2008). In wheat also, the effect is varied among tolerant and susceptible varieties. In maize, heat stress caused reduced light capture and radiation use efficiency and resulting biomass and grain yield as well as harvest index. It was observed that heat stress proved more detrimental at the flowering stage and then at the grain filling stage. High temperature (33°C–40°C) in maize is negatively affected light capture, Radiation-use efficiency (RUE), biomass and grain yield, harvest index although heat at the flowering stage results in higher yield reduction than that at grain filling period (Zhang et al. 2013). Among reproductive traits, grain filling duration was reduced when temperature was increased a few degrees above optimum temperature and hence resulted in reduced cereals' grain yield (Nahar et al. 2010). Similar trends were observed in sorghum and canola where filled seed weight, seed size, seeds per pod and number of fertile pods were reduced with induced heat stress (Sinsawat et al. 2004). Reduction in productivity noted in wheat (Cicchino et al. 2010) and maize (Högy et al. 2013) due to high temperature is due to reduced photosynthetic capacity and increased respiration costs which results in reduced assimilatory capacity and reduced radiation use efficiency. High temperature also affects grain quality. For instance, in barley grain, maltose and several proteinogenic amino acid concentrations were observed to be increased with increasing temperature which is anti-quality characteristics. On the other hand, concentrations of aluminum, lipids, fructose, starch and total non-structural carbohydrates were decreased (Högy et al. 2013).

#### 4.1.1.6 Oxidative Stress

Heat stress has been suggested to uncouple certain enzymes and metabolic pathways causing the accumulation of harmful reactive oxygen species leading to the development of oxidative stress. These ROS most commonly include singlet oxygen ($1O_2$), hydrogen peroxide ($H_2O_2$), superoxide radical ($O_2$•–) and hydroxyl radical (OH•) (Asada 2006; Vasseur et al. 2011). Majorly, these ROS are generated in the reaction centers of photosystems I and II in chloroplasts and some other organelles, e.g. mitochondria and peroxisomes (Soliman et al. 2011).

These ROS are proved to be detrimental to various physiological functions. Hydroxyl radicals may undergo a reaction with all biomolecules, like DNA, proteins, pigments, lipids and almost with all cell constituents. Similarly, singlet oxygen can cause the oxidation of DNA, protein and fatty acids. Heat stress may also cause the denaturation of proteins causing peroxidation of membrane lipids and disruption of cell membrane stability (Rodríguez et al. 2005). Increased electron leakage from the thylakoid membrane is yet another mechanism induced by heat stress which causes a decrease in photosynthetic light reaction (Halliwell 2006). Among cereal crops, heat stress has been observed to increase leaf temperature, reduce antioxidant enzyme activities and hence increase malondialdehyde content in leaves of rice plants (Hurkman et al. 2009). In wheat, oxidative stress induced by high temperature causes damage to cell membrane, resulting in enzyme deactivation and protein degradation, reducing membrane thermostability and increasing electrolyte leakage. Exposure of perennial ryegrass (*Lolium perenne* L.) to moderate (36°C) and severe high-temperature stress (40°C) caused increase in levels of $H_2O_2$ which was responsible for reduction in photosystem II efficiency, destruction of cell membrane stability and caused lipid peroxidation (Halliwell 2006). Cotton, soybean and sorghum were also reported to be affected by high temperature as heat stress caused membrane injury and lipid peroxidation (Adnan et al. 2016; Mohammed and Tarpley 2010; Tan et al. 2011; Young et al. 2004).

Although ROS are known mostly for their detrimental effects on plant metabolic processes they are hypothetically known for inducing tolerance toward heat stress by improving plant signaling behaviors and hence triggering heat shock responses. This research area is relatively untouched (Asada 2006).

## 4.2 Plants Adaptation to High-Temperature Stress

Plants try to adapt to high temperature via avoidance and tolerance mechanism.

### 4.2.1 Avoidance Mechanisms

Plants adapt various surviving mechanisms to survive under high-temperature conditions. Usually, these mechanisms are similar to those adopted to perform under water deficit conditions. These mechanisms may comprise of both long- and short-term ways of heat avoidance. In long run, plants adapt evolutionary phenological and morphological adaptations and hence finally become able to avoid heat stress to be listed as thermophiles. Short-term avoidance or acclimation mechanisms include transpirational cooling, changing leaf orientation and alteration of membrane lipid compositions. Plants adjust water balance by reducing water loss through closure of stomata and increased stomatal densities. On the other hand, to extract more water from the soil, plants evolve larger xylem vessels (Srivastava et al. 2012; Shah et al. 2017). Many crop plants adapt an escape mechanism by maturing earlier than normal and hence minimizing yield losses (Rodríguez et al. 2005). Many plant species form a thick waxy coat of small hairs (tomentose) which greatly reduce the absorption of solar radiation. Plants growing in a hot climate avoid heat stress by reducing the absorption of solar radiation. Paraheliotropism is yet another mechanism that is the turning of leaf blades away from sunlight and orient leaves parallel to sunlight. Leaf rolling is also a mechanism to absorb a minimum of incident solar radiation. Smaller leaves face less resistance to diffuse the heat to ambient air due to less resistance caused by smaller air boundary around leaves.

Plants undergo intensive transpiration to cool their leaves under well-hydrated conditions and surprisingly leaf temperature has been found 6°C–15°C lower than the ambient temperature. Other mechanisms employed by plants under heat stress conditions are leaf abscission leaving heat resistant buds and heat escaping by completing the life cycle during cooler months (Fitter and Hay 2012).

### 4.2.2 Tolerance Mechanisms

Heat tolerance points toward the ability of a plant to grow under high temperature and yield economically. The tolerance response to heat stress is highly variable among different crops, closely related species and even the same plant organs. The initial response toward heat stress is established by an initial stress signal in the form of changes in membrane fluidity or the establishment of an ionic or osmotic effect (Rodríguez et al. 2005). Many mechanisms have been adapted by evolving plants to tolerate heat stress. Both long-term evolutionary adaptations and short-term avoidance mechanisms are employed by plants in this regard. This includes mechanisms such as the use of osmoprotectants and ion transporters by plants, late embryogenesis abundant proteins and strengthening antioxidant defense (Rodríguez et al. 2005; Wang et al. 2004). Sudden heat incidence may trigger short-term acclimation mechanisms such as transpirational cooling, leaf orientation and changes in membrane lipid composition (Rodríguez et al. 2005). Some plant species also opt for an escape strategy which could lead to smaller grain losses (Adams et al. 2001; Rahman et al. 2016).

## 4.3 Summary and Conclusion

Extreme temperatures are becoming a matter of routine in a changing climate. Sudden rise in temperature and heat waves has disastrous effects on susceptible plant species from germination to yield and onward. The detrimental effects of high temperature begin to appear as soon as the seed is sown in the soil. High temperature results in low germination percentage, reduced plant emergence, emergence of abnormal seedlings, reduced plumule and radicle growth and poor seedling vigor. These effects result in poor crop stand which shows poor growth after emergence. The disastrous effects of high temperature go on and affect every plant part. Among vegetative parts, leaves are more susceptible to heat stress and signs of leaf senescence and abscission, rolling and drying of leaves, necrosis and leaf damage can be seen in heat-stressed plants resulting in poor photosynthetic performance with damaged chloroplast and hence poor assimilation and plant growth. Phenological durations such as days to anthesis, booting and maturity are reduced in plants under heat stress, thus reducing the total life period. After the completion of vegetative growth, plants experience impaired meiosis in their reproductive parts due to heat stress, resulting in reduced number of produced pollen grains, reduced pollen germination and impaired pollen tube growth, unfertilized embryo, reduced ovule viability, abnormal style and style positions and hence disturbed fertilization process. All these effects result in reduced plant yield. Heat stress also promotes the production of reactive oxygen species such as singlet oxygen ($1O_2$), hydrogen peroxide ($H_2O_2$), superoxide radical ($O_2$•–) and hydroxyl radical (OH•). These radicals may undergo reaction with all biomolecules, like DNA, proteins, pigments, lipids and almost all cell constituents causing alterations in them.

Indeed plants suffer from both these stresses, however, some plant species have adapted many avoidances and tolerance mechanisms to avoid and tolerate these extreme stresses. Avoidance mechanisms for heat stress include transpirational cooling, changing leaf orientation and alteration of membrane lipid compositions. Other mechanisms include maintaining water balance through closure of stomata, increased stomatal densities and evolving larger xylem vessels. For coping with low-temperature stress plants undergo alteration at a molecular level by the production of cryoprotectant compounds such as soluble sugars, sugar alcohols and low-molecular-weight nitrogenous compounds. Since extreme temperatures cannot be controlled anthropogenically, genetic breeding for the production of tolerant varieties is perhaps the ultimate need of time second only to climate change mitigation strategies.

# REFERENCES

Adams S, Cockshull K, Cave C (2001) Effect of temperature on the growth and development of tomato fruits. Ann Bot 88:869–877.

Adnan M, Shah Z, Arif M, Khan MJ, Mian IA, Sharif M, Alam M, Basir A, Ullah H, Rahman I, Saleem N (2016) Impact of rhizobial inoculum and inorganic fertilizers on nutrients (NPK) availability and uptake in wheat crop. Can J Soil Sci 96:169–176.

Ahamed K-U, Nahar K, Fujita M, Hasanuzzaman M (2010) Variation in plant growth, tiller dynamics and yield components of wheat (Triticum aestivum L.) due to high temperature stress. Adv Agric Bot 2:213–224.

Ali J, Muhammad H, Ullah I, Rashid JA, Adnan M, Ali M, Ahmad W, Rehman A, Khan J (2016) Mango seed germination in different media at different depth. J Nat Sci Res 6(1):56–59.

Asada K (2006) Production and scavenging of reactive oxygen species in chloroplasts and their functions. Plant Physiol 141:391–396.

Ashraf M, Hafeez M (2004) Thermotolerance of pearl millet and maize at early growth stages: Growth and nutrient relations. Biol Plant 48:81–86.

Casalegno S (2011) Global Warming Impacts: Case Studies on the Economy, Human Health, and on Urban and Natural Environments. BoD–Books on Demand.

Cheng L, Zou Y, Ding S, Zhang J, Yu X, Cao J, Lu G (2009) Polyamine accumulation in transgenic tomato enhances the tolerance to high temperature stress. J Integrative Plant Biol 51:489–499.

Cicchino M, Edreira J, Uribelarrea M, Otegui M (2010) Heat stress in field-grown maize: Response of physiological determinants of grain yield. Crop Sci 50:1438–1448.

Crafts-Brandner SJ, Salvucci ME (2002) Sensitivity of photosynthesis in a C4 plant, maize, to heat stress. Plant Physiol 129:1773–1780.

Essemine J, Ammar S, Bouzid S (2010) Impact of heat stress on germination and growth in higher plants: Physiological, biochemical and molecular repercussions and mechanisms of defence. J Biol Sci 10:565–572.

Fahad S, Hussain S, Saud S, Hassan S, Chauhan BS, Khan F, et al (2016a) Responses of rapid viscoanalyzer profile and other rice grain qualities to exogenously applied plant growth regulators under high day and high night temperatures. PLoS One 11(7):e0159590. https://doi.org/10.1371/journal.pone.0159590.

Fahad S, Hussain S, Saud S, Hassan S, Ihsan Z, Shah AN, Wu C, Yousaf M, Nasim W, Alharby H, Alghabari F, Huang J (2016c) Exogenously applied plant growth regulators enhance the morphophysiological growth and yield of rice under high temperature. Front Plant Sci 7:1250. https://doi.org/10.3389/fpls.2016.01250.

Fahad S, Hussain S, Saud S, Hassan S, Tanveer M, Ihsan MZ, Shah AN, Ullah A, Nasrullah KF, Ullah S, Alharby H, Nasim W, Wu C, Huang J (2016d) A combined application of biochar and phosphorus alleviates heat-induced adversities on physiological, agronomical and quality attributes of rice. Plant Physiol Biochem 103:191–198.

Fahad S, Hussain S, Saud S, Khan F, Hassan S, Jr A, Nasim W, Arif M, Wang F, Huang J (2016b) Exogenously applied plant growth regulators affect heat-stressed rice pollens. J Agron Crop Sci 202:139–150.

Fahad S, Hussain S, Saud S, Tanveer M, Bajwa AA, Hassan S, Shah AN, Ullah A, Wu C, Khan FA, Shah F, Ullah S, Chen Y, Huang J (2015a) A biochar application protects rice pollen from high-temperature stress. Plant Physiol Biochem 96:281–287.

Fahad S, Muhammad ZI, Abdul K, Ihsanullah D, Saud S, Saleh A, Wajid N, Muhammad A, Imtiaz AK, Chao W, Depeng W, Jianliang H (2018) Consequences of high temperature under changing climate optima for rice pollen characteristics-concepts and perspectives. Archives Agron Soil Sci. https://doi.org/10.1080/03650340.2018.1443213.

Fahad S, Adnan M, Hassan S, Saud S, Hussain S, Wu C, Wang D, Hakeem KR, Alharby HF, Turan V, Khan MA, Huang J (2019b) Rice responses and tolerance to high temperature. In: Hasanuzzaman M, Fujita M, Nahar K, Biswas JK (eds) Advances in rice research for abiotic stress tolerance. Woodhead Publ Ltd, Abington, pp 201–224.

Fahad S, Rehman A, Shahzad B, Tanveer M, Saud S, Kamran M, Ihtisham M, Khan SU, Turan V, Rahman MHU (2019a) Rice responses and tolerance to metal/metalloid toxicity. In: Hasanuzzaman M, Fujita M, Nahar K, Biswas JK (eds) Advances in rice research for abiotic stress tolerance. Woodhead Publ Ltd, Abington, pp 299–312.

Fitter AH, Hay RK (2012) Environmental physiology of plants. Academic press, London.

Foolad, M.R. (2005) Breeding for Abiotic Stress Tolerances in Tomato. In: Ashraf, M. and Harris, P.J.C., Eds., Abiotic Stresses: Plant Resistance through Breeding and Molecular Approaches, The Haworth Press Inc., New York, 613–684.

Greer DH, Weedon MM (2012) Modelling photosynthetic responses to temperature of grapevine (Vitis vinifera cv. Semillon) leaves on vines grown in a hot climate. Plant Cell Environ 35:1050–1064.

Halliwell B (2006) Oxidative stress and neurodegeneration: Where are we now? J Neurochem 97:1634–1658.

Högy P, Poll C, Marhan S, Kandeler E, Fangmeier A (2013) Impacts of temperature increase and change in precipitation pattern on crop yield and yield quality of barley. Food Chem 136:1470–1477.

Hurkman WJ, Vensel WH, Tanaka CK, Whitehand L, Altenbach SB (2009) Effect of high temperature on albumin and globulin accumulation in the endosperm proteome of the developing wheat grain. J Cereal Sci 49:12–23.

Khan I, Alam S, Hussain H, Shah B, Naeem A, Ullah W, Ali W, Adnan M, Junaid K, Shah SRA, Ahmed N, Iqbal M (2016) Study on the management of potato black scurf disease by using biocontrol agent and phytobiocides. J Ento Zool Studies 4(2):471–475.

Koini MA, Alvey L, Allen T, Tilley CA, Harberd NP, Whitelam GC, Franklin KA (2009) High temperature-mediated adaptations in plant architecture require the bHLH transcription factor PIF4. Curr Biol 19:408–413.

Kumar A, Omae H, Egawa Y, Kashiwaba K, Shono M (2005) Some physiological responses of snap bean (Phseolus vulgaris L.) to water stress during reproductive period. In: Proceedings of the International Conference on Sustainable Crop Production in Stress Environment: Management and Genetic Option, JNKVV, Jabarpur, India, pp 226–227.

Kumar S et al. (2011) Heat-stress induced inhibition in growth and chlorosis in mungbean (Phaseolus aureus Roxb.) is partly mitigated by ascorbic acid application and is related to reduction in oxidative stress. Acta Physiol Plantarum 33:2091.

Lobell DB, Field CB (2007) Global scale climate–crop yield relationships and the impacts of recent warming. Environ Res Lett 2:014002.

Lobell DB, Schlenker W, Costa-Roberts J (2011) Climate trends and global crop production since 1980. Science 333:616–620.

Maheswari M, Yadav S, Shanker AK, Kumar MA, Venkateswarlu B (2012) Overview of plant stresses: Mechanisms, adaptations and research pursuit. In: Venkateswarlu B, Shanker AK, Shanker C, Maheswari M (eds) Crop stress and its management: Perspectives and strategies. Springer, pp 1–18.

Marchand FL, Mertens S, Kockelbergh F, Beyens L, Nijs I (2005) Performance of High Arctic tundra plants improved during but deteriorated after exposure to a simulated extreme temperature event. Global Change Biol 11:2078–2089.

Mitra R, Bhatia C (2008) Bioenergetic cost of heat tolerance in wheat crop. Curr Sci 94:1049–1053.

Mohammed AR, Tarpley L (2010) Effects of high night temperature and spikelet position on yield-related parameters of rice (Oryza sativa L.) plants. Eur J Agron 33:117–123.

Naseer F, Baig J, Din SU, Nafees MA, Alam A, Hameed N, Adnan M, Arshad M, Khan MA, Romman M, Mian IA, Shah SRA (2016) Impact of water quality on distribution of macro-invertebrate in Jutialnallah, Gilgit-Baltistan, Pakistan. Int J Biosci 9(6):451–459.

Nahar K, Ahamed KU, Fujita M (2010) Phenological variation and its relation with yield in several wheat (Triticum aestivum L.) cultivars under normal and late sowing mediated heat stress conditio. Notulae Scientia Biol 2:51–56.

Omae H, Kumar A, Shono M (2012) Adaptation to high temperature and water deficit in the common bean (Phaseolus vulgaris L.) during the reproductive period. J Bot 2012.

Patel D, Franklin KA (2009) Temperature-regulation of plant architecture. Plant Signaling Behav 4:577–579.

Rahman I, Ali S, Rahman I, Adnan M, Ullah H, Basir A, Malik FA, Shah AS, Ibrahim M, Arshad M (2016) Effect of pre-storage seed priming on biochemical changes in okra seed. Pure Appl Biol 5(1):165–171.

Rodríguez M, Canales E, Borrás-Hidalgo O (2005) Molecular aspects of abiotic stress in plants. Biotecnol Aplicada 22:1–10.

Saitoh H (2008) Ecological and physiology of vegetable. Nousangyoson Bunka Kyoukai, Tokyo.

Sato S, Kamiyama M, Iwata T, Makita N, Furukawa H, Ikeda H (2006) Moderate increase of mean daily temperature adversely affects fruit set of Lycopersicon esculentum by disrupting specific physiological processes in male reproductive development. Ann Bot 97:731–738.

Savin R, Stone P, Nicolas M, Wardlaw I (1997) Effects of heat stress and moderately high temperature on grain growth and malting quality of barley. Aust J Agric Res 48:615–624.

Sinsawat V, Leipner J, Stamp P, Fracheboud Y (2004) Effect of heat stress on the photosynthetic apparatus in maize (Zea mays L.) grown at control or high temperature. Environ Exp Bot 52:123–129.

Soliman WS, Fujimori M, Tase K, Sugiyama SI (2011) Oxidative stress and physiological damage under prolonged heat stress in C3 grass Lolium perenne. Grassland Sci 57(2):101–106.

Srivastava S, Pathak AD, Gupta PS, Shrivastava AK, Srivastava AK (2012) Hydrogen peroxide-scavenging enzymes impart tolerance to high temperature induced oxidative stress in sugarcane. J Environ Biol 33:657.

Suwa R et al. (2010) High temperature effects on photosynthate partitioning and sugar metabolism during ear expansion in maize (Zea mays L.) genotypes. Plant Physiol Biochem 48:124–130.

Shah N, Khan A, Khisroon M, Adnan M, Jawad SM (2017) Seroprevalence and risk factors of toxoplasmosis in pregnant women of district Swabi. Pure Appl Biol 6(4):1306–1313.

Tan W, wei Meng Q, Brestic M, Olsovska K, Yang X (2011) Photosynthesis is improved by exogenous calcium in heat-stressed tobacco plants. J Plant Physiol 168:2063–2071.

Tubiello FN, Soussana J-F, Howden SM (2007) Crop and pasture response to climate change. Proc Natl Acad Sci USA 104:19686–19690.

Vasseur F, Pantin F, Vile D (2011) Changes in light intensity reveal a major role for carbon balance in Arabidopsis responses to high temperature. Plant Cell Environ 34:1563–1576.

Wahid A (2007) Physiological implications of metabolite biosynthesis for net assimilation and heat-stress tolerance of sugarcane (Saccharum officinarum) sprouts. J Plant Res 120:219–228.

Wang J, Cui L, Wang Y, Li J (2009) Growth, lipid peroxidation and photosynthesis in two tall fescue cultivars differing in heat tolerance. Biol Plant 53:237–242.

Wang W, Vinocur B, Shoseyov O, Altman A (2004) Role of plant heat-shock proteins and molecular chaperones in the abiotic stress response. Trends Plant Sci 9:244–252.

Yamamoto Y et al. (2008) Quality control of photosystem II: Impact of light and heat stresses. Photosynthesis Res 98:589–608.

Yang X et al. (2006) Tolerance of photosynthesis to photoinhibition, high temperature and drought stress in flag leaves of wheat: A comparison between a hybridization line and its parents grown under field conditions. Plant Sci 171:389–397.

Young LW, Wilen RW, Bonham-Smith PC (2004) High temperature stress of Brassica napus during flowering reduces micro-and megagametophyte fertility, induces fruit abortion, and disrupts seed production. J Exp Bot 55:485–495.

Yun-Ying C, Hua D, Li-Nian Y, Zhi-Qing W, Shao-Chuan Z, Jian-Chang Y (2008) Effect of heat stress during meiosis on grain yield of rice cultivars differing in heat tolerance and its physiological mechanism. Acta Agron Sinica 34:2134–2142.

Zhang X, Cai J, Wollenweber B, Liu F, Dai T, Cao W, Jiang D (2013) Multiple heat and drought events affect grain yield and accumulations of high molecular weight glutenin subunits and glutenin macropolymers in wheat. J Cereal Sci 57:134–140.

# 5

# *Molecular and Physiological Mechanism of Native Plants to Combat Drought Stress under Changing Climatic Conditions*

**Hasnain Alam**
*International Islamic University*
*United Arab Emirates University*

**Muhammad Zamin, Muhammad Adnan, and Rafiullah**
*University of Swabi*

**Shah Fahad**
*Hainan University*
*The University of Haripur*

**Muhammad Arif, Amanullah, and Ishaq Ahmad Mian**
*The University of Agriculture*

**Sahar Mumtaz**
*University of Education*

**Muhammad Hamzah Saleem and Abdul Saboor Khan**
*Huazhong Agricultural University*

**CONTENTS**



DOI: 10.1201/9781003286400-5

## 5.1 Introduction

Climate change had increased the global mean temperature and water scarcity throughout the world (Zamin et al., 2019; Alam et al., 2021) and is expected to become even more in the future (Haddeland et al., 2014). The rapid rate of urbanization, industrialization and agricultural expansion has resulted in the overexploitation and contamination of groundwater resources (Huq et al., 2020; Khattak et al., 2014). Global demand for water had increased threefold since the 1950s as the freshwater supply had been on the way out (Alam et al., 2017; Zamin and Khattak, 2017; Gleick, 2003). Water shortage is a serious issue as 1.1 billion people are lacking access to drinking water (Alam et al., 2021; Gleick, 1998). According to an estimate, an average of 90% of global fresh water is being used by agriculture and crop production (Zamin et al., 2018; Shiklomanov, 2000). Approaches employed for remediating water resources issues pose an adverse effect on water's quality and quantity (Scanlon et al., 2007). Agriculture growth had led to decreasing water quality and lowering water tables (Qureshi et al., 2010). Pakistan is also included among water deficit nations (Alam et al., 2017; Rasul et al., 2012).

Most of the introduced landscape plants are from tropical regions having much more water and maintenance requirements to sustain them in arid environments. Alternate of these exotic or introduced species are the native plant's species which are grown in a specific area without human aid or intervention. These plants are well adapted to local soil and environmental conditions, resistant to water stress and have minimum maintenance requirements. These plants can be introduced in landscapes instead of increasing the stress resistance of cultivated imported plants. Domesticating native plants will reduce the maintenance cost, conserve limited water and energy resources, reestablish and diverse native ecosystem while providing a natural look to parks and gardens (Zamin et al., 2019, Bhat et al., 2009). The successful approach is to screen the native plants for their genetic tolerance to water stress and evaluate their field performance in maintained landscapes.

## 5.2 Drought

American Heritage Dictionary defines drought as "a long period with no rain, especially during a planting season" (Morris and others 1969). Another definition of Random House Dictionary is "an extended period of dry weather, especially one injurious to crops" (House, 1987). Conceptual definitions provide little guidance to those who wish to apply them to current (i.e., real-time) drought assessments. Meteorological definitions consider drought solely based on the degree of dryness and the duration of the dry period. For example, meteorological drought has been defined as a "period of more than some particular number of days with precipitation less than some specified small amount" (McIntosh, 1972). Linsley et al. (1975) referred to it as a "sustained period of time without significant rainfall." Meteorological drought definitions are also often site specific. Some meteorological drought definitions developed for application in various countries are as follows: less than 2.5 mm of rainfall in 48 hours in the United States (Blumenstock, 1942), 15 days, none of which received as much as 0.25 mm in Britain (Organisation, 1936).

## 5.3 Drought Resistance Mechanisms

Plants have developed various resistance mechanisms to counter water stress. Water stress-avoiding refers to a range of morphological and physiological adaptations of plants to sustain suitable water status, either by preserving water during water stress periods or by ensuring an efficient water supply to above-ground organs (Clarke and Durley, 1981). These adaptations can be of three types: (1) enhanced water uptake, (2) reduced loss of water through transpiration and (3) storing water in plant tissues. Another approach to withstand water stress is water stress tolerance that includes physiological and biochemical mechanisms (Clarke and Durley, 1981). A combined mechanism of avoidance and tolerance had been found in most of the grasses for their survival under drought conditions (Zamin et al., 2019; Arraudeau, 1989).

## 5.4 Water Shortage and Possible Solutions for Green Sector

To solve the problems of water shortage many irrigation systems and equipment are being introduced (Shahin and Salem, 2014). Two kinds of approaches had been taken on so far to overcome the salinity problem. The first one is modifying the environment by managing the irrigation and drainage, and the second approach is genetically modifying the plants to enhance their stress tolerance. But, huge areas cannot be managed in this way, and the majority of the possible solutions are much expensive (Läuchli and Lüttge, 2002; Mahmood et al., 2003; Zamin et al., 2019). There are many abiotic factors which make genetic alteration in plants for tolerance (Wang et al., 2003). According to Saghir, the utilization of biotic approach can be the most feasible and economic solution. However, stress tolerance responses of plants are complex, and the functions of many genes controlling these mechanisms are unknown.

## 5.5 Native Plants and Landscaping

Amenity plantations in urban areas are generally inspired by water-rich European-style (Alam et al., 2017). According to Environment Agency Abu Dhabi, adopting native plants and increasing hardscape could save a significant amount of water and energy. This arid landscape policy can reduce half of the maintenance cost and energy requirements (Pitman et al., 2009).

Native plants generally refer to plants that are found to occur and grow naturally in a particular area without human aid or introduction (Al-Mashhanadi, 2015). Most of the introduced plants' species are difficult to acclimatize in the local environment, whereas native plants are most suitable for the local environment (Bhat et al., 2009). Bodle (2001), Hostetler et al. (2003) and Haehle and Brookwell (2004) argued that native species should perform better than exotic species in their indigenous habitat. Native plants also exhibit an advantage in rebuilding a suitable environment for wildlife by providing food and shelter (Fiedler, 2006; Anella, 2000).

Architects have promoted demand for native plants due to their eco-friendly nature in several states of North America with water restrictions for landscape use (Potts et al., 2002). Native plants are getting more importance in landscaping due to their associated environmental benefits. They are always well adapted to the local environment and need less maintenance. They also exhibit the ability to adjust or grow under stressful conditions in a more efficient manner than their cultivated relatives (Morales et al., 2001; Fiedler, 2006; Stephens et al., 2006; and Ochoa et al., 2009). Palm canopy was used to create a condition of microclimate. In conjunction with the *Arabian mashrabia* the species like *Phoenix dactylifera*, *Ziziphus spina Christi*, *Nerium oleander* and *Vinca rosea* were used as a natural air-conditioning and cooling method (Salama, 1990).

Native plants not only contribute to vernacular landscapes but also help in water conservation and solution for rising temperatures (Crewe, 2013). Arizona, the city of Tucson, was entirely dependent on groundwater. Groundwater supplies started to decline in the 1980s. The city government made it mandatory to use native plants with low-water requirements (McPherson and Haip, 1989) to combat the water scarcity.

Many reports had validated the importance of native plants in different countries including Saudi Arabia (Ricks, 1992), China (Zheng and Chen, 2008), Oman (Hopkins and Al-Yahyai, 2015) and different states of the United States (McPherson and Haip, 1989; Mee et al., 2003; Brzuszek et al., 2007; Love et al., 2009; Hilaire et al., 2010; Anonymous, 2011; Ricordi et al., 2014; Reid et al., 2008; Zamin et al., 2018).

## 5.6 Public Interest for Native Plants in Landscaping

Landscaping with native plants is becoming more popular due to their ecological and cultural functions (McMahan 2006). Native plants are well thought-out as a rising slot in the green industry (Hamill, 2005). Consumer demand for native plants has increased rapidly because of their eco-friendly nature, especially, in drought-affected areas. Moreover, native plants and desert landscaping (designed or natural landscape with desert plants) are preferred by the public (Hilaire et al., 2010; Yabiku et al., 2008; Yue et al., 2011).

As compared to invasive or nonnative plants, customers show more interest in plants labeled as native and agreeable to pay extra (Yue et al., 2011; 2012). There were different programs to promote native plants like "Nevada Grown" and "Utah's Choice (marketed by the Intermountain Native Plant Growers Association)." One-third of people were ready to give 20% more for plants labeled as native (Meyer, 2005; Curtis et al., 2009; Yue et al., 2011).

In the horticulture industry, the introduction and promotion of native plants had been slow, but consumers are preferring native plants (Gagliardi and Brand, 2007; Yue et al., 2011). Customers' perception about native plant's aesthetics is a limiting factor for the use of native plants. Another limiting factor is customers' lack of knowledge about specific native plant species' uses and care (Hooper, 2003). For increasing the native plant market, there is a need to educate consumers and agriculture based industry about native plants. In addition, various reports (Peppin et al., 2010; Woosaree, 2000) recommended focusing on consumer awareness and education about native plants.

## 5.7 Promotion of Native Plants Market

Several studies had been carried out to examine the customer's choice difficulties faced by producers and trends in native plant markets. Major issues in this regard included difficulty in seed treatments, seed viability and dormancy testing of seeds, threats posed to wild populations by the collection of wild native plants/seeds and above all, the most limiting factor is lack of sufficient scientific research on propagation techniques (Potts et al., 2002; Kauth and Pérez, 2011; Neufeld, 2010). According to Potts et al. (2002), specific propagation guidelines are very valuable for the green industry as many nurserymen are confronting issues in poor and slow germination rates. Many researchers mentioned some other issues that often limit the acceptance of landscape projects incorporating native plants such as lack of commercial seeds availability, unawareness and lack of knowledge of customer about native plants, unavailability of plant materials, unestablished maintenance requirements, selection of plants, unavailability of required size and species (Hooper, 2003; Potts et al., 2002; Ricordi et al., 2014; Tamimi, 1996). To sustain the native plant industry, there is a dire need to educate the consumer and grower about native plants (Meyer, 2005; Peppin et al., 2010; Woosaree, 2000).

## 5.8 Recommendation for Future Research

The landscaping sector utilizes a huge amount of freshwater resources. The best way to reduce landscape water consumption is to domesticate the wild species rather than to increase the salt or water stress tolerance of plant species. The successful approach is to select the wild species that have genetic tolerance to salt stress and have some economic and landscape potential. Most of the molecular investigation about salt and water stress response is done on cultivated crops. Studying native plants under stress conditions will help us to better understand the salt and water stress resistance mechanisms. The broad studies of the physiological and molecular response of native landscape plants to combine salt and water stress have not yet been done. Compared to exotic plants, halophytic and non-halophytic native plants have shown to perform well under limited water or saline conditions. These native plant species are well adapted to harsh environmental conditions and have developed salt and drought resistance mechanisms. These mechanisms may include stress avoidance or tolerance or combination of both. By utilizing these mechanisms, native plants not only withstand the harsh environmental conditions but also require less maintenance. Thus, we must conserve the limited energy and water resources. These salt- and drought-resistant native plants are not only used for fodder, medicines and stabilization, but many of these native plants are the best substitutes for exotic plants. These native plants are being used in different functional landscapes including roadside plantations, town houses and parks. In the context of increasing water scarcity and decreasing water quality, landscape architects should focus on introducing desert landscapes and incorporating more native plant species. Hence studies should be carried out to explore the native salt and drought resistance native plant species which can be used as a substitute for exotic species. Lab experiments can be conducted to study the salt and drought stress resistance mechanism and field studies should be carried out to study their performance in a maintained landscape under stress conditions. Moreover, introducing the identified species in landscaping will also preserve the biodiversity, wildlife habitats, horticulture heritage and national unique landscape of the country.

## REFERENCES

Alam H, Khattak JZ, Ksiksi TS, Saleem MH, Fahad S, Sohail H, Ali Q, Zamin M, El-Esawi MA, Saud S, Jiang X (2021) Negative impact of long-term exposure of salinity and drought stress on native *Tetraena mandavillei* L. Physiologia plantarum 172(2):1336–1351.

Alam H, Khattak JZK, Ppoyil SBT, et al (2017) Landscaping with native plants in the UAE: A review. Emir J Food Agric 29(10):729–741.

Al-Mashhanadi ASA (2015) Proposed mechanisms for introducing native plants in landscaping projects. In: Municipal Best Practices Forum, Abu Dhabi, UAE.

Anella LB (2000) Debunking native myths. Am Nurseryman Aug 15:39–40.

Anonymous (2011) Chapter Four: Landscaping with native plants: A gardener's guide for Missouri in the Native Landscaping Manual. Shaw Nature Reserve [a division of the Missouri Botanical Garden]: Gray Summit, MO.

Arraudeau M (1989) Breeding strategies for drought resistance. In: Baker FWC (ed) Drought resistance in cereals. CAB. International, Wallingford, UK, pp 107–116.

Bhat NR, Suleiman MK, Abdal M, et al (2009) Selection of crops for sustainable utilization of land and water resources in Kuwait. World J Agric Sci 5:201–206.

Blumenstock Jr G (1942) Drought in the United States analyzed by means of the theory of probability. US Department of Agriculture, USA.

Bodle M (2001) Waterwise south Florida landscapes. South Florida Water Management District, West Palm Beach, FL.

Brzuszek RF, Harkess RL, Mulley SJ (2007) Landscape architects' use of native plants in the southeastern United States. HortTechnology 17:78–81.

Clarke JM, Durley RC (1981) The responses of plants to drought stress. Water stress on plants. Praeger Publishers, New York, pp 89–139.

Crewe K (2013) Arizona native plants and the urban challenge. Landscape J 32:215–229.

Curtis KR, Cowee MW, Slocum SL (2009) Nevada wildland seed cooperative feasibility assessment. Technical Report UCED 2005/06-10. University Center for Economic Development, Department of Resource Economics, University of Nevada, Reno, Nevada.

Fiedler AK (2006) Evaluation of Michigan native plants to provide resources for natural enemy arthropods. MS thesis, Department of Entomology, Michigan State University.

Gagliardi JA, Brand MH (2007) Connecticut nursery and landscape industry preferences for solutions to the sale and use of invasive plants. HortTechnology 17:39–45.

Gleick PH (1998) The human right to water. Water Policy 1:487–503.

Gleick PH (2003) Global freshwater resources: Soft-path solutions for the 21st century. Science 302:1524–1528.

Haddeland I, Heinke J, Biemans H, et al (2014) Global water resources affected by human interventions and climate change. Proc Nat Acad Sci 111:3251–3256.

Haehle RG, Brookwell J (2004) Native Florida plants: Low maintenance landscaping and gardening. Taylor Trade Publishing, Maryland, USA

Hamill N (2005) Natives near and far. Ornamental Outlook 14:31–32.

Hilaire RS, VanLeeuwen DM, Torres P (2010) Landscape preferences and water conservation choices of residents in a high desert environment. HortTechnology 20:308–314.

Hooper VH (2003) Understanding Utah's native plant market: Coordinating public and private interest. MLA thesis, Utah State University.

Hopkins E, Al-Yahyai R (2015) Landscaping with Native Plants in Oman. In: VIII International Symposium on New Ornamental Crops and XII International Protea Research Symposium 1097, pp 181–192.

Hostetler ME, Klowden G, Miller SW, Youngentob KN (2003) Landscaping backyards for wildlife: Top ten tips for success. UF/IFAS EDIS Publication # UW175.

House R (1987) The random house dictionary of the English language. Random House, New York.

Huq ME, Fahad S, Shao Z, Sarven MS, Khan IA, Alam M, Saeed M, Ullah H, Adnan M, Saud S, Cheng Q(2020) Arsenic in a groundwater environment in Bangladesh: Occurrence and mobilization. J Environ Manage 262:110318.

Kauth PJ, Pérez HE (2011) Industry survey of the native wildflower market in Florida. HortTechnology 21:779–788.

Khattak A, Ahmed N, Hussain I, et al (2014) Spatial distribution of salinity in shallow ground water used for crop irrigation. Pak J Bot 46:531–537.

Läuchli A, Lüttge U (2002) Salinity: Environment-plants-molecules. Springer, Swiss.

Linsley Jr RK, Kohler MA, Paulhus JL (1975) Hydrology for engineers. McGraw-Hill, New York.

Love SS, Nobe K, Robbins JA, et al (2009) Landscaping with native plants. Short-season, High- altitude gardening. University of Idaho Extention Bulletin 862:1–24.

Mahmood K, Marcar NE, Naqvi MH, et al (2003) Genetic variation in Eucalyptus camaldulensis Dehnh. for growth and stem straightness in a provenance–family trial on saltland in Pakistan. For Ecol Manage 176:405–416.

McIntosh D.H. (1972) Meteorological glossary. Chemical Publishing Company, ISBN-10: 0-8206-0228-0/ 0820602280.

McMahan LR (2006) Understanding cultural reasons for the increase in both restoration efforts and gardening with native plants. Native Plants J 7:31–34.

McPherson EG, Haip RA (1989) Emerging desert landscape in Tucson. Geographical Review 79:435–449.

Mee W, Barnes J, Kjelgren R, Sutton R (2003) Water wise: Native plants for intermountain landscapes. Utah State University Press Publications, USA. p 143.

Meyer SE (2005) Intermountain Native Plant Growers Association: A nonprofit trade organization promoting landscape use of native plants. Native Plants J 6:104–107.

Morales MA, Olmos E, Torrecillas A, et al (2001) Differences in water relations, leaf ion accumulation and excretion rates between cultivated and wild species of *Limonium* sp. grown in conditions of saline stress. Flora 196:345–352.

Morris W, et al (1969) American heritage dictionary of the English language. American Heritage, USA.

Neufeld C (2010) Assessment of the Native Plant Materials Industry in western Canada and the northern United States–results from providers and users of native plant materials. Native Plant Society of Saskatchewan. Saskatoon, SK.

Ochoa J, Muñoz M, Vicente MJ, et al (2009) Native ornamental species for urban landscaping and xerogardening in semi-arid environments. In: II International Conference on Landscape and Urban Horticulture 881. pp 425–428.

Organisation BR (1936) British rainfall. Air Ministry, Meteorological Office, London.

Peppin DL, Fulé PZ, Lynn JC, et al (2010) Market perceptions and opportunities for native plant production on the southern Colorado plateau. Restor Ecol 18:113–124.

Pitman K, McDonnell R, Dawoud M (2009) Abu Dhabi water resources master plan. Environment Agency Abu Dhabi, Abu Dhabi.

Potts LE, Roll MJ, Wallner SJ (2002) Colorado native plant survey—Voices of the green industry. Native Plants J 3:121–125.

Qureshi AS, McCornick PG, Sarwar A, Sharma BR (2010) Challenges and prospects of sustainable groundwater management in the Indus Basin, Pakistan. Water Resour Manage 24:1551–1569.

Rasul G, Mahmood A, Sadiq A, Khan SI (2012) Vulnerability of Indus delta to climate change in Pakistan. Pak J Meteorol 8:89–107.

Reid S, Oki L, et al (2008) Field trials identify more native plants suited to urban landscaping. Calif Agric 62:97–104.

Ricks G (1992) Native & introduced species for naturalistic landscape in Saudi Arabia. Eng Sci 4:21–46.

Ricordi AH, Kaufman AJ, Cox LJ, et al (2014) Going native in Hawai 'i opportunities and barriers for using native plant material by landscape architects. Landscape J 33:127–139.

Salama MM (1990) Saudi Arabian flora and its application in landscape design projects. Doctoral dissertation, University of Edinburgh.

Scanlon BR, Jolly I, Sophocleous M, Zhang L (2007) Global impacts of conversions from natural to agricultural ecosystems on water resources: Quantity versus quality. Water Resour Res 43:1–18.

Shahin SM, Salem MA (2014) The five key solutions to rescue the landscaping sector in the United Arab Emirates (UAE). Discovery 25:41–47.

Shiklomanov IA (2000) Appraisal and assessment of world water resources. Water Int 25:11–32.

Stephens CJ, Schellhorn NA, Wood GM, Austin AD (2006) Parasitic wasp assemblages associated with native and weedy plant species in an agricultural landscape. Austral Entomol 45:176–184.

Tamimi LN (1996) The use of native Hawaiian plants by landscape architects in Hawaii. Master of Landscape Architecture, Virginia Polytechnic Institute and State University.

Wang W, Vinocur B, Altman A (2003) Plant responses to drought, salinity and extreme temperatures: Towards genetic engineering for stress tolerance. Planta 218:1–14.

Woosaree J (2000) Market assessment of the native plant industry in western Canada, Alberta Environment, Edmonton, Canada.

Yabiku ST, Casagrande DG, Farley-Metzger E (2008) Preferences for landscape choice in a southwestern desert city. Environ Behav 40:382–400.

Yue C, Hurley T, Anderson NO (2012) Heterogeneous consumer preferences for native and invasive plants: Evidence from experimental auctions. HortScience 47:1091–1095.

Yue C, Hurley TM, Anderson N (2011) Do native and invasive labels affect consumer willingness to pay for plants? Evidence from experimental auctions. Agric Econ 42:195–205.

Zamin M, Khattak AM (2017) Performance of Sporobolus spicatus ecotypes, UAE native grass, under various salinity levels. Pure Appl Biol (PAB) 6(2):595–604.

Zamin M, Khattak AM, Alyafei MA, Sajid M, Shakur M, Shah S, Khan I, Rashid, HU (2018). Sporobolus spicatus, a potential turf grass under the climatic conditions of United Arab Emirates. J Sci Agric 2:1–8.

Zamin M, Khattak AM, Salim AM, Marcum KB, Shakur M, Shah S, Jan I, Fahad, S. (2019). Performance of Aeluropus lagopoides (mangrove grass) ecotypes, a potential turfgrass, under high saline conditions. Environ Sci Pollut Res 26(13):13410–13421.

Zheng W, Chen L (2008) Investigation and analysis of landscaping utilization potentiality of native plant species diversity in Huangshi city. Chin Landsc Archit 7:10.

# 6

# Climate Change-Oriented Elevation of Carbon Dioxide: Implications on Plant Physiology and Global Food Security

**Hamayun Shaheen**
*University of Azad Jammu and Kashmir*

**Anila Ulfat**
*University of Poonch*

**CONTENTS**



## 6.1 An Introduction to Climate Change and Elevated Carbon Dioxide: Past, Present and Future

Over the ancient century, the rapid rise in population and the increasing intensity and scale of human initiatives have made it possible to alter the earth on a large scale. One significant measure of human action is the rate of use of energy. This rate has accelerated amazingly in the past hundred years as of rapidly increasing human population coupled with increasing per capita energy consumption resulting in a significant climate change. Broadly speaking, climatic change is triggered by exchanges of energy, momentum and chemicals between land surfaces, the atmosphere and the oceans (Ainsworth and Long 2005). Oceanic and atmospheric circulation, radiative transfer, turbulent mixing and photochemistry are all involved. These processes are chiefly natural, but more or less are susceptible to human influence.

DOI: 10.1201/9781003286400-6

Among all processes, greenhouse gases (GHGs) are perhaps the most critical entrants responsible for bringing about climate change. Carbon dioxide is the major GHG along with halocarbons, methane, and oxides of sulfur and nitrogen. These gases enter the air mainly as byproducts of the combustion of coal, natural gas and petroleum, and also concerned with other industrial and agricultural activities (Ainsworth and Rogers 2007; Deepranjan et al. 2021; Haider et al. 2021; Huang Li et al.2021; Ikram et al. 2021; Jabborova et al. 2021; Khadim et al. 2021a, b; Manzer et al. 2021; Muzammal et al. 2021; Abdul et al. 2021a, b; Ashfaq et al. 2021; Amjad et al. 2021; Atif et al. 2021; Athar et al. 2021; Adnan et al. 2018a,b; Adnan et al. 2019; Akram et al. 2018; Aziz et al. 2017a, b; Chang et al. 2021; Chen et al. 2021;Emre et al. 2021; Habib et al. 2017; Hafiz et al. 2016; Hafiz et al. 2019; Ghulam et al. 2021; Guofu et al. 2021; Hafeez et al. 2021; Khan et al. 2021; Kamran et al. 2017; Muhammad et al. 2019; Safi et al. 2021; Sajjad et al. 2019; Saud et al. 2013; Saud et al. 2014; Saud et al. 2017; Saud et al. 2016; Shah et al. 2013; Saud et al. 2020; Qamar et al. 2017; Hamza et al. 2021; Irfan et al. 2021;Wajid et al. 2017; Yang et al. 2017; Zahida et al. 2017; Depeng et al. 2018; Hussain et al. 2020; Hafiz et al. 2020a, b; Shafi et al. 2020; Wahid et al. 2020; Subhan et al. 2020; Zafar-ul-Hye et al. 2020a, b; Zafar et al. 2021; Adnan et al. 2020; Ilyas et al. 2020; Saleem et al. 2020a, b, c; Rehman et al. 2020; Farhat et al. 2020; Wu et al. 2020; Mubeen et al. 2020; Farhana et al. 2020; Jan et al. 2019; Wu et al. 2019; Ahmad et al. 2019; Baseer et al. 2019; Hafiz et al. 2018;Tariq et al. 2018). Their rates of emission into the air are roughly proportional to the global rate of energy consumption arising from human activity. Thus, as human population and per capita energy consumption have increased, the demand and consumption of the GHGs have abruptly risen resulting in their increased concentration in the atmosphere. As they rise, these gases trap radiations re-radiated from the earth's surface. These GHGs are responsible for rising temperature at the earth's surface unless some compensating process cancels out this tendency (Bencze et al. 2014).

Natural rises in carbon dioxide concentrations have periodically warmed the earth's temperature in ice age cycles during the past million years or even more. The warm periods began with a small increase in sunlight due to a tiny vibrate in the earth's axis of rotation or the path of its orbit around the sun. That slight bit of extra sunlight initiated a little bit of warming. By way of the oceans warmed, they outgassed carbon dioxide (Hocking and Meyer 1985). The additional carbon dioxide in the atmosphere improved the initial warming. Based on air bubbles ensnared in mile-thick ice burgs, we come to know that during the past million years at ice age cycles, carbon dioxide not ever exceeded 300 ppm (Ainsworth 2008). Before the industrial revolution ongoing in the mid-1700s, the global average sum of carbon dioxide was about 280 ppm (Jones et al. 1985; Brown and Higginbotham 1986). In fact, the last time the atmospheric $CO_2$ amounts were this high was more than 3 million years ago, when temperature was 2°C–3°C (3.6°F–5.4°F) higher than during the pre-industrial era, and sea level was 15–25 m (50–80 feet) higher than today. Carbon dioxide concentrations are rising mostly because of the fossil fuels that people are burning for energy (Aranjueloa et al. 2014). Fossil fuels like coal and oil contain carbon that plants pulled out of the atmosphere through photosynthesis over the span of many millions of years; we are returning that carbon to the atmosphere in just a few hundred years (Bazzaz and Fajer 1992).

A significant development in the $CO_2$ level monitoring happened with the initiation of interval continuous observations at Mauna Loa Volcanic Observatory in Hawaiian Islands, USA from 1958. At that time, the global atmospheric carbon dioxide was 315 ppm. On May 9, 2013, the daily average carbon dioxide measured at Mauna Loa surpassed 400 ppm for the first time on record (Bloom et al. 2014). In the 1960s, the global growth rate of atmospheric carbon dioxide was roughly $0.6 \pm 0.1$ ppm per year (Berryman et al. 1993). Over the past decade, however, the growth rate has been closer to 2.3 ppm per year. The annual rate of increase in atmospheric carbon dioxide over the past 60 years is about 100 times faster than previous natural increases, such as those that occurred at the end of the last ice age 11,000–17,000 years ago (Figure 6.1).

According to a report from NOAA and the American meteorological society, global atmospheric carbon dioxide was $405.0 \pm 0.1$ ppm in 2017, a new record higher than at any point in the past 800,000 years. Between 2016 and 2017, global annual mean carbon dioxide increased $2.2 \pm 0.1$ ppm, which was slightly less than the increase between 2015 and 2016 (3.0 ppm per year).

The human population has just crossed the mark of seven billion, and by the middle of this century it is expected to exceed nine billion (Godfray et al., 2010). This huge population explosion synchronized with rapid industrialization, deforestation and immense fossil fuel consumption is predicted to cause massive

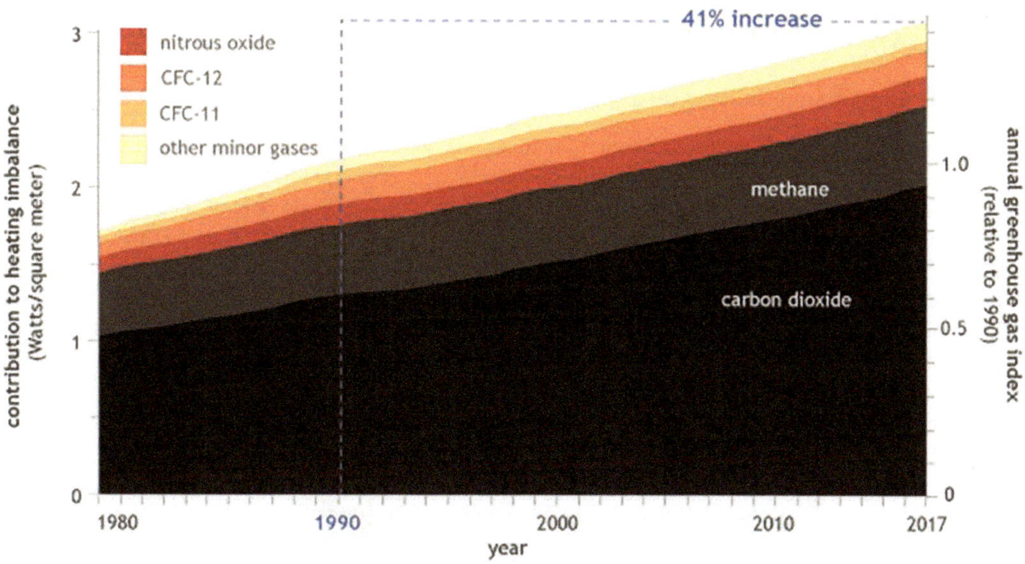


**FIGURE 6.1** Atmospheric carbon dioxide concentrations in parts per million (ppm) for the past 800,000 years, based on EPICA (ice core) data. The peaks and valleys in carbon dioxide levels track the coming and going of ice ages (low carbon dioxide) and warmer interglacial (higher levels). Throughout these cycles, atmospheric carbon dioxide was never higher than 300 ppm; in 2017, it reached 405.0 ppm (black dot).

**Source: NOAA Climate.gov, based on Epica Dome C data (Lüthi, et al., 2008; Ali et al. 2018)**

$CO_2$ inputs in the atmosphere having disastrous consequences. The rate of change of $CO_2$ concentration has accelerated with models predicting that the $CO_2$ concentration will increase to 550 mL/L by the middle of this century and climb up to 815 mL/L by the year 2100 (Long and Ort 2010; Feng et al. 2014).

## 6.2 Why Carbon Dioxide Got More Consideration?

Carbon dioxide got attention as a greenhouse gas because it is indubitably rising in concentration. Among GHGs, $CO_2$ is a gas that absorbs heat, warmed by sunlight as a result earth's land and ocean surfaces constantly radiate thermal infrared energy. Without this natural greenhouse effect, the earth's average annual temperature would be below freezing instead of close to 60°F. But increases in GHGs have tipped the earth's energy budget out of balance, trapping surplus heat and raising the earth's average temperature. Though $CO_2$ absorbs less heat per molecule than the other GHGs like methane or nitrous oxide it is more abundant and stays in the atmosphere much longer (Bowes 1996). While carbon dioxide is less abundant and less powerful than water vapor on a molecule per molecule basis, it absorbs wavelengths of thermal energy that water vapor does not, which means it adds to the greenhouse effect in a unique way. Increases in atmospheric $CO_2$ are accountable for about two-thirds of the total energy imbalance that is causing the earth's temperature to rise (Budak et al. 2013).

Another reason carbon dioxide is important in the earth's system is that it dissolves into the ocean like the fizz in a can of soda. It reacts with water molecules, producing carbonic acid and lowering the ocean's pH. Since the start of the industrial revolution, the pH of the ocean's surface waters has dropped from 8.21 to 8.10. This drop in pH is called ocean acidification. A drop of 0.1 may not seem like a lot, but the pH scale is logarithmic; a 1-unit drop in pH means a tenfold increase in acidity (Fangmeier et al. 1999; Altaf et al. 2016). A change of 0.1 means a roughly 30% increase in acidity. Increasing acidity interferes with the ability of marine life to extract calcium from the water to build their shells and skeletons. If global energy demand continues to grow and to be met mostly with fossil fuels, atmospheric carbon dioxide will likely exceed 900 ppm by the end of this century.

## 6.3 Impacts of Climate Change

### 6.3.1 Climate Change Is Responsible in Driving Global Hunger

Climate change is among the leading causes of rising global hunger according to a UN report released by the FAO. Overall, the number of hungry people grew per year reaching a total of 821 million worldwide. Experts warn that this number will continue to rise if countries fail to tackle climate change and to build resilience to its unavoidable impacts (Hikosaka et al. 2011; Fahad and Bano 2012; Fahad et al. 2017; Fahad et al. 2013; Fahad et al. 2014a, b; Fahad et al. 2016 a, b, c, d; Fahad et al. 2015a, b; Fahad et al. 2018a, b; Fahad et al. 2019a, b; Fahad et al. 2020; Fahad et al. 2021a, b, c, d, e, f; Hesham and Fahad 2020. Iqra et al. 2020; Akbar et al. 2020; Mahar et al. 2020; Noor et al. 2020; Bayram et al. 2020; Amanullah and Fahad 2018a, b; Amanullah and Fahad 2017; Amanullah et al. 2020; Amanullah et al. 2021; Rashid et al. 2020; Arif et al. 2020; Amir et al. 2020; Saman et al. 2020; Muhammad Tahir et al. 2020; Md Jakir and Allah 2020; Mahmood et al. 2021; Farah et al. 2020; Sadam et al. 2020; Unsar et al. 2020; Fazli et al. 2020; Md. Enamul et al. 2020; Gopakumar et al. 2020; Zia-ur-Rehman 2020; Ayman et al. 2020; Al-Wabel et al. 2020a, b; Senol 2020; Amjad et al. 2020; Ibrar et al. 2020; Sajid et al. 2020; Muhammad et al. 2021; Sidra et al. 2021;Zahir et al. 2021; Sahrish et al. 2022). The number of extreme climate-related disasters, including extreme heat, droughts, floods and storms, has doubled since the early 1990s, with an average of 213 of these events occurring every year during the period of 1990–2016. These disasters harm agricultural productivity of major crops such as wheat, rice and maize causing food price hikes and income losses that reduce people's access to food (Adnan et al. 2018a, b; Akram et al. 2018; Aziz et al. 2017a; Habib et al. 2017; Hafiz et al. 2016; Hafiz et al. 2019; Kamran et al. 2017; Muhammad et al. 2019; Sajjad et al. 2019; Saud et al. 2013; Saud et al. 2014; Saud et al. 2017; Saud et al. 2016; Shah et al. 2013; Qamar et al. 2017; Wajid et al. 2017; Yang et al. 2017; Zahida et al.2017; Depeng et al. 2018).

### 6.3.2 Climate, Weather and Crops

Among all, climate change emerges as a troubling problem that influences all others. Earth's climate has swung into and out of ice ages since the dawn of time. In the last 50 years, however, things have changed (Feng et al. 2014; Adnan et al. 2018; Akram et al. 2018; Aziz et al. 2017a; Habib et al. 2017; Hafiz et al. 2016; Hafiz et al. 2019; Kamran et al. 2017; Muhammad et al. 2019; Sajjad et al. 2019; Saud et al. 2013; Saud et al. 2014; Saud et al. 2017; Saud et al. 2016; Shah et al. 2013; Qamar et al. 2017; Wajid et al. 2017; Yang et al. 2017; Zahida et al. 2017; Depeng et al. 2018). Average global temperatures have increased ever more quickly, with new recorded highs in 2014, then again in 2015, and again in 2016. Climate change is also increasing the severity and frequency of extreme weather events, such as powerful storms and droughts. As a result, some regions of the world are getting wetter, while others are becoming drier. Agriculture is one of the industries most exposed and vulnerable to climate change (Fahad and Bano 2012; Fahad et al.2017; Fahad et al. 2013; Fahad et al. 2014a, b; Fahad et al. 2016a, b, c, d; Fahad et al. 2015a, b; Fahad et al. 2018a,b; Fahad et al. 2019a, b). Crops and livestock are extremely sensitive to temperature and precipitation. A late spring frost can be devastating, and a heat wave during the flowering stage can result in sharply reduced yields. In short, agriculture is the "Goldilocks industry" the weather should not be too hot or too cold, and rainfall must be "just right." Producing enough food for everyone in the world depends heavily on climate. This means that it will be impossible to curb hunger without preparing for and adapting to climate change (Ferrario et al. 1997; Adnan et al. 2018a, b; Akram et al. 2018; Aziz et al. 2017a; Habib et al. 2017; Hafiz et al. 2016; Hafiz et al. 2019; Kamran et al. 2017; Muhammad et al. 2019; Sajjad et al. 2019; Saud et al. 2013; Saud et al. 2014; Saud et al. 2017; Saud et al. 2016; Shah et al. 2013; Qamar et al. 2017; Wajid et al. 2017; Yang et al. 2017; Zahida et al. 2017; Depeng et al. 2018).

### 6.3.3 Climate Change's Pervasive Influence

Climate change also intensifies other stresses on global food production change (Fahad and Bano 2012; Fahad et al. 2017; Fahad et al. 2013; Fahad et al. 2014a, b; Fahad et al. 2016a, b, c, d; Fahad et al. 2015a, b;

Fahad et al. 2018a, b; Fahad et al. 2019a, b). Consider the critical role of water. Meat consumption alone accounts for an estimated 22% of global water use, and this need will increase in a hotter world. Climate change also alters rainfall patterns: Some places will have too little water to farm, while others may have enough but find that it falls at the wrong time, or arrives less frequently but in larger rainfall events. Even seemingly disparate factors like international trade are affected by climate change, with serious ramifications for food security. As climate change drives permanent shifts in the geography of world agricultural production zones, international trade will emerge as an important resiliency mechanism for reducing hunger and for enhancing equal access to food.

## 6.4 Adapting New Changes to Combat Climate Change

Climate change is anticipated to force more than 100 million people into extreme poverty by 2030 (Fahad and Bano 2012; Fahad et al. 2017; Fahad et al. 2013; Fahad et al. 2014a, b; Fahad et al. 2016a, b, c, d; Fahad et al. 2015a, b; Fahad et al. 2018; Fahad et al. 2019a, b). Adapting to climate change is a key way to combat this and technology can help. For instance, precision agriculture can leverage computers, global positioning systems, geographic information systems and sensors to provide the data necessary to give each tiny parcel of land on a field exactly the inputs it needs. And a resurgent interest in the use of the time-honored technology of cover crops may mitigate climate change impacts. We can go even smaller in our measurements with the emergence of nanotechnology. Aside from making field sensors smaller and more compact, nanotechnologies can also help improve how fertilizers and pesticides are released. By putting chemical inputs into tiny capsules or in gels, it is possible to control when and how these inputs are released to make them more effective, reducing chemical emissions and runoff. But ultimately, it is up to individuals (De Graaff et al. 2006).

Around the world, people must wield their social power to encourage mitigation of climate change and promote investments in technologies for adaptation. We need everyone at the table contributing to a food-secure future. While a great deal of media and public attention has focused on the effects that such higher concentrations of $CO_2$ are likely to have on global climate, so this rising $CO_2$ can be utilized to cope with the different environmental stresses triggered by climate change that influence plants productivity.

## 6.5 Ecosystem Controls on Carbon Sequestering

In natural ecosystems molecular, several physiological and ecological controls on carbon sequestering exist which regulate the balance of $CO_2$ and maintain the ecosystem functioning. Molecular and physiological controls on carbon sequestering include molecular responses like gene transcription along with diverse physiological responses. The primary physiological responses include photosynthesis, photorespiration, dark respiration, stomatal regulation whereas the secondary physiological responses are photosynthate concentration, photosynthate translocation, plant water status, transpiration, tissue water potential, water-use efficiency (WUE) and leaf temperature. The tertiary whole-plant responses include growth rate (mass, height, leaf area, node formation), growth form (branch number, leaf area and number, root architecture, root versus shoot mass, leaf specific mass), reproduction (flower and fruit number and size, nectar production, seed size and number, seed germination) and phenology (time to germinate, reproduction, leaf senescence and whole-plant senescence) (Taub et al. 2008).

The controls of ecological origin include primary organism interaction such as competition, herbivory, pollination, shelter and plant microbial interactions such as disease, decomposition and symbiosis. Secondary interactions include evolutionary responses and genetic differentiation, whereas the tertiary ecosystem responses involve integration of all effects through time.

Experiments have been performed in a wide variety of settings, including greenhouses and chambers of a variety of sizes and designs. However, plants grown in chambers may not experience the effects of increasing $CO_2$ the same way as plants growing in more natural settings. For this reason, techniques of Free-Air Carbon dioxide Enrichment (FACE) have been developed that allow natural or agricultural ecosystems to be fumigated with elevated concentrations of $CO_2$ in the field without the use of chambers

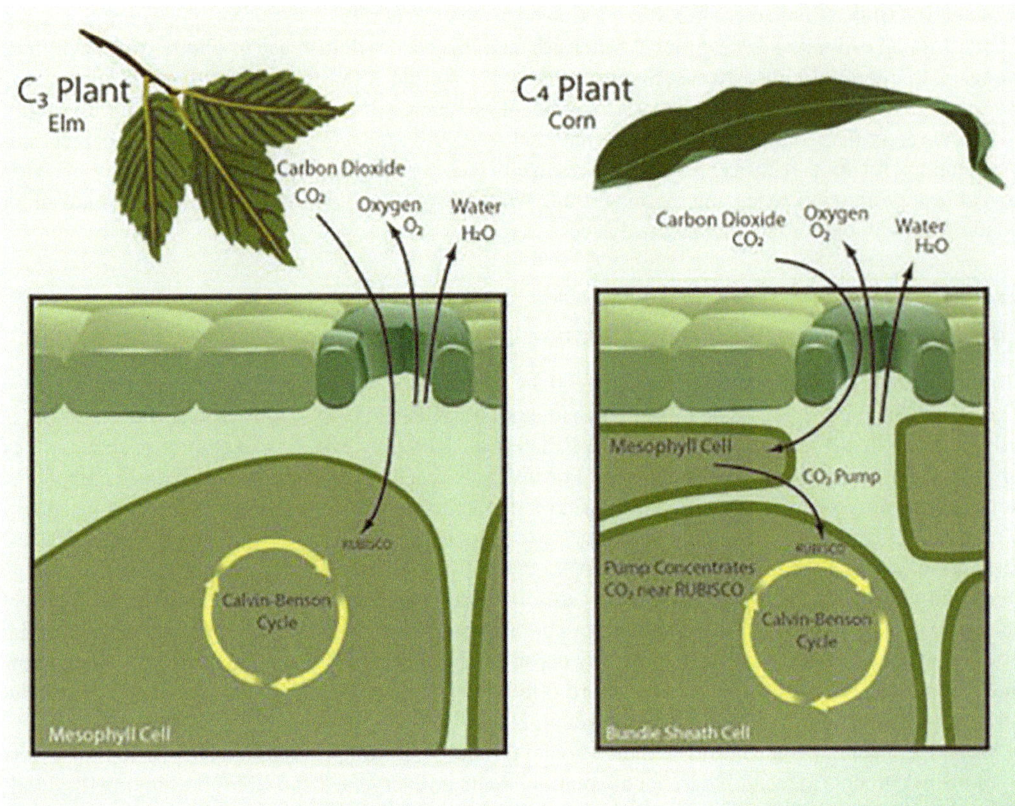


**FIGURE 6.2** Photosynthetic variation in $C_3$ and $C_4$ photosynthesis mechanism.

(Figure 6.2). As these experiments are the most naturalistic, they should provide the best indication of the responses of plants to increased $CO_2$ under the real-world conditions of the future.

### 6.5.1 Plant Community Interactions under Elevated $CO_2$

A number of experiments have found that some plant species that respond positively to elevated $CO_2$ when grown alone experience decreased growth under elevated $CO_2$ when grown in mixed plant communities (Poorter and Navas 2003). This effect likely results because the direct positive effects of elevated $CO_2$ are outweighed by negative effects due to stimulation of the growth of competitors. Rising atmospheric concentrations of $CO_2$ may therefore lead to changes in the composition of plant communities, as some species reap more of an advantage from the increased $CO_2$ than do others. In mixed-species experiments under high-fertility conditions, $C_4$ plants decrease as a proportion of the biomass of plant communities under elevated $CO_2$. Similarly, under low-fertility conditions, legumes increase as a proportion of the biomass of plant communities under elevated $CO_2$ (Poorter and Navas 2003).

### 6.5.2 Differences among Plant Functional Types in Response to Elevated $CO_2$

One of the most important determinants of species differences in response to elevated $CO_2$ is photosynthetic type. Most plant species (~90%) utilize a photosynthetic process known as $C_3$ photosynthesis. Other species use either of two physiologically distinct processes known as $C_4$ and Crassulacean acid metabolism (CAM) photosynthesis (Figure 6.3). $C_4$ plants include most tropical and sub-tropical grasses and several important crops, including maize (corn), sugar cane, sorghum and millets. There has been considerably more research on the responses to elevated $CO_2$ in $C_4$ than in CAM plants.

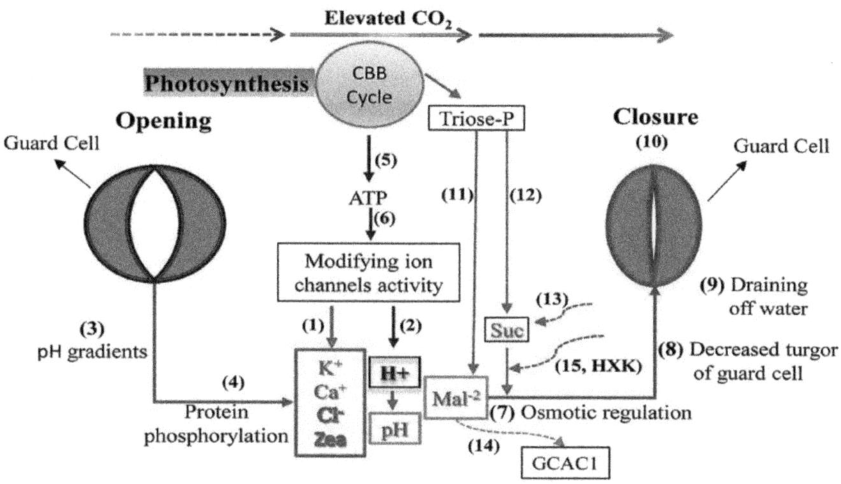


**FIGURE 6.3** Gaseous exchange mechanism under elevated carbon.

$C_4$ plants use a biochemical pump to concentrate $CO_2$ at the locations within the leaf where the Rubisco enzyme mediates incorporation of $CO_2$ by the Calvin–Benson photosynthetic cycle. Since $CO_2$ concentrations are already high within the bundle sheath cells, increasing atmospheric $CO_2$ concentrations above current levels has little direct effect on photosynthetic rates for $C_4$ species. $C_4$ species do respond to elevated $CO_2$ by decreasing stomatal conductance; this may lead to some indirect enhancement of photosynthesis by helping avoid water stress under drought conditions (Leakey 2009). In FACE experiments, stimulation of photosynthesis by elevated $CO_2$ in $C_4$ plants is only about one-third of that experienced by $C_3$ species. $C_4$ plants also show little or no enhancement of growth in some studies (Ainsworth and Long 2005). The very limited data available also shows no increase in $C_4$ crop yield in FACE studies (Long et al. 2006). While there is little FACE data available on the effects of elevated $CO_2$ on plant nitrogen and protein concentrations, data from chamber experiments shows $C_4$ plants to be much less responsive than $C_3$ plants in this regard (Cotrufo et al. 1998a,b). The picture that emerges is that $C_4$ plants are in general relatively unresponsive to the elevation of atmospheric $CO_2$ above current ambient levels. In contrast to $C_4$ species, another group of plants, legumes may be especially capable of responding to elevated $CO_2$ with increased photosynthesis and growth. For most plants, growth under elevated $CO_2$ can alter the internal balance between carbon and nitrogen.

Many studies in controlled environments have shown that, compared to other plant species, legumes show greater enhancement of photosynthesis and growth by elevated $CO_2$. Decreases in tissue nitrogen concentrations under elevated $CO_2$ are also smaller for legumes than for other $C_3$ species (Jablonski et al. 2002). In FACE experiments, soybeans show a greater response to elevated $CO_2$ than wheat and rice in photosynthesis and overall growth, although not in harvestable yield (Long et al. 2006).

## 6.6 Overall Effects of Elevated $CO_2$ on Agricultural Crops and Plants

As a result of climate change, the resultant increase in $CO_2$ will lead to a mosaic of both positive and negative effects on major crops used to feed the global population, many of which may yet be unknown.

### 6.6.1 Short- and Long-Term Exposure

Experiments have been done on short-term (36 hours) and long-term (16 days) periods of $CO_2$ enrichment imposed on vegetative, growth chamber-grown soybean plants (Glycine mar. IL.I Merr. cv. "Clay"). Results revealed that neither total nor specific nodule activities were increased during exposure to short-term (36 hours) $CO_2$ enrichment. Dry weight of the leaves and stems plus petioles showed an increase within 36 hours of $CO_2$ enrichment whereas the dry weights of the roots and nodules were not altered by short-term $CO_2$ enrichment. Short-term $CO_2$ enrichment increased the total nonstructural carbohydrates in the leaves and stems plus petioles, but not in the roots and nodules. Analyses of the separate pools of carbohydrate reserves indicated that the majority of the additional carbohydrate provided by short-term $CO_2$ enrichment was stored as leaf starch with relatively little being partitioned to the roots and nodules. Long-term $CO_2$ enrichment (16 days) did not enhance specific nodule activity. Shoot, root and nodule dry weights were increased 109, 34% and 56%, respectively (Fujita et al. 2013). Total nodule activity per plant was significantly enhanced only after 16 days of treatment and was related to increased nodule mass. These results indicate that the increased total nodule activity in response to $CO_2$ enrichment is a consequence of a general growth response of the plant. Results of both studies indicate that nodule activity was not directly limited by current photosynthesis but rather by the partitioning and utilization of photosynthetic in the plant (Godfray et al. 2010).

### 6.6.2 Impact of Elevated $CO_2$ on Gaseous Exchange Mechanism

Crops sense and respond directly to rising elevated $CO_2$ through photosynthesis and stomatal conductance, and this is the basis for the fertilization effect on yield (Long et al. 2004). In $C_3$ plants, mesophyll cells containing ribulose-1, 5-bisphosphate carboxylase-oxygenase (RuBisCO) are in direct contact with the intercellular air space that is connected to the atmosphere via stomatal pores in the epidermis. Hence, in $C_3$ crops, rising $CO_2$ increases net photosynthetic $CO_2$ uptake because RuBisCO is not $CO_2$-saturated in today's atmosphere and because $CO_2$ inhibits the competing oxygenation reaction leading to photorespiration. RuBisCO is highly conserved across terrestrial plants, so instantaneous responses to increased elevated $CO_2$ may be generalized across $C_3$ plants, including rice, soybeans and wheat. At 25°C, an increase in e[$CO_2$] from the present-day value of 380 ppm to that of 550 ppm, projected for the year 2050, would increase $C_3$ photosynthesis by 38% (Long et al. 2004).

In contrast, in $C_4$ crops such as maize and sorghum, RuBisCO is localized to bundle sheath cells in which $CO_2$ is concentrated to three to six times atmospheric e[$CO_2$] (von Caemmerer 2003). This concentration is sufficient to saturate RuBisCO and in theory would prevent any increase in $CO_2$ uptake with rising e[$CO_2$]. Although $C_4$ crops may not show a direct response in photosynthetic activity, an indirect increase in the efficiency of water use via reduction in stomatal conductance may still increase yield. Several *Arabidopsis thaliana* mutants showing impaired stomatal responses to $CO_2$ have been isolated using a high-throughput technique employing thermography of rosette leaves, and analyzed (Hashimoto et al. 2006; Negi et al. 2008; 2013, 2014; Hashimoto-Sugimoto et al. 2013). In addition, the mesophylls of *Commelina communis* plants were shown to greatly accelerate guard cell responses to $CO_2$ (Fujita et al. 2013).

One of the most consistent effects of elevated atmospheric $CO_2$ on plants is an increase in the rate of photosynthetic carbon fixation by leaves. Across a range of FACE experiments, with a variety of plant species, the growth of plants at elevated $CO_2$ concentrations of 475–600 ppm increases leaf photosynthetic rates by an average of 40% (Ainsworth and Rogers 2007). Carbon dioxide concentrations are also important in regulating the openness of stomata, pores through which plants exchange gases, with the external environment. Open stomata allow $CO_2$ to diffuse into leaves for photosynthesis, but also provide a pathway for water to diffuse out of leaves. Therefore, plants regulate the degree of stomatal opening as a compromise between the goals of maintaining high rates of photosynthesis and low rates of water loss. As $CO_2$ concentrations increase, plants can maintain high photosynthetic rates with relatively low stomatal conductance and decreases stomatal conductance of water by an average of 22% (Ainsworth 2008).

It is expected to decrease overall plant water use, although the magnitude of the overall effect of $CO_2$ will depend on how it affects other determinants of plant water use, such as plant size, morphology, and leaf temperature and decreases in whole-plant water use of 5%–20%. This in turn can have consequences for the hydrological cycle of entire ecosystems, with soil moisture levels and runoff both increasing under elevated $CO_2$ (Leakey et al. 2009). Since photosynthesis and stomatal behavior are central to plant carbon and water metabolism, the growth of plants under elevated $CO_2$ leads to a large variety of secondary effects on plant physiology. The additional photosynthate enables plants to grow faster under elevated $CO_2$, with dry matter production increased on average by 17% and 30% for the above- and below-ground portions respectively (Ainsworth and Long 2005; de Graaff et al. 2006). This increased growth is also reflected in the harvestable yield of crops, with wheat, rice and soybean all showing increases in yield of 12%–14% under elevated $CO_2$ in FACE experiments (Long et al. 2006).

### 6.6.3 Impact of Elevated $CO_2$ on Different Agricultural Crops to Cope Different Stresses by Altering Metabolic Pathway

Due to the decline in groundwater resources, it is the main concern to make efficient use of groundwater and to indorse less water demanding forms of crop production. In most of the world, drought is one of the most important limiting factors for cereals. Significant differences have been observed by investigating the influence of water shortage on WUE of plants. Simulated drought stress by the shooting decreases WUE compared to the well-watered plants with meaningful differences among the varieties at the different $CO_2$ levels. The enriched $CO_2$ results in a significant increase of WUE under stress conditions simulates by the heading compared to the plants grown under atmospheric $CO_2$ level.

e[$CO_2$] is also reported to reduce the stomatal conductance, which conserved soil moisture and plant capability under drought in sorghum, a $C_4$ grass with protein content in grains by 60% grown under e[$CO_2$] (Amanda 2015).

A doubling of $CO_2$ level initially accelerates carbon fixation in $C_3$ plants by about 30%, yet after days to weeks of exposure to high $CO_2$ concentrations, depending on species, carbon fixation declines until it stabilizes at a rate that averages 12% above ambient controls. This general phenomenon, known as $CO_2$ acclimation, is correlated with a decline in the activity of Rubisco and other enzymes in the Calvin cycle (Bowes 1993; Moore et al. 1998). The change in Calvin cycle enzyme activities is not necessarily selective; rather, it often follows a decline in overall shoot protein and N contents (Makino and Mae 1999). Shoot N contents diminish by an average of 14% with a doubling of $CO_2$ (Cotrufo et al. 1998a, b), a difference that exceeds what would be expected if a given amount of N were diluted by additional biomass (Makino and Mae 1999).

Experiments on winter wheat grown under drought stress in elevated $CO_2$ concentrations showed decreased soil water humidity from 20 to 25 volumetric water content (VWC%) to 3–5 VWC%. Significant differences were found between the varieties and treatments in the antioxidant enzyme activity and a number of phenological parameters. Elevated $CO_2$ concentration improved stress tolerance increasing biomass production and the grain number in both varieties. High $CO_2$ concentration also reduced the negative effect of drought stress on yield parameters.

Similar experiments on yield of winter wheat under 0°C±2°C, 4°C and 6°C in daily maximum and minimum temperatures, and 380, 420, 460 and 500 ppm elevated $CO_2$ levels showed that variation in both temperatures affected yield and all simulated physiological parameters. Results indicated that when both temperatures increased by 6°C, yield decreased 30%, while 6°C decrease in temperatures increased yield by 37%. A similar effect was perceived on the other simulated parameters such as biomass and 1000 kernel weight. Elevated $CO_2$ positively affected yield, grain number, leaf area and biomass. However, elevated $CO_2$ reduced harvest index and evapotranspiration and non-significantly affected flowering date, maturity and 1000 kernel weight. Every 40 ppm increase in $CO_2$ level increased yields about 150 kg/ha in all elevated $CO_2$ treatments.

Trials on crop yield and grain quality spring wheat (*Triticum aestivum* L.) showed promoted aboveground biomass (+11.8%) and grain yield (+10.4%). However, adverse effects were predominantly observed on whole-grain quality characteristics. Although the thousand-grain weight remained unchanged, size distribution was significantly shifted toward smaller grains, which may directly relate to lower market value. Total grain protein concentration decreased significantly by 7.4% under elevated $CO_2$, and protein and amino acid composition were altered. Corresponding to the decline in grain protein concentration, $CO_2$ enrichment resulted in an overall decrease in amino acid concentrations, with greater reductions in non-essential than essential amino acids. Minerals such as potassium, molybdenum and lead increased, while manganese, iron, cadmium and silicon decreased, suggesting that adjustments of agricultural practices may be required to retain current grain quality standards (Harnos et al. 2002; Arshad et al. 2016). The concentration of fructose and fructan, as well as amounts per area of total and individual nonstructural carbohydrates, except for starch, significantly increased in the grain. The same holds for the number of lipids. With regard to the mixing and rheological properties of the flour, a significant increase in gluten resistance under elevated $CO_2$ was observed. $CO_2$ enrichment affected grain quality characteristics that are important for consumer nutrition and health, and for industrial processing and marketing, which have to date received little attention.

Short- and long-term growth responses of *Poa annua* L. at ambient and elevated (ambient +200 mmol mol$^{-1}$) atmospheric $CO_2$ showed enhanced growth during the early phase, but this was reversed in the reproductive phase. Seed mass and percentage germination were significantly reduced in elevated $CO_2$. Soybean (*Glycine max* L. Merr. cv. Bragg) grown in elevated $CO_2$ stress showed that in the absence of other climatic stresses, photosynthesis and carbohydrate metabolism would perform well under rising atmospheric [$CO_2$] and temperature predicted for the 21st century.

Experiments on sugar cane focusing on Leaf $CO_2$ exchange rate, primary enzymes involved in leaf photosynthesis and sucrose metabolism, chlorophyll contents and soluble protein grown at ambient (360 mmol/mol) and double-ambient (elevated, 720 mmol/mol) $CO_2$ revealed that elevated-$CO_2$ plants had up to 51% lower stomatal conductance and 39% less transpiration, which resulted in 26%–52% greater WUE than ambient-$CO_2$ plants. For the elevated-$CO_2$ plants, total Chlorophyll (Chl) and soluble protein were 31% and 15% greater. The up-regulation of the key photosynthesis and sucrose metabolism enzymes at early stages of leaf development would indicate an acclimation to elevated $CO_2$ for the $C_4$ sugarcane plant. An up-regulation of the enzymes, together with a reduction in leaf stomatal conductance and transpiration and an improvement in leaf WUE and plant water status, could lead to an enhancement in leaf area, plant biomass accumulation and sucrose production for the $CO_2$-enriched sugarcane plants.

Rice crop (*Oryza sativa* L.) was subjected to elevated $CO_2$ to investigate the relationship between plant growth, grain yield, nutrient acquisition and partitioning. Elevated $CO_2$ increased biomass at tillering, and this was largely due to an increase in root mass by 160%. However elevated $CO_2$ had no effect on total nutrient uptake (N, P, K, Mg and Ca). But nutrient partitioning among organs was significantly altered. N partitioning to leaf blades was significantly decreased, whereas the N partitioning into the leaf sheaths and roots was increased. Nutrient-use efficiency of N, P, K and Mg in all organs was significantly increased at elevated $CO_2$. At harvest maturity, grain yield was increased by 27% at elevated $CO_2$ while grain protein concentration was decreased by a similar magnitude (28%), suggesting that critical nutrient requirements for rice might need to be reassessed with global climate change.

Effect of different $CO_2$ concentrations in radish (*Raphanus sativus* L., cv. *Žara)* on carbohydrate, chlorophyll contents and net photosynthetic productivity were measured. The results showed that elevated $CO_2$ increases total carbohydrate content and changes in hexoses/sucrose ratio. It was also found that leaf carbohydrate contents affect the intensity of photosynthetic pigment synthesis (Adnan et al. 2018a, b).

Maize is one of the major cultivated crops having a central role in ensuring the food security of the country. The open-top chamber (OTC) experiment was designed in China to control contrasting [$CO_2$], i.e., 390, 450 and 550 mmol and the experiment with 15% increasing precipitation levels was also set based on the average monthly precipitation of 5–9 months from 1981 to 2010 and controlled by irrigation. Thus, six treatments, C550W+15%, C550W0, C450W+15%, C450W0, C390W+15% and C390W0, were included in the study. The results showed that the irrigation under elevated [$CO_2$] levels increased the leaf net photosynthetic rate (Pn) and intercellular $CO_2$ concentration (Ci) of maize. Similarly, the stomatal conductance (gs) and transpiration rate (Tr) decreased with elevated [$CO_2$], but irrigation has a positive effect on increased of them at each [$CO_2$] level, resulting in WUE higher in natural precipitation treatment than irrigation treatment at elevated [$CO_2$] levels. Irradiance-response parameters, e.g., maximum net photosynthetic rate (Pnmax) and light saturation points (LSP), were increased under elevated [$CO_2$] and irrigation, and dark respiration (Rd) was increased as well. The growth characteristics, e.g., plant height, leaf area and aboveground biomass were enhanced, resulting in an improved yield and ear characteristics except axle diameter. The study concluded by reporting that, future elevated [$CO_2$] may favor maize when coupled with an increasing amount of precipitation in Northeast China.

The effects of nitrogen [75 and 150 kg (N)/ha] and elevated $CO_2$ on growth, photosynthetic rate, contents of soluble leaf proteins and activities of ribulose-1,5-bisphosphate carboxylase/oxygenase (Rubisco) and nitrate reductase were studied on wheat (*Triticum aestivum* L.) grown in open-top chambers under either ambient or elevated $CO_2$ concentration ($350\pm50$, $600\pm50$ μmol/mol) and analyzed at 40, 60 and

90 d after sowing. Plants grown under elevated $CO_2$ showed greater photosynthetic rate and were taller and attained greater leaf area along with higher total plant dry mass at all growth stages than those grown under ambient carbon. Total soluble and Rubisco protein contents decreased under elevated $CO_2$ but the activation of Rubisco was higher at elevated $CO_2$ with a higher N supply. Nitrogen increased the nitrate reductase activity whereas elevated $CO_2$ reduced it. Thus, elevated $CO_2$ causes increased growth and PN ability per unit uptake of N in wheat plants, even if N is limiting.

The response of seed production to $CO_2$ concentration [$CO_2$] is known to vary considerably among $C_3$ annual species. Meta-analysis reveals that the increase in seed mass per plant under elevated [$CO_2$] is mainly due to increase in seed nitrogen per plant rather than seed [N] dilution. Nitrogen-fixing legumes enhanced nitrogen acquisition more than non-nitrogen fixers, resulting in a large increase in seed mass per plant. In Poaceae, an increase in seed mass per plant was also caused by a decrease in seed [N]. Greater carbon allocation to albumen (endosperm and/or perisperm) than the embryo may account for [N] reduction in grass seeds. These differences in $CO_2$ response of seed production among functional groups may affect their fitness, leading to changes in species composition in the future high-[$CO_2$] ecosystem (Hikosaka et al. 2011).

Long-term growth of Arabidopsis at high $CO_2$ (1000 mL L21) resulted in a twofold increase in nonstructural carbohydrates, a large decrease in the expression of Rubisco protein and the transcript of rbcL, the gene encoding the large subunit of Rubisco (approximately 35%–40%), and an even greater decline in mRNA of rbcS, the gene encoding the small subunit (approximately 60%). This differential response of protein and mRNAs suggests that transcriptional/posttranscriptional processes and protein turnover may determine the final amount of leaf Rubisco protein at high $CO_2$. Analysis of mRNA levels of individual rbcS genes indicated that reduction in total rbcS transcripts was caused by decreased expression of all four rbcS genes. Short-term transfer of Arabidopsis plants grown at ambient $CO_2$ to high $CO_2$ resulted in a decrease in total rbcS mRNA by day 6, whereas Rubisco content and rbcL mRNA decreased by day 9. Transfer to high $CO_2$ reduced the maximum expression level of the primary rbcS genes (1A and, particularly, 3B) by limiting their normal pattern of accumulation through the night period. The decreased nighttime levels of rbcS mRNA were associated with a nocturnal increase in leaf hexoses. We suggest that prolonged nighttime hexose metabolism resulting from exposure to elevated $CO_2$ affects rbcS transcript accumulation and, ultimately, the level of Rubisco protein.

### 6.6.4 Impact of Elevated $CO_2$ on Other Trees and Fruits

Several studies have shown that elevated $CO_2$ affects the growth rhythm of forest trees by altering the timing of bud burst and growth cessation (Cannell and Smith 1986; Murray et al. 1994; Ceulemans et al. 1995). In the boreal region, early bud burst results in the early onset of growth thereby prolonging the short growing season and potentially increasing wood production (Beuker 1994). Earlier bud bursts, however, may increase the risk of frost damage from late spring frosts (Murray et al. 1989). Increased atmospheric $CO_2$ concentrations may also affect bud phenology directly through changes in tree biochemistry and physiology. For example, changes in starch or hormonal concentrations may alter dormancy status and growth patterns, by shifting the timing and duration of the vegetative season (Cannell 1990).

In trees, elevated $CO_2$ can increase total leaf area (Koch et al. 1986), leaf weight (Brown and Higginbotham 1986; Norby and Neill 1989), leaf weight to area ratio (Conroy et al. 1986; Berryman et al. 1993; Pettersson et al. 1993) and branching frequency (Sionit et al. 1985; Samuelson and Seiler 1993). Root biomass, root length, root branching and lateral root production are also reported to increase in response to elevated $CO_2$ (Rogers et al. 1994; Day et al. 1996; Janssens et al. 1998). Research was done on deciduous *Quercus robur* and *Pinus pinaster* to investigate the effects of e[$CO_2$] concentration showed significant reductions in the activities of superoxide dismutases. The activities of ascorbate peroxidase and catalase were also reduced in most cases. The activities of dehydroascorbate reductase, monodehydroascorbate radical reductase, glutathione reductase and guaiacol peroxidase were affected little or not at all by elevated $CO_2$. The results suggest that growth in elevated $CO_2$ might reduce oxidative stress to which leaf tissues are normally exposed and enhance metabolic flexibility to encounter increased stress by increases in antioxidative capacity.

Suspension-cultured pear fruit (*Pyrus communis* L.) cells in aging media exposed to elevated $CO_2$ exhibited reduced respiration. Ethylene production of both air and $CO_2$-treated cells also declined to barely discernible levels by day 3. Fructose 6-phosphate (F6P) accumulated, while levels of fructose 1, 6-bisphosphate (F1, 6-P2), and activities of ATP and PPi phosphofructokinases (PFK and PFP) declined in response to elevated $CO_2$. These results indicate an inhibitory effect of $CO_2$ at the site of action of both phosphofructokinases in the glycolytic pathway, which could account, at least in part, for the observed reduction in respiration. The responses to elevated $CO_2$ levels of the cell suspension system and intact pear fruit ventilated with air + 10% $CO_2$ are compared, revealing a close similarity. Elevated $CO_2$ significantly improved the date of bud burst during both years in Scots pine (*Pinus sylvestris* L.). The total needle area, stem diameter, number and length of shoots were increased in response to elevated $CO_2$ (Zvereva and Kozlov 2006).

## 6.7 Conclusion

It was predicted from above literature provided that interaction of e[$CO_2$] and different stresses have an effect on the adaptation of plants to future climate scenarios. From a recent literature survey, it is concluded that increased $CO_2$ concentrations are expected to enhance the photosynthetic production of carbohydrates and other organic compounds. It was also depicted that nitrogen level decline under elevated $CO_{2.}$ It might be due to decrease in N under elevated [$CO_2$] is the results from dilution due to accumulation of nonstructural carbohydrates. Another possible mechanism for decreased N under elevated $CO_2$ is dilution by plant secondary compounds. The precise fact is still not achieved about this dilution but researchers observed that it might be functional and biomass dilution. It was concluded that coordinated efforts to increase photosynthetic efficiency, carbohydrate metabolism, high antioxidant potential, increased sink capacity, and improvement for nitrogen uptake would potentially improve grain yield and quality under rising $CO_2$ levels.

## 6.8 Future Prospective

With an emerging issue of climate change, that brings severe abiotic stress for plants of the next generation, Plant should be grown under elevated $CO_2$. It would stimulate growth but is often coupled with low quality. This is mainly due to relaxation of photosynthesis to a lower rate when exposed to elevated $CO_2$ over longer periods. A sustained stimulation for 22 century requires plant growth at elevated $CO_2$ along with acquisition of additional nitrogen in order to maximize increased carbon assimilation. This research identified that nutrient management strategies are required under elevated $CO_2$. The new combinations of the nutrient source with elevated $CO_2$ will be needed, along with more efficient cultivars. However, increased use of nitrogenous fertilizers is neither environmentally nor economically favorable. These fertilizers are too costly for poor farmers and their application also causes pollution of nitrate. Nitrogen-limiting conditions, therefore, reduce the responsiveness of plants to elevated [$CO_2$] and decrease the photosynthetic rate. In order to sustain productivity under changing environmental factors, there is a need to look for plants that can utilize the positive effect of $CO_2$, even at low level of nitrogen.

## REFERENCES

Abdul S, Muhammad AA, Shabir H, Hesham A. El E, Sajjad H, Niaz A, Abdul G, Sayyed RZ, Fahad S, Subhan D, Rahul D (2021a) Zinc nutrition and arbuscular mycorrhizal symbiosis effects on maize (Zea mays L.) growth and productivity. J Saudi Soc Agric Sci. https://doi.org/10.1016/j.sjbs.2021.06.096.

Abdul S, Muhammad AA, Subhan D, Niaz A, Fahad S, Rahul D, Mohammad JA, Omaima N, Muhammad Habib ur R, Bernard RG (2021b) Effect of arbuscular mycorrhizal fungi on the physiological functioning of maizeunder zinc-deficient soils. Sci Rep 11:18468.

Adnan M, Fahad S, Khan IA, Saeed M, Ihsan MZ, Saud S, Riaz M, Wang D, Wu C (2019) Integration of poultry manure and phosphate solubilizing bacteria improved availability of Ca bound P in calcareous soils. 3 Biotech 9(10):368.

Adnan M, Fahad S, Muhammad Z, Shahen S, Ishaq AM, Subhan D, Zafar-ul-Hye M, Martin LB, Raja MMN, Beena S, Saud S, Imran A, Zhen Y, Martin B, Jiri H, Rahul D (2020) Coupling phosphate-solubilizing bacteria with phosphorus supplements improve maize phosphorus acquisition and growth under lime induced salinity stress. Plants 9(900). https://doi.org/10.3390/plants9070900.

Adnan M, Shah Z, Sharif M, Rahman H (2018b) Liming induces carbon dioxide ($CO_2$) emission in PSB inoculated alkaline soil supplemented with different phosphorus sources. Environ Sci Poll Res 25(10):9501–9509.

Adnan M, Zahir S, Fahad S, Arif M, Mukhtar A, Imtiaz AK, Ishaq AM, Abdul B, Hidayat U, Muhammad A, Inayat-Ur R, Saud S, Muhammad ZI, Yousaf J, Amanullah Hafiz MH, Wajid N (2018a) Phosphate-solubilizing bacteria nullify the antagonistic effect of soil calcification on bioavailability of phosphorus in alkaline soils. Sci Rep 8:4339. https://doi.org/10.1038/s41598-018-22653-7.

Ahmad S, Kamran M, Ding R, Meng X, Wang H, Ahmad I, Fahad S, Han Q (2019) Exogenous melatonin confers drought stress by promoting plant growth, photosynthetic capacity and antioxidant defense system of maize seedlings. PeerJ 7:e7793. http://doi.org/10.7717/peerj.7793.

Ainsworth EA (2008) Rice production in a changing climate: A meta-analysis of responses to elevated carbon dioxide and elevated ozone concentrations. Global Change Biol 14:1642–1650

Ainsworth EA, Long SP (2005) What have we learned from 15 years of free air $CO_2$ enrichment (FACE)? A meta-analytic review of the responses of photosynthesis, canopy properties and plant production to rising $CO_2$. New Phytol 165:351–372.

Ainsworth EA, Rogers A (2007) The response of photosynthesis and stomatal conductance to rising [$CO_2$]: Mechanisms and environmental interactions. Plant Cell Environ 30:258–270.

Akbar H, Timothy JK, Jagadish T, M. Golam M, Apurbo KC, Muhammad F, Rajan B, Fahad S, Hasanuzzaman M (2020) Agricultural land degradation: Processes and problems undermining future food security. In: Fahad S, Hasanuzzaman M, Alam M, Ullah H, Saeed M, Khan AK, Adnan M (eds) Environment, climate, plant and vegetation growth. Springer Publ Ltd, Springer Nature Switzerland AG. Part of Springer Nature, pp 17–62. https://doi.org/10.1007/978-3-030-49732-3.

Akram R, Turan V, Hammad HM, Ahmad S, Hussain S, Hasnain A, Maqbool MM, Rehmani MIA, Rasool A, Masood N, Mahmood F, Mubeen M, Sultana SR, Fahad S, Amanet K, Saleem M, Abbas Y, Akhtar HM, Waseem F, Murtaza R, Amin A, Zahoor SA, ul Din MS, Nasim W (2018) Fate of organic and inorganic pollutants in paddy soils. In: Hashmi MZ, Varma A (eds) Environmental pollution of paddy soils, soil biology. Springer International Publishing AG, Switzerland, pp 197–214.

Altaf M, Raziq F, Khan I, Hussain H, Shah B, Ullah W, Naeem A, Adnan M, Junaid K, Shah SRA, Attaullah, Iqbal M (2016) Study on the response of different maize cultivars to various inoculum levels of Bipolarismaydis(Y. Nisik & C. Miyake) shoemaker under field conditions. J Ento Zool Stud 4(2):533–537.

Al-Wabel MI, Abdelazeem S, Munir A, Khalid E, Adel RAU (2020b) Extent of climate change in Saudi Arabia and its impacts on agriculture: A case study from Qassim Region. In: Fahad S, Hasanuzzaman M, Alam M, Ullah H, Saeed M, Khan AK, Adnan M (eds) Environment, climate, plant and vegetation growth. Springer Publ Ltd, Springer Nature Switzerland AG. Part of Springer Nature. pp 635–658. https://doi.org/10.1007/978-3-030-49732-3.

Al-Wabel MI, Ahmad M, Usman AR, Akanji M, Rafique MI (2020a) Advances in pyrolytic technologies with improved carbon capture and storage to combat climate change. In: Fahad S, Hasanuzzaman M, Alam M, Ullah H, Saeed M, Khan AK, Adnan M (eds) Environment, climate, plant and vegetation growth. Springer Publ Ltd, Springer Nature Switzerland AG. Part of Springer Nature. pp 535–576. https://doi.org/10.1007/978-3-030-49732-3.

Amanullah, Fahad S (eds) (2017) Rice - technology and production. IntechOpen, Croatia. http://dx.doi.org/10.5772/64480.

Amanullah, Fahad S (eds) (2018a) Corn - Production and human Health in changing climate. IntechOpen, United Kingdom. http://dx.doi.org/10.5772/intechopen.74074.

Amanullah, Fahad S (eds) (2018b) Nitrogen in agriculture – updates. IntechOpen, Croatia. http://dx.doi.org/10.5772/65846.

Amanullah, Muhammad I, Haider N, Shah K, Manzoor A, Asim M, Saif U, Izhar A, Fahad S, Adnan M, et al. (2021) Integrated foliar nutrients application improve wheat (Triticum Aestivum L.) productivity under calcareous soils in drylands. Commun Soil Sci Plant Anal. https://doi.org/10.1080/00103624.2021.1956521.

Amanullah, Shah K, Imran, Hamdan AK, Muhammad A, Abdel RA, Muhammad A, Fahad S, Azizullah S, Brajendra P (2020) Effects of climate change on irrigation water quality. In: Fahad S, Hasanuzzaman M, Alam M, Ullah H, Saeed M, Khan AK, Adnan M (eds) Environment, climate, plant and vegetation growth. Springer Publ Ltd, Springer Nature Switzerland AG. Part of Springer Nature. pp 123–132. https://doi.org/10.1007/978-3-030-49732-3.

Amir M, Muhammad A, Allah B, Sevgi Ç, Haroon ZK, Muhammad A, Emre A (2020) Bio fortification under climate change: The fight between quality and quantity. In: Fahad S, Hasanuzzaman M, Alam M, Ullah H, Saeed M, Khan AK, Adnan M (Ed.), Environment, climate, plant and vegetation growth. Springer Publ Ltd, Springer Nature Switzerland AG. Part of Springer Nature. pp 173–228. https://doi.org/10.1007/978-3-030-49732-3.

Amjad I, Muhammad H, Farooq S, Anwar H (2020) Role of plant bioactives in sustainable agriculture. In: Fahad S, Hasanuzzaman M, Alam M, Ullah H, Saeed M, Khan AK, Adnan M (eds) Environment, climate, plant and vegetation growth. Springer Publ Ltd, Springer Nature Switzerland AG. Part of Springer Nature. pp 591–606. https://doi.org/10.1007/978-3-030-49732-3.

Amjad SF, Mansoora N, Din IU, Khalid IR, Jatoi GH, Murtaza G, Yaseen S, Naz M, Danish S, Fahad S, et al. (2021) Application of zinc fertilizer and mycorrhizal inoculation on physio- biochemical parameters of wheat grown under water-stressed environment. Sustainability 13:11007. https://doi.org/10. 3390/su131911007.

Aranjueloa I, Cabrerizoa PM, Aparicio-Tejoa PM, Arrese-Igor C (2014) Unravelling the mechanisms that improve photosynthetic performance of $N_2$-fixing pea plants exposed to elevated [$CO_2$]. Environ Exp Bot 99:167–174. https://doi.org/10.1016/j.envexpbot.2013.10.020.

Arif M, Talha J, Muhammad R, Fahad S, Muhammad A, Amanullah, Kawsar A, Ishaq AM, Bushra K, Fahd R (2020) Biochar; a remedy for climate change. In: Fahad S, Hasanuzzaman M, Alam M, Ullah H, Saeed M, Khan AK, Adnan M (eds) Environment, climate, plant and vegetation growth. Springer Publ Ltd, Springer Nature Switzerland AG. Part of Springer Nature. pp 151–172. https://doi.org/10.1007/978-3-030-49732-3.

Arshad M, Adnan M, Ali A, Khan AK, Khan F, Khan A, Kamal MA, Alam M, Ullah H, Saleem A, Hussain A, Shahwar D (2016) Integrated effect of phosphorous and zinc on wheat quality and soil properties. Adv Environ Biol 10(2):40–45.

Ashfaq AR, Uzma Y, Niaz A, Muhammad AA, Fahad S, Haider S, Tayebeh Z, Subhan D, Süleyman T, Hesham AElE, Pramila T, Jamal MA, Sulaiman AA, Rahul D (2021) Toxicity of Cadmium and nickel in the context of applied activated carbon biochar for improvement in soil fertility. Saudi Soc Agric Sci. https://doi.org/10.1016/j.sjbs.2021.09.035.

Athar M, Masood IA, Sana S, Ahmed M, Xiukang W, Sajid F, Sher AK, Habib A, Faran M, Zafar H, Farhana G, Fahad S (2021) Bio-diesel production of sunflower through sulphur management in a semi-arid subtropical environment. Environ Sci Pollut Res. https://doi.org/10. 1007/s11356-021-16688-z.

Atif B, Hesham A, Fahad S (2021) Biochar coupling with phosphorus fertilization modifies antioxidant activity, osmolyte accumulation and reactive oxygen species synthesis in the leaves and xylem sap of rice cultivars under high-temperature stress. Physiol Mol Biol Plants. https://doi.org/10. 1007/s12298-021-01062-7.

Ayman EL Sabagh, Hossain A, Barutçular C, Iqbal MA, Islam MS, Fahad S, Sytar O, Çig F, Meena RS, Erman M (2020) Consequences of salinity stress on the quality of crops and its mitigation strategies for sustainable crop production: An outlook of arid and semi- arid regions. In: Fahad S, Hasanuzzaman M, Alam M, Ullah H, Saeed M, Khan AK, Adnan M (eds) Environment, climate, plant and vegetation growth. Springer Publ Ltd, Springer Nature Switzerland AG. Part of Springer Nature. pp 503–534. https://doi.org/10.1007/978-3-030-49732-3.

Aziz K, Daniel KYT, Fazal M, Muhammad ZA, Farooq S, Fan W, Fahad S, Ruiyang Z (2017a) Nitrogen nutrition in cotton and control strategies for greenhouse gas emissions: A review. Environ Sci Pollut Res 24:23471–23487. https://doi.org/10.1007/s11356-017-0131-y.

Aziz K, Daniel KYT, Muhammad ZA, Honghai L, Shahbaz AT, Mir A, Fahad S (2017b) Nitrogen fertility and abiotic stresses management in cotton crop: A review. Environ Sci Pollut Res 24:14551–14566. https://doi.org/10.1007/s11356-017-8920-x.

Baseer M, Adnan M, Fazal M, Fahad S, Muhammad S, Fazli W, Muhammad A, Jr. Amanullah, Depeng W, Saud S, Muhammad N, Muhammad Z, Fazli S, Beena S, Mian AR, Ishaq AM (2019) Substituting urea by organic wastes for improving maize yield in alkaline soil. J Plant Nutr. https://doi.org/10.1080/01904167.2019.1659344.

Bayram AY, Seher Ö, Nazlican A (2020) Climate change forecasting and modeling for the year of 2050. In: Fahad S, Hasanuzzaman M, Alam M, Ullah H, Saeed M, Khan AK, Adnan M (eds) Environment, climate, plant and vegetation growth. Springer Publ Ltd, Springer Nature Switzerland AG. Part of Springer Nature. pp 109–122. https://doi.org/10.1007/978-3-030-49732-3.

Bazzaz FA, Fajer ED (1992) Plant life in a $CO_2$-rich world. Sci Am 266:68–74.

Bencze S, Bamberger Z, Janda T, Balla K, Varga B, Bedö Z, et al. (2014). Physiological response of wheat varieties to elevated atmospheric $CO_2$ and low water supply levels. Photosynthetica 52:71–82. https://doi.org/10.1007/s11099-014-0008-y.

Berryman CA, Eamus D, Duff GA (1993) The influence of $CO_2$ enrichment on growth, nutrient content and biomass allocation of Marantes corymbosa. Aust J Bot 41:195–209.

Beuker E (1994) Adaptation to climatic changes of the timing of bud burst in populations of Pinus sylvestris L. and Picea abies (L.) Karrst. Tree Physiol 14:961–970.

Bloom N, Lemos R, Sadun R, Scur D, Reenen JV (2014) The new empirical economics of management. J Eur Econ Assoc 12(4):835–876.

Bowes G (1993) Facing the inevitable: Plants and increasing atmospheric $CO_2$. Annu Rev Plant Physiol Plant Mol Biol 44:309–332.

Bowes G (1996) Elevated carbon dioxide. In: Baker N (ed) Advances in photosynthesis, environmental stress and photosynthesis, 3. Academic Press, Dordrecht, pp 387–407.

Brown K, Higginbotham KO (1986) Effects of carbon dioxide enrichment and nitrogen supply on growth of boreal tree seedlings. Tree Physiol 2:223–232.

Budak H, Kantar M, Yucebilgili Kurtoglu K (2013) Drought tolerance in modern and wild wheat. Sci World J 54:82–46.

Cannell MGR (1990) Modelling the phenology of trees. Silva Carelica 15:11–27.

Cannell MGR, Smith RI (1986). Climatic warming, spring budburst, and forest damage on trees. J Appl Ecol 23:177–191. https://doi.org/10.2307/2403090.

Ceulemans R, Jiang XN, Shao BY (1995) Effects of elevated atmospheric $CO_2$ on growth, biomass production and nitrogen allocation of two Populus clones. J Biogeogr 22:261–268.

Chang W, Qiujuan J, Evgenios A, Haitao L, Gezi L, Jingjing Z, Fahad S, Ying J (2021) Hormetic effects of zinc on growth and antioxidant defense system of wheat plants. Sci Total Environ. https://doi.org/10.1016/j.scitotenv.2021.150992.

Chen Y, Guo Z, Dong L, Fu Z, Zheng Q, Zhang G, Qin L, Sun X, Shi Z, Fahad S, Xie F, Saud S (2021) Turf performance and physiological responses of native Poa species to summer stress in Northeast China. PeerJ 9:e12252. http://doi.org/10.7717/peerj.12252.

Conroy JP, Barlow EWR, Bevege DI (1986) Response of Pinus radiata seedlings to carbon dioxide enrichment at different levels of water and phosphorus: Growth, morphology and anatomy. Ann Bot 7:165–177.

Cotrufo MF, Berg B, Kratz W (1998a) Increased atmospheric $CO_2$ and litter quality. Environ Rev 6:1–12.

Cotrufo MF, Ineson P, Scott A (1998b) Elevated $CO_2$ reduces the nitrogen concentration of plant tissues. Global Change Biol 4:43–54.

Day FP, Weber EP, Hinckle CR, Drake BG (1996) Effects of elevated atmospheric $CO_2$ on fine root length and distribution in an oak-palmetto scrub ecosystem in central Florida. Global Change Biol 2:143–148.

De Graaff MA, Van Groeningen KJ, Six J, Hungate BA, Van Kessel C (2006) Interactions between plant growth and nutrient dynamics under elevated $CO_2$: A meta analysis. Global Change Biol 12:1–15.

Deepranjan S, Ardith S, O. Siva D, Sonam S, Shikha, Manoj P, Amitava R, Sayyed RZ, Abdul G, Mohammad JA, Subhan D, Fahad S, Rahul D (2021) Optimizing nutrient use efficiency, productivity, energetics, and economics of red cabbage following mineral fertilization and biopriming with compatible rhizosphere microbes. Sci Rep 11:15680.

Depeng W, Fahad S, Saud S, Muhammad K, Aziz K, Mohammad NK, Hafiz MH, Wajid N (2018) Morphological acclimation to agronomic manipulation in leaf dispersion and orientation to promote "Ideotype" breeding: Evidence from 3D visual modeling of "super" rice (Oryza sativa L.). Plant Physiol Biochem. https://doi.org/10.1016/j.plaphy.2018.11.010.

Emre B, Ömer SU, Martín LB, Andre D, Fahad S, Rahul D, Muhammad Z-ul-H, Ghulam SH, Subhan D (2021) Studying soil erosion by evaluating changes in physico-chemical properties of soils under different land-use types. J Saudi Soc Agric Sci 20:190–197.

Fahad S, Abdul B, Adnan M. (eds) (2018b) Global wheat production. IntechOpen, United Kingdom. http://dx.doi.org/10.5772/intechopen.72559.

Fahad S, Adnan M, Hassan S, Saud S, Hussain S, Wu C, Wang D, Hakeem KR, Alharby HF, Turan V, Khan MA, Huang J (2019b) Rice responses and tolerance to high temperature. In: Hasanuzzaman M, Fujita M, Nahar K, Biswas JK (eds) Advances in rice research for abiotic stress tolerance. Woodhead Publ Ltd, Abington Hall Abington, Cambridge, UK, pp 201–224.

Fahad S, Bajwa AA, Nazir U, Anjum SA, Farooq A, Zohaib A, Sadia S, NasimW, Adkins S, Saud S, Ihsan MZ, Alharby H, Wu C, Wang D, Huang J (2017) Crop production under drought and heat stress: Plant responses and management options. Front Plant Sci 8:1147. https://doi.org/10.3389/fpls.2017.01147.

Fahad S, Bano A (2012) Effect of salicylic acid on physiological and biochemical characterization of maize grown in saline area. Pak J Bot 44:1433–1438.

Fahad S, Chen Y, Saud S, Wang K, Xiong D, Chen C, Wu C, Shah F, Nie L, Huang J (2013) Ultraviolet radiation effect on photosynthetic pigments, biochemical attributes, antioxidant enzyme activity and hormonal contents of wheat. J Food, Agric Environ 11(3&4):1635–1641.

Fahad S, Hasanuzzaman M, Alam M, Ullah H, Saeed M, Ali Khan I, Adnan M. (eds) (2020) Environment, climate, plant and vegetation growth. Springer Nature Switzerland AG. https://doi.org/10.1007/978-3-030-49732-3

Fahad S, Hussain S, Bano A, Saud S, Hassan S, Shan D, Khan FA, Khan F, Chen Y, Wu C, Tabassum MA, Chun MX, Afzal M, Jan A, Jan MT, Huang J (2014a) Potential role of phytohormones and plant growth-promoting rhizobacteria in abiotic stresses: Consequences for changing environment. Environ Sci Pollut Res 22(7):4907–4921. https://doi.org/10.1007/s11356-014-3754-2.

Fahad S, Hussain S, Matloob A, Khan FA, Khaliq A, Saud S, Hassan S, Shan D, Khan F, Ullah N, Faiq M, Khan MR, Tareen AK, Khan A, Ullah A, Ullah N, Huang J (2014b) Phytohormones and plant responses to salinity stress: A review. Plant Growth Regul 75(2):391–404. https://doi.org/10. 1007/s10725-014-0013-y.

Fahad S, Hussain S, Saud S, Hassan S, Chauhan BS, Khan F, et al. (2016a) Responses of rapid viscoanalyzer profile and other rice grain qualities to exogenously applied plant growth regulators under high day and high night temperatures. PLoS One 11(7):e0159590. https://doi.org/10.1371/journal.pone.0159590.

Fahad S, Hussain S, Saud S, Hassan S, Ihsan Z, Shah AN, Wu C, Yousaf M, Nasim W, Alharby H, Alghabari F, Huang J (2016c) Exogenously applied plant growth regulators enhance the morphophysiological growth and yield of rice under high temperature. Front Plant Sci 7:1250. https://doi.org/10.3389/fpls.2016.01250.

Fahad S, Hussain S, Saud S, Hassan S, Tanveer M, Ihsan MZ, Shah AN, Ullah A, Nasrullah KF, Ullah S, Alharby H, Wu C, Huang J (2016d) A combined application of biochar and phosphorus alleviates heat-induced adversities on physiological, agronomical and quality attributes of rice. Plant Physiol Biochem 103:191–198.

Fahad S, Hussain S, Saud S, Khan F, Hassan S, Nasim W, Arif M, Wang F, Huang J (2016b) Exogenously applied plant growth regulators affect heat-stressed rice pollens. J Agron Crop Sci 202:139–150.

Fahad S, Hussain S, Saud S, Tanveer M, Bajwa AA, Hassan S, Shah AN, Ullah A, Wu C, Khan FA, Shah F, Ullah S, Chen Y, Huang J (2015a) A biochar application protects rice pollen from high-temperature stress. Plant Physiol Biochem 96:281–287

Fahad S, Muhammad ZI, Abdul K, Ihsanullah D, Saud S, Saleh A, Wajid N, Muhammad A, Imtiaz AK, Chao W, Depeng W, Jianliang H (2018a) Consequences of high temperature under changing climate optima for rice pollen characteristics-concepts and perspectives, Archives Agron Soil Sci. https://doi.org/10.1080/03650340.2018.1443213.

Fahad S, Nie L, Chen Y, Wu C, Xiong D, Saud S, Hongyan L, Cui K, Huang J (2015b) Crop plant hormones and environmental stress. Sustain Agric Rev 15:371–400.

Fahad S, Rehman A, Shahzad B, Tanveer M, Saud S, Kamran M, Ihtisham M., Khan SU, Turan V, Rahman MHU (2019a) Rice responses and tolerance to metal/metalloid toxicity. In: Hasanuzzaman M, Fujita M, Nahar K, Biswas JK (eds) Advances in rice research for abiotic stress tolerance. Woodhead Publ Ltd, Abington Hall Abington, Cambridge, UK, pp 299–312.

Fahad S, Saud S, Yajun C, Chao W, Depeng W (eds) (2021f) Abiotic stress in plants. IntechOpen, United Kingdom. http://dx.doi.org/10.5772/intechopen.91549.

Fahad S, Sönmez O, Saud S, Wang D, Wu C, Adnan M, Arif M, Amanullah (eds) (2021e) Engineering tolerance in crop plants against abiotic stress, 1st edn. ed, Footprints of climate variability on plant diversity. CRC Press, Boca Raton, FL.

Fahad S, Sönmez O, Saud S, Wang D, Wu C, Adnan M, Turan V (eds) (2021a) Plant growth regulators for climate-smart agriculture, 1st edn. ed, Footprints of climate variability on plant diversity. CRC Press, Boca Raton, FL.

Fahad S, Sönmez O, Saud S, Wang D, Wu C, Adnan M, Turan V (eds) (2021b) Climate change and plants: Biodiversity, growth and interactions, 1st edn. ed, Footprints of climate variability on plant diversity. CRC Press, Boca Raton, FL.

Fahad S, Sönmez O, Saud S, Wang D, Wu C, Adnan M, Turan V (eds) (2021c) Developing climate resilient crops: Improving global food security and safety, 1st edn. ed, Footprints of climate variability on plant diversity. CRC Press, Boca Raton, FL.

Fahad S, Sönmez O, Turan V, Adnan M, Saud S, Wu C, Wang D (eds) (2021d) Sustainable soil and land management and climate change, 1st edn. ed, Footprints of climate variability on plant diversity. CRC Press, Boca Raton, FL.

Fangmeier DD, Clemmens AJ, El-Ansary M, Strelkoff TS, Osman HE (1999) Influence of land leveling precision on level-basin advance and performance. Trans ASAE 42(4):1019–1025.

Farah R, Muhammad R, Muhammad SA, Tahira Y, Muhammad AA, Maryam A, Shafaqat A, Rashid M, Muhammad R, Qaiser H, Afia Z, Muhammad AA, Muhammad A, Fahad S (2020) Alternative and non-conventional soil and crop management strategies for increasing water use efficiency. In: Fahad S, Hasanuzzaman M, Alam M, Ullah H, Saeed M, Khan AK, Adnan M (eds) Environment, climate, plant and vegetation growth. Springer Publ Ltd, Springer Nature Switzerland AG. Part of Springer Nature. pp 323–338. https://doi.org/10.1007/978-3-030-49732-3.

Farhana G, Ishfaq A, Muhammad A, Dawood J, Fahad S, Xiuling L, Depeng W, Muhammad F, Muhammad F, Syed AS (2020) Use of crop growth model to simulate the impact of climate change on yield of various wheat cultivars under different agro-environmental conditions in Khyber Pakhtunkhwa, Pakistan. Arabian J Geosci 13:112 https://doi.org/10.1007/s12517-020-5118-1.

Farhat A, Hafiz MH, Wajid I, Aitazaz AF, Hafiz FB, Zahida Z, Fahad S, Wajid F, Artemi C (2020) A review of soil carbon dynamics resulting from agricultural practices. J Environ Manage 268:110319.

Fazli W, Muhmmad S, Amjad A, Fahad S, Muhammad A, Muhammad N, Ishaq AM, Imtiaz AK, Mukhtar A, Muhammad S, Muhammad I, Rafi U, Haroon I, Muhammad A (2020) Plant-microbes interactions and functions in changing climate. In: Fahad S, Hasanuzzaman M, Alam M, Ullah H, Saeed M, Khan AK, Adnan M (eds) Environment, climate, plant and vegetation growth. Springer Publ Ltd, Springer Nature Switzerland AG. Part of Springer Nature. pp 397–420. https://doi.org/10.1007/978-3-030-49732-3.

Feng GD, Yang SZ, Wang YH, Deng MR, Zhu HH (2014) Acinetobacter guangdongensis sp. nov., isolated from abandoned lead-zinc ore. Int J Syst Evol Microbiol 64:3417–3421.

Ferrario-Mery S, Thibaud MC, Betsche T, Valadier MH, Foyer CH (1997) Modulation of carbon and nitrogen metabolism, and of nitrate reductase, in untransformed Nicotiana plumbaginifolia during $CO_2$ enrichment of plants grown in pots and hydroponic culture. Planta 202:510–521.

Fujita T, Noguchi K, Terashima I (2013) Apoplastic mesophyll signals induce rapid stomatal responses to $CO_2$ in Commelina communis. New Phytol 199:395–406.

Ghulam M, Muhammad AA, Donald LS, Sajid M, Muhammad FQ, Niaz A, Ateeq ur R, Shakeel A, Sajjad H, Muhammad A, Summia M, Aqib HAK, Fahad S, Rahul D, Mazhar I, Timothy DS (2021) Formalin fumigation and steaming of various composts differentially influence the nutrient release, growth and yield of muskmelon (Cucumis melo L.). Sci Rep 11:21057.

Godfray HCJ, Beddington JR, Crute IR, Haddad L, Lawrence D, Muir JF, Pretty J, Robinson S, Thomas SM, Toulmin C (2010) Food security: The challenge of feeding 9 billion people. Science 327:812–818.

Gopakumar L, Bernard NO, Donato V (2020) Soil microarthropods and nutrient cycling. In: Fahad S, Hasanuzzaman M, Alam M, Ullah H, Saeed M, Khan AK, Adnan M (eds) Environment, climate, plant and vegetation growth. Springer Publ Ltd, Springer Nature Switzerland AG. Part of Springer Nature. pp 453–472. https://doi.org/10.1007/978-3-030-49732-3.

Gries C, Kimball BA, Idso SB (1993) Nutrient uptake during the course of a year by sour orange trees growing in ambient and elevated carbon dioxide concentrations. J Plant Nutr 16:129–147.

Guofu L, Zhenjian B, Fahad S, Guowen C, Zhixin X, Hao G, Dandan L, Yulong L, Bing L, Guoxu J, Saud S (2021) Compositional and structural changes in soil microbial communities in response to straw mulching and plant revegetation in an abandoned artificial pasture in Northeast China. Global Ecol Conserv 31:e01871.

Habib ur R, Ashfaq A, Aftab W, Manzoor H, Fahd R, Wajid I, Md. Aminul I, Vakhtang S, Muhammad A, Asmat U, Abdul W, Syeda RS, Shah S, Shahbaz K, Fahad S, Manzoor H, Saddam H, Wajid N (2017) Application of CSM-CROPGRO-Cotton model for cultivars and optimum planting dates: Evaluation in changing semi-arid climate. Field Crops Res. http://dx.doi.org/10.1016/j.fcr.2017.07.007.

Hafeez M, Farman U, Muhammad MK, Xiaowei L, Zhijun Z, Sakhawat S, Muhammad I, Mohammed AA, G. Mandela FG, Nicolas D, Muzammal R, Fahad S, Yaobin L (2021) Metabolic-based insecticide resistance mechanism and ecofriendly approaches for controlling of beet armyworm Spodoptera exigua:a review. Environ Sci Pollut Res. https://doi.org/10.1007/s11356-021-16974-w.

Hafiz MH, Abdul K, Farhat A, Wajid F, Fahad S, Muhammad A, Ghulam MS, Wajid N, Muhammad M, Hafiz FB (2020b) Comparative effects of organic and inorganic fertilizers on soil organic carbon and wheat productivity under arid region. Commun Soil Sci Plant Anal. https://doi.org/10.1080/00103624.2020.1763385.

Hafiz MH, Farhat A, Ashfaq A, Hafiz FB, Wajid F, Carol Jo W, Fahad S, Gerrit H (2020a) Predicting kernel growth of maize under controlled water and nitrogen applications. Int J Plant Prod. https://doi.org/10.1007/s42106-020-00110-8.

Hafiz MH, Farhat A, Shafqat S, Fahad S, Artemi C, Wajid F, Chaves CB, Wajid N, Muhammad M, Hafiz FB (2018) Offsetting land degradation through nitrogen and water management during maize cultivation under arid conditions. Land Degrad Dev:1–10. https://doi.org/10.1002/ldr.2933.

Hafiz MH, Muhammad A, Farhat A, Hafiz FB, Saeed AQ, Muhammad M, Fahad S, Muhammad A (2019) Environmental factors affecting the frequency of road traffic accidents: A case study of sub-urban area of Pakistan. Environ Sci Pollut Res. https://doi.org/10.1007/s11356-019-04752-8.

Hafiz MH, Wajid F, Farhat A, Fahad S, Shafqat S, Wajid N, Hafiz FB (2016) Maize plant nitrogen uptake dynamics at limited irrigation water and nitrogen. Environ Sci Pollut Res 24(3):2549–2557. https://doi.org/10.1007/s11356-016-8031-0.

Haider SA, Lalarukh I, Amjad SF, Mansoora N, Naz M, Naeem M, Bukhari SA, Shahbaz M, Ali SA, Marfo TD, Subhan D, Rahul D, Fahad S (2021) Drought stress alleviation by potassium-nitrate-containing chitosan/montmorillonite microparticles confers changes in Spinacia oleracea L. Sustainability 13:9903. https://doi.org/10.3390/su13179903.

Hamza SM, Xiukang W, Sajjad A, Sadia Z, Muhammad N, Adnan M, Fahad S, et al. (2021) Interactive effects of gibberellic acid and NPK on morpho-physio-biochemical traits and organic acid exudation pattern in coriander (Coriandrum sativum L.) grown in soil artificially spiked with boron. Plant Physiol Biochem 167:884–900.

Harnos N, Bencze S, Janda T, Juhász A, Veisz O (2002). Interactions between elevated $CO_2$ and water stress in two winter wheat cultivars differing in drought resistance. Cereal Res Commun 30:359–366.

Hashimoto M, Negi J, Young J, Israelsson M, Schroeder J, Iba K (2006) Arabidopsis HT1 kinase controls stomatal movements in response to $CO_2$. Nat Cell Biol 8:391–397.

Hashimoto-Sugimoto M, Higaki T, Yaeno T, Nagami A, Irie M, Fujimi M, et al. (2013) A Munc13-like protein in Arabidopsis mediated H+-ATPase translocation that is essential for stomatal responses. Nat Commun 4:2215.

Hesham FA, Fahad S (2020) Melatonin application enhances biochar efficiency for drought tolerance in maize varieties: Modifications in physio-biochemical machinery. Agron J 112(4):1–22.

Hikosaka K, Kinugasa T, Oikawa S, Onoda Y, Hirose T (2011) Effects of elevated $CO_2$ concentration on seed production in C3 annual plants. J Exp Bot 62:1523–1530. https://doi.org/10.1093/jxb/erq401.

Hocking PJ, Meyer, CP (1985) Responses of Noogoora burr (Xanthium occidentale Bertol.) to nitrogen supply and carbon dioxide enrichment. Ann Bot (N. S.) 55:835–844.

Huang Li Y, Li Xiao X, Zhang Yun B, Fahad S, Wang F (2021) dep1 improves rice grain yield and nitrogen use efficiency simultaneously by enhancing nitrogen and dry matter translocation. J Integrative Agric. https://doi.org/10.1016/S2095-3119(21)63795-4.

Hussain MA, Fahad S, Rahat S, Muhammad FJ, Muhammad M, Qasid A, Ali A, Husain A, Nooral A, Babatope SA, Changbao S, Liya G, Ibrar A, Zhanmei J, Juncai H (2020) Multifunctional role of brassinosteroid and its analogues in plants. Plant Growth Regul. https://doi.org/10.1007/s10725-020-00647-8.

Ibrahim M, Jamal Y, Basir A, Adnan M (2016) Response of Sesame (Sesamumindicuml.) to various levels of Nitrogen and Phosphorus in agro-climatic condition of Peshawar. Pure Appl Biol 5(1):121–126.

Ibrar K, Aneela R, Khola Z, Urooba N, Sana B, Rabia S, Ishtiaq H, Mujaddad Ur Rehman, Salvatore M (2020) Microbes and environment: Global warming reverting the frozen zombies. In: Fahad S, Hasanuzzaman M, Alam M, Ullah H, Saeed M, Khan AK, Adnan M (eds) Environment, climate, plant and vegetation growth. Springer Publ Ltd, Springer Nature Switzerland AG. Part of Springer Nature. pp 607–634. https://doi.org/10.1007/978-3-030-49732-3.

Ikram U, Khadim D, Muhammad T, Muhammad S, Fahad S (2021) Gibberellic acid and urease inhibitor optimize nitrogen uptake and yield of maize at varying nitrogen levels under changing climate. Environ Sci Pollut Res. https://doi.org/10. 1007/s11356-021-16049-w.

Ilyas M, Mohammad N, Nadeem K, Ali H, Aamir HK, Kashif H, Fahad S, Aziz K, Abid U (2020) Drought tolerance strategies in plants: A mechanistic approach. J Plant Growth Regul. https://doi.org/10.1007/s00344-020-10174-5.

Iqra M, Amna B, Shakeel I, Fatima K, Sehrish L, Hamza A, Fahad S (2020) Carbon cycle in response to global warming. In: Fahad S, Hasanuzzaman M, Alam M, Ullah H, Saeed M, Khan AK, Adnan M (eds) Environment, climate, plant and vegetation growth. Springer Publ Ltd, Springer Nature Switzerland AG. Part of Springer Nature. pp 1–16. https://doi.org/10.1007/978-3-030-49732-3.

Irfan M, Muhammad M, Muhammad JK, Khadim MD, Dost M, Ishaq AM, Waqas A, Fahad S, Saud S, et al. (2021) Heavy metals immobilization and improvement in maize (Zea mays L.) growth amended with biochar and compost. Sci Rep 11:18416.

Jabborova D, Sulaymanov K, Sayyed RZ, Alotaibi SH, Enakiev Y, Azimov A, Jabbarov Z, Ansari MJ, Fahad S, Danish S, et al. (2021) Mineral fertilizers improves the quality of turmeric and soil. Sustainability 13:9437. https://doi.org/10.3390/su13169437.

Jablonski LM, Wang X, et al. (2002) Plant reproduction under elevated $CO_2$ conditions: A meta-analysis of reports on 79 crop and wild species. New Phytol 156:9–26.

Jan M, Muhammad Anwar-ul-Haq, Adnan NS, Muhammad Y, Javaid I, Xiuling L, Depeng W, Fahad S (2019) Modulation in growth, gas exchange, and antioxidant activities of salt-stressed rice (Oryza sativa L.) genotypes by zinc fertilization. Arab J Geosci 12:775. https://doi.org/10.1007/s12517-019-4939-2.

Janssens IA, Crookshanks M, Taylor G, Ceulemans R (1998) Elevated $CO_2$ increases fine root production, respiration, rhi-zosphere respiration and soil $CO_2$ efflux in Scots pine seedlings. Global Change Biol 4:871–878.

Jones P, Allen LH, Jr, Jones JW (1985) Responses of a soybean canopy photosynthesis and transpiration to whole-day temperature changes in different $CO_2$ environments. Agron J 77:242–249.

Kamran M, Wenwen C, Irshad A, Xiangping M, Xudong Z, Wennan S, Junzhi C, Shakeel A, Fahad S, Qingfang H, Tiening L (2017) Effect of paclobutrazol, a potential growth regulator on stalk mechanical strength, lignin accumulation and its relation with lodging resistance of maize. Plant Growth Regul 84:317–332. https://doi.org/10.1007/s10725-017-0342-8.

Khadim D, Fahad S, Jahangir MMR, Iqbal M, Syed SA, Shah AK, Ishaq AM, Rahul D, et al. (2021a) Biochar and urease inhibitor mitigate $NH_3$ and $N_2O$ emissions and improve wheat yield in a urea fertilized alkaline soil. Sci Rep 11:17413.

Khadim D, Saif-ur-R, Fahad S, Syed SA, Shah AK, et al. (2021b) Influence of variable biochar concentration on yield-scaled nitrous oxide emissions, Wheat yield and nitrogen use efficiency. Sci Rep 11:16774.

Khan MMH, Niaz A, Umber G, Muqarrab A, Muhammad AA, Muhammad I, Shabir H, Shah F, Vibhor A, Shams H. A-H, Reham A, Syed MBA, Nadiyah MA, Ali TKZ, Subhan D, Rahul D (2021) Synchronization of Boron application methods and rates is environmentally friendly approach to improve quality attributes of Mangifera indica L. On sustainable basis. Saudi J Bio Sci. https://doi.org/10.1016/j.sjbs.2021.10.036.

Koch KE, Jones PH, Avigne WT, Allen LH (1986) Growth and dry matter partitioning and diurnal activities of RuBP carboxylase in citrus seedlings maintained at two levels of $CO_2$. Physiol Plant 67:477–484.

Leakey ADB (2009) C4 crop responses to rising [$CO_2$]. Proc Royal Soc B: Biol Sci. https://doi.org/10.1098/rspb.2008.1517.

Leakey ADB, Xu F, Gillespie KM, McGrath JM, Ainsworth EA, Ort DR (2009) The genomic basis for stimulated respiratory carbon loss to the atmosphere by plants growing under elevated [$CO_2$]. Proc Nat Acad Sci USA, 106:3597–3602.

Long SP, Ainsworth EA, Leakey ADB, Ort DR (2006) Food for thought: Lower-than-expected crop yield stimulation with rising $CO_2$ conditions. Science 312:1918–1921.

Long SP, Ainsworth EA, Rogers A, Ort DR (2004) Rising atmospheric carbon dioxide: Plants FACE the future. Ann Rev Plant Biol 55:591–628.

Long SP, Ort DR (2010) More than taking the heat: Crops and global change. Curr Opin Plant Biol 13:1–8.

Lüthi D, et al. (2008) High-resolution carbon dioxide concentration record 650,000–800,000 years before present. Nat 453:379–382.

Mahar A, Amjad A, Altaf HL, Fazli W, Ronghua L, Muhammad A, Fahad S, Muhammad A, Rafiullah, Imtiaz AK, Zengqiang Z (2020) Promising technologies for Cd-contaminated soils: Drawbacks and possibilities. In: Fahad S, Hasanuzzaman M, Alam M, Ullah H, Saeed M, Khan AK, Adnan M (eds) Environment, climate, plant and vegetation growth. Springer Publ Ltd, Springer Nature Switzerland AG. Part of Springer Nature. pp 63–92. https://doi.org/10.1007/978-3-030-49732-3.

Mahmood Ul H, Tassaduq R, Chandni I, Adnan A, Muhammad A, Muhammad MA, Muhammad H-ur-R, Mehmood AN, Alam S, Fahad S (2021) Linking plants functioning to adaptive responses under heat stress conditions: A mechanistic review. J Plant Growth Regul. https://doi.org/10.1007/s00344-021-10493-1.

Makino A, Mae T (1999) Photosynthesis and plant growth at elevated levels of $CO_2$. Plant Cell Physiol 40:999–1006.

Manzer HS, Saud A, Soumya M, Abdullah A. Al-A, Qasi DA, Bander MA. Al-M, Hayssam MA, Hazem MK, Fahad S, Vishnu DR, Om PN (2021) Molybdenum and hydrogen sulfide synergistically mitigate arsenic toxicity by modulating defense system, nitrogen and cysteine assimilation in faba bean (Vicia faba L.) seedlings. Environ Pollut 290:117953.

Md Jakir H, Allah B (2020) Development and applications of transplastomic plants; a way towards eco-friendly agriculture. In: Fahad S, Hasanuzzaman M, Alam M, Ullah H, Saeed M, Khan AK, Adnan M (eds) Environment, climate, plant and vegetation growth. Springer Publ Ltd, Springer Nature Switzerland AG. Part of Springer Nature. pp 285–322. https://doi.org/10.1007/978-3-030-49732-3.

Md. Enamul H, Shoeb AZM, Mallik AH, Fahad S, Kamruzzaman MM, Akib J, Nayyer S, K. M. Mehedi A, Swati AS, Md Yeamin A, Most SS (2020) Measuring vulnerability to environmental hazards: Qualitative to quantitative. In: Fahad S, Hasanuzzaman M, Alam M, Ullah H, Saeed M, Khan AK, Adnan M (eds) Environment, climate, plant and vegetation growth. Springer Publ Ltd, Springer Nature Switzerland AG. Part of Springer Nature. pp 421–452. https://doi.org/10.1007/978-3-030-49732-3.

Moore BD, Cheng SH, Seemann JR (1998) Sucrose cycling, Rubisco expression, and prediction of photosynthetic acclimation to elevated atmospheric $CO_2$. Plant Cell Environ 21:905–915.

Mubeen M, Ashfaq A, Hafiz MH, Muhammad A, Hafiz UF, Mazhar S, Muhammad Sami ul Din, Asad A, Amjed A, Fahad S, Wajid N (2020) Evaluating the climate change impact on water use efficiency of cotton-wheat in semi-arid conditions using DSSAT model. J Water Climate Change. https://doi.org/10.2166/wcc.2019.179/622035/jwc2019179.pdf.

Muhammad N, Muqarrab A, Khurram S, Fiaz A, Fahim N, Muhammad A, Shazia A, Omaima N, Sulaiman AA, Fahad S, Subhan D, Rahul D (2021) Kaolin and Jasmonic acid improved cotton productivity under water stress conditions. J Saudi Soc Agric Sci 28:6606–6614. https://doi.org/10.1016/j.sjbs.2021.07.043.

Muhammad Z, Abdul MK, Abdul MS, Kenneth BM, Muhammad S, Shahen S, Ibadullah J, Fahad S (2019) Performance of Aeluropus lagopoides (mangrove grass) ecotypes, a potential turfgrass, under high saline conditions. Environ Sci Pollut Res. https://doi.org/10.1007/s11356-019-04838-3.

Muhammad Tahir ul Q, Amna F, Amna B, Barira Z, Xitong Z, Ling-Ling C (2020) Effectiveness of conventional crop improvement strategies vs. omics. In: Fahad S, Hasanuzzaman M, Alam M, Ullah H, Saeed M, Khan AK, Adnan M (eds) Environment, climate, plant and vegetation growth. Springer Publ Ltd, Springer Nature Switzerland AG. Part of Springer Nature. pp 253–284. https://doi.org/10.1007/978-3-030-49732-3.

Murray MB, Cannell MGR, Smith RI (1989) Date of bud burst of fifteen tree species in Britain following climatic warming. J Appl Ecol 26:693–700.

Murray MB, Smith RI, Leith ID, Fowler D, Lee HSJ, Friend AD, et al. (1994) Effects of elevated $CO_2$, nutrition and climate warming on bud phenology in Sitka spruce (Picea sitchensis) and their impact on the risk of frost damage. Tree Physiol 14:691–706. https://doi.org/10.1093/treephys/14.7-8-9.691.

Muzammal R, Fahad S, Guanghui D, Xia C, Yang Y, Kailei T, Lijun L, Fei-Hu L, Gang D (2021) Evaluation of hemp (Cannabis sativa L.) as an industrial crop: A review. Environ Sci Pollut Res. https://doi.org/10.1007/s11356-021-16264-5.

Negi J, Hashimoto-Sugimoto M, Kusumi K, Iba K (2014) New approaches to the biology of stomatal guard cells. Plant Cell Phyisol 55:241–250.

Negi J, Matuda O, Nagasawa T, Oba Y, Takahashi H, Kawai-Yamada M, et al. (2008) $CO_2$ regulator SLAC1 and its homologues are essential for anion homeostasis in plant cells. Nature 452:483–486.

Negi J, Morikawa K, Konishi M, Yokoyama R, Nakano T, Kusumi K, et al. (2013) A Dof transcription factor, SCAP1, is essential for the development of functional stomata in Arabidopsis. Curr Biol 23:479–486.

Noor M, Naveed ur R, Ajmal J, Fahad S, Muhammad A, Fazli W, Saud S, Hassan S (2020) Climate change and costal plant lives. In: Fahad S, Hasanuzzaman M, Alam M, Ullah H, Saeed M, Khan AK, Adnan M (eds) Environment, climate, plant and vegetation growth. Springer Publ Ltd, Springer Nature Switzerland AG. Part of Springer Nature. pp 93–108. https://doi.org/10.1007/978-3-030-49732-3.

Norby RJ, Neill EGO (1989) Growth dynamics and water use of seedlings of Quercus alba L. in $CO_2$-enriched atmospheres. New Phytol 111:491–500.

Nowak RS, Ellsworth DS, Smith SD (2004) Functional responses of plants to elevated atmospheric $CO_2$-do photosynthetic and productivity data from FACE experiments support early predictions? New Phytol 162:253–280.

Pettersson R, McDonald AJS, Stadenberg I (1993) Response of small birch plants (Betula pendula Roth.) to elevated $CO_2$ and nitrogen supply. Plant Cell Environ 16:1115–1121.

Poorter H, Navas ML (2003) Plant growth and competition at elevated $CO_2$: On winners, losers and functional groups. New Phytol 157:175–198.

Qamar uz Z, Zubair A, Muhammad Y, Muhammad ZI, Abdul K, Fahad S, Safder B, Ramzani PMA, Muhammad N (2017) Zinc biofortification in rice: Leveraging agriculture to moderate hidden hunger in developing countries. Arch Agron Soil Sci 64:147–161. https://doi.org/10.1080/03650340.2017.1338343.

Rashid M, Qaiser H, Khalid SK, Al-Wabel MI, Zhang A, Muhammad A, Shahzada SI, Rukhsanda A, Ghulam AS, Shahzada MM, Sarosh A, Muhammad FQ (2020) Prospects of biochar in alkaline soils to mitigate climate change. In: Fahad S, Hasanuzzaman M, Alam M, Ullah H, Saeed M, Khan AK, Adnan M (eds) Environment, climate, plant and vegetation growth. Springer Publ Ltd, Springer Nature Switzerland AG. Part of Springer Nature. pp 133–150. https://doi.org/10.1007/978-3-030-49732-3.

Rehman M, Fahad S, Saleem MH, Hafeez M, Muhammad Habib ur R, Liu F, Deng G (2020) Red light optimized physiological traits and enhanced the growth of ramie (Boehmeria nivea L.). Photosynthetica 58(4):922–931.

Rogers HH, Runion GB, Krupa SV (1994) Plant responses to atmospheric $CO_2$ enrichment with emphasis on roots and the rhi-zosphere. Environ Pollut 83:155–189.

Sadam M, Muhammad Tahir ul Q, Ghulam M, Muhammad SK, Faiz AJ (2020) Role of biotechnology in climate resilient agriculture. In: Fahad S, Hasanuzzaman M, Alam M, Ullah H, Saeed M, Khan AK, Adnan M (eds) Environment, climate, plant and vegetation growth. Springer Publ Ltd, Springer Nature Switzerland AG. Part of Springer Nature. pp 339–366. https://doi.org/10.1007/978-3-030-49732-3.

Safi UK, Ullah F, Mehmood S, Fahad S, Ahmad Rahi A, Althobaiti F, et al. (2021) Antimicrobial, anti-oxidant and cytotoxic properties of Chenopodium glaucum L.. PLoS One 16(10):e0255502. https://doi.org/10.1371/journal.pone.0255502.

Sahrish N, Shakeel A, Ghulam A, Zartash F, Sajjad H, Mukhtar A, Muhammad AK, Ahmad K, Fahad S, Wajid N, Sezai E, Carol Jo W, Gerrit H (2022) Modeling the impact of climate warming on potato phe-nology. Eur J Agron 132:126404.

Sajid H, Jie H, Jing H, Shakeel A, Satyabrata N, Sumera A, Awais S, Chunquan Z, Lianfeng Z, Xiaochuang C, Qianyu J, Junhua Z (2020) Rice production under climate change: Adaptations and mitigating strategies. In: Fahad S, Hasanuzzaman M, Alam M, Ullah H, Saeed M, Khan AK, Adnan M (eds) Environment, Climate, Plant and Vegetation Growth. Springer Publ Ltd, Springer Nature Switzerland AG. Part of Springer Nature. pp 659–686. https://doi.org/10.1007/978-3-030-49732-3.

Sajjad H, Muhammad M, Ashfaq A, Waseem A, Hafiz MH, Mazhar A, Nasir M, Asad A, Hafiz UF, Syeda RS, Fahad S, Depeng W, Wajid N (2019) Using GIS tools to detect the land use/land cover changes during forty years in Lodhran district of Pakistan. Environ Sci Pollut Res. https://doi.org/10.1007/s11356-019-06072-3.

Saleem MH, Fahad S, Adnan M, Mohsin A, Muhammad SR, Muhammad K, Qurban A, Inas AH, Parashuram B, Mubassir A, Reem MH (2020a) Foliar application of gibberellic acid endorsed phytoextraction of copper and alleviates oxidative stress in jute (Corchorus capsularis L.) plant grown in highly copper-contaminated soil of China. Environ Sci Pollut Res. https://doi.org/10.1007/s11356-020-09764-3.

Saleem MH, Fahad S, Shahid UK, Mairaj D, Abid U, Ayman ELS, Akbar H, Analía L, Lijun L (2020c) Copper-induced oxidative stress, initiation of antioxidants and phytoremediation potential of flax (Linum usitatissimum L.) seedlings grown under the mixing of two different soils of China. Environ Sci Poll Res. https://doi.org/10.1007/s11356-019-07264-7.

Saleem MH, Rehman M, Fahad S, Tung SA, Iqbal N, Hassan A, Ayub A, Wahid MA, Shaukat S, Liu L, Deng G (2020b) Leaf gas exchange, oxidative stress, and physiological attributes of rapeseed (Brassica napus L.) grown under different light-emitting diodes. Photosynthetica 58(3):836–845.

Saman S, Amna B, Bani A, Muhammad Tahir ul Qamar, Rana MA, Muhammad SK (2020) QTL mapping for abiotic stresses in cereals. In: Fahad S, Hasanuzzaman M, Alam M, Ullah H, Saeed M, Khan AK, Adnan M (eds) Environment, climate, plant and vegetation growth. Springer Publ Ltd, Springer Nature Switzerland AG. Part of Springer Nature. pp 229–252. https://doi.org/10.1007/978-3-030-49732-3.

Samuelson LJ, Seiler JR (1993) Interactive role of elevated $CO_2$, nutrient limitations, and water stress in the growth responses of red spruce seedlings. For Sci 39:348–358

Saud S, Chen Y, Fahad S, Hussain S, Na L, Xin L, Alhussien SA (2016) Silicate application increases the photosynthesis and its associated metabolic activities in Kentucky bluegrass under drought stress and post-drought recovery. Environ Sci Pollut Res 23(17):17647–17655. https://doi.org/10.1007/s11356-016-6957-x.

Saud S, Chen Y, Long B, Fahad S, Sadiq A (2013) The different impact on the growth of cool season turf grass under the various conditions on salinity and drought stress. Int J Agric Sci Res 3:77–84.

Saud S, Fahad S, Cui G, Chen Y, Anwar S (2020) Determining nitrogen isotopes discrimination under drought stress on enzymatic activities, nitrogen isotope abundance and water contents of Kentucky bluegrass. Sci Rep 10:6415. https://doi.org/10.1038/s41598-020-63548-w.

Saud S, Fahad S, Yajun C, Ihsan MZ, Hammad HM, Nasim W, Amanullah Jr, Arif M, Alharby H (2017) Effects of nitrogen supply on water stress and recovery mechanisms in Kentucky bluegrass plants. Front Plant Sci 8:983. https://doi.org/10.3389/fpls.2017.00983.

Saud S, Li X, Chen Y, Zhang L, Fahad S, Hussain S, Sadiq A, Chen Y (2014) Silicon application increases drought tolerance of Kentucky bluegrass by improving plant water relations and morph physiological functions. SciWorld J 2014:1–10. https://doi.org/10.1155/2014/368694.

Senol C (2020) The effects of climate change on human behaviors. In: Fahad S, Hasanuzzaman M, Alam M, Ullah H, Saeed M, Khan AK, Adnan M (eds) Environment, climate, plant and vegetation growth. Springer Publ Ltd, Springer Nature Switzerland AG. Part of Springer Nature. pp 577–590. https://doi.org/10.1007/978-3-030-49732-3

Shafi MI, Adnan M, Fahad S, Fazli W, Ahsan K, Zhen Y, Subhan D, Zafar-ul-Hye M, Martin B, Rahul D (2020) Application of single superphosphate with humic acid improves the growth, yield and phosphorus uptake of wheat (Triticum aestivum L.) in calcareous soil. Agron 10:1224. https://doi.org/10.3390/agronomy10091224.

Shah F, Lixiao N, Kehui C, Tariq S, Wei W, Chang C, Liyang Z, Farhan A, Fahad S, Huang J (2013) Rice grain yield and component responses to near 2°C of warming. Field Crop Res 157:98–110.

Sidra K, Javed I, Subhan D, Allah B, Syed IUSB, Fatma B, Khaled DA, Fahad S, Omaima N, Ali TKZ, Rahul D (2021) Physio-chemical characterization of indigenous agricultural waste materials for the development of potting media. J Saudi Soc Agric Sci. https://doi.org/10.1016/j.sjbs.2021.08.058.

Sionit N, Strain BR, Hellmers H, Riechers GH, Jaeger CH (1985) Long-term atmospheric $CO_2$ enrichment affects the growth and development of Liquidambar styraciflua and Pinus taeda seed-lings. Can J For Res 15:468–471.

Subhan D, Zafar-ul-Hye M, Fahad S, Saud S, Martin B, Tereza H, Rahul D (2020) Drought stress alleviation by ACC deaminase producing Achromobacter xylosoxidans and Enterobacter cloacae, with and without timber waste biochar in maize. Sustainability 12(6286). https://doi.org/10.3390/su12156286.

Tariq M, Ahmad S, Fahad S, Abbas G, Hussain S, Fatima Z, Nasim W, Mubeen M, ur Rehman MH, Khan MA, Adnan M (2018) The impact of climate warming and crop management on phenology of sunflower-based cropping systems in Punjab, Pakistan. Agric For Meterol 256:270–282.

Taub DR, Miller B, Allen H (2008) Effects of elevated $CO_2$ on the protein concentration of food crops: A meta-analysis. Glob Change Biol 14:565–575.

Unsar NU, Muhammad R, Syed HMB, Asad S, Mirza AQ, Naeem I, Muhammad H ur R, Fahad S, Shafqat S (2020) Insect pests of cotton crop and management under climate change scenarios. In: Fahad S, Hasanuzzaman M, Alam M, Ullah H, Saeed M, Khan AK, Adnan M (eds) Environment, climate, plant and vegetation growth. Springer Publ Ltd, Springer Nature Switzerland AG. Part of Springer Nature. pp 367–396. https://doi.org/10.1007/978-3-030-49732-3.

von Caemmerer S (2003) C4 photosynthesis in a single C3 cell is theoretically inefficient but may ameliorate internal $CO_2$ diffusion limitations of C3 leaves. Plant Cell Environ 26:1191–1197.

Wahid F, Fahad S, Subhan D, Adnan M, Zhen Y, Saud S, Manzer HS, Martin B, Tereza H, Rahul D (2020) Sustainable management with mycorrhizae and phosphate solubilizing bacteria for enhanced phosphorus uptake in calcareous soils. Agriculture 10(334). https://doi.org/10.3390/agriculture10080334.

Wajid N, Ashfaq A, Asad A, Muhammad T, Muhammad A, Muhammad S, Khawar J, Ghulam MS, Syeda RS, Hafiz MH, Muhammad IAR, Muhammad ZH, Muhammad Habib ur R, Veysel T, Fahad S, Suad S, Aziz K, Shahzad A (2017) Radiation efficiency and nitrogen fertilizer impacts on sunflower crop in contrasting environments of Punjab. Pak Environ Sci Pollut Res 25:1822–1836. https://doi.org/10.1007/s11356-017-0592-z.

Wu C, Kehui C, She T, Ganghua L, Shaohua W, Fahad S, Lixiao N, Jianliang H, Shaobing P, Yanfeng D (2020) Intensified pollination and fertilization ameliorate heat injury in rice (Oryza sativa L.) during the flowering stage. Field Crops Res 252:107795.

Wu C, Tang S, Li G, Wang S, Fahad S, Ding Y (2019) Roles of phytohormone changes in the grain yield of rice plants exposed to heat: A review. PeerJ 7:e7792. https://doi.org/10.7717/peerj.7792.

Yang Z, Zhang Z, Zhang T, Fahad S, Cui K, Nie L, Peng S, Huang J (2017) The effect of season-long temperature increases on rice cultivars grown in the central and southern regions of China. Front Plant Sci 8:1908. https://doi.org/10.3389/fpls.2017.01908.

Zafar-ul-Hye M, Akbar MN, Iftikhar Y, Abbas M, Zahid A, Fahad S, Datta R, Ali M, Elgorban AM, Ansari MJ, et al. (2021) Rhizobacteria inoculation and Caffeic acid alleviated drought stress in lentil plants. Sustainability 13:9603. https://doi.org/10.3390/su13179603.

Zafar-ul-Hye M, Muhammad N, Subhan D, Fahad S, Rahul D, Mazhar A, Ashfaq AR, Martin B, Jiˇrí H, Zahid HT, Muhammad N (2020a) Alleviation of cadmium adverse effects by improving nutrients uptake in bitter gourd through cadmium tolerant rhizobacteria. Environments 7(54). https://doi.org/10.3390/environments7080054.

Zafar-ul-Hye M, Tahzeeb-ul-Hassan M, Muhammad A, Fahad S, Martin B, Tereza D, Rahul D, Subhan D (2020b) Potential role of compost mixed biochar with rhizobacteria in mitigating lead toxicity in spinach. Sci Rep 10:12159. https://doi.org/10.1038/s41598-020-69183-9.

Zahida Z, Hafiz FB, Zulfiqar AS, Ghulam MS, Fahad S, Muhammad RA, Hafiz MH, Wajid N, Muhammad S (2017) Effect of water management and silicon on germination, growth, phosphorus and arsenic uptake in rice. Ecotoxicol Environ Saf 144:11–18.

Zahir SM, Zheng HG, Ala Ud D, Amjad A, Ata Ur R, Kashif J, Shah F, Saud S, Adnan M, Fazli W, Saud A, Manzer HS, Shamsher A, Wajid N, Hafiz MH, Fahad S (2021) Synthesis of silver nanoparticles using Plantago lanceolata extract and assessing their antibacterial and antioxidant activities. Sci Rep 11:20754.

Zia-ur-Rehman M (2020) Environment, climate change and biodiversity. In: Fahad S, Hasanuzzaman M, Alam M, Ullah H, Saeed M, Khan AK, Adnan M (eds) Environment, climate, plant and vegetation growth. Springer Publ Ltd, Springer Nature Switzerland AG. Part of Springer Nature. pp 473–502. https://doi.org/10.1007/978-3-030-49732-3.

Zvereva EL, Kozlov MV (2006) Consequences of simultaneous elevation of carbon dioxide and temperature for plant–herbivore interactions: A metaanalysis. Global Change Biol 12:27–41.

# 7

# *Effect of Environmental Pollution on Plant Growth*

**Amanullah, Shah Khalid, Imran, Hamdan Ali Khan, Muhammad Arif, Hamida Bibi, Nadia, Gohar Ayub, and Muhammad Sajid**
*The University of Agriculture*

**Abdel Rahman Altawaha**
*Al-Hussein Bin Talal University*

**Muhammad Adnan and Muhammad Zamin**
*University of Swabi*

**Shah Fahad**
*Hainan University*
*The University of Haripur*

**Azizullah Shah**
*Pakistan Agriculture Research Council*

**Brajendra Parmar**
*ICAR-Indian Institute of Rice Research, Soil Science*

**Sahar Mumtaz**
*University of Education*

## CONTENTS



DOI: 10.1201/9781003286400-7

## 7.1 Introduction

Air pollution is one of the most serious issues. It is harmful to the ecosystem (Tripathi and Gautam 2007). Recently, air pollutants cause injury to vegetation and decrease the yield of crops (Joshi and Swami 2007; Ilahi et al. 2020). Air pollution in urban areas has a deleterious effect throughout the world. The growth in the number of automobiles and industries is responsible for the release of toxic gases and other air pollutants into the atmosphere (Jahan and Iqbal 1992). Different sources, their percent contribution and their ultimate effect on crop productivity are given in Figures 7.1–7.3, respectively. Toxic gases are released in all the combustion processes including oxide of nitrogen, sulfur and oxide of carbon as well as other toxic heavy metals and isotopes of radioactive elements (Agbaire and Esiefarienrhe 2009). Increase in the world population results in increased road transportation and industries which leads to air pollutants (Adnan et al. 2016). Deleterious air pollution effects on plants and animals and their ecosystems have been widely studied. Most of the work has been done on the effects of air pollutants on plants growth at different levels ranging from biochemical to ecosystem levels. It is concluded that the concentration of ozone is higher in rural areas than urban areas, as concentrations of $NO_2$ and $SO_2$ are high in urban areas. Air pollution is one of the environmental stresses which mostly affect plant productivity (Woo and Lee 2007). Industrialization is mainly responsible for air pollution. Pollutants could be classified as either primary or secondary. Primary pollutants are those which are directly released into the air and pollute the air directly, while ones formed by the interaction of primary pollutants are called secondary pollutants (Agbaire 2009; Wang et al. 2021). Plants grown in urban areas are negatively affected by different air pollutants including $NO_2$, $SO_2$ and other toxic gases (Jahan and Iqbal 1992). Growth of plants is directly affected by air pollutants through leaves and indirectly by soil contaminations. Plants prior to the visible damage show physiological changes when exposed to air pollutants.

$SO_2$ in the air badly affect the different vital physiological and morphological process of plants. Increase in soil moisture and humidity intensified $SO_2$ plants injury. Release of toxic gases from automobiles and industries is the main source of air pollution, and plants are very susceptible to air pollution (Joshi and Swami 2009). Trees have a very important role in enhancing air quality in urban environments, by absorbing these toxic gases (Woo and Je 2006; Khan et al. 2017). Plants absorb various toxic air pollutants by providing large-surface leaf area and decreasing the concentration of air pollutants and are used as monitors of air pollution (Joshi and Swami 2009).

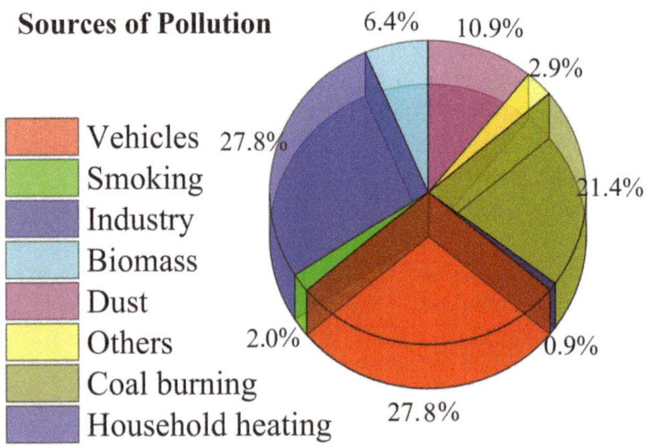


**FIGURE 7.1** Percent contribution of different sources of pollution (Rajper et al., 2018).

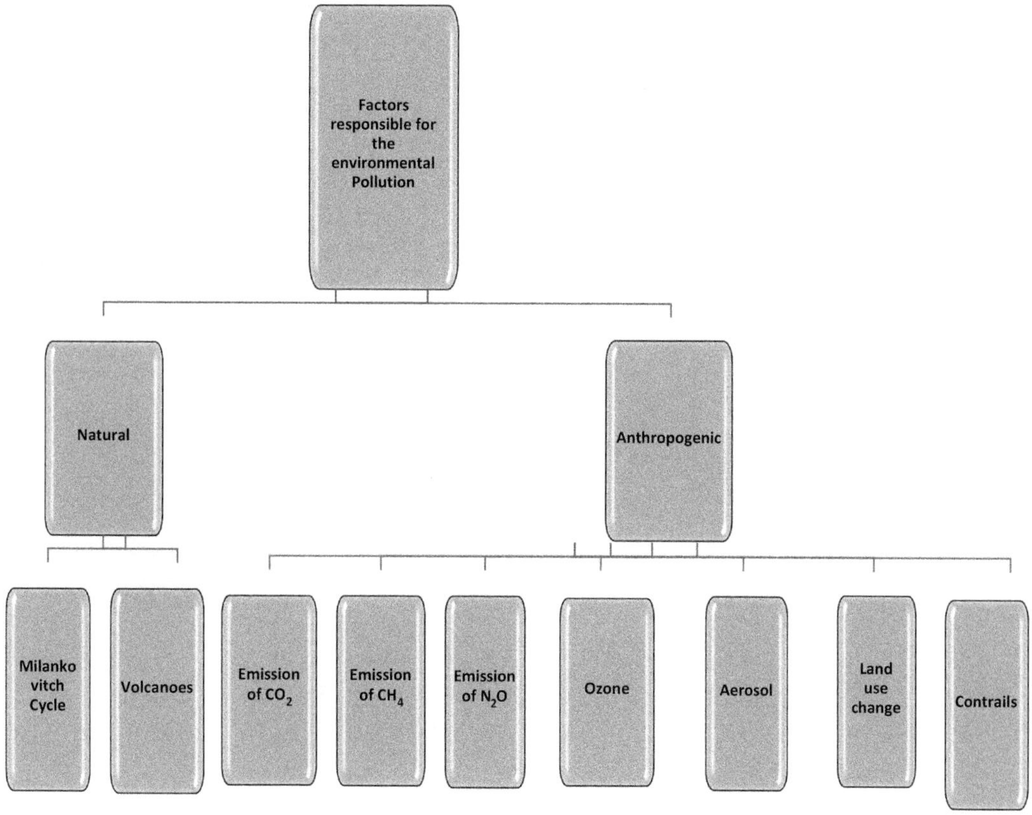


**FIGURE 7.2** Factors responsible for environmental pollution.

## 7.2 Deleterious Effects of Air Pollution on Plants

### 7.2.1 Leaf Morphology of Plants and Air Pollution

Air pollution is responsible for stomatal damage, leaf injury, decreased photosynthetic activity, premature senescence and reduced growth and yield in plant species. Decrease in the number of leaves and leaf area might be due to less production of leaves and an increased rate of senescence. Low leaf area decreases absorption of radiation and rate of photosynthesis (Hammad et al. 2015). Air pollution also decreases the length of petiole and leaf area (Dineva 2004).

Decrease in leaf area in the polluted air as compared with a clean environment is reported by previous researchers (Jahan and Iqbal 1992). Considerable decrease in area and length of petiole of *G. Officinale* was recorded under polluted air. $SO_2$ and heavy dust decrease leaf blade dimension of five trees of different species (Jahan and Iqbal 1992). Bhatti and Iqbal (1988) reported the exhaustive effect of the automobile on roadside trees. Seyyednejad et al. also reported a decrease in leaf area of *Albizia lebbeck* under air pollution conditions.

Growth of *Cassia siamea* plants at two sites—polluted and non-polluted—considerably exhibited the phonology of flowering and morphology of flowers (Chauhan et al. 2004). Increase in photosynthetic rates under $CO_2$ elevated conditions was abolished by leaf area reduction.

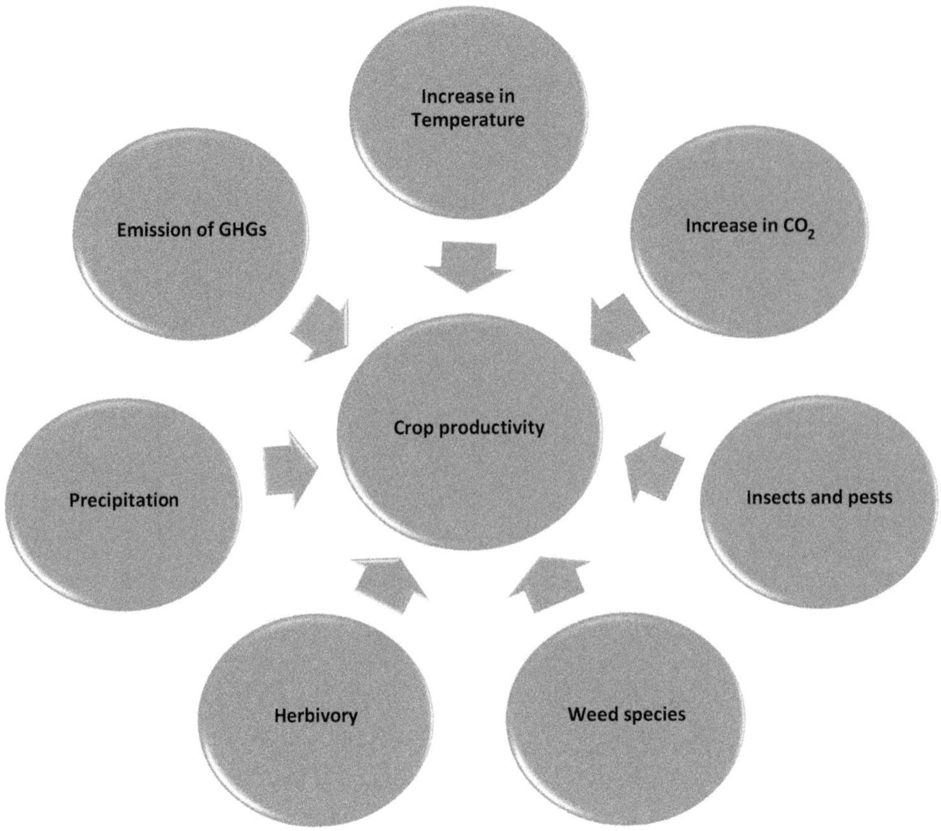


**FIGURE 7.3** Factors responsible for decreasing crop productivity.

### 7.2.2 Pigments Content and Air Pollution

Air pollution increases the closure of stomata, which decreases $CO_2$ concentration in the leaves and prevents the fixation of carbon. Net photosynthetic rate is mostly used as an indicator of air pollutants on the growth of the tree (Woo and Lee 2007). Plants under polluted conditions continuously absorb and integrate different toxic pollutants, which changes plant biochemical processes and integration of different metabolites in their system (Agbaire and Esiefarienrhe 2009). $SO_2$, NOX, $CO_2$ and other pollutants when captivated by the leaves may cause a decrease in chlorophyll and carotenoid concentration of leaves, which affect directly the productivity of plants (Joshi and Swami 2009; Honour et al. 2009; Arif et al. 2015).

Air pollution decreases the chlorophyll content of the leaves gradually which consequently decreases the photosynthetic rate of a plant (Joshi and Swami 2007; Joshi and Swami 2009; Basir et al. 2016). Chlorophyll concertation is the productivity index of a plant. Some of the condiments have chlorophyll content, while others decrease it (Agbaire and Esiefarienrhe 2009). Leaf surface shad by deposition of suspended pollutants decrease chlorophyll concentration, decrease photosynthesis rate, stimulate stomatal closer, increase leaf temperature and decrease stomatal gases exchange (Joshi and Swami 2009). Air pollutants enter the plant through the stomata which leads to chloroplast denaturation in polluted leaves. High concentration of $SO_2$ gases leads to chlorophyll demolition (Tripathi and Gautam 2007; Joshi and Swami 2007, 2009) most of the researches showed a decrease in contents of chlorophyll under air pollution (Joshi and Swami 2007, 2009; Tripathi and Gautam 2007). Yun exposed decrease in photosynthesis due to damage of PSII function in tobacco species. Free radical oxidative is responsible for the carotenoid reduction in plants (Fleschin et al. 2003).

Under air pollution several investigators have described decrease in carotenoid concentration (Joshi and Swami, 2009; Tripathi and Gautam, 2007).

### 7.2.3 Leaf Structure and Air Pollution

Air pollutants such as $SO_2$, heavy metals and ozone substantially damage leaves of plants by triggering chlorosis or yellowing of plant leaves abnormally which leads to deficiency of chlorophyll (Joshi and Swami 2009; Tripathi and Gautam 2007). Chlorophyll is an important component of the photosynthetic process. Chlorophyll prepared food for the plant by absorbing energy from the sun, without this molecule plants cannot assemble their food or energy. Increase in photosynthetic rates under $CO_2$ elevated conditions was abolished by leaf area reduction.

### 7.2.4 Flowering and Air Pollution

An exhaustive impact of air pollution on plants is delayed flowering. Growing of *Cassia siamea* plants at two sites, polluted and non-polluted exhibited considerable in phonology of flowering and morphology of flower (Chauhan et al. 2004). Plants under stress will not produce flowers but, somehow, will use their survival resources to retrieve the threat.

### 7.2.5 Root Damage and Air Pollution

Air pollutants such as oxides of sulfur and nitrogen cause acid rain, thus creating a very complex situation for the plant roots to uptake nutrients from the soil and affecting the availability of essential macro- and micronutrients. Mobility of the aluminum increases in acidic soil, which is commonly present in inert state, as a result of increased aluminum mobility to stimulate the system of roots and decrease absorption of calcium. Increase in the concentration of aluminum and other toxic heavy metals indirectly impacts on plant roots' absorbing capacity by decreasing the activity of soil microorganisms, which decreases the breakdown of organic matter and reduces the availability of plant nutrients.

### 7.2.6 Stomata Damage and Air Pollution

Each leaf has very small pores on either ventral or dorsal or both sides depending on plant species for the exchange of gases ($CO_2$ and $O_2$). During the processes of photosynthesis, $CO_2$ is taken up and $O_2$ is released as a by-product to the environment. Air pollution has negatively affected stomatal aperture size and density by decreasing the size of stomata and leaf area. As a result, the exchange of gases becomes slow during photosynthesis which leads to yield reduction in plants.

### 7.2.7 Plant Sugar and Air Pollution

All living organisms need soluble sugar for structure and energy purposes. Plants prepare carbohydrates during the process of photosynthesis and breakdown for the release of energy during the process of respiration (Tripathi and Gautam 2007; Tzvetkova and Kolarov 1996). Air pollution considerably decreases soluble sugar contents of plants (Tripathi and Gautam 2007). Decrease in soluble sugar might be due to increase in respiration and decrease in $CO_2$ absorption in polluted air because of chlorophyll corrosion. Air pollutants like $NO_2$, $H_2S$ and $SO_2$ under stress conditions considerably decrease leaf-soluble sugars, and the reaction of ketones and aldehydes with air pollutant sulfate also reduces air-soluble sugar contents (Tripathi and Gautam 2007).

### 7.2.8 Proline and Air Pollution

Stress like low and high temperatures, air and soil pollutions, and drought can increase reactive oxygen species in cells of plants, which are tremendously reactive. More exposition plants to the air pollution pushed chloroplast high level of energy, which increases reactive oxygen species generation oxidation rate (Woo and Lee 2007). As a result of air pollutants, the production of reactive oxygen species increases, which causes increased cellular prooxidative destruction. Proline act as a scavenger to shelter plants from oxidative damage (Wang et al. 2009). Seyyednejad et al. reported a considerable increase in proline contents in *Albizia lebbeck* cultivated in polluted areas. It is acting a defensive reaction of plants in response to many stresses. Apparently, proline has a chief character in defense of various types of stresses. Proline accretion in plants is a functional comeback to osmotic stress, salt (Woodward and Bennett 2005) and heavy metals (Wang et al. 2009).

## 7.3 Conclusion

It is concluded that air pollution is one serious issue which is a result of an increasing world population, transportation and industries. Air pollution is responsible for stomatal damage, leaf injury, decreased photosynthetic activity, premature senescence, reduced growth and yield in plant species, decreased leaf area, increased rate of senescence and closure of stomata, and decreased chlorophyll and carotenoid concentration of leaves. Air pollution decreases the uptake of plant nutrients from the soil and increases reactive oxygen species.

## REFERENCES

Adnan M, Shah Z, Arif M, Khan MJ, Mian IA, Sharif M, Alam M, Basir A, Ullah H, Rahman I, Saleem N (2016) Impact of rhizobial inoculum and inorganic fertilizers on nutrients (NPK) availability and uptake in wheat crop. Can J Soil Sci 96:169–176.

Agbaire PO (2009) Air Pollution Tolerance Indices (APTI) of some plants around Erhoike-Kokori oil exploration site of Delta State, Nigeria. Int J Phy Sci 4:366–368.

Agbaire PO, Esiefarienrhe E (2009) Air Pollution Tolerance Indices of some plants around Otorogun gas plant in Delta state, Nigeria. J Appl Sci Environ Manage 13:11–14.

Arif M, Shah T, Ilyas M, Ahmad W, Mian AA, Jadoon MA, Adnan M (2015) Effect of organic manures and their levels on weeds density and maize yield. Pak J Weed Sci Res 21(4):517–522.

Basir A, Jan MT, Alam M, Shah AS, Afridi K, Adnan M, Ali K, Mian IA (2016) Impacts of tillage, stubble management and nitrogen on wheat production and soil properties. Can J Soil Sci 97(2):133–140.

Bhatti GH, Iqbal MZ (1988) Investigations into the effect of automobile exhausts on the phenology, periodicity and productivity of some roadside trees. Acta Soc Bot Bol 57:395–399.

Chauhan SVS, Bharati C, Anita R (2004) Impact of air pollution on floral morphology of Cassia siamea Lamk. J Envi Biol 25:291–297.

Dineva SB (2004) Comparative studies of the leaf morphology and structure of white ash *Fraxinus americana* L. and London plane tree Platanus acerifolia Willd growing in polluted area. J Den Bio 52:3–8.

Fleschin S, Fleschin M, Nita S, et al. (2003) Free radicals mediate protein oxidation in biochemistry. Roum Biotec Lett 5:479–495.

Hammad M, Ahmed N, Khan IA, Shah B, Khan A, Ali M, Rasheed MT, Junaid K, Adnan M, Zaki AB, Ahmed S (2015) Studies on the efficacy of selected insecticides against anopheles mosquitoes of village Goth. Bhoorji (Sindh) Pakistan. J Ento Zool Stud 3(6):169–173.

Honour SL, Bell JN, Ashenden TW et al. (2009) Responses of herbaceous plants to urban air pollution: Effects on growth, phenology and leaf surface characteristics. Environ Pollut 157:1279–1286.

Ilahi H, Wahid F, Ullah R, Adnan M, Ahmad J, Azeem M, Amin I, Amin H (2020) Evaluation of bacterial and fertility status of Rawalpindi cultivated soil with special emphasis on fluoride content. Int J Agric Environ Res 6(3):177–184.

Jahan S, Iqbal MZ (1992) Morphological and anatomical studies on leaves of different plants affected by motor vehicle exhusted. J Isl Acad Sci 5:21–23.

Joshi PC, Swami A (2007) Physiological responses of some tree species under roadside automobile pollution stress around city of Haridwar, India. Environmentalist 27:365–374.

Joshi PC, Swami A (2009) Air pollution induced changes in the photo synthetic pigments of selected plant species. J Environ Bio 30:295–298.

Khan A, Yousafzai AM, Shah N, Adnan M, Jehan S, Jawad SM (2017) Scrutinizing impact of atrazine on stress biomarker: Plasma glucose concentration of fresh water fish, grass carp, (*Ctenopharyngodonidella*). Pure Appl Biol 6(4):1319–1327.

Tripathi AK, Gautam M (2007) Biochemical parameters of plants as indicators of air pollution. J Environ Bio 28:127–132.

Tzvetkova N, Kolarov D (1996) Effect of air pollution on carbohydrate and nutrient concentrations in some deciduous tree spicies. J Plant Phy 22:53–63.

Wang F, Zeng B, Sun Z, et al (2009) Relationship between proline and Hg2+-induced oxidative stress in a tolerant rice mutant. Arch Environ Contam Toxicol 56:723–731.

Wang X, Saleem MH, Parveen A, Mumtaz S, Hassan A, Adnan M, Fiaz S, Ali S, Iqbal Khan Z, Ali S, Yasin G (2021) Proximate composition and nutritive value of some leafy vegetables from Faisalabad, Pakistan. Sustainability 13(15):8444.

Woo SY, Je SM (2006) Bhotosynthetic rates and antioxidant enzyme activity of Platanus occidentalis growing under two levels of air pollution along the streets of Seoul. J Plant Biol 49:315–319.

Woo SY, Lee DK (2007) Net photosynthetic rate, ascorbate peroxidase and glutathione reductase activities of Erythrina orientalis in polluted and non-polluted areas. Photosynthetica 45:293–295.

Woodward AJ, Bennett AT (2005) The effect of salt stress and abscisic acid on proline production, chlorophyll content and growth of in vitro propagated shoots of Eucalyptus camaldulensis. Plant Tis Cult 82:189–200.

# 8

# *Phyto-Ecological Studies of Genus* Phoenix *(Linn.) (Date Palms) from Various Zones of Pakistan*

**Abdullah, Shujaul Mulk Khan, Zahoor ul Haq, and Saifullah**
*Quaid-i-Azam University*

**CONTENTS**



## 8.1 Introduction

### 8.1.1 Historical Background

Date palm is the oldest known fruit crop and has been domesticated for about 5000 years in the Middle East and North Africa (Zohary and Hopf, 2000). The oldest record from Mesopotamia (Iraq) shows that probably date palm culture was established before 3000 BCE. Due to wide distribution and long history of date palm culture, the main origination point is unknown, but the most probable places of its origination are southern Iraq (ancient Mesopotamia) or western India (Wrigley, 1995). According to a well-known French-Swiss botanist Alphonse Pyrame de Candolle (1806–1893), the date palm was cultivated in prehistoric time from Western Africa (Senegal) to South Asia (Indus river basin). Genus *Phoenix* comprises 17 species, distributed in different areas of the globe such as southern areas of Mediterranean region, Canary Island, Kingdom of Arabia, throughout Africa, Middle East, Iran, Mesopotamia, Pakistan, Afghanistan, southern slopes of Himalaya, Assam, South to the Malayan Peninsula, China and N. Samatra Senegal, Mali, Egypt, Mauritania, Sudan, Algeria, Nigeria, Morocco, Chad, Libya, Iraq, Tunisia, Yemen, UAE and Oman since the ancient times. It is a dioecious palm involved in the economic and social aspects of people in different areas of the globe. In Pakistan, date palm plays an important role in the livelihoods and economy of the local inhabitants. It is grown in different areas of Baluchistan, Punjab, Khyber Pakhtunkhwa, Suleman range and semi-desert-like zones and plains of Sindh Province

DOI: 10.1201/9781003286400-8 

**FIGURE 8.1** (a) Date palm harvesting. (b) A scenic view of the date palm habitat. (c) Fruit bunches of date palm. (d) Date palm distribution in salt range. (e) Date tree with unripen fruit bunches. (f) Date palm unripen bunches in the Western Himalayan province.

(Markhand et al., 2010; Haider et al., 2013). Pakistan has unique territory and four seasons and great suitability to various native and exotic date palm varieties. Dhakki, Aseel, Koharba and Begum Jungi are among the native and highly significantly grown cultivars across the various ecological zones of the country. Date palm industry plays an important role, is well-adapted all over the country and gives 10.3% of the world's total date production which is 70 thousand tons (Iqbal et al., 2012; Khan et al. 2016). Pakistan is ranked the world's fifth-highest dates producer with a production of 70 thousand tons/year and contributes 9% shares to the world's total dates production. The Arabian rulers spread the date palm in Southern Europe during the 12–15th centuries. Furthermore, in the 18th century. date palms were domesticated from the West in the continent of America by the Spaniards. Currently, the Muslim countries are leading in the world in production of dates (Anonymous, 2011; Ahmad et al., 2015).

### 8.1.2 Taxonomic Detail of Genus Phoenix

#### *8.1.2.1 Genus Phoenix*

It is a solitary or clump-forming dioecious palm having feather-like leaves. The genus is characterized by the stout trunk, covered with petiole bases, uniformly thickened and rarely branched.

#### *8.1.2.2 Leaves*

Leaves are pinnate having numerous pinnae, unequal in length and width, elongate, lanceolate, coriaceous, induplicate, ensiform and folded longitudinally and attached to the petiole with folded bases (Figure 8.1). Usually, the lowest pinnae is modified into stout and sharp spines.

#### *8.1.2.3 Inflorescence*

*Phoenix* species have simply branched and of axillary nature inflorescence. Flowers are unisexual, small and sessile on the drooping or erect peduncle. Male flowers are ovoid or oblong, cup-shaped calyx, three-toothed. The calyx is shorter than petals, oblong or valvate. Stamens are usually six subulate filaments at the base united while anthers are erect. Female flowers have three imbricate petals, rounded, usually longer than calyx. Carpels are three apocarpous, stigmas are sessile; staminodes six in number or fused into a six-lobed cup; each carpel has one lobe.

#### *8.1.2.4 Fruit*

The fruit is a one-seeded berry, oblong or ovoid, pericarp fleshy with a membranous endocarp. Seeds are straight, ventrally grooved with homogeneous endosperm.

In Pakistan, the genus *Phoenix* is represented by four species, e.g. *Phoenix dactylifera* L., *Phoenix canariensis* Hort. ex Chabaud, *Phoenix sylvestris* (Linn.) Roxb and *Phoenix loureirii* Kunth, Enum.

### 8.1.3 Study Area Pakistan

#### *8.1.3.1 Geography and Climate*

Pakistan lies on a unique geographical position at 23°–37° N longitude and 61°–81° E latitude with an area of 796,096 km$^2$. The altitude of the country ranges from sea level to the second highest peak of the globe K2 with an altitude of 8611 m. Different zones of the country also have a great variability in the range of the temperature starting from well below freezing point up to 52°C in the district of Sibi Baluchistan. Annual mean rainfall ranges from 50 mm at the point of Nok Kundi in the province of Baluchistan up to a very maximum of 2032 mm in the uplands of Kashmir Himalaya during monsoon (Ali, 1978; Ali & Qaiser, 1986). The diverse flora and fauna is the result of the great variability in temperature, elevation, rainfall and other environmental variables (Ali, 1978) (Figure 8.2).

#### *8.1.3.2 Geology of Pakistan*

Pakistan is blessed with diverse mountainous ranges namely Himalaya, Hindukush, Karakoram, Toba Kakar, Sulaiman, Kirthar and Salt ranges. These mountain ranges host remarkable glaciers, peaks, gorges and biodiversity hotspots which play a significant role in the socio-economic growth of the country. Historically, these are very popular for their potential in the formation and deposition of metals, nonmetals or minerals diversity. Few of these hidden resources have been discovered and are being used for the well-being of the humans. On the basis of tectonic environment, few of the tectonostratigraphic/mineralogic zones have been developed in the region. Different types of metals are excavated from the hidden treasures of these mountains across the country such as chromite, gold, copper, antimony, iron, zinc, silver, manganese and other radioactive metals. On the other hand, graphite, fluorite, feldspar, limestone, mica, quartz, soapstone, rock phosphate, Kaoline, magnesite, talc, nepheline syenite and barite. Gemstone and decorative materials (ruby, peridote, tourmaline, emerald, aquamarine, epidote, marble, salt and coal) resources are found in the maximum amount in Pakistan. The districts of Buner, Mohmand, Bajaur and some other areas in northern Pakistan host different marble verities in a considerable amount. Khewra, Warcha and Kalabagh are great salt mines of Pakistan. These mines are not only a source of salts but also provide recreational values for tourists (Figure 8.3). From these ranges, a vast amount of salts are excavated. A medium quality of salts are also found in these salt ranges but the most popular places in Sind province are Lakhra, Thatta, Thar, Sonda, Jherruck, and Haji Coal, Central Salt Range, Eastern Salt Range and Makerwal in the Punjab province and Sor Range/Degari, Khost-Sharig-Harnai, Duki, Musakhel, Mach-Kingri, Pir Ismail Ziarat and Chamalong coal mine in the province of Baluchistan. Most parts of the country, in general, and tribal districts and Gilgit Baltistan, in particular, are still unexplored and need to be explored for the well-being and mineral wealth of the country (Shah and Jan, 2002; Khan et al., 2015; Khan et al., 2016).

### 8.1.4 Date Palm and Monsoon

Date palm mostly adopts dry and arid zones where the summer is characterized by blazing heat and winter with a little precipitation. Still, in some countries of the world, the climate in summer is not fully

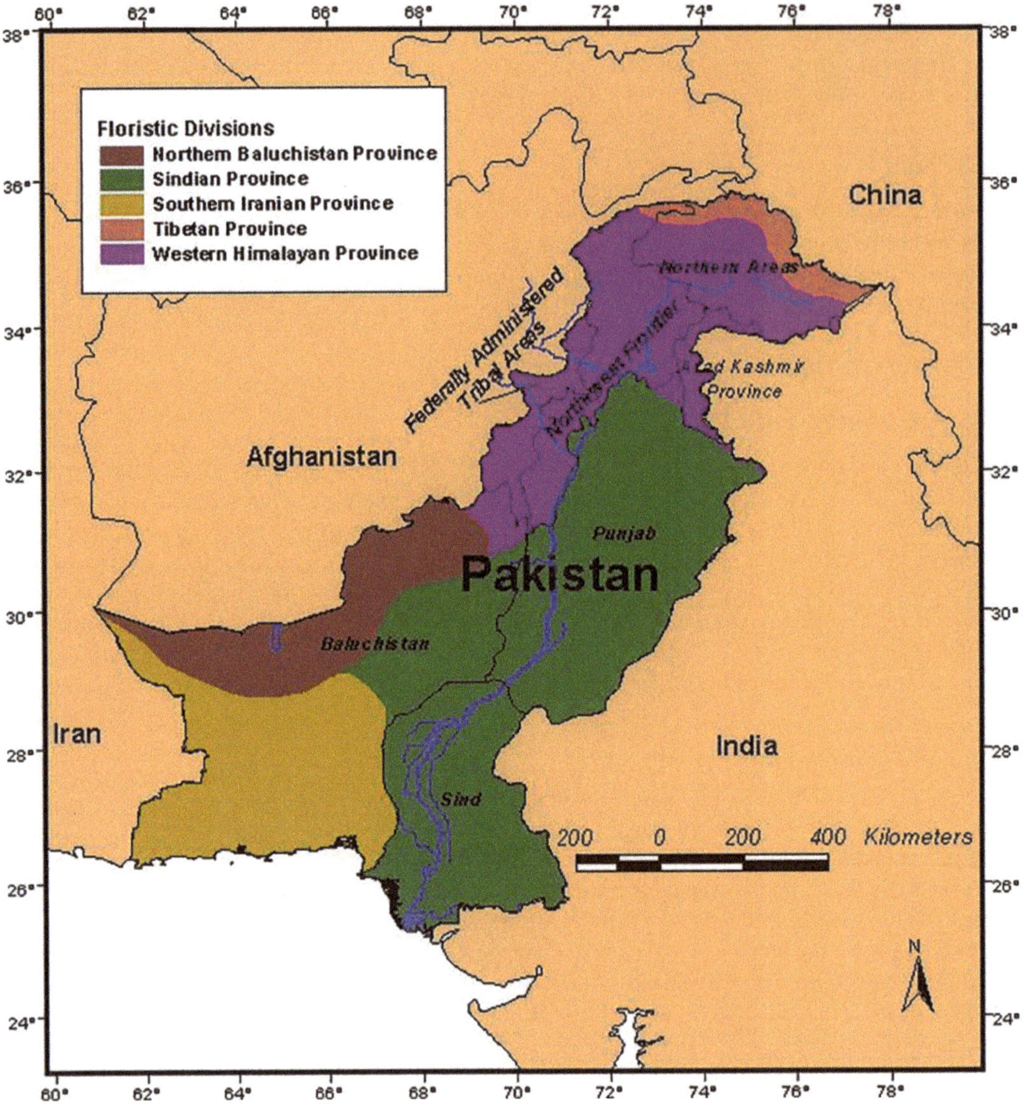


FIGURE 8.2 Floristic provinces of Pakistan (flora of Pakistan).

dried and monsoon rains take place, for example, in Oman, Pakistan and the coast of the Sahara Desert. Monsoon rains take place in July, August and September almost in the fruit-harvesting season of date palm. Due to precipitation, cracks form on the fruit within 1–2 days (Zaid and Dewet, 1999) and destroy all varieties in the stage of harvesting, while those varieties that are not mature and at green stage show resistance to precipitation. The date fruits are highly damaged by rain when they ripen late or if the rain occurs in the early season.

### 8.1.5 Temperature Requirements of Date Palm

The maximum highest temperatures of the areas is popular for dates growing having low humidity, long days and great insolation in the summer season. According to Zaid and de wet (1999), different varieties of the date palm are cultivated in semi-arid and arid zones in which the summer season is long and hot, with no or low precipitation and humidity especially during the ripening season of the date palm. Some

**FIGURE 8.3** (a) Scenic view of Khewra salt range. (b) Khewra mine internal view. (c) Various items processed from salt rocks. (d) Administration center of Khewra salt mines. (e) Historical view of the salt range. (f) Route for tourist train to the Khewra mine.

palm genera such as *Phoenix* and *Nannorrhops* can tolerate temperature above 50°C, while in winter, both genera can tolerate temperature below freezing point. According to Zaid and Dewet (1999), date is a thermophilic palm species and shows more flexibility and elasticity to fluctuations in temperature. Throughout the arid and semi-arid regions of the globe, date palm cultivation constitutes significant and highly successful agricultural practices. Leaves of date palm having a very thick cuticle and on the base of leaves appendages like spines take place which shows that it is highly suitable and native species of dry areas. It was also expected that the date palm is one of the first fruit trees domesticated in dry areas. Date palms have a great capacity to trap water from the soil that drainage rapidly due to their deep roots. Its root growth and development depend upon the characteristics of soil, genotype depth and availability of underground water. Zaid and de Wet give an example of date palm roots that spread 25 m from the plant and more than 6 m deep in the soil. These characteristics of the date palms contribute to the reduction of evaporation and the increase of water uptake. These features of date palms are very important for their adaptation to dry and hot weathers. Date palms are used by the local inhabitants throughout the year due to which its seeds are distributed all over the ecological zones and hence germinate almost in all districts of Pakistan, but fruit formation occurs only in some areas. In different areas of Khyber Pakhtunkhwa such as Hazara (Abbottabad, Mansehra) and Malakand division, due to cold climatic conditions, fruit formation does not take place. These areas are characterized by the well-established monsoon, maximum humidity and cold wind. On the other hand, some areas of Pakistan such as Islamabad and Rawalpindi receive monsoon, but the climate is relatively dry. Therefore, fruit formation occurs in date palm but does not reach the ripening stage at the right time.

### 8.1.6 Dates City, Khairpur

Khairpur of the Sind province is the leading district in the production of date palms. Every year in Khairpur during the harvesting season, many people are involved in the processing of date palms and a new city for about two months is developed. Everywhere in the area, it looks like a big festival. People

**TABLE 8.1**
Durations of Different Stages of Date Palm Life Cycle

| S. No. | Main Stages in the Date Palm Life Cycle | Duration |
|---|---|---|
| 1 | Suitable age of suckers for transplanting | 2–3 years |
| 2 | Time to start bearing fruits | 4–5 years |
| 3 | Time to full production | 6–8 years |
| 4 | Full productivity but no more offshoots formation | 30–35 years |
| 5 | Normal economic bearing life | 50 years |
| 6 | Declining fruit yields after | 50 years |
| 7 | Time of transplanting | Whole year, especially from May to July |
| 8 | Time of flowering | February to March |
| 9 | Time of harvest | July to October |

*Source:* Jatoi et al. (2009).

from other areas across the country come to the city of dates to earn money and achieve livelihood facilities. They are involved in fruit picking, harvesting and making Chohara while trucks loader are used to transfer the crop from the city of dates to other cities of the country.

### 8.1.7 Pollination in Date Palm

All varieties of dates are pollinated naturally by wind pollination method in the local oasis. On the other hand, in modern orchards formed for commercial purposes, it is pollinated by manual pollination methodologies. Furthermore, in the case of manual pollination, 40–50 female date palms are pollinated with the flowers of a single male date palm (Rani et al., 2007; Zeeshan et al., 2016). From pollination stage to fruit maturation of dates, it takes 6–7 months in a season (Saleem, 2004) (Table 8.1).

## 8.2 Conclusion

This study certifies that date palms remain massively used fruit plant species all over the globe. Distributions of date palms are strongly influenced by different ecological factors. Temperature, rainfall and humidity are the various ecological drivers responsible for the distribution of date palms in different ecological zones in Pakistan. The Sindian province provides suitable climatic conditions for the growth and production of date palm followed by the South Iranian floristic province. The remaining three provinces are characterized by monsoon rains and cold climatic conditions which are not suitable for the growth and ripening in dates. Finally, it is concluded that date palms are the native and suitable plant species of hot areas where the temperature ranges from 40°C to 50°C. We believe that our current study will be helpful for researchers and scholars working on date palms all over the world.

## REFERENCES

Ahmad K, Adnan M, Khan MA, Hussain Z, Junaid K, Saleem N, Ali M, Basir A (2015) Bioactive neem leaf powder enhances the shelf life of stored mungbean grains and extends protection from pulse beetle. Pak J Weed Sci Res 21(1):71–81.

Ali SI (1978) The Flora of Pakistan—some general and analytical remarks. Notes Roy Bot Gard Edinburgh 36:427–439.

Ali SI, Qaiser M (1986) A phytogeographical analysis of the phanerogams of Pakistan and Kashmir. Proc R Soc Edinburgh Sec B—Biol Sci 89:89–101.

Anonymous (2011) SBI, "Dates Drying & De-Hydration Plant. 5–42 Sindh Board of Investment Government of Sindh 108-F, PECHSII Karachi, Pakistan".

Haider MS, Khan IA, Naqvi SA, Jaskani MJ, Khan RW, Nafees M, Pasha I (2013) Fruit developmental stages effects on biochemical attributes in date palm. Pak J Agri Sci 50(4):577–583.

Iqbal M, Rahman UU, Niamatullah M, Khan SM, Akhtar A (2012) Periodic growth and development of fruits of different date cultivars grown under the agro-climatic conditions of DI Khan. Pak J Agric Sci 64(3):259–264.

Jatoi MA, Markhand Z, Solangi N (2009) Dates in Sindh—facts and figures. In: Proceedings of the international dates seminar, vol. 28, pp 59–70.

Khan A, Yousafzai AM, Shah N, Ahmad MS, Farooq M, Aziz F, Adnan M, Rizwan M, Jawad SM (2016) Enzymatic profile activity of grass carp (*Ctenopharyngodonldella*) after exposure to the pollutant named atrazine (herbicide). Pol J Environ Stud 25(5):2001–2006.

Khan MA, Basir A, Adnan M, Saleem N, Khan A, Shah SRA, Shah JA, Ali K (2015) Effect of tillage, organic and inorganic nitrogen on maize yield. Am Eurasian J Agric Environ Sci 15(12):2489–2494.

Khan N, Shah Z, Adnan M, Ali M, Khan B, Mian IA, Ali A, Zahoor M, Roman M, Ullah L, Khaliq A, Khan WA, Alam A (2016) Evaluation of soil for important properties and chromium concentration in the basin of chromite hills in lower Malakand. Adv Environ Biol 10(7):141–147.

Markhand GS, Abul-Soad AA, Mirbahar AA, Kanhar NA (2010) Fruit characterization of Pakistani dates. Pak J Bot 42(6):3715–3722.

Rani CI, Kalaiselvi T, Jegadeeswari V (2007) The date palm. Accessed at www.techno-preneur.net/informationdesk/.../2007/.../Date_palm.

Saleem SA (2004) Aspects of ripening of Dhakki dates (*Phoenix dactylifera* L.) and Post-harvest stability employing hurdle technology. Ph.D thesis, Food Technology Department, Gomal University, Dera Ismail Khan, Pakistan.

Shah M, Jan M (2002). Mineral resource potential of the Himalaya, Karakoram and Hindu Kush mountains, northern Pakistan.

Wrigley G (1995) Date palm. In: Smartt J, Simmonds, NW (eds) Evolution of crop plants, 2nd edn. Longman Group, Essex, UK, pp 399–403.

Zaid A, De Wet PF (1999) Chapter IV Climatic requirements of Date Palm. FAO Plant Production and Protection Papers 58–73.

Zeeshan MA, Khan I, Shah B, Naeem A, Khan N, Ullah W, Adnan M, Shah SRA, Junaid K, Iqbal M (2016) Study on the management of *Ralstoniasolanacearum*(Smith) with spent mushroom compost. J Ento Zool Stud 4(3):114–121.

Zohary D, M Hopf (2000) Domestication of plants in the old world—The origin and spread of cultivated plants in West Asia, Europe, and the Nile Valley. Oxford University Press, Oxon, UK.

Zohary M (1973) Geobotanical foundations of the middle east, 2 vols. Gustav Fischer Verlag Press, Stuttgart, Swets & Zeitlinger, Amsterdam.

# 9

## *Soil Biodiversity in Changing Climate*

**Ahmed Ragab Henawy and Asmaa A. Halema**
*Cairo University*

**CONTENTS**



## 9.1 Introduction

Climate change is an inconstant variation in the climate condition and preserving for a prolonged duration (Fahad and Bano 2012; Fahad et al. 2017; Fahad et al. 2013; Fahad et al. 2014a, b; Fahad et al. 2016a, b, c, d; Fahad et al. 2015a, b; Fahad et al. 2018a, b; Fahad et al. 2019a, b; Fahad et al. 2020; Fahad et al. 2021a, b, c, d, e, f; Hesham and Fahad 2020). Climate change causes due to the natural internal or external processes or continuous changes in land use or in the atmosphere (Karl and Trenberth 2003; Mora et al. 2013; Adnan et al. 2018a, b; Akram et al. 2018a, b; Aziz et al. 2017a; Habib-Ur-Rahman et al. 2017; Hafiz et al. 2016, 2019; Kamran et al. 2017; Muhammad et al. 2019; Sajjad et al. 2019; Saud et al. 2013; Saud et al. 2014; Saud et al. 2017; Saud et al. 2016; Shah et al. 2013; Qamar et al. 2017; Wajid et al. 2017; Yang et al. 2017; Zahida et al. 2017; Depeng et al. 2018). The greenhouse gases prevalence within the atmosphere may be a natural characterization of the climate system and assists to take care of our planet (Mora et al. 2013). Greenhouse gases enable the energy of the sun to pass across the atmosphere to the earth's surface because of its transparency to forthcoming solar radiation. Then, the surface of the earth can absorb it and use it in important processes. Apart from the emitted radiation passes across the

DOI: 10.1201/9781003286400-9 

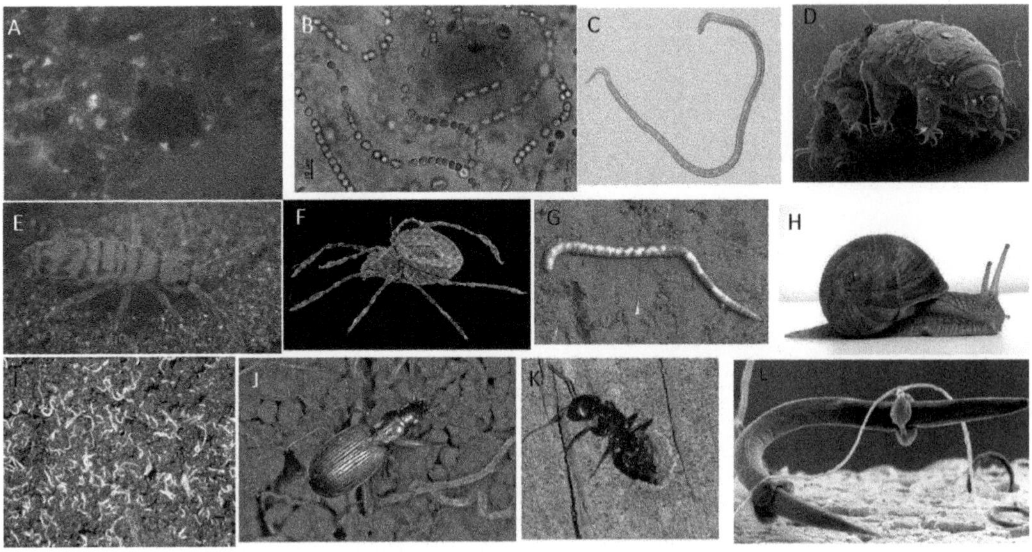

**FIGURE 9.1** A selection of organisms of soil biota. A–L, The selection of organisms includes bacteria (A) filamentous cyanobacterium (B), nematode (C), tardigrade (D), collembolan (E), mite (F), earthworm (G), snail (H), pot worms (I), carabids (J), ants (K), the carnivorous fungi Drechslerella anchonia which captures nematodes in rings which grow along its hyphae then penetrates the skin and consumes the nematode from the inside out (L). All photographs are from the European Soil Biodiversity Atlas, courtesy of A. Jones; individual photo credits are K. Ritz (A); K. Wolowsk (B); H. van Megen (C); Eye of Science/Science Photo Library (D); U. Tartes (E); J. Mourek (F); D. Cluzeau (G); J. Schoner (H); J. Römbke(I); G. Colm(J); A. Mori / D. Grasso (K); G. Barron / N. Allin.

atmosphere that emits back to space, however, the part is absorbed by greenhouse gases and then emitted back in all directions (Schmidt et al. 2010). The impact of this event is warming the surface of the earth and the lower atmosphere. These gases have heat-trapping capabilities and longer residence time in the atmosphere, so it has a robust warming effect. If the greenhouse effect does not exist, the average temperature of the earth would be 0°F (−18°C), rather than the present average of 59°F (15°C) (Jacobson 2005; Karl and Trenberth 2003; Adnan et al. 2016; Iqra et al. 2020; Akbar et al. 2020; Mahar et al. 2020; Noor et al. 2020; Bayram et al. 2020; Amanullah and Fahad 2018a, b; Amanullah and Fahad 2017; Amanullah et al. 2020; Amanullah et al. 2021; Rashid et al. 2020; Arif et al. 2020; Amir et al. 2020; Saman et al. 2020; Tahir ul Qamar et al. 2020; Md Jakir and Allah 2020; Mahmood et al. 2021; Farah et al. 2020; Sadam et al. 2020; Unsar et al. 2020; Fazli et al. 2020; Huq et al. 2020; Gopakumar et al. 2020; Zia-ur-Rehman 2020; Sabagh et al. 2020; Al-Wabel et al. 2020a, b; Senol 2020; Amjad et al. 2020; Ibrar et al. 2020; Sajid et al. 2020; Muhammad et al. 2021; Sidra et al. 2021; Zahir et al. 2021; Sahrish et al. 2022).

Soil biodiversity alludes to all soil living organisms. The soil has an incredible diversity of organisms and these organisms are extremely abundant in the soil. It varies and differs from soil to soil, relying on some factors related to the soil condition such as soil texture, pH, organic matter content, and environmental conditions. Soil communities found within the same ecosystem range from microscopic organisms (bacteria, fungi, and protozoa) to larger organisms (ants, earthworms, woodlice, and moles), as shown in Figure 9.1. Soil biota can be categorized into three groups depending on their size: microfauna, mesofauna, and macrofauna (Wallwork 1970; Swift et al. 1979). Microfauna's size range is 1–100 µm and includes bacteria, protozoa, fungi, and nematodes. Mesofauna's size range is from 100 µm to 2 mm and includes tardigrades, collembola, and mites. Macrofauna's size range is >2 mm and includes earthworms, ants, woodlice, centipedes, amphibians, reptiles, mammals, and birds, as shown in Figure 9.2. In another classification, it can be classified into detritivores: fungivores and bacterivores (Hunt et al. 1987; Chen et al. 1996; Briones et al. 1999; Ingham et al. 1986). Also, it can be distinguished into above and below-ground according to the preferable environment (Jing et al. 2015).

Soil organisms play primary roles in the following ecological functions: (1) driving the essential components of nutrient cycles, (2) controlling soil organic matter dynamics, (3) soil carbon sequestration

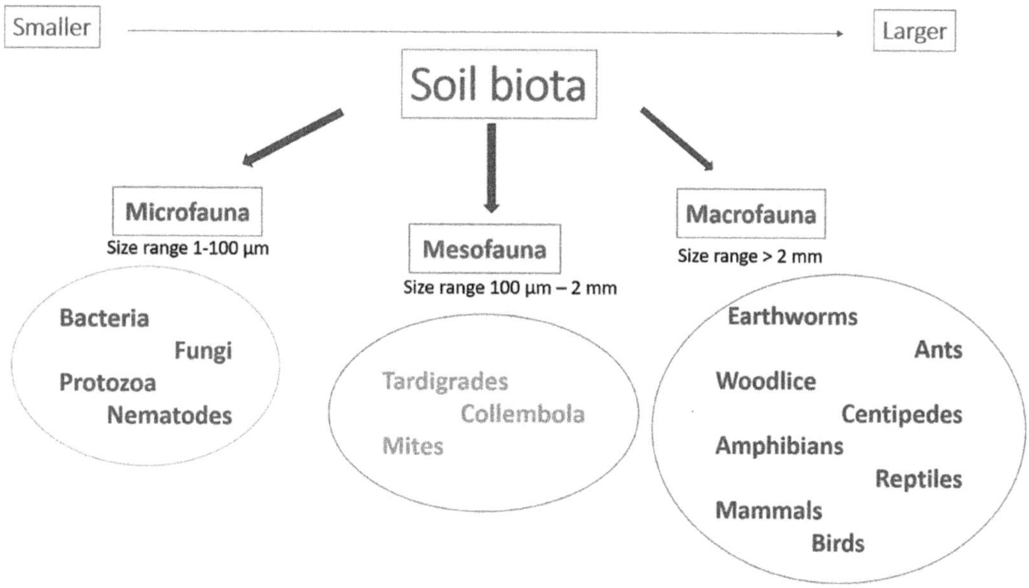


**FIGURE 9.2** Soil biota can be categorized into three groups according to size.

and greenhouse gases emission, (4) adjusting the control of soil and water physical properties, And (5) promoting the amount of nutrient acquisition and efficiency by promoting plant health and vegetation (Lavelle and Spain 2001; Swift et al. 1979; Duxbury 1989; Ingham et al. 1985; Brady et al. 2010).

Soil organisms participate in the soil food web. Generally, the roots and dead organic matters degradation is the main ax in the soil food web. Also, the stability of soil food web is based on the performance stability of the ecosystem (Killham 1994; Adl 2003). Soil microorganisms perform a wide range of functions as follows: (1) organic matter decomposition, (2) degradation of toxic residues and release of nutrients into the plant, (3) symbiotic associations with plant roots, (4) antagonist to pathogens, and (5) solubilization of minerals and contribution to soil aggregation and structure (Swift et al. 1979; Coleman 1984; Adl 2003). So, soil biodiversity plays a very essential role in the global system.

## 9.2 Climate Change Effects on Soil Properties

Climate change has an important impact on soil biological, physical, and chemical properties such as soil pH, electrical conductivity, rate of acidification or alkalization, and cation-exchange capacity through some climate change drivers such as $CO_2$, temperature, N deposition and rainfall. Thus, it affects critical processes in soil (Burns et al. 2013), and these properties are considered as indicators for soil health under climate change stress (Allen et al. 2011).

### 9.2.1 Physical Properties

#### *9.2.1.1 Soil Temperature*

Soil temperature is the main effector in most soil functions. There is a positive relationship between soil temperature and air temperature. So, when the air temperature increases, the soil temperature increases. The soil temperature system is controlled by evaporation, radiation gains and losses at the surface, and heat conduction. Also, it is associated with snow cover depth over a period of 30 years (Qian et al. 2011). With higher soil temperature, soil functions will be accelerated, organic matter decomposition will be rapid, the microbiological activity will be increased, nutrients release will be quicker, and chemical weathering of minerals and nitrification rate will be increased. Nevertheless, the type of vegetation on soil surface will affect its temperature (DEFRA 2005).

#### 9.2.1.2 Soil Water

Precipitation and temperature can affect water content of the soil. Precipitation leads to fast changes in soil water, and temperature leads to boost water evapotranspiration loss from the soil. Water balance and soil moisture reflected the integral influence of climate-hydrology vegetation-land use changes (Várallyay 1990; Várallyay and Farkas 2008). High temperature leads to an increase in the evaporation and transpiration if the plant is not suffering from limited water supply decreases surface runoff, water infiltration, groundwater flow, and water storage, especially if associated with low precipitation, it would give more opportunity for water penetration (Várallyay 2010). Additionally, soil moisture content significantly affects various soil functions including structure formation, porosity, aggregation, organic matter turnover, weathering, and clay translocation (Karmakar et al. 2016). With high rainfall, the decomposition of organic matters reduces because high rainfall increases the susceptibility of water erosion, temporary flooding, leaching rate, and water saturation. In contrast, low rainfall increases the susceptibility of wind erosion, represses root growth, suppresses the organic matter deposition, and boosts the accumulation of salt in soil (Anjali and Dhananjaya 2017).

#### 9.2.1.3 Soil Structure

The organic and inorganic matters and the cultivation method significantly influence the soil structure. When soil organic matter levels decrease, the rates of soil aggregate infiltration and stability decrease, and the susceptibility to erosion and compaction increases (Bot and Benites 2005).

#### 9.2.1.4 Bulk Density

Bulk density has a negative relationship with soil organic matter and soil organic carbon (Magdoff and Weil 2004). Bulk density increases due to loss of organic carbon due to increase of the organic matter decomposition and makes the soil more susceptible to compaction (Davidson and Janssens 2006; Birkás et al. 2009).

### 9.2.2 Chemical Properties

#### 9.2.2.1 Soil pH

Reth et al. (2005) reported that climate change drivers will affect nutrient and carbon cycling, organic matter status, and plant available water which has an impact on soil pH. The soil pH decreases if the elevation gradient increases. The soil pH decreases due to increase in basic cations leaching in the higher elevations, increased nitrification, and greater precipitation. So the soil pH decreases when moving from low to high elevation (Smith et al. 2002). While it was proposed that most soils would not be subjected to rapid pH changes resulting from drivers of climate change such as $CO_2$ fertilization, variable precipitation elevated temperatures, and atmospheric N deposition (Brinkman and Sombroek 1996; De Vries and Breeuwsma 1987; McCarthy et al. 2001), it is likely, however, that these climate change drivers will affect plant available water, C and nutrient cycling, organic matter status and hence plant productivity, which in turn soil pH will be affected (Reth et al. 2005). Also, it was reported that there is a relationship between pH and electrical conductivity.

#### 9.2.2.2 Electrical Conductivity

Increase in electrical conductivity with elevation is a contrast to the hypothesis that the basic cations leaching leads to decrease soil pH with increasing elevation. However, in the higher elevations, a significant amount of nitrate participates in the EC increasing over the 500 m elevation because the lower pH leads to increase $H^+$ ion concentrations (Smith et al. 2002). The soluble salts concentration dynamics in soils from different climatic areas were examined and found a nonlinear relationship between the rainfall and soluble salts content, with sites that received less than 200 mm rainfall having high soluble contents (Pariente 2001).

#### 9.2.2.3 Cation-Exchange Capacity

Elevated temperatures increase the loss of soil organic matter, which results in the cation-exchange capacity loss of soils. The low cation-exchange capacity of soil increases basic cations leaching as a result of high rainfall, leads to transport of alkalinity from soil to watercourses, and makes the soil acidified (Davidson and Janssens 2006; Rengel 2011).

### 9.2.3 Biological Properties

#### 9.3.3.1 Soil Organic Matter

Soil organic matter is the most complex living and non-living component in the soil. The decrease in Soil Organic matters (SOM) results in soil structure loss and decline in soil biodiversity and fertility, so it leads to an increase in soil compaction through it increases both bulk density and erosion and also reduced water-holding capacity (Magdoff and Weil 2004).

#### 9.2.3.2 Soil Carbon

If the $CO_2$ concentration is high, it will increase the organic matter supplies to the soil because high $CO_2$ concentration will raise the rates of photosynthesis and boost the water use efficiency of crops. The temperature has a significant impact on the soil. With high temperature, the soil degradation and natural decomposition of organic matters speed up. Additionally, at higher temperatures, soil moisture variation is more significant (Borken et al. 2003). The organic matter decomposition rate decreases with high moisture contents because of the low oxygen supply in soil (Zhou et al. 2002; Schjønning et al. 2003). Also, the carbon, phosphorus, sulfur, and nitrogen cycles speed up in the soil, plant, and atmosphere systems. In the soil, increase in root and microbial activities require higher $CO_2$ pressure in soil air and $CO_2$ activity in soil water so it would boost the plant nutrient release rates from the soil mineral weathering) Hu et al. 2001; Drigo et al. 2008; Shah et al. 2016). High C/N ratios under high $CO_2$ result in slow remobilization of plant nutrients from the litter and uptake by the root (Drigo et al. 2008). Also, soil respiration is sensitive to changes in the seasonal timing of rainfall, which is predicted to change according to global and regional climate models (Chou et al. 2008)

#### 9.2.3.3 Soil Microbial Biomass

Soil microbial biomass significantly declined during the experiments of long-term activated climatic warming (Rinnan et al. 2007). Also, it was reported that soil microbial biomass responds to short-term environmental changes (Haynes 2008; Qadeem et al. 2015). Also, it was reported that metabolic and microbial quotients have a significant sensitivity to changes under N fertilization and elevated $CO_2$ conditions (Moscatelli et al. 2005).

#### 9.2.3.4 Enzyme Activities

Microbial enzyme activities may be stimulated by elevated $CO_2$. Also, microbial enzymes abundance and C turnover affect the soil microbial community (Dorodnikov et al. 2009). This enzyme activity may rely on the size of soil aggregate. Additionally, the extracellular enzymes which are responsible for nutrient cycling processes and soil organic carbon decomposition may be affected by N deposition in the atmosphere (Frey et al. 2004).

## 9.3 Climate Change Effects on Soil Biodiversity

### 9.3.1 Effect of Changes in Soil Temperature on Soil Biodiversity

Generally, the effect of climate on soil microorganisms was recognized by the seasonal dynamics of microbial populations. Composition, growth and activity of microbes are influenced by changes in moisture and temperature, which are considered as the main factors controlled by climate. If the temperature is high, the microbial activity increases. So, in the winter season, activity and growth of microbes decrease (Schadt et al. 2003). So, temperature positively affects the activity and growth of microbes. High temperatures can enhance the breakdown of organic pollutants due to the high microbial activity. It was reported that high temperature can differentially activate functional genes involved in the degradation of carbon in soil (Gouin et al. 2013; Zhou et al. 2012; Wu et al. 2011; Bardgett et al. 2013). It was reported that this activation is not permeant, and in the long term, the microbes will adapt to the higher temperature (Hartley et al. 2008; Bradford et al. 2008). It was observed that with prolonged high temperature, the soil carbon decomposition rate was accelerated then declined because of the reduction of soil carbon, microbial activity, and biomass over time (Davidson and Janssens 2006; Knorr et al. 2005; Frey et al. 2008; Steinweg et al. 2008). Additionally, it was observed that microbial biomass, soil respiration, and organic matter decomposition increased with temperature in short-term studies (Bradford et al. 2008). So the substrates of labile soil carbon are exhausted by increased microbial activity (Bradford 2013; Allison and Martiny 2008; Classen et al. 2015). These warming effects are extremely complicated and accompanied by adaptation, acclimation, and evolution (Classen et al. 2015). However, it was not abolished the unexpected outcome from higher soil temperature on the bioremediation process. Also, toxicity of pollutants can rise with increase in temperature and thus can result in low levels of soil quality (Noyes et al. 2009; Rohr et al. 2013).

Generally, the increasing precipitation was preferred by the fungi; on the other hand, enrichment of $CO_2$ was preferred by the bacteria (Blankinship et al. 2011). The abundance of fungi boosted in warmed areas, as well as the abundance of bacteria raised in warmed areas with elevated $CO_2$ but declined in warmed areas with ambient $CO_2$. The abundance of Proteobacteria and Acidobacteria changed according to the changes in precipitation, where as Acidobacteria decreased along with the Proteobacteria increasing in wet areas comparing to dry areas (Mandal 2012).

Additionally, the temperature has an important impact on nematodes and microarthropods. Nematodes are sensitive to soil moisture and temperature depending on their metabolic condition (Hoschitz and Kaufmann 2004; Ruess et al. 1999). These organisms have a distinguished technique to survive under extreme environmental conditions through cysts formation or entering dormant stages to survive under the most extreme changes of soil temperature and moisture (McSorley 2003; Wall and Virginia 1999). Effects of high temperatures and droughts on nematodes are mainly dependent on how they influence soil moisture. In arid zones like deserts, survival of nematode relies on soil moisture, while in temperate areas, their survival is unlikely to be at stake unless soils dry out completely (Papatheodorou et al. 2004; Strong et al. 2004). The high temperature increases the abundance and biomass of microarthropod only under wet conditions (Harte et al. 1996). Soil moisture and temperature can control microarthropods' biology like springtails and mites and influence their population abundance (Roy and Roy 2006; Cassagne et al. 2003). It can survive above 20°C and is limited to 50°C (Vannier 1994). Generally, species that live on the litter surface can tolerate higher temperatures than species living further down in the soil. Choudhuri (1963) reported that most mites and springtails have been reported to limit their lethal temperature around 35°C–40°C. So, species living in warm areas are more resistant to high temperature than those living in temperate and cold areas. Temperature can also influence springtails reproduction and development rates with important impacts on population growth (Dlekkruger and Roske 1995; Choi and Ryoo 2003). Closely related microarthropod species can differ in soil moisture sensitivity and temperature tolerance; each species seems to require quite specific conditions of temperature and moisture (Christiansen 1964). Additionally, thermotolerance differs according to the developmental stages (Chown and Nicolson 2004).

The soil moisture and temperature changes can strongly influence the earthworm's physiology. When soils are too dry, earthworms often boost their burrowing activity, lose weight, or enter into diapause (Booth et al. 2000; Holmstrup 2001; Saleem et al. 2015). In contrast, growth is preferred under high temperatures and high moisture. The abundance of Collembola tends to decline with high temperature (Dollery et al. 2006).

### 9.3.2 Effect of Changes in $CO_2$ on Soil Biodiversity

Higher atmospheric $CO_2$ has an indirect impact on soil microbial activity through changes in litter deposition, root exudation, and higher plant growth because $CO_2$ levels in the soil are higher than atmospheric $CO_2$ (Phillips et al. 2011; Burns et al. 2013).

The interactions of $CO_2$ and $N_2$ differ between soil biota groups (Reich 2009; Eisenhauer et al. 2012). The richness of soil microarthropod taxa declined at elevated N, but only at ambient $CO_2$. This suggested that elevated $CO_2$ negatively affects N enrichment. The same interaction for the richness of plant species was found and ascribed the changes to various effects of $CO_2$ and N on plant traits and soil resources altering competitive interactions between plant species (Reich 2009). Also, rhizodeposition is reduced at elevated N at ambient $CO_2$, suggesting that elevated $CO_2$ boosts soil microbial biomass, rhizodeposition, and shoot litter inputs. In contrast, the richness of nematode taxa was at a minimum at elevated $CO_2$ and elevated N (Eisenhauer et al. 2012). This suggests that elevated N has a negative effect on the richness of nematode, reduced rhizodeposition, and predatory nematodes.

Also, elevated $CO_2$ has a negative impact on the richness of nematode species through boosting the biomass of shoot litter and changing plant stoichiometry (Novotny et al. 2007; Reich, 2009). Also, the density of Prostigmata, Astigmata, and fungal-feeding nematodes was high at elevated N and ambient $CO_2$ compared to ambient N. This means distinct changes in the bacteria–fungi ratio and the composition of the microbial community in response to global change drivers (Ruess et al. 1999; Chung et al. 2007). The abundance of Ciliates differed significantly in response to summer drought and elevated N, this supported that the active ciliates abundance is sensitive to drought (Foissner 1997; Deepranjan et al. 2021; Haider et al. 2021; Li-ying et al. 2021; Ikram et al. 2021; Jabborova et al. 2021; Khadim et al. 2021a, b; Siddiqui et al. 2021; Muzammal et al. 2021; Abdul et al. 2021a, b; Ashfaq et al. 2021; Amjad et al. 2021; Atif et al. 2021; Athar et al. 2021; Adnan et al. 2018a, b; Adnan et al. 2019; Akram et al. 2018a, b; Aziz et al. 2017a, b; Chang et al. 2021; Chen et al. 2021; Emre et al. 2021; Habib-Ur-Rahman et al. 2017; Hafiz et al. 2016; Hafiz et al. 2019; Ghulam et al. 2021; Guofu et al. 2021; Hafeez et al. 2021; Khan et al. 2021; Kamran et al. 2017; Muhammad et al. 2019; Safi et al. 2021; Sajjad et al. 2019; Saud et al. 2013; Saud et al. 2014; Saud et al. 2017; Saud et al. 2016; Shah et al. 2013; Saud et al. 2020; Qamar et al. 2017; Hamza et al. 2021; Irfan et al. 2021;Wajid et al. 2017; Yang et al. 2017; Zahida et al. 2017; Depeng et al. 2018; Hussain et al. 2020; Hafiz et al. 2020a, b; Shafi et al. 2020; Wahid et al. 2020; Subhan et al. 2020; Zafar-ul-Hye et al. 2020a, b; Zafar et al. 2021; Adnan et al. 2020; Ilyas et al. 2020; Saleem et al. 2020a, b, c; Rehman 2020; Frahat et al. 2020; Wu et al. 2020; Mubeen et al. 2020; Farhana 2020; Jan et al. 2019; Wu et al. 2019; Ahmad et al,. 2019; Baseer et al. 2019; Hafiz et al. 2018; Tariq et al. 2018).

## 9.4 Conclusion

Climate change can affect the chemical, physical, and biological properties of soil which leads to changes in soil biodiversity. These changes are due to the effect of climate change on temperature, moisture, and precipitation. All these effects change the richness of soil organisms and lead to discovering new features for these organisms or new species that can adapt to these climate changes.

## REFERENCES

Abdul S, Muhammad AA, Shabir H, Hesham A. El E, Sajjad H, Niaz A, Abdul G, RZ Sayyed, Fahad S, Subhan D, Rahul D (2021a) Zinc nutrition and arbuscular mycorrhizal symbiosis effects on maize (Zea mays L.) growth and productivity. J Saudi Soc Agric Sci https://doi.org/10.1016/j.sjbs.2021.06.096.

Abdul S, Muhammad AA, Subhan D, Niaz A, Fahad S, Rahul D, Mohammad JA, Omaima N, ur Rehman MH, Bernard RG (2021b) Effect of arbuscular mycorrhizal fungi on the physiological functioning of maize under zinc-deficient soils. Sci Rep 11:18468.

Adl SM (2003) The ecology of soil decomposition. Cabi Publishing, Wallingford.

Adnan M, Fahad S, Khan IA, Saeed M, Ihsan MZ, Saud S, Riaz M, Wang D, Wu C (2019). Integration of poultry manure and phosphate solubilizing bacteria improved availability of Ca bound P in calcareous soils. 3 Biotech 9(10):1–10.

Adnan M, Fahad S, Muhammad Z, Shahen S, Ishaq AM, Subhan D, Zafar-ul-Hye M, Martin LB, Raja MMN, Beena S, Saud S, Imran A, Zhen Y, Martin B, Jiri H, Rahul D (2020) Coupling phosphate-solubilizing bacteria with phosphorus supplements improve maize phosphorus acquisition and growth under lime induced salinity stress. Plants 9(900). https://doi.org/10.3390/plants9070900.

Adnan M, Shah Z, Arif M, Khan MJ, Mian IA, Sharif M, Alam M, Basir A, Ullah H, Rahman I, Saleem N (2016). Impact of rhizobial inoculum and inorganic fertilizers on nutrients (NPK) availability and uptake in wheat crop. Can J Soil Sci 96:169–176.

Adnan M, Shah Z, Sharif M, Rahman H (2018b). Liming induces carbon dioxide ($CO_2$) emission in PSB inoculated alkaline soil supplemented with different phosphorus sources. Environ Sci Pollut Res 25(10):9501–9509.

Adnan M, Zahir S, Fahad S, Arif M, Mukhtar A, Imtiaz AK, Ishaq AM, Abdul B, Hidayat U, Muhammad A, Inayat-Ur R, Saud S, Muhammad ZI, Yousaf J, Amanullah, Hafiz MH, Wajid N (2018a) Phosphate-solubilizing bacteria nullify the antagonistic effect of soil calcification on bioavailability of phosphorus in alkaline soils. Sci Rep 8:4339. https://doi.org/10.1038/s41598-018-22653-7.

Ahmad S, Kamran M, Ding R, Meng X, Wang H, Ahmad I, Fahad S, Han Q (2019) Exogenous melatonin confers drought stress by promoting plant growth, photosynthetic capacity and antioxidant defense system of maize seedlings. PeerJ 7:e7793. https://doi.org/10.7717/peerj.7793.

Ajlan A (2016) A review on bioremediation. Microbiol 1:103.

Akbar H, Timothy JK, Jagadish T, M. Golam M, Apurbo KC, Muhammad F, Rajan B, Fahad S, Hasanuzzaman M (2020) Agricultural land degradation: processes and problems undermining future food security. In: Fahad S, Hasanuzzaman M, Alam M, Ullah H, Saeed M, Khan AK, Adnan M (eds) Environment, climate, plant and vegetation growth. Springer, pp 17–62. https://doi.org/10.1007/978-3-030-49732-3.

Akram R, Turan V, Hammad HM, Ahmad S, Hussain S, Hasnain A, Maqbool MM, Rehmani MIA, Rasool A, Masood N, Mahmood F, Mubeen M, Sultana SR, Fahad S, Amanet K, Saleem M, Abbas Y, Akhtar HM, Waseem F, Murtaza R, Amin A, Zahoor SA, ul Din MS, Nasim W (2018a) Fate of organic and inorganic pollutants in paddy soils. In: Hashmi MZ, Varma A (eds) Environmental pollution of paddy soils, soil biology. Springer, Cham, pp 197–214.

Akram R, Turan V, Wahid A, Ijaz M, Shahid MA, Kaleem S, Hafeez A, Maqbool MM, Chaudhary HJ, Munis, MFH, Mubeen M, Sadiq N, Murtaza R, Kazmi DH, Ali S, Khan N, Sultana SR, Fahad S, Amin A, Nasim W (2018b) Paddy land pollutants and their role in climate change. In: Hashmi MZ, Varma A (eds) Environmental pollution of paddy soils, soil biology. Springer, Cham, pp 113–124.

Allen DE, Singh BP, Dalal RC (2011) Soil health indicators under climate change: a review of current knowledge. In: Singh BP et al. (eds) Soil health and climate change. Springer, Berlin, Heidelberg, pp 25–45.

Allison SD, Martiny JB (2008) Resistance, resilience, and redundancy in microbial communities. Proc Natl Acad Sci U S A 105:11512–11519.

Al-Wabel MI, Abdelazeem S, Munir A, Khalid E, Adel RAU (2020b) Extent of Climate Change in Saudi Arabia and Its Impacts on Agriculture: A Case Study from Qassim Region. In: Fahad S, Hasanuzzaman M, Alam M, Ullah H, Saeed M, Khan AK, Adnan M (Ed.), Environment, Climate, Plant and Vegetation Growth. Springer Publ Ltd, Springer Nature Switzerland AG. Part of Springer Nature. PP. 635–658. https://doi.org/10.1007/978-3-030-49732-3

Al-Wabel MI, Munir Ahmad, Adel R. A. Usman, Mutair Akanji, and Muhammad Imran Rafique (2020a) Advances in Pyrolytic Technologies with Improved Carbon Capture and Storage to Combat Climate Change. In: Fahad S, Hasanuzzaman M, Alam M, Ullah H, Saeed M, Khan AK, Adnan M (Ed.), Environment, Climate, Plant and Vegetation Growth. Springer Publ Ltd, Springer Nature Switzerland AG. Part of Springer Nature. PP. 535–576. https://doi.org/10.1007/978-3-030-49732-3

Amanullah MI, Fahad S (Eds.) (2017) Rice – technology and production. IntechOpen. https://doi.org/10.5772/64480.

Amanullah MI, Fahad S (Eds.) (2018a) Corn – production and human health in changing climate. IntechOpen. https://doi.org/10.5772/intechopen.74074.

Amanullah MI, Fahad S (Eds.) (2018b) Nitrogen in agriculture – updates. IntechOpen. https://doi.org/10.5772/65846.

Amanullah MI, Haider N, Shah K, Manzoor A, Asim M, Saif U, Izhar A, Fahad S, Adnan M et al. (2021) Integrated foliar nutrients application improve wheat (Triticum aestivum L.) productivity under calcareous soils in drylands. Commun Soil Sci Plant Anal. https://doi.org/10.1080/00103624.2021.1956521.

Amanullah MI, Shah K, Imran, Hamdan AK, Muhammad A, Abdel RA, Muhammad A, Fahad S, Azizullah S, Brajendra P (2020) Effects of climate change on irrigation water quality. In: Fahad S, Hasanuzzaman M, Alam M, Ullah H, Saeed M, Khan AK, Adnan M (eds) Environment, climate, plant and vegetation growth. Springer, pp 123–132. https://doi.org/10.1007/978-3-030-49732-3.

Amir M, Muhammad A, Allah B, Sevgi Ç, Haroon ZK, Muhammad A, Emre A (2020) Bio fortification under climate change: the fight between quality and quantity. In: Fahad S, Hasanuzzaman M, Alam M, Ullah H, Saeed M, Khan AK, Adnan M (eds) Environment, climate, plant and vegetation growth. Springer, pp 173–228. https://doi.org/10.1007/978-3-030-49732-3.

Amjad I, Muhammad H, Farooq S, Anwar H (2020) Role of plant bioactives in sustainable agriculture. In: Fahad S, Hasanuzzaman M, Alam M, Ullah H, Saeed M, Khan AK, Adnan M (eds) Environment, climate, plant and vegetation growth. Springer, pp 591–606. https://doi.org/10.1007/978-3-030-49732-3.

Amjad SF, Mansoora N, Din IU, Khalid IR, Jatoi GH, Murtaza G, Yaseen S, Naz M, Danish S, Fahad S et al. (2021) Application of zinc fertilizer and mycorrhizal inoculation on physio-biochemical parameters of wheat grown under water-stressed environment. Sustainability 13:11007. https://doi.org/10.3390/su131911007.

Anjali MC, Dhananjaya BC (2017) Effect of phosphorus levels with or without PSB seed treatment on dynamics of P in soil under groundnut (Arachis hypogaea L.). Environ Ecol 35:3217–3221.

Arif M, Talha J, Muhammad R, Fahad S, Muhammad A, Amanullah MI, Kawsar A, Ishaq AM, Bushra K, Fahd R (2020) Biochar; a remedy for climate change. In: Fahad S, Hasanuzzaman M, Alam M, Ullah H, Saeed M, Khan AK, Adnan M (eds) Environment, climate, plant and vegetation growth. Springer, pp 151–172. https://doi.org/10.1007/978-3-030-49732-3.

Ashfaq AR, Uzma Y, Niaz A, Muhammad AA, Fahad S, Haider S, Tayebeh Z, Subhan D, Süleyman T, Hesham AElE, Pramila T, Jamal MA, Sulaiman AA, Rahul D (2021) Toxicity of Cadmium and nickel in the context of applied activated carbon biochar for improvement in soil fertility. Saudi Soc Agric Sci. https://doi.org/10.1016/j.sjbs.2021.09.035.

Athar M, Masood IA, Sana S, Ahmed M, Xiukang W, Sajid F, Sher AK, Habib A, Faran M, Zafar H, Farhana G, Fahad S (2021) Bio-diesel production of sunflower through sulphur management in a semi-arid subtropical environment. Environ Sci Pollut Res https://doi.org/10.1007/s11356-021-16688-z.

Atif B, Hesham A, Fahad S (2021) Biochar coupling with phosphorus fertilization modifies antioxidant activity, osmolyte accumulation and reactive oxygen species synthesis in the leaves and xylem sap of rice cultivars under high-temperature stress. Physiol Mol Biol Plants. https://doi.org/10.1007/s12298-021-01062-7.

Aziz K, Daniel KYT, Fazal M, Muhammad ZA, Farooq S, Fan W, Fahad S, Ruiyang Z (2017a) Nitrogen nutrition in cotton and control strategies for greenhouse gas emissions: a review. Environ Sci Pollut Res 24:23471–23487. https://doi.org/10.1007/s11356-017-0131-y.

Aziz K, Daniel KYT, Muhammad ZA, Honghai L, Shahbaz AT, Mir A, Fahad S (2017b) Nitrogen fertility and abiotic stresses management in cotton crop: a review. Environ Sci Pollut Res 24:14551–14566. https://doi.org/10.1007/s11356-017-8920-x.

Banerjee A, Roy A, Dutta S, Mondal S (2016) Bioremediation of hydrocarbon-a review. Int J Adv Res 4:1303–1313.

Bardgett RD, Manning P, Morriën E, De Vries FT (2013) Hierarchical responses of plant–soil interactions to climate change: consequences for the global carbon cycle. J Ecol 101:334–343.

Baseer M, Adnan M, Fazal M, Fahad S, Muhammad S, Fazli W, Muhammad A, Jr. Amanullah MI, Depeng W, Saud S, Muhammad N, Muhammad Z, Fazli S, Beena S, Mian AR, Ishaq AM (2019) Substituting urea by organic wastes for improving maize yield in alkaline soil. J Plant Nutr. https://doi.org/10.1080/01904167.2019.1659344.

Bayram AY, Seher Ö, Nazlican A (2020) Climate change forecasting and modeling for the year of 2050. In: Fahad S, Hasanuzzaman M, Alam M, Ullah H, Saeed M, Khan AK, Adnan M (eds) Environment, climate, plant and vegetation growth. Springer, pp 109–122. https://doi.org/10.1007/978-3-030-49732-3.

Birkás M, Dexter A, Szemők A (2009) Tillage-induced soil compaction, as a climate threat increasing stressor. Cereal Res Commun 37:379–382.

Blankinship JC, Niklaus PA, Hungate BA (2011) A meta-analysis of responses of soil biota to global change. Oecologia 165:553–565.

Booth L, Heppelthwaite V, Mcglinchy A (2000) The effect of environmental parameters on growth, cholinesterase activity and glutathione S-transferase activity in the earthworm (Apporectodea caliginosa). Biomarkers 5:46–55.

Borken W, Davidson EA, Savage K, Gaudinski J, Trumbore SE (2003) Drying and wetting effects on carbon dioxide release from organic horizons. Soil Sci Soc Am J 67:1888–1896.

Bot A, Benites J (2005) The importance of soil organic matter: key to drought-resistant soil and sustained food production. FAO. No. 80.

Bradford MA (2013). Thermal adaptation of decomposer communities in warming soils. Front Microbiol 4:333.

Bradford MA, Davies CA, Frey SD, Maddox TR, Melillo JM, Mohan JE, Wallenstein MD (2008) Thermal adaptation of soil microbial respiration to elevated temperature. Ecol Lett 11:1316–1327.

Brady NC, Weil RR, Brady NC (2010) Elements of the nature and properties of soils (No. 631.4 B733E.). Pearson Educational International, Upper Saddle River, NJ.

Brinkman R, Sombroek WG (1996) The effects of global change on soil conditions in relation to plant growth and food production. Global Clim Change Agri Prod 49–63.

Briones MJI, Ineson P, Sleep D (1999) Use of δ13C to determine food selection in collembolan species. Soil Biol Biochem 31:937–940.

Burns RG, DeForest JL, Marxsen J, Sinsabaugh RL, Stromberger ME, Wallenstein MD, Zoppini A (2013) Soil enzymes in a changing environment: current knowledge and future directions. Soil Biol Biochem 58:216–234.

Cassagne N, Gers C, Gauquelin T (2003) Relationships between Collembola, soil chemistry and humus types in forest stands (France). Biol Fertil Soils 37:355–361.

Chang W, Qiujuan J, Evgenios A, Haitao L, Gezi L, Jingjing Z, Fahad S, Ying J (2021) Hormetic effects of zinc on growth and antioxidant defense system of wheat plants. Sci Total Environ. https://doi.org/10.1016/j.scitotenv.2021.150992.

Chen B, Snider RJ, Snider RM (1996) Food consumption by Collembola from northern Michigan deciduous forest. Pedobiologia 40:149–161.

Chen Y, Guo Z, Dong L, Fu Z, Zheng Q, Zhang G, Qin L, Sun X, Shi Z, Fahad S, Xie F, Saud S (2021) Turf performance and physiological responses of native Poa species to summer stress in Northeast China. PeerJ 9:e12252. http://doi.org/10.7717/peerj.12252.

Cho YG, Rhee SK, Lee ST (2000) Effect of soil moisture on bioremediation of chlorophenol-contaminated soil. Biotechnol Lett 22:915–919.

Choi WI, Ryoo MI (2003) A matrix model for predicting seasonal fluctuations in field populations of Paronychiurus kimi (Collembola: Onychiruidae). Ecol Model 162:259–265.

Chou WW, Silver WL, Jackson RD, Thompson AW, Allen-Diaz BA (2008) The sensitivity of annual grassland carbon cycling to the quantity and timing of rainfall. Glob Change Biol 14:1382–1394.

Choudhuri DK (1963) Temperature and its effect on the three species of the genus Onychiurus collembola. Proc Zool Soc 16:97–98.

Chowdhury N, Marschner P, Burns RG (2011) Soil microbial activity and community composition: impact of changes in matric and osmotic potential. Soil Biol Biochem 43:1229–1236.

Chown SL, Nicolson S (2004) Insect physiological ecology: mechanisms and patterns. Oxford University Press, Oxford.

Christiansen K (1964) Bionomics of collembola. Ann Rev Entomol 9:147–178.

Chung H, Zak DR, Reich PB, Ellsworth DS (2007). Plant species richness, elevated $CO_2$, and atmospheric nitrogen deposition alter soil microbial community composition and function. Glob Change Biol 13:980–989.

Classen AT, Sundqvist MK, Henning JA, Newman GS, Moore JA, Cregger MA, Patterson CM (2015) Direct and indirect effects of climate change on soil microbial and soil microbial-plant interactions: what lies ahead? Ecosphere 6:1–21.

Coleman DC (1984) Soil nutrient transformations in the rhizosphere via animal-microbial interactions. Inver Microbe Interact 35–58.

Davidson EA, Janssens IA (2006) Temperature sensitivity of soil carbon decomposition and feedbacks to climate change. Nat 440(7081):165.

De Vries W, Breeuwsma A (1987) The relation between soil acidification and element cycling. Water Air Soil Pollut 35:293–310.

Deepranjan S, Ardith S, O. Siva D, Sonam S, Shikha, Manoj P, Amitava R, Sayyed RZ, Abdul G, Mohammad JA, Subhan D, Fahad S, Rahul D (2021) Optimizing nutrient use efficiency, productivity, energetics, and economics of red cabbage following mineral fertilization and biopriming with compatible rhizosphere microbes. Sci Rep 11:15680.

DEFRA (2005) Impact of climate change on soil functions. Final Project Report, Research and Development. DEFRA, London.

Depeng W, Fahad S, Saud S, Muhammad K, Aziz K, Mohammad NK, Hafiz MH, Wajid N (2018) Morphological acclimation to agronomic manipulation in leaf dispersion and orientation to promote "Ideotype" breeding: evidence from 3D visual modeling of "super" rice (*Oryza sativa* L.). Plant Physiol Biochem 135:499–510. https://doi.org/10.1016/j.plaphy.2018.11.010.

Dlekkruger B, Roske H (1995) Modelling the population dynamics of Isotoma notabilis (Collembola) on sites of different agricultural usage. Pedobiologia 39:58–73.

Dollery R, Hodkinson ID, Jónsdóttir IS (2006) Impact of warming and timing of snow melt on soil microarthropod assemblages associated with Dryas-dominated plant communities on Svalbard. Ecography 29:111–119.

Dorodnikov M, Blagodatskaya E, Blagodatsky S, Marhan S, Fangmeier A, Kuzyakov Y (2009) Stimulation of microbial extracellular enzyme activities by elevated $CO_2$ depends on soil aggregate size. Glob Change Biol 15:1603–1614.

Drigo B, Kowalchuk GA, Van Veen JA (2008) Climate change goes underground: effects of elevated atmospheric $CO_2$ on microbial community structure and activities in the rhizosphere. Biol Fertil Soils 44:667–679.

Duxbury JM (1989) Soil organic matter as a source and a sink of plant nutrients. Dynamics of soil organic matter in tropical ecosystems. In: Coleman DC, Oades JM, Uehara G (eds) Dynamics of soil organic matter in tropical ecosystems, NifTAL Project. University of Hawaii, Honolulu, pp 33–67.

Eisenhauer N, Cesarz S, Koller R, Worm K, Reich PB (2012) Global change belowground: impacts of elevated $CO_2$, nitrogen, and summer drought on soil food webs and biodiversity. Glob Change Biol 18:435–447.

Emre B, Ömer SU, Martín LB, Andre D, Fahad S, Rahul D, Zafar-ul-Hye M, Ghulam SH, Subhan D (2021) Studying soil erosion by evaluating changes in physico-chemical properties of soils under different land-use types. J Saudi Soc Agric Sci 20:190–197.

Fahad S, Abdul B, Adnan M. (Eds.) (2018b) Global wheat production. IntechOpen. http://doi.org/10.5772/intechopen.72559.

Fahad S, Bajwa AA, Nazir U, Anjum SA, Farooq A, Zohaib A, Sadia S, NasimW, Adkins S, Saud S, Ihsan MZ, Alharby H, Wu C, Wang D, Huang J (2017) Crop production under drought and heat stress: plant responses and management options. Front Plant Sci 8:1147. https://doi.org/10.3389/fpls.2017.01147.

Fahad S, Bano A (2012) Effect of salicylic acid on physiological and biochemical characterization of maize grown in saline area. Pak J Bot 44:1433–1438.

Fahad S, Chen Y, Saud S, Wang K, Xiong D, Chen C, Wu C, Shah F, Nie L, Huang J (2013) Ultraviolet radiation effect on photosynthetic pigments, biochemical attributes, antioxidant enzyme activity and hormonal contents of wheat. J Food Agric Environ 11(3&4):1635–1641.

Fahad S, Hussain S, Bano A, Saud S, Hassan S, Shan D, Khan FA, Khan F, Chen Y, Wu C, Tabassum MA, Chun MX, Afzal M, Jan A, Jan MT, Huang J (2014a) Potential role of phytohormones and plant growth-promoting rhizobacteria in abiotic stresses: consequences for changing environment. Environ Sci Pollut Res 22(7):4907–4921. https://doi.org/10.1007/s11356-014-3754-2.

Fahad S, Hussain S, Matloob A, Khan FA, Khaliq A, Saud S, Hassan S, Shan D, Khan F, Ullah N, Faiq M, Khan MR, Tareen AK, Khan A, Ullah A, Ullah N, Huang J (2014b) Phytohormones and plant responses to salinity stress: a review. Plant Growth Regul 75(2):391–404. https://doi.org/10.1007/s10725-014-0013-y.

Fahad S, Hussain S, Saud S, Hassan S, Chauhan BS, Khan F et al. (2016a) Responses of rapid viscoanalyzer profile and other rice grain qualities to exogenously applied plant growth regulators under high day and high night temperatures. PLoS One 11(7):e0159590. https://doi.org/10.1371/journal.pone.0159590.

Fahad S, Hussain S, Saud S, Hassan S, Ihsan Z, Shah AN, Wu C, Yousaf M, Nasim W, Alharby H, Alghabari F, Huang J (2016c) Exogenously applied plant growth regulators enhance the morphophysiological growth and yield of rice under high temperature. Front Plant Sci 7:1250. https://doi.org/10.3389/fpls.2016.01250.

Fahad S, Hussain S, Saud S, Hassan S, Tanveer M, Ihsan MZ, Shah AN, Ullah A, Nasrullah KF, Ullah S, Alharby HNW, Wu C, Huang J (2016d) A combined application of biochar and phosphorus alleviates heat-induced adversities on physiological, agronomical and quality attributes of rice. Plant Physiol Biochem 103:191–198.

Fahad S, Hussain S, Saud S, Khan F, Hassan S, Jr A, Nasim W, Arif M, Wang F, Huang J (2016b) Exogenously applied plant growth regulators affect heat-stressed rice pollens. J Agron Crop Sci 202:139–150.

Fahad S, Hussain S, Saud S, Tanveer M, Bajwa AA, Hassan S, Shah AN, Ullah A, Wu C, Khan FA, Shah F, Ullah S, Chen Y, Huang J (2015a) A biochar application protects rice pollen from high-temperature stress. Plant Physiol Biochem 96:281–287.

Fahad S, Muhammad ZI, Abdul K, Ihsanullah D, Saud S, Saleh A, Wajid N, Muhammad A, Imtiaz AK, Chao W, Depeng W, Jianliang H (2018a) Consequences of high temperature under changing climate optima for rice pollen characteristics-concepts and perspectives. Arch Agron Soil Sci. https://doi.org/10.1080/03650340.2018.1443213.

Fahad S, Nie L, Chen Y, Wu C, Xiong D, Saud S, Hongyan L, Cui K, Huang J (2015b) Crop plant hormones and environmental stress. Sustain Agric Rev 15:371–400.

Fahad S, Saud S, Yajun C, Chao W, Depeng W (Eds.) (2021f) Abiotic stress in plants. IntechOpen. https://doi.org/10.5772/intechopen.91549.

Fahad S, Hasanuzzaman M, Alam M, Ullah H, Saeed M, Ali Khan I, Adnan M. (Eds.) (2020) Environment, climate, plant and vegetation growth. Springer. https://doi.org/10.1007/978-3-030-49732-3.

Fahad S, Adnan M, Hassan S, Saud S, Hussain S, Wu C, Wang D, Hakeem KR, Alharby HF, Turan V, Khan MA, Huang J (2019b). Rice responses and tolerance to high temperature. In: Hasanuzzaman M, Fujita M, Nahar K, Biswas JK (eds) Advances in rice research for abiotic stress tolerance. Woodhead Publications Ltd, Abington, Cambridge, pp 201–224.

Fahad S, Rehman A, Shahzad B, Tanveer M, Saud S, Kamran M, Ihtisham M, Khan SU, Turan V, Rahman MHU (2019a) Rice responses and tolerance to metal/metalloid toxicity. In: Hasanuzzaman M, Fujita M, Nahar K, Biswas JK (eds) Advances in rice research for abiotic stress tolerance. Woodhead Publications Ltd, Abington, Cambridge, pp 299–312.

Fahad S, Sönmez O, Saud S, Wang D, Wu C, Adnan M, Arif M. (Eds.) (2021e). Engineering tolerance in crop plants against abiotic stress, First edn, Footprints of climate variability on plant diversity. CRC Press, Boca Raton, FL.

Fahad S, Sönmez O, Saud S, Wang D, Wu C, Adnan M, Turan V. (Eds.) (2021b). Climate change and plants: biodiversity, growth and interactions, First edn, Footprints of climate variability on plant diversity. CRC Press, Boca Raton, FL.

Fahad S, Sönmez O, Saud S, Wang D, Wu C, Adnan M, Turan V. (Eds.) (2021c). Developing climate resilient crops: improving global food security and safety, First edn, Footprints of climate variability on plant diversity. CRC Press, Boca Raton, FL.

Fahad S, Sönmez O, Saud S, Wang D, Wu C, Adnan M, Turan V. (Eds.) (2021a). Plant growth regulators for climate-smart agriculture, First edn, Footprints of climate variability on plant diversity. CRC Press, Boca Raton, FL.

Fahad S, Sönmez O, Turan V, Adnan M, Saud S, Wu C, Wang D. (Eds.) (2021d). Sustainable soil and land management and climate change, First edn, Footprints of climate variability on plant diversity. CRC Press, Boca Raton, FL.

Farah R, Muhammad R, Muhammad SA, Tahira Y, Muhammad AA, Maryam A, Shafaqat A, Rashid M, Muhammad R, Qaiser H, Afia Z, Muhammad AA, Muhammad A, Fahad S (2020) Alternative and non-conventional soil and crop management strategies for increasing water use efficiency. In: Fahad S, Hasanuzzaman M, Alam M, Ullah H, Saeed M, Khan AK, Adnan M (eds) Environment, climate, plant and vegetation growth. Springer, pp 323–338. https://doi.org/10.1007/978-3-030-49732-3.

Farhana G, Ishfaq A, Muhammad A, Dawood J, Fahad S, Xiuling L, Depeng W, Muhammad F, Muhammad F, Syed AS (2020) Use of crop growth model to simulate the impact of climate change on yield of various wheat cultivars under different agro-environmental conditions in Khyber Pakhtunkhwa, Pakistan. Arab J Geosci 13:112. https://doi.org/10.1007/s12517-020-5118-1.

Farhat A, Hafiz MH, Wajid I, Aitazaz AF, Hafiz FB, Zahida Z, Fahad S, Wajid F, Artemi C (2020) A review of soil carbon dynamics resulting from agricultural practices. J Environ Manage 268:110319.

Fazli W, Muhammad S, Amjad A, Fahad S, Muhammad A, Muhammad N, Ishaq AM, Imtiaz AK, Mukhtar A, Muhammad S, Muhammad I, Rafi U, Haroon I, Muhammad A (2020) Plant-microbes interactions and functions in changing climate. In: Fahad S, Hasanuzzaman M, Alam M, Ullah H, Saeed M, Khan AK, Adnan M (eds) Environment, climate, plant and vegetation growth. Springer, pp 397–420. https://doi.org/10.1007/978-3-030-49732-3.

Foissner W (1997) Global soil ciliate (Protozoa, Ciliophora) diversity: a probability-based approach using large sample collections from Africa, Australia and Antarctica. Biodivers Conserv 6:1627–1638.

Franzluebbers AJ (1999). Microbial activity in response to water-filled pore space of variably eroded southern Piedmont soils. Appl Soil Ecol 11:91–101.

Frey SD, Drijber R, Smith H, Melillo J (2008) Microbial biomass, functional capacity, and community structure after 12 years of soil warming. Soil Biol Biochem 40:2904–2907.

Frey SD, Knorr M, Parrent JL, Simpson RT (2004). Chronic nitrogen enrichment affects the structure and function of the soil microbial community in temperate hardwood and pine forests. Forest Ecol Manag 196:159–171.

Ghulam M, Muhammad AA, Donald LS, Sajid M, Muhammad FQ, Niaz A, Ateeq ur Rehman AR, Shakeel A, Sajjad H, Muhammad A, Summia M, Aqib HAK, Fahad S, Rahul D, Mazhar I, Timothy DS (2021) Formalin fumigation and steaming of various composts differentially influence the nutrient release, growth and yield of muskmelon (*Cucumis melo* L.). Sci Rep 11:21057.

Gopakumar L, Bernard NO, Donato V (2020) Soil microarthropods and nutrient cycling. In: Fahad S, Hasanuzzaman M, Alam M, Ullah H, Saeed M, Khan AK, Adnan M (eds) Environment, climate, plant and vegetation growth. Springer, pp 453–472. https://doi.org/10.1007/978-3-030-49732-3.

Gouin T, Armitage JM, Cousins IT, Muir DC, Ng CA, Reid L, Tao S (2013) Influence of global climate change on chemical fate and bioaccumulation: the role of multimedia models. Environ Toxicol Chem 32:20–31.

Guofu L, Zhenjian B, Fahad S, Guowen C, Zhixin X, Hao G, Dandan L, Yulong L, Bing L, Guoxu J, Saud S (2021) Compositional and structural changes in soil microbial communities in response to straw mulching and plant revegetation in an abandoned artificial pasture in Northeast China. Glob Ecol Conserv 31:e01871.

Habib-Ur-Rahman M, Ashfaq A, Aftab W, Manzoor H, Fahd R, Wajid I, Md. Aminul I, Vakhtang S, Muhammad A, Asmat U, Abdul W, Syeda RS, Shah S, Shahbaz K, Fahad S, Manzoor H, Saddam H, Wajid N (2017) Application of CSM-CROPGRO-cotton model for cultivars and optimum planting dates: evaluation in changing semi-arid climate. Field Crops Res. http://doi.org/10.1016/j.fcr.2017.07.007.

Hafeez M, Farman U, Muhammad MK, Xiaowei L, Zhijun Z, Sakhawat S, Muhammad I, Mohammed AA, Mandela F-G, Nicolas D, Muzammal R, Fahad S, Yaobin L (2021) Metabolic-based insecticide resistance mechanism and e approaches for controlling of beet armyworm Spodoptera exigua: a review. Environ Sci Pollut Res. https://doi.org/10.1007/s11356-021-16974-w.

Hafiz MH, Abdul K, Farhat A, Wajid F, Fahad S, Muhammad A, Ghulam MS, Wajid N, Muhammad M, Hafiz FB (2020b) Comparative effects of organic and inorganic fertilizers on soil organic carbon and wheat productivity under arid region. Commun Soil Sci Plant Anal. https://doi.org/10.1080/00103624.2020.1763385.

Hafiz MH, Farhat A, Ashfaq A, Hafiz FB, Wajid F, Carol Jo W, Fahad S, Gerrit H (2020a) Predicting kernel growth of maize under controlled water and nitrogen applications. Int J Plant Prod. https://doi.org/10.1007/s42106-020-00110-8.

Hafiz MH, Farhat A, Shafqat S, Fahad S, Artemi C, Wajid F, Chaves CB, Wajid N, Muhammad M, Hafiz FB (2018) Offsetting land degradation through nitrogen and water management during maize cultivation under arid conditions. Land Degrad Dev 1–10. https://doi.org/10.1002/ldr.2933.

Hafiz MH, Muhammad A, Farhat A, Hafiz FB, Saeed AQ, Muhammad M, Fahad S, Muhammad A (2019) Environmental factors affecting the frequency of road traffic accidents: a case study of sub-urban area of Pakistan. Environ Sci Pollut Res https://doi.org/10.1007/s11356-019-04752-8.

Hafiz MH, Wajid F, Farhat A, Fahad S, Shafqat S, Wajid N, Hafiz FB (2016) Maize plant nitrogen uptake dynamics at limited irrigation water and nitrogen. Environ Sci Pollut Res 24(3):2549–2557. https://doi.org/10.1007/s11356-016-8031-0.

Haider SA, Lalarukh I, Amjad SF, Mansoora N, Naz M, Naeem M, Bukhari SA, Shahbaz M, Ali SA, Marfo TD, Subhan D, Rahul D, Fahad S (2021) Drought stress alleviation by potassium-nitrate-containing chitosan/montmorillonite microparticles confers changes in Spinacia oleracea L. Sustainability 13:9903. https://doi.org/10.3390/su13179903.

Hamza SM, Xiukang W, Sajjad A, Sadia Z, Muhammad N, Adnan M, Fahad S et al. (2021) Interactive effects of gibberellic acid and NPK on morpho-physio-biochemical traits and organic acid exudation pattern in coriander (Coriandrum sativum L.) grown in soil artificially spiked with boron. Plant Physiol Biochem 167:884–900.

Harte J, Rawa A, Price V (1996) Effects of manipulated soil microclimate on mesofaunal biomass and diversity. Soil Biol Biochem 28:313–322.

Hartley IP, Hopkins DW, Garnett MH, Sommerkorn M, Wookey PA (2008) Soil microbial respiration in arctic soil does not acclimate to temperature. Ecol lett 11:1092–1100.

Haynes RJ (2008) Soil organic matter quality and the size and activity of the microbial biomass: their significance to the quality of agricultural soils. In: Huang Q, Pan Ming H, Violante A (eds) Soil mineral microbe-organic interactions. Springer, Berlin, Heidelberg, pp 201–231.

Hesham FA, Fahad S (2020) Melatonin application enhances biochar efficiency for drought tolerance in maize varieties: modifications in physio-biochemical machinery. Agron J 112(4):1–22.

Holmstrup M (2001) Sensitivity of life history parameters in the earthworm Aporrectodea caliginosa to small changes in soil water potential. Soil Biol Biochem 33:1217–1223.

Hoschitz M, Kaufmann R (2004) Soil nematode communities of Alpine summits–site differentiation and microclimatic influences. Pedobiologia 48:313–320.

Howard DM, Howard PJA (1993) Relationships between $CO_2$ evolution, moisture content and temperature for a range of soil types. Soil Biol Biochem 25:1537–1537.

Hu S, Chapin III FS, Firestone MK, Field CB, Chiariello NR (2001) Nitrogen limitation of microbial decomposition in a grassland under elevated $CO_2$. Nature 409(6817):188.

Huang Li Y, Li X-X, Zhang Y-B, Fahad S, Wang F (2021) depl improves rice grain yield and nitrogen use efficiency simultaneously by enhancing nitrogen and dry matter translocation. J Integr Agric. https://doi.org/10.1016/S2095-3119(21)63795-4.

Hunt HW, Coleman DC, Ingham ER, Ingham RE, Elliott ET, Moore JC, Morley CR (1987). The detrital food web in a shortgrass prairie. Biol Fertil Soils 3:57–68.

Hussain MA, Fahad S, Rahat S, Muhammad FJ, Muhammad M, Qasid A, Ali A, Husain A, Nooral A, Babatope SA, Changbao S, Liya G, Ibrar A, Zhanmei J, Juncai H (2020) Multifunctional role of brassinosteroid and its analogues in plants. Plant Growth Regul. https://doi.org/10..1007/s10725-020-00647-8.

Huq M, Shoeb AZ, Hossain MA, Fahad S, Kamruzzaman MM, Javed A, Saleem N, Adnan KM, Sarker SA, Ali MY, Sarven M (2020) Measuring vulnerability to environmental hazards: qualitative to quantitative. In: Fahad S, Hasanuzzaman M, Alam M, Ullah H, Saeed M, Khan AK, Adnan M (eds) Environment, climate, plant and vegetation growth. Springer, pp 421–452. https://doi.org/10.1007/978-3-030-49732-3.

Ibrar K, Aneela R, Khola Z, Urooba N, Sana B, Rabia S, Ishtiaq H, Ur Rehman M, Salvatore M (2020) Microbes and environment: global warming reverting the frozen zombies. In: Fahad S, Hasanuzzaman M, Alam M, Ullah H, Saeed M, Khan AK, Adnan M (eds) Environment, climate, plant and vegetation growth. Springer, pp 607–634. https://doi.org/10.1007/978-3-030-49732-3.

Ikram U, Khadim D, Muhammad T, Muhammad S, Fahad S (2021) Gibberellic acid and urease inhibitor optimize nitrogen uptake and yield of maize at varying nitrogen levels under changing climate. Environ Sci Pollut Res. https://doi.org/10.1007/s11356-021-16049-w.

Ilyas M, Mohammad N, Nadeem K, Ali H, Aamir HK, Kashif H, Fahad S, Aziz K, Abid U, (2020) Drought tolerance strategies in plants: a mechanistic approach. J Plant Growth Regul. https://doi.org/10.1007/s00344-020-10174-5.

Ingham ER, Trofymow JA, Ames R, Hunt HW, Morley CR, Moore JC, Coleman DC (1986) Trophic interactions and nitrogen cycling in a semi-arid grassland soil. I. Seasonal dynamics of the natural populations, their interactions and effects on nitrogen cycling. J Appl Ecol 23:597–614.

Ingham RE, Trofymow JA, Ingham ER, Coleman DC (1985). Interactions of bacteria, fungi, and their nematode grazers: effects on nutrient cycling and plant growth. Ecol Monogr 55:119–140.

Iqra M, Amna B, Shakeel I, Fatima K, Sehrish L, Hamza A, Fahad S (2020) Carbon cycle in response to global warming. In: Fahad S, Hasanuzzaman M, Alam M, Ullah H, Saeed M, Khan AK, Adnan M (eds) Environment, climate, plant and vegetation growth. Springer, pp 1–16. https://doi.org/10.1007/978-3-030-49732-3.

Irfan M, Muhammad M, Muhammad JK, Khadim MD, Dost M, Ishaq AM, Waqas A, Fahad S, Saud S et al. (2021) Heavy metals immobilization and improvement in maize (Zea mays L.) growth amended with biochar and compost. Sci Rep 11:18416.

Jabborova D, Sulaymanov K, Sayyed RZ, Alotaibi SH, Enakiev Y, Azimov A, Jabbarov Z, Ansari MJ, Fahad S, Danish S et al. (2021) Mineral fertilizers improves the quality of turmeric and soil. Sustainability 13:9437. https://doi.org/10.3390/su13169437.

Jacobson MZ (2005) Correction to "Control of fossil-fuel particulate black carbon and organic matter, possibly the most effective method of slowing global warming". J Geo Res Atmos 110:D14.

Jan M, Anwar-ul-Haq M, Adnan NS, Muhammad Y, Javaid I, Xiuling L, Depeng W, Fahad S, (2019) Modulation in growth, gas exchange, and antioxidant activities of salt-stressed rice (Oryza sativa L.) genotypes by zinc fertilization. Arab J Geosci 12:775. https://doi.org/10.1007/s12517-019-4939-2.

Jing X, Sanders NJ, Shi Y, Chu H, Classen AT, Zhao K, He JS (2015) The links between ecosystem multi functionality and above-and belowground biodiversity are mediated by climate. Nat Commun 6:8159.

Kamarn M, Wenwen C, Irshad A, Xiangping M, Xudong Z, Wennan S, Junzhi C, Shakeel A, Fahad S, Qingfang H, Tiening L (2017) Effect of paclobutrazol, a potential growth regulator on stalk mechanical strength, lignin accumulation and its relation with lodging resistance of maize. Plant Growth Regul 84:317–332. https://doi.org/10.1007/s10725-017-0342-8.

Karl TR, Trenberth KE (2003) Modern global climate change. Sci 302:1719–1723.

Karmakar R, Das I, Dutta D, Rakshit A (2016) Potential effects of climate change on soil properties: a review. Sci Int 4:51–73.

Khadim D, Fahad S, Jahangir MMR, Iqbal M, Syed SA, Shah AK, Ishaq AM, Rahul D et al. (2021a) Biochar and urease inhibitor mitigate $NH_3$ and $N_2O$ emissions and improve wheat yield in a urea fertilized alkaline soil. Sci Rep 11:17413.

Khadim D, Saif-ur-R M, Fahad S, Syed SA, Shah AK et al. (2021b) Influence of variable biochar concentration on yield-scaled nitrous oxide emissions, wheat yield and nitrogen use efficiency. Sci Rep 11:16774.

Khan MMH, Niaz A, Umber G, Muqarrab A, Muhammad AA, Muhammad I, Shabir H, Shah F, Vibhor A, Shams HA, Reham A, Syed MBA, Nadiyah M, Ali TKZ, Subhan D, Rahul D (2021) Synchronization of Boron application methods and rates is environmentally friendly approach to improve quality attributes of Mangifera indica L. On sustainable basis. Saudi J Bio Sci. https://doi.org/10.1016/j.sjbs.2021.10.036.

Killham K (1994) Soil ecology. Cambridge University Press, Cambridge.

Knorr W, Prentice IC, House JI, Holland EA (2005) Long-term sensitivity of soil carbon turnover to warming. Nature 433:298.

Lavelle P, Spain AV (2001) Soil ecology. Springer Science & Business Media, London.

Magdoff F, Weil RR (2004) Soil organic matter in sustainable agriculture. CRC Press, Boca Raton, FL.

Mahar A, Amjad A, Altaf HL, Fazli W, Ronghua L, Muhammad A, Fahad S, Muhammad A, Rafiullah, Imtiaz AK, Zengqiang Z (2020) Promising technologies for Cd-contaminated soils: drawbacks and possibilities. In: Fahad S, Hasanuzzaman M, Alam M, Ullah H, Saeed M, Khan AK, Adnan M (eds) Environment, climate, plant and vegetation growth. Springer, pp 63–92. https://doi.org/10.1007/978-3-030-49732-3.

Mahmood UH, Tassaduq R, Chandni I, Adnan A, Muhammad A, Muhammad MA, Habib-ur-Rehman M, Mehmood AN, Alam S, Fahad S (2021) Linking plants functioning to adaptive responses under heat stress conditions: a mechanistic review. J Plant Growth Regul https://doi.org/10.1007/s00344-021-10493-1.

Mandal A (2012) Impact of climate change on soil biodiversity – a review. Agric Rev 33:283–292.

McCarthy JJ, Canziani OF, Leary NA, Dokken DJ, White KS (Eds.) (2001) Climate change 2001: impacts, adaptation, and vulnerability: contribution of Working Group II to the third assessment report of the Intergovernmental Panel on Climate Change, Vol. 2. Cambridge University Press, Cambridge.

McSorley R (2003). Adaptations of nematodes to environmental extremes. Fla Entomol 86:138–143.

Md Jakir H, Allah B (2020) Development and applications of transplastomic plants; a way towards eco-friendly agriculture. In: Fahad S, Hasanuzzaman M, Alam M, Ullah H, Saeed M, Khan AK, Adnan M (eds) Environment, climate, plant and vegetation growth. Springer, pp 285–322. https://doi.org/10.1007/978-3-030-49732-3.

Mora C, Frazier AG, Longman RJ, Dacks RS, Walton MM, Tong EJ, Ambrosino CM (2013) The projected timing of climate departure from recent variability. Nature 502:183.

Moscatelli MC, Lagomarsino A, Marinari S, De Angelis P, Grego S (2005) Soil microbial indices as bioindicators of environmental changes in a poplar plantation. Ecol Indic 5:171–179.

Mubeen M, Ashfaq A, Hafiz MH, Muhammad A, Hafiz UF, Mazhar S, Sami ul Din M, Asad A, Amjed A, Fahad S, Wajid N (2020) Evaluating the climate change impact on water use efficiency of cotton-wheat in semi-arid conditions using DSSAT model. J Water Clim Change. https://doi.org/10.2166/wcc.2019.179/622035/jwc2019179.pdf.

Muhammad N, Muqarrab A, Khurram S, Fiaz A, Fahim N, Muhammad A, Shazia A, Omaima N, Sulaiman AA, Fahad S, Subhan D, Rahul D (2021) Kaolin and Jasmonic acid improved cotton productivity under water stress conditions. J Saudi Soc Agric Sci 28:6606–6614. https://doi.org/10.1016/j.sjbs.2021.07.043.

Muhammad Z, Abdul MK, Abdul MS, Kenneth BM, Muhammad S, Shahen S, Ibadullah J, Fahad S (2019) Performance of Aeluropus lagopoides (mangrove grass) ecotypes, a potential turfgrass, under high saline conditions. Environ Sci Pollut Res https://doi.org/10.1007/s11356-019-04838-3.

Muzammal R, Fahad S, Guanghui D, Xia C, Yang Y, Kailei T, Lijun L, Fei-Hu L, Gang D (2021) Evaluation of hemp (Cannabis sativa L.) as an industrial crop: a review. Environ Sci Pollution Res. https://doi.org/10.1007/s11356-021-16264-5.

Noor M, ur Rehman N, Ajmal J, Fahad S, Muhammad A, Fazli W, Saud S, Hassan S (2020) Climate change and costal plant lives. In: Fahad S, Hasanuzzaman M, Alam M, Ullah H, Saeed M, Khan AK, Adnan M (eds) Environment, climate, plant and vegetation growth. Springer, pp 93–108. https://doi.org/10.1007/978-3-030-49732-3.

Novotny AM, Schade JD, Hobbie SE, Kay AD, Kyle M, Reich PB, Elser JJ (2007) Stoichiometric response of nitrogen-fixing and non-fixing dicots to manipulations of $CO_2$, nitrogen, and diversity. Oecologia 151:687–696.

Noyes PD, McElwee MK, Miller HD, Clark BW, Van Tiem LA, Walcott KC, Levin ED (2009) The toxicology of climate change: environmental contaminants in a warming world. Environ Int 35:971–986.

Papatheodorou EM, Argyropoulou MD, Stamou GP (2004) The effects of large-and small-scale differences in soil temperature and moisture on bacterial functional diversity and the community of bacterivorous nematodes. Appl Soil Ecol 25: 37–49.

Pariente S (2001) Soluble salts dynamics in the soil under different climatic conditions. Catena 43:307–321.

Phillips RP, Finzi AC, Bernhardt ES (2011) Enhanced root exudation induces microbial feedbacks to N cycling in a pine forest under long-term $CO_2$ fumigation. Ecol lett 14: 187–194.

Qadeem W, Sattar S, Adnan M, Zaman M, Ali I, Shah SRA, Junaid K, Said F, Rahman IU, Saleem N (2015) Efficacy of botanical and microbial extracts against Angoumois grain moth *Sitotrogacerealella* (Olivier) (Lepidoptera: Gelechiidae) under laboratory conditions. J Entomol Zool Studies 3(5):451–454.

Qamar Z, Zubair A, Muhammad Y, Muhammad ZI, Abdul K, Fahad S, Safder B, Ramzani PMA, Muhammad N (2017) Zinc biofortification in rice: leveraging agriculture to moderate hidden hunger in developing countries. Arch Agron Soil Sci 64:147–161. https://doi.org/10.1080/03650340.2017.1338343.

Qian B, Gregorich EG, Gameda S, Hopkins DW, Wang XL (2011). Observed soil temperature trends associated with climate change in Canada. J Geo Res Atmos 116:3–25.

Rahman UI, Ali S, Rahman LU, Adnan M, Ibrahim M, Saleem N, Irshad M (2016) Effect of seed priming on growth parameters of Okra (Abelmoschusesculentus L.). Pure Appl Biol 5(1):165–171.

Rashid M, Qaiser H, Khalid SK, Mohammad I. Al-Wabel, Zhang A, Muhammad A, Shahzada SI, Rukhsanda A, Ghulam AS, Shahzada MM, Sarosh A, Muhammad FQ (2020) Prospects of biochar in alkaline soils to mitigate climate change. In: Fahad S, Hasanuzzaman M, Alam M, Ullah H, Saeed M, Khan AK, Adnan M (eds) Environment, climate, plant and vegetation growth. Springer, pp 133–150. https://doi.org/10.1007/978-3-030-49732-3.

Rehman M, Fahad S, Saleem MH, Hafeez M, Habib-Ur-Rahman M, Liu F, Deng G (2020) Red light optimized physiological traits and enhanced the growth of ramie (Boehmeria nivea L.). Photosynthetica 58(4):922–931.

Reich PB (2009) Elevated $CO_2$ reduces losses of plant diversity caused by nitrogen deposition. Science 326:1399–1402.

Rengel Z (2011) Soil pH, soil health and climate change. In: Soil health and climate change. Springer, Berlin, Heidelberg, pp 69–85.

Reth S, Reichstein M, Falge E (2005) The effect of soil water content, soil temperature, soil pH-value and the root mass on soil $CO_2$ efflux–a modified model. Plant Soil 268:21–33.

Rinnan R, Michelsen A, Bååth E, Jonasson S (2007) Fifteen years of climate change manipulations alter soil microbial communities in a subarctic heath ecosystem. Glob Change Biol 13:28–39.

Rohr JR, Johnson P, Hickey CW, Helm RC, Fritz A, Brasfield S (2013) Implications of global climate change for natural resource damage assessment, restoration, and rehabilitation. Environ Toxicol Chem 32:93–101.

Roy S, Roy MM (2006) Spatial distribution and seasonal abundance of soil mites and collembola in grassland and Leucaena plantation in a semi-arid region. Trop Ecol 47:57–62.

Ruess L, Michelsen A, Schmidt IK, Jonasson S (1999) Simulated climate change affecting microorganisms, nematode density and biodiversity in subarctic soils. Plant Soil 212:63–73.

Sabagh AEL, Hossain A, Barutçular C, Iqbal MA, Sohidul Islam M, Fahad S, Sytar O, Çig F, Meena RS, Erman M (2020) Consequences of salinity stress on the quality of crops and its mitigation strategies for sustainable crop production: an outlook of arid and semi-arid regions. In: Fahad S, Hasanuzzaman M, Alam M, Ullah H, Saeed M, Khan AK, Adnan M (eds) Environment, climate, plant and vegetation growth. Springer, pp 503–534. https://doi.org/10.1007/978-3-030-49732-3.

Sadam M, Tahir ul Qamar M, Ghulam M, Muhammad SK, Faiz AJ (2020) Role of biotechnology in climate resilient agriculture. In: Fahad S, Hasanuzzaman M, Alam M, Ullah H, Saeed M, Khan AK, Adnan M (eds) Environment, climate, plant and vegetation growth. Springer, pp 339–366. https://doi.org/10.1007/978-3-030-49732-3.

Safi UK, Ullah F, Mehmood S, Fahad S, Ahmad Rahi A, Althobaiti F, et al. (2021) Antimicrobial, antioxidant and cytotoxic properties of Chenopodium glaucum L. PLoS One 16(10):e0255502. https://doi.org/10.1371/journal.pone.0255502.

Sahrish N, Shakeel A, Ghulam A, Zartash F, Sajjad H, Mukhtar A, Muhammad AK, Ahmad K, Fahad S, Wajid N, Sezai E, Carol Jo W, Gerrit H (2022) Modeling the impact of climate warming on potato phenology. Eur J Agron 132:126404.

Sajid H, Jie H, Jing H, Shakeel A, Satyabrata N, Sumera A, Awais S, Chunquan Z, Lianfeng Z, Xiaochuang C, Qianyu J, Junhua Z (2020) Rice production under climate change: adaptations and mitigating strategies. In: Fahad S, Hasanuzzaman M, Alam M, Ullah H, Saeed M, Khan AK, Adnan M (eds) Environment, climate, plant and vegetation growth. Springer, pp 659–686. https://doi.org/10.1007/978-3-030-49732-3.

Sajjad H, Muhammad M, Ashfaq A, Waseem A, Hafiz MH, Mazhar A, Nasir M, Asad A, Hafiz UF, Syeda RS, Fahad S, Depeng W, Wajid N (2019) Using GIS tools to detect the land use/land cover changes during forty years in Lodhran district of Pakistan. Environ Sci Pollut Res. https://doi.org/10.1007/s11356-019-06072-3.

Saleem MH, Fahad S, Adnan M, Mohsin A, Muhammad SR, Muhammad K, Qurban A, Inas AH, Parashuram B, Mubassir A, Reem MH (2020a) Foliar application of gibberellic acid endorsed phytoextraction of copper and alleviates oxidative stress in jute (Corchorus capsularis L.) plant grown in highly copper-contaminated soil of China. Environ Sci Pollut Res. https://doi.org/10.1007/s11356-020-09764-3.

Saleem MH, Fahad S, Shahid UK, Mairaj D, Abid U, Ayman ELS, Akbar H, Analía L, Lijun L (2020c) Copper-induced oxidative stress, initiation of antioxidants and phytoremediation potential of flax (Linum usitatissimum L.) seedlings grown under the mixing of two different soils of China. Environ Sci Poll Res https://doi.org/10.1007/s11356-019-07264-7.

Saleem MH, Rehman M, Fahad S, Tung SA, Iqbal N, Hassan A, Ayub A, Wahid MA, Shaukat S, Liu L, Deng G (2020b) Leaf gas exchange, oxidative stress, and physiological attributes of rapeseed (Brassica napus L.) grown under different light-emitting diodes. Photosynthetica 58(3):836–845.

Saleem N, Adnan M, Khan NA, Zaheer S, Jalal F, Amin M, Khan WM, Arif M, Rahman I, Ibrahim M, Jamal Y, Shah SRA, Junaid K, Ali M (2015) Dual purpose canola: grazing and grains options. Pak J Weed Sci Res 21(2):295–304.

Saleh J, Maftoun M (2008) Interactive effect of NaCl levels and Zinc sources and levels on the growth and mineral composition of rice. J Agric Sci Tech 10:325–336.

Saman S, Amna B, Bani A, Tahir ul Qamar M, Rana MA, Muhammad SK (2020) QTL mapping for abiotic stresses in cereals. In: Fahad S, Hasanuzzaman M, Alam M, Ullah H, Saeed M, Khan AK, Adnan M (eds) Environment, climate, plant and vegetation growth. Springer, pp 229–252. https://doi.org/10.1007/978-3-030-49732-3.

Saud S, Chen Y, Fahad S, Hussain S, Na L, Xin L, Alhussien SA (2016) Silicate application increases the photosynthesis and its associated metabolic activities in Kentucky bluegrass under drought stress and post-drought recovery. Environ Sci Pollut Res 23(17):17647–17655. https://doi.org/10.1007/s11356-016-6957-x.

Saud S, Chen Y, Long B, Fahad S, Sadiq A (2013) The different impact on the growth of cool season turf grass under the various conditions on salinity and drought stress. Int J Agric Sci Res 3:77–84.

Saud S, Fahad S, Cui G, Chen Y, Anwar S (2020) Determining nitrogen isotopes discrimination under drought stress on enzymatic activities, nitrogen isotope abundance and water contents of Kentucky bluegrass. Sci Rep 10:6415. https://doi.org/10.1038/s41598-020-63548-w.

Saud S, Fahad S, Yajun C, Ihsan MZ, Hammad HM, Nasim W, Arif M, Alharby H (2017) Effects of nitrogen supply on water stress and recovery mechanisms in Kentucky bluegrass plants. Front Plant Sci 8:983. https://doi.org/10.3389/fpls.2017.00983.

Saud S, Li X, Chen Y, Zhang L, Fahad S, Hussain S, Sadiq A, Chen Y (2014) Silicon application increases drought tolerance of Kentucky bluegrass by improving plant water relations and morph physiological functions. SciWorld J 2014:1–10. https://doi.org/10.1155/2014/368694.

Schadt CW, Martin AP, Lipson DA, Schmidt SK (2003) Seasonal dynamics of previously unknown fungal lineages in tundra soils. Science 301:1359–1361.

Schimel J, Balser TC, Wallenstein M (2007) Microbial stress-response physiology and its implications for ecosystem function. Ecology 88:1386–1394.

Schjønning P, Thomsen IK, Moldrup P, Christensen BT (2003) Linking soil microbial activity to water-and air-phase contents and diffusivities. Soil Sci Soc Am J 67:156–165.

Schmidt GA, Ruedy RA, Miller RL, Lacis AA (2010) Attribution of the present-day total greenhouse effect. J Geo Res Atmos 27:115(D20).

Senol C (2020) The effects of climate change on human behaviors. In: Fahad S, Hasanuzzaman M, Alam M, Ullah H, Saeed M, Khan AK, Adnan M (eds) Environment, climate, plant and vegetation growth. Springer, pp 577–590. https://doi.org/10.1007/978-3-030-49732-3.

Shafi MI, Adnan M, Fahad S, Fazli W, Ahsan K, Zhen Y, Subhan D, Zafar-ul-Hye M, Martin B, Rahul D (2020) Application of single superphosphate with humic acid improves the growth, yield and phosphorus uptake of wheat (Triticum aestivum L.) in calcareous soil. Agronomy 10:1224. https://doi.org/10.3390/agronomy10091224.

Shah B, Khan IA, Akbar R, Fayaz W, Zaman M, Din MMU, Ahmad N, Adnan M, Junaid K, Shah SRA, Rahman IU (2016) Population dynamics of corn flea beetle and fourspotted leaf beetle on different maize cultivars at Peshawar. J Entomol Zool Studies 4(1):308–311.

Shah F, Lixiao N, Kehui C, Tariq S, Wei W, Chang C, Liyang Z, Farhan A, Fahad S, Huang J (2013) Rice grain yield and component responses to near 2°C of warming. Field Crop Res 157:98–110.

Siddiqui MH, Alamri S, Mukherjee S, Al-Amri AA, Alsubaie QD, Al-Munqedhi BM, Ali HM, Kalaji HM, Fahad S, Rajput VD, Narayan OP (2021) Molybdenum and hydrogen sulfide synergistically mitigate arsenic toxicity by modulating defense system, nitrogen and cysteine assimilation in faba bean (*Vicia faba* L.) seedlings. Environ Pollut 290:117953.

Sidra K, Javed I, Subhan D, Allah B, Syed IUSB, Fatma B, Khaled DA, Fahad S, Omaima N, Ali TKZ, Rahul D (2021) Physio-chemical characterization of indigenous agricultural waste materials for the development of potting media. J Saudi Soc Agric Sci. https://doi.org/10.1016/j.sjbs.2021.08.058.

Silva CC, Guido ML, Ceballos JM, Marsch R, Dendooven L (2008) Production of carbon dioxide and nitrous oxide in alkaline saline soil of Texcoco at different water contents amended with urea: a laboratory study. Soil Biol Biochem 40:1813–1822.

Smith JL, Halvorson JJ, Bolton Jr, H (2002) Soil properties and microbial activity across a 500 m elevation gradient in a semi-arid environment. Soil Biol Biochem 34:1749–1757.

Steinweg JM, Plante AF, Conant RT, Paul EA, Tanaka DL (2008) Patterns of substrate utilization during long-term incubations at different temperatures. Soil Biol Biochem 40: 2722–2728.

Strong DT, Wever HD, Merckx R, Recous S (2004) Spatial location of carbon decomposition in the soil pore system. Eur J Soil Sci 55: 739–750.

Subhan D, Zafar-ul-Hye M, Fahad S, Saud S, Martin B, Tereza H, Rahul D (2020) Drought stress alleviation by ACC deaminase producing *Achromobacter xylosoxidans* and *Enterobacter cloacae*, with and without timber waste biochar in maize. Sustainability 12:6286. https://doi.org/10.3390/su12156286.

Swift MJ, Heal OW, Anderson JM, Anderson JM (1979) Decomposition in terrestrial ecosystems, Vol. 5. University of California Press, Berkeley.

Tahir ul Qamar M, Amna F, Amna B, Barira Z, Xitong Z, Ling-Ling C (2020) Effectiveness of conventional crop improvement strategies vs. omics. In: Fahad S, Hasanuzzaman M, Alam M, Ullah H, Saeed M, Khan AK, Adnan M (eds) Environment, climate, plant and vegetation growth. Springer, pp 253–284. https://doi.org/10.1007/978-3-030-49732-3.

Tariq M, Ahmad S, Fahad S, Abbas G, Hussain S, Fatima Z, Nasim W, Mubeen M, Habib-Ur-Rahman, Khan MA, Adnan M (2018) The impact of climate warming and crop management on phenology of sunflower-based cropping systems in Punjab, Pakistan. Agric For Metereol 256:270–282.

Unsar N-U, Muhammad R, Syed HMB, Asad S, Mirza AQ, Naeem I, Habib-Ur-Rahman M, Fahad S, Shafqat S (2020) Insect pests of cotton crop and management under climate change scenarios. In: Fahad S, Hasanuzzaman M, Alam M, Ullah H, Saeed M, Khan AK, Adnan M (eds) Environment, climate, plant and vegetation growth. Springer, pp 367–396. https://doi.org/10.1007/978-3-030-49732-3.

Vannier G (1994) The thermobiological limits of some freezing intolerant insects: the supercooling and thermostupor points. Acta Oecol 15:31–42.

Várallyay G (1990) Influence of climatic change on soil moisture regime, texture, structure and erosion. In: Scharpenseel HW, Schomaker M, Ayoub A (eds) Developments in soil science, Vol. 20. Elsevier, pp 39–49.

Várallyay G (2010) The impact of climate change on soils and on their water management. Agron Res 8:385–396.

Várallyay G, Farkas CS (2008) Impact of climate change on soils in Hungary. In: Harnos Z, Csete L (eds) Climate change: environment-risk-society. Research results. Szaktudás Kiado Haz, Budapest, pp 91–129.

Wahid F, Fahad S, Subhan D, Adnan M, Zhen Y, Saud S, Manzer HS, Martin B, Tereza H, Rahul D (2020) Sustainable management with mycorrhizae and phosphate solubilizing bacteria for enhanced phosphorus uptake in calcareous soils. Agriculture 10(334). https://doi.org/10.3390/agriculture10080334.

Wajid N, Ashfaq A, Asad A, Muhammad T, Muhammad A, Muhammad S, Khawar J, Ghulam MS, Syeda RS, Hafiz MH, Muhammad IAR, Muhammad ZH, Habib ur Rahman M, Veysel T, Fahad S, Suad S, Aziz K, Shahzad A (2017) Radiation efficiency and nitrogen fertilizer impacts on sunflower crop in contrasting environments of Punjab. Pak Environ Sci Pollut Res 25:1822–1836. https://doi.org/10.1007/s11356-017-0592-z.

Wall DH, Virginia RA (1999) Controls on soil biodiversity: insights from extreme environments. Appl Soil Ecol 13:137–150.

Wallwork JA (1970) Ecology of soil animals. Ecology of soil animals. McGraw-Hill, London.

Wu C, Kehui C, She T, Ganghua L, Shaohua W, Fahad S, Lixiao N, Jianliang H, Shaobing P, Yanfeng D (2020) Intensified pollination and fertilization ameliorate heat injury in rice (Oryza sativa L.) during the flowering stage. Field Crops Res 252:107795.

Wu C, Tang S, Li G, Wang S, Fahad S, Ding Y (2019) Roles of phytohormone changes in the grain yield of rice plants exposed to heat: a review. PeerJ 7:e7792. https://doi.org/10.7717/peerj.7792.

Wu Z, Dijkstra P, Koch GW, Peñuelas J, Hungate BA (2011) Responses of terrestrial ecosystems to temperature and precipitation change: a meta-analysis of experimental manipulation. Glob Change Biol 17:927–942.

Yang Z, Zhang Z, Zhang T, Fahad S, Cui K, Nie L, Peng S, Huang J (2017) The effect of season-long temperature increases on rice cultivars grown in the central and southern regions of China. Front Plant Sci 8:1908. https://doi.org/10.3389/fpls.2017.01908.

Zafar-ul-Hye M, Akbar MN, Iftikhar Y, Abbas M, Zahid A, Fahad S, Datta R, Ali M, Elgorban AM, Ansari MJ et al. (2021) Rhizobacteria inoculation and caffeic acid alleviated drought stress in lentil plants. Sustainability 13:9603. https://doi.org/10.3390/su13179603.

Zafar-ul-Hye M, Muhammad N, Subhan D, Fahad S, Rahul D, Mazhar A, Ashfaq AR, Martin B, Jiří H, Zahid HT, Muhammad N (2020a) Alleviation of cadmium adverse effects by improving nutrients uptake in bitter gourd through cadmium tolerant rhizobacteria. Environments 7(54). https://doi.org/10.3390/environments7080054.

Zafar-ul-Hye M, Tahzeeb-ul-Hassan M, Muhammad A, Fahad S, Martin B, Tereza D, Rahul D, Subhan D (2020b) Potential role of compost mixed biochar with rhizobacteria in mitigating lead toxicity in spinach. Sci Rep 10:12159. https://doi.org/10..1038/s41598-020-69183-9.

Zahida Z, Hafiz FB, Zulfiqar AS, Ghulam MS, Fahad S, Muhammad RA, Hafiz MH, Wajid N, Muhammad S (2017) Effect of water management and silicon on germination, growth, phosphorus and arsenic uptake in rice. Ecotoxicol Environ Saf 144:11–18.

Zahir SM, Zheng HG, Ala Ud D, Amjad A, Ata Ur R, Kashif J, Shah F, Saud S, Adnan M, Fazli W, Saud A, Manzer HS, Shamsher A, Wajid N, Hafiz MH, Fahad S (2021) Synthesis of silver nanoparticles using Plantago lanceolata extract and assessing their antibacterial and antioxidant activities. Sci Rep 11:20754

Zhou J, Xia B, Treves DS, Wu LY, Marsh TL, O'Neill RV, Tiedje JM (2002) Spatial and resource factors influencing high microbial diversity in soil. Appl Environ Microbiol 68:326–334.

Zhou J, Xue K, Xie J, Deng Y, Wu L, Cheng X, Luo Y (2012) Microbial mediation of carbon-cycle feedbacks to climate warming. Nat Clim Change 2:106.

Zia-ur-Rehman M (2020) Environment, climate change and biodiversity. In: Fahad S, Hasanuzzaman M, Alam M, Ullah H, Saeed M, Khan AK, Adnan M (eds) Environment, climate, plant and vegetation growth. Springer, pp 473–502. https://doi.org/10.1007/978-3-030-49732-3.

# 10

## *Biochar Magic against Abiotic Stresses*

**Mona H. Soliman**
*Taibah University*
*Cairo University*

**Haifa Abdulaziz S. Alhaithloul**
*Jouf University*

**Hossam Eldean A. Awad**
*Cairo University*

**Munazzam Jawad Shahid**
*Government College University Faisalabad*

**Shafaqat Ali**
*Government College University Faisalabad*
*China Medical University (CMU)*

**Amr A. El Kelish**
*Suez Canal University*

**CONTENTS**



DOI: 10.1201/9781003286400-10

## 10.1 Introduction

### 10.1.1 What Is "Biochar"?

Biochar is a solid carbonaceous material remaining after thermal decomposition of agricultural biomass and forest wastes to temperatures typically between 300°C and 700°C under oxygen-deprived conditions, a process known as "pyrolysis" (Fahad et al. 2015a; Fahad et al. 2016d). Moreover, gasification and hydrothermal carbonization, are alternative methods through which biochar can be obtained as a by-product by releasing the volatile gases leaving behind carbon-rich biochar. In distinction to the initial biomass feedstock that principally contains polysaccharide, hemicellulose, and lignin, biochar falls into the spectrum of materials referred to as "charcoal," or "black carbon"; however, it excludes black carbon derived from fossil fuels or non-biomass wastes. Biochar production systems can be ranged from small scale such as household cook stoves, which are low-cost and relatively simple technologies, to large industrial pyrolysis plants, which are more capital cost-intensive (Pratt and Moran 2010).

Understanding the efficiency of different biochar production systems and how these chars interact with different types of soils is the main practice of amending soils with organic C-rich by-products for fertility management goes back millennia. New instances of traditional practice using biochar are still being discovered around the world. Some of the most remarkable and currently best-understood properties of biochars include their effects on soil nutrient dynamics and the high stability of the carbon of which they are composed. Nowadays, biochar is best utilized as a tool to fight climate change and improve soil fertility. Biochar can store carbon in the soil for hundreds to thousands of years (Zimmerman et al. 2011; Shah et al. 2015; Fahad et al. 2015a; Fahad et al. 2016d). The potential use of organic C-rich by-products from the bioenergy industries improves soil fertility by retaining nutrients and the soil water content, potentially enhances nutrient bioavailability, and could also result in increased crop yields and replenish soil C in previously degraded soils by smallholder farmers and thereby, reducing the need for additional use of fertilizers (Chan and Xu 2009; El-Naggar et al. 2019; Forbes et al. 2006; Fahad et al.2015a; Fahad et al.2016d). Biochar systems are particularly relevant in developing country contexts and could be leveraged to address global challenges associated with food production and climate change.

### 10.1.2 Biochar Systems Technology: New Contribution to Safe Environment

Biochar systems are inherently complex, and further research and life-cycle analysis are needed to better understand their associated opportunities and risks in developing countries. Four main factors need to be taken into account:

1. **Improving soil and agricultural productivity**. This includes soil pH, nutrient availability, soil moisture, soil organic matter, and the amount of biochar applied (Figure 10.1).
2. **The impacts on climate change**. Carbon storage and stabilization are probably the most direct and important quality for climate change mitigation efforts based on biochar. Indirect factors contributing to emission reductions include the following: (a) reduced greenhouse gas emission and reduced use of fertilizer results in reduced emissions and use of other fertilizer products; (b) retention of nutrients like nitrogen in the soil limits consequent emission of nitrous oxide into the atmosphere; (c) as an agricultural waste diversion, thus avoiding methane emissions; and (d) biochar can be used as a promising potential substance in rehabilitating low fertility soils to improve crop yield, sustainability, and food security by increasing the microbial life in the soil and increases carbon storage in the soil (El-Naggar et al. 2019; Fahad et al. 2015a; Fahad et al. 2016d). Biochar exhibits a great potential to efficiently tackle water contaminants such as nitrate, phosphate, and toxic organic compounds as a result of agriculture techniques (Kookana 2010; Mizuta et al. 2004; Mohan et al. 2014; Tan et al. 2015). Biochar offers a simple and sustainable solution to waste management by converting agricultural waste into a powerful soil enhancer, and cropland diversity can be preserved, and deforestation discouraged

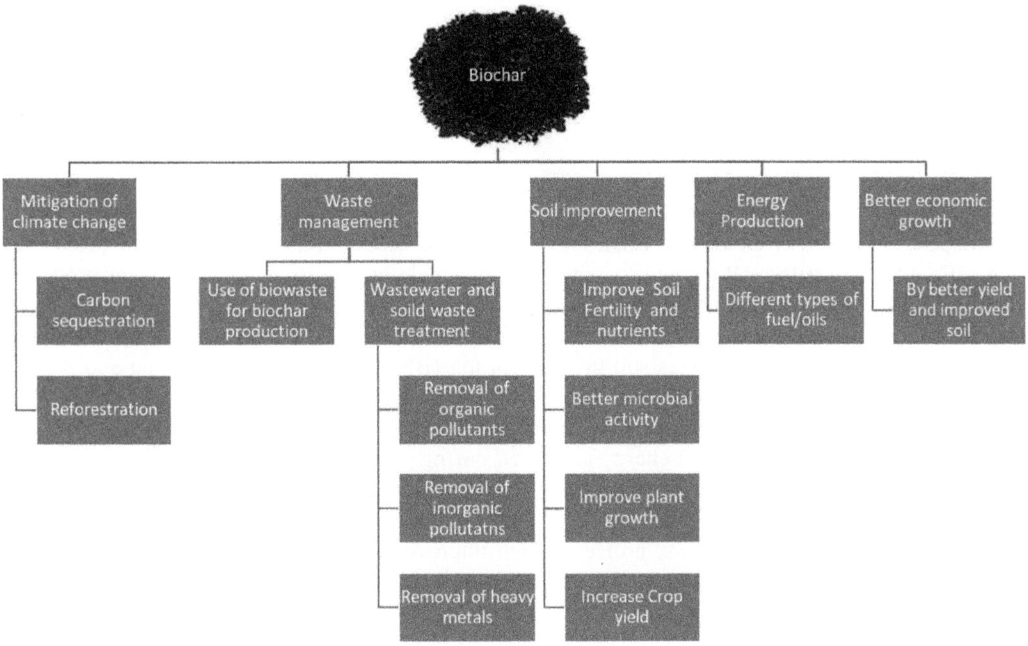


**FIGURE 10.1** Benefits associated with biochar.

(Lehmann and Joseph 2015). On the other hand, the risks of negative climate impact related to biochar lie primarily in the negative feedbacks that may occur directly or indirectly during biochar production and application. Such risks include emissions of methane and nitrous oxide during inefficient pyrolysis and degradation of soil organic matter after biochar application on unsuitable soils. Also, biomass cookstoves are significant sources of black carbon, which could have a global impact on climate and human health. The great advantage of biochar is that it is one of the few greenhouse gas reduction strategies that can withdraw carbon dioxide from the atmosphere, but the overall net climate impact of biochar could only be assessed through a full life-cycle assessment.

3. **The social impacts**. Biochar can help to improve food security by contributing to sustainable production systems and maintaining an eco-friendly environment (Qambrani et al. 2017) through affecting energy, health, and economics. The energy produced from biochar can be regarded as being the environmentally friendly energy storage technology obtained from renewable sources (Gao et al. 2017; Gu et al. 2015) and can potentially be used for such critical functions as refrigeration of vaccines, water pumping, and lighting after sunset. Biochar utilization as soil amendment is a proposed mechanism to repair degraded soil and increase fertility, further improve crop production, and increase conventional agricultural productivity to cope with global food demand in the next 30 years (Atkinson et al. 2010; Biederman and Harpole 2013; Lal 2015; Spokas et al. 2012). Biochar use can reduce pressure on wooded ecosystems and decrease the burden (mainly on women) of fuel gathering. Improved cookstoves can also reduce indoor air pollution throughout the developing world, especially for women and children. Risks of biochar use can include potential emission of toxins and inhalation of dust and small particulate matter; therefore, it is necessary to overcome multiple risks and constraints by both researchers and consumers to better public health.
4. **Competing uses of biomass**. The use of dedicated energy crops for biochar production could be strongly recommended as one of the best management practices to meet the challenges of upland agriculture. For example, the costs and benefits must, therefore, be weighed, of leaving biomass in situ versus using it to produce biochar that is then added to the soil.

## 10.2 Biochar for Sustainable Crop Growth, Soil Management and Yield Development in Agriculture

### 10.2.1 Introduction

Soil amendment with biochar significantly affects the physical characteristics of soil and directly affect the plant growth and development as growth survival is principally correlated to soil physical composition (Chan et al. 2008a; Chan and Xu 2009; Lehmann and Rondon 2006; Fahad et al. 2015a; Fahad et al. 2016d). Biochar is a potentially valuable as soil organic amendment produced from biomass by pyrolysis. Biochar enhances soil aggregation, increases the capacity for soil nutrient retention and water-holding capacity, provides soil organisms with habitats, modulates microbial and biodiversity activities, and stabilizes soil organic carbon (Blanco-Canqui 2017; Sandhu and Kumar 2017). Pyrolysis also produces biofuels as fuel gases or oils. Biochar is designed to alleviate climate change by providing renewables and soil amendment that can considerably improve the sequestration of net soil carbon. The biochar and energy generated during pyrolysis can be both useful as a renewable source of energy and improvement of soil as by-products. More than 1.6 million children per year die from biomass exposure (Edelstein et al. 2008) in the developing world. The International Biochar Initiative has been assisting in many projects which improve cook store efficiency while producing biochar for soil application and in plantation systems. More recent studies of the biochar production process focused on the role it plays in increasing demand for biomass-based energy resources that use greenhouse gas instead of fossil fuels emissions and slow climate change. Biochar improves soil characteristics as increasing bio-factor soil system activity (Lehmann et al. 2011; Paz-Ferreiro et al. 2014) declines emission of soil greenhouses and hence, increases the content of carbon recalcitrant in soil (Gascó et al. 2016).

### 10.2.2 Biochar Soil Amendment Significance to Farmers and Ranchers

Farmers and ranchers may be concerned with biochar as a soil modification that can improve fertility and reduce the need for other fertilizers with low environmental costs. Supplementation of biochar to the soil is concomitant with an increase in the availability of basic cations, and mineral nutrients such as phosphorus and total nitrogen (Glaser et al. 2002; Lehmann et al. 2003). Soil physiochemical properties, as well as fertility, can primarily be enhanced including pH, electrical conductivity, cation, and anion exchange capacity (Liang et al. 2006; Major et al. 2010; B. Singh et al. 2010). Moreover, the bioavailability of organic (Yu et al. 2009) and inorganic pollutants (Hua et al. 2014) and soil porosity and agglomeration have also been modified and facilitated oxygen diffusion leading consequently to microbial biodegradation (Amonette and Joseph 2009; Yaghoubi and Reddy 2011). In general, alkaline pH nature and mineral components from biochar such as Nitrogen, Phosphorus, Potash (NPK) and other trace elements can be significant agronomic benefits for many soils, at least in the short to medium term, in turn, root growth and reducing $Al^{3+}$ toxicity (Chan et al. 2008a; Cheng et al. 2006; Lehmann and Joseph 2015; Lehmann and Rondon 2006). Biochar with higher pH could modify the soil to be less acidic (Yuan et al. 2017, 2011). In addition to biochar's also play main role in the climate change mitigation by storing carbon in the soil in a stable form (Alburquerque et al. 2013; Atkinson et al. 2010; Gomez et al. 2014; Lehmann 2007; Atif et al. 2021) (Table 10.1).

Biochar can be mixed with other soil amendments and inputs such as cover cropping and compost or lime before application because it enhanced the availability of nutrients to crops, and also increased the wheat yield, reported by the Machado et al. (2017) Similarly, a mixture of biochar and dairy manure can contribute to improving soil water retention capabilities (Sandhu and Kumar 2017; Atif et al. 2021). Zhang et al. (2010) have also shown that the emissions of powerful greenhouse gases such as methane and nitrous oxide from agricultural soils are reduced, and may have an additional climate mitigation impact. Spokas et al. (2009) reported a reduction in the production of carbon dioxide by the addition of a different biochar concentration of 2%–60% (w/w), a decelerated production of nitrous oxides at levels

**TABLE 10.1**

Effects of Biochar Application on Soil Properties

| Biochar | Soil Type | Effect on Soil Properties | References |
|---|---|---|---|
| Wood charcoal | Ferralsol and Anthrosol | Increase the, pH, carbon content, phosphorus availability, reduced the leaching of applied fertilizers and nutrients | |
| Charcoal mixed soil | Haplic Acrisols | Increase the soil total porosity, soil hydraulic conductivity, and reduced bulk density | Oguntunde et al. (2008) |
| Hard wood | Not given | Increased carbon sequestration and reduced the soil nitrogen leaching | Zavalloni et al. (2011) |
| Food waste and corn stover | Alfisol | Increase in total potassium and sodium contents | Rajkovich et al. (2012) |
| Wheat straw | Chernozemic soil | Reduced $N_2O$ and $CO_2$ emission and changed the enzyme's activity | Wu et al. (2013) |
| Wheat straw | Fimi-Orthic Anthrosols | Increase the soil organic carbon content, pH, total nitrogen, and reduced $N_2O$ emission | Li et al. (2015) |
| Biochar from multiple sources | Various soil types | Improved soil pH, potassium, total nitrogen contents and increase the yield of crops and bacterial population | Thies et al. (2015) |
| Wood and manure | Different soil types | Improved the soil's saturated hydraulic conductivity, increase total nitrogen concentration, and reduced $NH_4$-N leaching | Ajayi et al. (2016) |
| Wheat Straw | Rice paddy soil | Increase the soil pH and $N_2O$ emission and *amoA* gene abundance | Lin et al. (2017) |
| Paper and wood waste | Different soil | Increased soil pH, carbon sequestration, electrical conductivity, and soil fertility | Randolph et al. (2017) |
| Banana peels waste | Not given | Decreased $NH4^+$-N and $N_2O$ and $CO_2$ emission, increased soil ammonium nitrogen, soil microbial biomass and microbial biomass nitrogen, and enzyme activity | Sial et al. (2019) |
| Crop residue | Not given | Increase the availability of silicon (Si) and nutrients to crops | Li et al. (2019) |
| Rice husk | Alfisol and Ultisol | Increased soil pH, soil organic carbon, total nitrogen and available phosphorus and potassium contents and nutrients availability and reduced the soil nitrate leaching | Oladele et al. (2019) |

above 20%, and environmental oxidation of methanol at all levels over unfinished terrestrial levels. Several reports documented that the control of certain pathogens may be possible in soils amended with biochar (Bonanomi et al. 2015; Elmer and Pignatello 2011; Thies et al. 2015). Furthermore, greater surface area, porous structure, and hydrophobicity of biochar are considered as beneficial habitat for soil nitrogen-fixing bacteria, useful fungi, and microorganisms enhancing their activities, which in turn can make some nutrients more available to crops (El-Ramady et al. 2018; Lehmann et al. 2011). Also, biochar decline expected harmful damages owing to soilborne pathogenic microorganisms (Elmer and Pignatello 2011; Zwart and Kim 2012).

### 10.2.3 Biochar as an Economic Potential Tool in Agriculture

For agriculture and ranchers, there are three sources of biochar economic potential: as a soil amendment that could replace partially fertilized materials; as a source of heat, bio-oil, and agricultural and ranch gas; and as a carbon off in future market cap and trade markets. For example, a farm or ranch with significant sources for renewable biomass available to be collected could hide these biomasses to other forms like heat, liquid, or gas fuel by machinery operation and also return to the fields to enhance fertility and collect a carbon-offset payment. Note, however, that before such a project is fully optimized, several economic, institutional, and regulatory questions must be answered. Further research is, however, necessary before biochar appliances can be validated for farms and grazing lands as a carbon-off and potential income source for farmers and farmers.

## 10.3 Various Abiotic Stresses Threaten Crop Production

### 10.3.1 Salinity Stress "Impacts on Plant Growth and Development"

Abiotic stresses such as drought, salinity, metals, and temperature adversely affect crop productivity worldwide causing more than 50% loss in yield potential of major crop plants (Ke et al. 2018; Ahanger and Agarwal 2017; Buchanan et al. 2000; Calderón et al. 2018; Adnan et al. 2018a,b; Akram et al. 2018a,b; Aziz et al. 2017a; Habib et al. 2017; Hafiz et al. 2016; Hafiz et al. 2019; Kamran et al. 2017; Muhammad et al. 2019; Sajjad et al. 2019; Saud et al. 2013; Saud et al. 2014; Saud et al. 2017; Saud et al. 2016; Shah et al. 2013; Qamar et al. 2017; Wajid et al. 2017; Yang et al. 2017; Zahida et al. 2017; Depeng et al. 2018; Fahad and Bano 2012; Fahad et al. 2017; Fahad et al. 2013; Fahad et al. 2014a,b; Fahad et al. 2016a,b,c,d; Fahad et al. 2015a,b; Fahad et al. 2018a,b; Fahad et al. 2019a,b; Fahad et al. 2020; Fahad et al. 2021a,b,c,d,e,f; Hesham and Fahad 2020). Therefore, the needs for prevention of crop losses and the production of more food and feed to meet the demands of over-increasing human populations have gained more and more attention (Wani et al. 2017; Deepranjan et al. 2021; Haider et al. 2021; Huang Li et al. 2021; Ikram et al. 2021; Jabborova et al. 2021; Khadim et al. 2021a,b; Manzer et al. 2021; Muzammal et al. 2021; Abdul et al. 2021a,b; Ashfaq et al. 2021; Amjad et al. 2021; Atif et al. 2021; Athar et al. 2021; Adnan et al. 2018; Adnan et al. 2019; Akram et al. 2018a,b; Aziz et al. 2017a,b; Chang et al. 2021; Chen et al. 2021; Emre et al. 2021; Habib et al. 2017; Hafiz et al. 2016; Hafiz et al. 2019; Ghulam et al. 2021; Guofu et al. 2021; Hafeez et al. 2021; Khan et al. 2021; Kamran et al. 2017; Muhmmad et al. 2019; Safi et al. 2021; Sajjad et al. 2019; Saud et al. 2013; Saud et al. 2014; Saud et al. 2017; Saud et al. 2016; Shah et al. 2013; Saud et al. 2020; Qamar et al. 2017; Hamza et al. 2021; Irfan et al. 2021;Wajid et al. 2017; Yang et al. 2017; Zahida et al. 2017; Depeng et al. 2018; Hussain et al. 2020; Hafiz et al. 2020a,b; Shafi et al. 2020; Wahid et al. 2020; Subhan et al. 2020; Zafar-ul-Hye et al. 2020a,b; Zafar et al. 2021; Adnan et al. 2020; Ilyas et al. 2020; Saleem et al 2020a,b,c; Rehman et al. 2020; Frahat et al. 2020; Wu et al. 2020; Mubeen et al. 2020; Farhana et al. 2020; Jan et al. 2020; Wu et al. 2019a; Ahmad et al. 2019; Baseer et al. 2019; Hafiz et al. 2018; Tariq et al. 2018; Iqra et al. 2020; Akbar et al. 2020; Mahar et al. 2020; Noor et al. 2020; Bayram et al. 2020; Amanullah and Fahad 2018a,b; Amanullah and Fahad 2017; Amanullah et al. 2020; Amanullah et al. 2021; Rashid et al. 2020; Arif et al. 2020; Amir et al. 2020; Saman et al. 2020; Muhammad Tahir et al. 2020; Md Jakir and Allah 2020; Mahmood et al. 2021; Farah et al. 2020; Sadam et al. 2020; Unsar et al. 2020; Fazli et al. 2020; Md. Enamul et al. 2020; Gopakumar et al. 2020; Zia-ur-Rehman 2020; Ayman et al. 2020; Al-Wabel et al. 2020a,b; Senol 2020; Amjad et al. 2020; Ibrar et al. 2020; Sajid et al. 2020; Muhammad et al. 2021; Sidra et al. 2021; Zahir et al. 2021; Sahrish et al. 2022). On the one hand, climate change has exacerbated the situation, and on the other hand, shortage of quality water for irrigation has compelled agriculturalists to use saline waters for irrigation purposes, resulting in conversion of arable lands into unproductive waste lands (Hu and Xiong 2014; Ke et al. 2016). Globally high salinity affects more than 20% of irrigated land, thereby affecting the net agricultural productivity (Ke et al. 2016; Qadir et al. 2014; Wani et al. 2016). It has been confirmed that the effect of salinity stress depends on the duration and severity, and high salinity hampers plant growth and development, inducing physiological drought and ion toxicity reflecting on the altered physiological and morphological processes (Munns 2005; James et al. 2011; Rahnama et al. 2010; Ahanger and Agarwal 2017; Soliman et al. 2018; Ahmad et al. 2018). Salinity-mediated physiological dryness reduces seed germination and root vitality (Banerjee and Roychoudhury 2018; Ibrahim 2016; Mechergui et al. 2017; Zhang et al. 2014). Like other stresses, high salinity increases the generation of reactive oxygen species (ROS) damaging the cellular molecules like DNA, RNA, proteins, carbohydrates, and lipids because of their highly toxic and reactive nature, thereby intensifying the oxidative stress effects (Ashraf et al. 2018a; Ke et al. 2016, 2018; Sharif et al. 2018). For averting the stress-induced oxidative damage, plants have an endogenous antioxidant scavenging system consisting of enzymes such as superoxide dismutase, catalase, and ascorbate peroxidase, which work together to limit the uncontrolled oxidation by scavenging the ROS and protecting the plant from the oxidative damage (Li et al. 2018; Ahanger et al. 2017; Galano et al. 2013; Gill and Tuteja 2010). It is quite obvious that plant species differ in tolerance level to high salinity and the same trend has been observed in different varieties of the same crop species (Ahanger et al. 2017; Munns 2005; Munns and Tester 2008; Sharif et al. 2018).

### 10.3.2 Water Stress

Water is a central molecule, which comprises 80%–95% of most herbaceous plants, and over 50% of woody plant biomass, in all physiological processes of plants (Kramer and Boyer 1995). If plants confront an imbalance in the water status, then plants develop various mechanisms to adapt their growth to this water condition. Water imbalance includes either shortage (drought) or increase (waterlogging or flooding) of water. Water stress affects many aspects of plant performance including several physiological and biochemical processes in particular photosynthesis. When stress continues, plant growth and productivity will decrease severely. A suitable water condition is contributed to enhancing crop growth and yields in all parts of the world. Recently, plants to maintain survival and counteract these adverse environmental conditions have developed different molecular mechanisms for reducing their resource consumption and adapting their growth and development safely (Ahuja et al. 2010; Osakabe et al. 2014; Skirycz and Inzé 2010; Van Ha et al. 2014; Yamaguchi-Shinozaki and Shinozaki 2006).

#### *10.3.2.1 Drought Stress*

Drought stress limits agronomic sustainability and results in a pronounced crop yield loss by inducing physiological and biochemical processes that hinder the plant growth and development (Anjum et al. 2017; Bano et al. 2012; Niu et al. 2016; Shao et al. 2009; Zandalinas et al. 2018). Furthermore, oxidative stress occurs in the plant cell under drought conditions resulting in a higher leakage of electrons toward $O_2$ during photosynthetic and respiratory processes which leads to enhancement and accumulation of ROS in terms of superoxide anions ($O_2^{\cdot-}$), singlet oxygen ($O_2^{1}$), hydroxyl or perhydroxyl radicals ($OH^-$), and hydrogen peroxide (Apel and Hirt 2004; Asada 2006; Waraich et al. 2011). Consequently, enhancing plant growth performance and sustaining high yield/quality under variable environmental stresses are needed to meet challenges of the rapid increase in population and climate change impacts on agriculture (Bisbis et al. 2018; Hajar et al. 2014; Lesk et al. 2016; Pachauri et al. 2014). Improvements in a synthesis of osmoprotectants (i.e., proline and soluble sugar) and secondary metabolites like phenolic compounds, as a metabolic adaptation, enables plants to cope with water stress (Ashraf et al. 2018a; Gupta and De 2017; Munns 2005; Petridis et al. 2012; Pinheiro et al. 2004; Sairam and Tyagi 2004). In particular, the balance between the formation and consumption of ROS under optimal growth conditions is firmly controlled by an efficient antioxidant system. Minimizing oxidative damage contributes to a standard cellular function enabling continuous plant growth under stress conditions (Ashraf et al. 2015; Chandrakar et al. 2018; El-Mashad and Mohamed 2012; Jiang and Huang 2001; Mahatma et al. 2009; Mittler 2002; Mittler et al. 2004). In the recent decades, the identification of proteomic tools in several traits has proved that the expression and balance of genes encoding antioxidant enzymes (i.e., superoxide dismutase, catalase, and ascorbate peroxidase) are critical regulators of biochemical pathways and sophisticated response mechanisms for drought stress tolerance through controlling the ROS production (Apel and Hirt 2004; Ashraf 2010; Scandalios 2005, 2002; Zhou et al. 2015; Adnan et al. 2016). For instance, catalase gene expression is downregulated while GPX gene expression showed a differential change in the activity under different abiotic stress conditions (e.g., salt treatment) in wheat plants (Castorina et al. 2018; Dat et al. 2000; Faghani et al. 2015). Furthermore, proteomic analysis of wheat showed a great abundance of proteins related to defense and oxidative stress in response to water stress. However, developing applications to diminish the deleterious effects of drought stress ensures optimal crop growth and productivity.

#### *10.3.2.2 Flooding Stress*

Like other factors, waterlogging is considered a significant growth determining factor. Waterlogging has been a global problem for agriculture. Climate change causing the overflow of rivers and heavy rainfall has been considered a primary reason for waterlogging (Kim et al. 2015; Milly et al. 2002; Wright et al. 2017). The soil is considered waterlogged if the standing waters are 20% higher than the field capacity (Agarwal et al. 2006). The resulting soil-hypoxic or anoxia conditions due to waterlogging have been considered as the resultant main effect affecting the plant developmental stages, survival, and hence the crop production (Kozdrój and van Elsas 2000; Jackson and Colmer 2005; Paul et al. 2016). Growth and development of plants with an optimum yield under the anaerobic condition is the most intensive management strategy of breeding crops for sustainable agriculture (Osakabe et al. 2014; Khan et al. 2016; Pedersen et al. 2017). Plants under flood stress develop a set of morphological, physiological, anatomical, and biochemical adaptations to resist the damaging effects of low oxygen (Chen et al. 2016; Ravanbakhsh et al. 2017; Striker and Colmer 2017).

Moreover, developmental programs are elicited via regulation of gene expression and adaptive mechanisms such as the development of adventitious roots (Chen et al. 2016; Song et al. 2017) or the regulation of ion uptake (Carter et al. 2006; Colmer and Flowers 2008) to mitigate the flooding stress (Bailey-Serres and Voesenek 2008; Li et al. 2018; Osakabe et al. 2014). Modifications in the root system play a key role in withstanding the logging effects. Accumulation of lignin and suberin at the cell level as well as the formation of the adventitious root system in tomato and aerenchyma development in wheat has been reported (Colmer and Voesenek 2009; Ezin et al. 2010; Herzog et al. 2016). Soil waterlogging generally generates declined (hypoxic) or complete (anoxic) lack of oxygen conditions which are very detrimental to normal plant development (Armstrong et al. 2000). Oxygen deficiency negatively influences the aerobic respiration resulting in reduced plant growth by affecting photosynthesis, hormonal balance, and nutrient uptake hence resulting in stunted growth and yield reduction (Terazawa et al. 1992; Valliyodan et al. 2016). Waterlogging-induced stomatal closure limits the level of $CO_2$ in plant cells which consequently induces oxidative stress due to free radical accumulation (Sairam et al. 2009). It has been reported that waterlogging-induced programmed cell death is related to generation of ethylene in aerenchyma (Lenochová et al. 2009).

### 10.3.3 Temperature Stress

Since the nineteenth century, temperature has been one of the important factors for the proper growth and development of plants and affects plant metabolism at several points. Plants as sessile in nature are confronted with environmental variables without escaping trials. However, both very high or very low temperatures destroy and interfere with the basis of the thermodynamic principles which regulate the rates of biological reactions (Franklin and Wigge 2014; Fahad et al. 2015a; Fahad et al. 2016d; Mahmood et al. 2021; Fahad et al. 2021a, b, c, d, e, f; Hesham and Fahad 2020; Atif et al. 2021). In the context of agricultural food production, the increase or decrease in temperature due to global climate change is a major concern that leads to a decrease in crop productivity. In this section, a brief illustration of high and low temperatures and their effects on plant responses and metabolism as well as adaption mechanisms are reviewed separately.

#### *10.3.3.1 Heat Stress*

Global warming currently has an enormous impact on the growth, survival and phenology of plant communities (Fahad et al. 2016a,b,c,d; Fahad et al. 2015a,b; Fahad et al. 2018a,b; Fahad et al. 2019a,b). These include increasing average surface air temperature as a result of increased atmospheric carbon dioxide levels due to the consumption of fossil fuels and an associated increase in erratic temperature events and severity (Bokhorst et al. 2009). It is anticipated that the expected temperature changes over the next 30–50 years will be within the governmental Panel Climate Change of 2°C–3°C (IPCC). From an environmental perspective, by 2080, about half the European species and some plants will be vulnerable or extinct, if they cannot adapt to temperate changes and could not complete life cycles and adjust floral transition timing (Willis and Reid 2008). Several agronomic traits will be affected by increasing temperature over optimum values, including seed germination, seedling growth, stem elongation, leaf development, and final yield (Parent and Tardieu 2012; Probert 2000; Schmuths et al. 2006). Therefore, a major threat to both agricultural productivity and future food security is high-temperature stress due to global warming as controlling the timing of the crop growing season (Battisti and Naylor 2009; Ye et al. 2015). It is estimated that an increase in temperature of 1°C would increase the demand for crop water by 4%–4.5% (Ye et al. 2015).

The link between the temperature response of plants and the thermal stability of metabolic enzymes (Burke and orzech usda-ars 1988) is an interesting development for our understanding of plant environmental responses. Heat stress interferes with several important processes at the molecular, physiological, and cellular levels as this directly affects the stability of proteins, membranes, RNA species, and the cytoskeleton structure which can have negative effects on growth and development that may trigger plant death (Hasanuzzaman et al. 2013a, 2013b; Howarth 2005). High temperatures also increase damage to the thylakoid membrane and result in reduced photosystem II activity (P. V. Prasad et al. 2008; P. V. V. Prasad et al. 2008; Prasad and Djanaguiraman 2011; Ristic et al. 2008). Major metabolome, transcriptome, and proteome changes were initiated to counteract thermal stress (Kotak et al. 2007; Mittler 2002; Penfield 2008). Heat waves, particularly during the anthesis and grain-filling stages, are detrimental to the photosynthetic system and significantly reduce yields (Feng et al. 2015).

#### *10.3.3.2 Cold Stress*

Chilling stress is one of the biggest abiotic stresses reducing crop production (Chinnusamy et al. 2007). Many reports examine the reduction in plant growth by cold stress (Kidokoro et al. 2017). Cold temperatures hamper the potential crop production in the mid- and high-latitude regions (Tuteja et al. 2011). The geographical distribution of plant species is generally restricted by low temperatures and associated with decreasing in crop yield and quality (Ogbomida et al. 2018). Moreover, climate change severely disturbs the biochemistry, quantity, and quality of crop yields (Cramer et al. 2011; Jaleel et al. 2009; Miller et al. 2010; Moraes de Freitas et al. 2016). Chilling injury can occur at temperatures between 0°C and 15°C (Allen and Ort 2001). Various phenotypic symptoms of plants in response to chilling stress are clear, including reductions in leaf chlorophyll content, wilting, chlorosis, and necrosis. Moreover, different metabolic modifications are induced, such as reductions in unsaturated fatty acid contents and increases in the permeability of cell membranes, which collectively reduce plant performance (Bracale and Coraggio 2003; Miura and Furumoto 2013; Tong 2005). One of the essential cellular responses throughout a freeze-thaw cycle is the maintenance of plasma membrane integrity. This maintenance reduces the efficiency of photosystem II, damages photosystem I, alters the carbon reduction cycle, $CO_2$ assimilation, and photosynthetic pigment complex systems, and accelerates ROS formation (Allen and Ort 2001; Guy 1990; Saibo et al. 2008). Overall photosynthesis processes and their various structures, including the chloroplast and the two photosystems with the electron transport chain as well as the carbon reduction cycle, and control of stomatal conductance are affected directly or indirectly under chilling stress (Lambers et al. 2008). Thus, temperatures below critical trigger the generation of ROS and further disturbances the net photosynthetic rate and physiological functions as a result of the accumulation of photo-energy (Alam and Jacob 2002). Plants of temperate origin have evolved unique abilities/traits that enable them to combat and survive with chilling/freezing temperatures. Improvement of persistence in harsh winters remains a key objective of breeding programs developing suitable cultivars. The effects of cold stress are usually reflected as the failure of germination and poor growth vigor (Law and Crafts-Brandner 2001; Mwanamwenge et al. 1999). Cold-stress injuries cause irreversible damage to cellular components and metabolisms. As a result, the plant fails to germinate or exhibits poor seedling vigor. Cold stress affects all aspects of cellular functions (Chinnusamy et al. 2007). Altered membrane integrity is also a common effect of cold stress (Kratsch and Wise 2000).

### 10.3.4 Heavy Metal Stress

Heavy metal pollution, due to its non-biodegradability, remains persistent in the environment, and their concentrations continue to increase as they enter the food chain, posing a serious threat to plant growth and human health, thus has become one of the most severe environmental problems today, which pose a potential health risk to water, air and soil (Ogbomida et al. 2018). The term "heavy metals" refers to any metallic element that has a relatively high density and is toxic or poisonous even at low concentration. The main contributors to heavy metal contamination of soil include human activities such as smelting and melting, sludge dumping, intensive agriculture, power transmission, and fuel and electricity generation (Dembitsky 2003; Igwe and Abia 2006). Besides these, heavy metals are added to soil and water through natural processes like excessive rock weathering, soil erosion, and volcanic eruptions. Some minerals like As, Pb, Cd, Cd, Hg, and Se are considered to be non-essential among heavy metals because they do not participate in any known physiological function of plants (Gaur and Adholeya 2004; Subhashini and Swamy 2013). Metals can also substitute for essential elements which disrupt their function and cause indirect toxic effects in enzyme cascades (Asati et al. 2016). In this case, contamination by heavy metals renders the soil unfit for crop growth and is detrimental to biodiversity (Etesami and Jeong 2018).

Soils often have a heavy metal contamination, including lead (Pb), zinc (Zn), nickel (Ni), iron (Fe), chromium (Cr), copper (Cu), arsenic (As), silver (Ag), and the platinum group elements (Kabata-Pendias 1993; Kabata-Pendias and Pendia 1984; Sharma et al. 2014, 2006, 2016; Sharma and Dubey 2005). Not only does heavy metal soil contamination reduce agricultural output, but it also adversely affects wildlife and human health (Guo et al. 2016; Jaishankar et al. 2014; Kloke et al. 1984). Also, they are highly toxic because they accumulate in plants and animals that cause metabolism to inhibit and kill (Table 10.2).

When plants absorb these heavy metals above their limits, they can cause growth and cell death. Certain types of plants have been developing physiological and biochemical mechanisms and strategies for tolerating toxic metal levels and tolerating high metal accumulation, without altering the plant's metabolic properties (Revathi and Venugopal 2013).

### 10.3.5 Mineral Nutrition Imbalance and Constraints to Plant Growth

Production and quality of healthy crops require a wide range of mineral elements indispensable as nutrients from the soil for growth and development. Plant nutrient applications must be optimized in order to reach the goal of the availability of soil nutrients. Hence, plant nutrition has impacted both the growth of plants and other living organisms (Maathuis 2009). The deficiency and excessive quantities of nutrients are both detrimental to plant growth. A proper supply of nutrients can alleviate some of the severe productivity restrictions associated with many stress conditions (Dwivedi et al. 2016). Plants have developed highly flexible and finely balanced mechanisms to be acclimated to variable soil nutrient concentrations (Buet et al. 2019). The sustainable agricultural production requires management strategies that improve nutrient phytoavailability and impaired deficiencies in critical mineral nutrients (macro- and micronutrients) (Branzini et al. 2012; Hasanuzzaman et al. 2018). All mineral nutrients have a specific range of plant sufficiency. An imbalance within this range could influence crop growth and development, and it might lead to either a nutrient deficiency or toxicity (Hasanuzzaman et al. 2018). Toxicity occurs when a nutrient exceeds its plant requirements, resulting in a decrease in plant growth or quality (McCauley 2011; McCauley et al. 2012, 1989). Moreover, pH of the soil solution has a pivotal role in limiting crop production and determines the availability of mineral nutrients and exposure to toxic elements (Mueller et al. 2012; Rengel 2015; White and Brown 2010).

Remarks by Z. Ahmad et al. (2018) and Ali et al. (2018) show that potassium (K) supplementation potentially increases the grain yield in wheat by improving physiological functions such as cell osmoregulation, plant photosynthesis, and antioxidative status. Boron deficiency is one of the most significant abiotic problems affecting a range of crop plants with special impact on citrus causing a significant negative impact on metabolism and productivity (Cakmak and Römheld 1997; Camacho-Cristóbal et al. 2018; Landi et al. 2019; Liu et al. 2018), while the remedial effects of B deficiency on root growth in citrus tissue can be alleviated by restored central metabolites such as proline, myo-inositol, and 6-deoxy-D glucose (Wu et al. 2019a). Plant differentiation is mainly affected by an increase or decrease in specific minerals such as nitrogen concentration as an increase in the root/shoot ratio is owing to nitrogen deficiency (Cakmak and Kirkby 2008). Magnesium (Mg) is considered an essential mineral for plant crops and classified as macronutrients as mainly functioning in the chlorophyll molecule; therefore, lack of Mg would induce photosynthesis inhibition and reduce export sugars from leaves to roots (Hosseini et al. 2019) as a result reduced root growth (Marschner and Marschner 2012; Rengel and Marschner 2005). Among nutrients, nitrogen stress has the greatest impact on the continuity of plant growth (Drenovsky et al. 2012).

**TABLE 10.2**
Biochar for Heavy Metal Removal

| Type of Biochar | Metal | Summary | Reference |
|---|---|---|---|
| Digested dairy waste biochar and digested whole sugar beet biochar | $Pb^{2+}$, $Cu^{2+}$, $Ni^{2+}$ and $Cd^{2+}$ | All the metals were removed 100%. Removal was mainly by surface precipitation mechanism | Inyang et al. (2012) |
| Hydrochar (hydrothermal carbonization of peanut hull) | $Pb^{2+}$, $Cu^{2+}$, $Ni^{2+}$ and $Cd^{2+}$ | Lead sorption capacity was 22.82 mg/g, all other metals were removed successfully. The $H_2O_2$ modification increased the oxygen-containing functional group and as result increased the sorption ability of biochar | Xue et al. (2012) |
| Biochar produced from switch grass via hydrothermal carbonization | $Cu^{2+}$ and $Cd^{2+}$ | 100% removal of copper and cadmium achieved. Metals were mainly removed by the sorption process | Regmi et al. (2012) |
| Rice husk and dairy manure–derived biochar | Pb, Cu, Zn, and Cd | Dairy manure biochar was more effective in metals removal (486 mmol/kg) than rice husk-derived biochar (65.5–140 mmol/kg) for all four metals.<br>Precipitation with $CO_3^{2-}$ or $PO_4^{3-}$ was mainly responsible for metal removal by manure biochar and rice husk biochar metals removed by their complexation with phenolic-OH | Xu et al. (2013a) |
| Dairy manure biochar (DMB) prepared at 350°C and 200°C | Cu, Zn, and Cd | Sorption capacities were 54.4, 32.8, and 51.4 mg/g for Cu, Zn, and Cd, respectively, for DMB 350 and 48.4, 31.6, and 31.9 mg/g for Cu, Zn, and Cd, respectively, of DMB200. The main removal process was precipitation with $PO_4^{3-}$ or $CO_3^{2-}$ origination in biochar | Xu et al. (2013b) |
| Magnetic biochar composite fabricated from pine bark waste and $CoFe_2O_4$ | $Pb^{2+}$ and $Cd^{2+}$ | Successful sorption of metals ion observed. Metals were removed by adsorption of metals ions to biochar | Reddy and Lee (2014) |
| Rice husk, organic solid waste, and sewage sludge–derived biochar | Cr (VI) and As (V) | Removal rate was 95% for Cr (III), 89% for Cr (VI), and 53% for As (V). High $Fe_2O_3$ contents enhanced metal adsorption | Agrafioti et al. (2014) |
| Polyethylenimine modified biochar | Cr | Adsorption capacity was 435.7 mg/g. Removal of Cr (VI) depended on pH and low pH favored the removal process | Ma et al. (2014) |
| Engineered biochar | Pb (II), Cu (II) and Cd (II) | Sorption rate were 153.1, 34.2, and 28.1 mg/g for Pb (II), Cu (II), and Cd (II) respectively. The removal of metals was mainly by surface adsorption; the dosage, initial solution pH, and ionic strength affected the removal of heavy metals by biochar | Wang et al. (2015a,b) |
| Pine wood waste biochar | Cr and Pb | Adsorption capacities were 680 and 520 μmol/g for Cr and Pb, respectively. Biochar has a high cation exchange capacity and high adsorption for metals | Abdel-Fattah et al. (2015) |
| Sesame straw biochar | Pb, Cd, Cr, Cu, and Zn | Adsorption capacities (mg/g) were Pb (102), Cd (86), Cr (65), Cu (55) and Zn (34). High interference among the metals was observed under multi-metal sorption condition | Park et al. (2016) |
| Hickory wood modified with NaOH | $Pb^{2+}$, $Cd^{2+}$, $Cu^{2+}$, $Zn^{2+}$, and $Ni^{2+}$ | Showed a strong ability to filter heavy metals from aqueous solution. The modified biochar exhibited high metal adsorption capacity | Ding et al. (2016) |
| Anaerobic digestion sludge–derived biochar | $Pb^{2+}$ | Adsorption capacity was 51.20 mg/g. Electrostatic attraction, precipitation, surface complexation, and ion exchange were possible $Pb^{2+}$ removal mechanisms | Ho et al. (2017) |
| Poplar catkins–derived biochar | U (VI) and Co (II) | Adsorptions rates were 71.85 and 110.17 mg/g for U (VI) and Co (II), respectively. Metals removed by adsorption process including surface complexes and hydrogen-binding interaction | Li et al. (2016) |

*(Continued)*

**TABLE 10.2 (*Continued*)**
Biochar for Heavy Metal Removal

| Type of Biochar | Metal | Summary | Reference |
|---|---|---|---|
| Waste marine macro-algal biomass–derived magnetic biochar | Cd, Cu, and Zn | Removal rate was 63.52–69.37 mg/g for copper. This high metal removal performance was due to the presence of various oxygen-containing functional groups in abundance | Son et al. (2018) |
| Cow manure biochar modified with $HNO_3$ | U (VI) | The maximum adsorption capacity was 355.6 mg/g which is higher than most of carbon-based adsorbents. The highly efficient adsorption of U (VI) was due to the carboxyl groups on the biochar surface | Jin et al. (2018) |
| Rice straw biochar derived by higher temperature | Pb | The immobilized lead fraction on biochar was 11.34 %. Higher pH value, surface area resulted in higher metal removal. Precipitation was the key mechanism for lead removal | Shen et al. (2019) |
| Biochar of anaerobically digested sludge | Pb (II) and Cd (II) | The adsorption capacities were 0.75 and 0.55 mmoL/g for Pb (II) and Cd (II) respectively. Adsorption was the main removal mechanism, preference of biochar for Pb (II) compared to Cd (II) was observed | Ni et al. (2019) |

# REFERENCES

Abdul S, Muhammad AA, Shabir H, Hesham A. El E, Sajjad H, Niaz A, Abdul G, Sayyed RZ, Fahad S, Subhan D, Rahul D (2021a) Zinc nutrition and arbuscular mycorrhizal symbiosis effects on maize (Zea mays L.) growth and productivity. J Saudi Soc Agric Sci. https://doi.org/10.1016/j.sjbs.2021.06.096.

Abdul S, Muhammad AA, Subhan D, Niaz A, Fahad S, Rahul D, Mohammad JA, Omaima N, Muhammad Habib ur R, Bernard RG (2021b) Effect of arbuscular mycorrhizal fungi on the physiological functioning of maizeunder zinc-deficient soils. Sci Rep 11:18468.

Adnan M, Fahad S, Khan IA, Saeed M, Ihsan MZ, Saud S, Riaz M, Wang D, Wu C (2019) Integration of poultry manure and phosphate solubilizing bacteria improved availability of Ca bound P in calcareous soils. 3 Biotech 9(10):368.

Adnan M, Fahad S, Muhammad Z, Shahen S, Ishaq AM, Subhan D, Zafar-ul-Hye M, Martin LB, Raja MMN, Beena S, Saud S, Imran A, Zhen Y, Martin B, Jiri H, Rahul D (2020) Coupling phosphate-solubilizing bacteria with phosphorus supplements improve maize phosphorus acquisition and growth under lime induced salinity stress. Plants 9(900). https://doi.org/10.3390/plants9070900.

Adnan M, Shah Z, Saleem N, Basir A, Rahman IU, Ullah H, Ibrahim M, Shah JA, Muhammad AK, Ali Shah SR (2016) Isolation and evaluation of summer legumes Rhizobia as PGPR. Pure Appl Bio 5(1):127–133.

Adnan M, Shah Z, Sharif M, Rahman H (2018b) Liming induces carbon dioxide ($CO_2$) emission in PSB inoculated alkaline soil supplemented with different phosphorus sources. Environ Sci Pollut Res 25(10):9501–9509.

Adnan M, Zahir S, Fahad S, Arif M, Mukhtar A, Imtiaz AK, Ishaq AM, Abdul B, Hidayat U, Muhammad A, Inayat-Ur R, Saud S, Muhammad ZI, Yousaf J, Amanullah, Hafiz MH, Wajid N (2018a) Phosphate-solubilizing bacteria nullify the antagonistic effect of soil calcification on bioavailability of phosphorus in alkaline soils. Sci Rep 8:4339. https://doi.org/10.1038/s41598-018-22653-7.

Agarwal PK, Agarwal P, Reddy MK, Sopory SK (2006) Role of DREB transcription factors in abiotic and biotic stress tolerance in plants. Plant Cell Rep 25:1263–1274.

Ahanger MA, Agarwal RM (2017) Salinity stress induced alterations in antioxidant metabolism and nitrogen assimilation in wheat (*Triticum aestivum* L) as influenced by potassium supplementation. Plant Physiol Biochem 115:449–460.

Ahanger MA, Tomar NS, Tittal M, Argal S, Agarwal RM (2017) Plant growth under water/salt stress: ROS production; antioxidants and significance of added potassium under such conditions. Physiol Mol Biol Plants 23:731–744.

Ahmad P, Ahanger MA, Alyemeni MN, Wijaya L, Alam P (2018) Exogenous application of nitric oxide modulates osmolyte metabolism, antioxidants, enzymes of ascorbate-glutathione cycle and promotes growth under cadmium stress in tomato. Protoplasma 255:79–93.

Ahmad S, Kamran M, Ding R, Meng X, Wang H, Ahmad I, Fahad S, Han Q (2019) Exogenous melatonin confers drought stress by promoting plant growth, photosynthetic capacity and antioxidant defense system of maize seedlings. PeerJ 7:e7793. http://doi.org/10.7717/peerj.7793.

Ahmad Z, Anjum S, Waraich EA, Ayub MA, Ahmad T, Tariq RMS, Ahmad R, Iqbal MA (2018) Growth, physiology, and biochemical activities of plant responses with foliar potassium application under drought stress – a review. J Plant Nutr 41:1734–1743.

Ahuja I, de Vos RC, Bones AM, Hall RD (2010) Plant molecular stress responses face climate change. Trends Plant Sci 15:664–674.

Akbar H, Timothy JK, Jagadish T, M. Golam M, Apurbo KC, Muhammad F, Rajan B, Fahad S, Hasanuzzaman M (2020) Agricultural land degradation: processes and problems undermining future food security. In: Fahad S, Hasanuzzaman M, Alam M, Ullah H, Saeed M, Khan AK, Adnan M (eds) Environment, climate, plant and vegetation growth. Springer Publ Ltd, Springer Nature Switzerland AG. Part of Springer Nature, pp 17–62. https://doi.org/10.1007/978-3-030-49732-3.

Akram R, Turan V, Hammad HM, Ahmad S, Hussain S, Hasnain A, Maqbool MM, Rehmani MIA, Rasool A, Masood N, Mahmood F, Mubeen M, Sultana SR, Fahad S, Amanet K, Saleem M, Abbas Y, Akhtar HM, Waseem F, Murtaza R, Amin A, Zahoor SA, ul Din MS, Nasim W (2018a) Fate of organic and inorganic pollutants in paddy soils. In: Hashmi MZ, Varma A (eds) Environmental pollution of paddy soils, soil biology. Springer International Publishing AG, Switzerland, pp 197–214.

Akram R, Turan V, Wahid A, Ijaz M, Shahid MA, Kaleem S, Hafeez A, Maqbool MM, Chaudhary HJ, Munis, MFH, Mubeen M, Sadiq N, Murtaza R, Kazmi DH, Ali S, Khan N, Sultana SR, Fahad S, Amin A, Nasim W (2018b) Paddy land pollutants and their role in climate change. In: Hashmi MZ, Varma A (eds) Environmental pollution of paddy soils, soil biology. Springer International Publishing AG, Switzerland, pp 113–124.

Alam B, Jacob J (2002) Overproduction of photosynthetic electrons is associated with chilling injury in green leaves. Photosynthetica 40:91–95.

Alburquerque JA, Salazar P, Barrón V, Torrent J, del Campillo M, del C, Gallardo A, Villar R (2013) Enhanced wheat yield by biochar addition under different mineral fertilization levels. Agron Sustain Dev 33:475–484.

Ali S, Rizwan M, Qayyum MF, Ok YS, Ibrahim M, Riaz M, Arif MS, Hafeez F, Al-Wabel MI, Shahzad AN (2017) Biochar soil amendment on alleviation of drought and salt stress in plants: a critical review. Environ Sci Pollut Res 24:12700–12712.

Allen DJ, Ort DR (2001) Impacts of chilling temperatures on photosynthesis in warm-climate plants. Trends Plant Sci 6:36–42.

Al-Wabel MI, Abdelazeem S, Munir A, Khalid E, Adel RAU (2020b) Extent of climate change in Saudi Arabia and its impacts on agriculture: a case study from Qassim Region. In: Fahad S, Hasanuzzaman M, Alam M, Ullah H, Saeed M, Khan AK, Adnan M (eds) Environment, climate, plant and vegetation growth. Springer Publ Ltd, Springer Nature Switzerland AG. Part of Springer Nature. pp 635–658. https://doi.org/10.1007/978-3-030-49732-3.

Al-Wabel MI, Ahmad M, Usman AR, Akanji M, Rafique MI (2020a) Advances in pyrolytic technologies with improved carbon capture and storage to combat climate change. In: Fahad S, Hasanuzzaman M, Alam M, Ullah H, Saeed M, Khan AK, Adnan M (eds) Environment, climate, plant and vegetation growth. Springer Publ Ltd, Springer Nature Switzerland AG. Part of Springer Nature. pp 535–576. https://doi.org/10.1007/978-3-030-49732-3.

Amanullah, Fahad S (eds) (2017) Rice - technology and production. IntechOpen, Croatia. http://dx.doi.org/10.5772/64480.

Amanullah, Fahad S (eds) (2018a) Corn - Production and human Health in changing climate. IntechOpen, United Kingdom. http://dx.doi.org/10.5772/intechopen.74074.

Amanullah, Fahad S (eds) (2018b) Nitrogen in agriculture – updates. IntechOpen, Croatia. http://dx.doi.org/10.5772/65846.

Amanullah, Muhammad I, Haider N, Shah K, Manzoor A, Asim M, Saif U, Izhar A, Fahad S, Adnan M, et al (2021) Integrated foliar nutrients application improve wheat (Triticum Aestivum L.) productivity under calcareous soils in drylands. Commun Soil Sci Plant Anal. https://doi.org/10.1080/00103624.2021.1956521.

Amanullah, Shah K, Imran, Hamdan AK, Muhammad A, Abdel RA, Muhammad A, Fahad S, Azizullah S, Brajendra P (2020) Effects of climate change on irrigation water quality. In: Fahad S, Hasanuzzaman M, Alam M, Ullah H, Saeed M, Khan AK, Adnan M (eds) Environment, climate, plant and vegetation growth. Springer Publ Ltd, Springer Nature Switzerland AG. Part of Springer Nature. pp 123–132. https://doi.org/10.1007/978-3-030-49732-3.

Amir M, Muhammad A, Allah B, Sevgi Ç, Haroon ZK, Muhammad A, Emre A (2020) Bio fortification under climate change: the fight between quality and quantity. In: Fahad S, Hasanuzzaman M, Alam M, Ullah H, Saeed M, Khan AK, Adnan M (eds) Environment, climate, plant and vegetation growth. Springer Publ Ltd, Springer Nature Switzerland AG. Part of Springer Nature. pp 173–228. https://doi.org/10.1007/978-3-030-49732-3.

Amjad I, Muhammad H, Farooq S, Anwar H (2020) Role of plant bioactives in sustainable agriculture. In: Fahad S, Hasanuzzaman M, Alam M, Ullah H, Saeed M, Khan AK, Adnan M (eds) Environment, climate, plant and vegetation growth. Springer Publ Ltd, Springer Nature Switzerland AG. Part of Springer Nature. pp 591–606. https://doi.org/10.1007/978-3-030-49732-3.

Amjad SF, Mansoora N, Din IU, Khalid IR, Jatoi GH, Murtaza G, Yaseen S, Naz M, Danish S, Fahad S, et al (2021) Application of zinc fertilizer and mycorrhizal inoculation on physio- biochemical parameters of wheat grown under water-stressed environment. Sustainability 13:11007. https://doi.org/10.3390/su131911007.

Amonette JE, Joseph S (2009) Physical properties of biochar. In: Lehmann J, Joseph S (eds) Biochar for environmental management. Science and technology, pp 33–53. Earthscan, London.

Anjum SA, Ashraf U, Tanveer M, Khan I, Hussain S, Zohaib A, Abbas F, Saleem MF, Wang L (2017) Drought tolerance in three maize cultivars is related to differential osmolyte accumulation, antioxidant defense system, and oxidative damage. Front Plant Sci 8:69.

Apel K, Hirt H (2004) Reactive oxygen species: metabolism, oxidative stress, and signal transduction. Annu Rev Plant Biol 55:373–399.

Arif M, Talha J, Muhammad R, Fahad S, Muhammad A, Amanullah, Kawsar A, Ishaq AM, Bushra K, Fahd R (2020) Biochar; a remedy for climate change. In: Fahad S, Hasanuzzaman M, Alam M, Ullah H, Saeed M, Khan AK, Adnan M (eds) Environment, climate, plant and vegetation growth. Springer Publ Ltd, Springer Nature Switzerland AG. Part of Springer Nature. pp 151–172. https://doi.org/10.1007/978-3-030-49732-3.

Armstrong W, Cousins D, Armstrong J, Turner DW, Beckett PM (2000) Oxygen distribution in wetland plant roots and permeability barriers to gas-exchange with the rhizosphere: a microelectrode and modelling study with *Phragmites australis*. Ann Bot 86:687–703.

Asada K (2006) Production and scavenging of reactive oxygen species in chloroplasts and their functions. Plant Physiol 141:391–396.

Asati A, Pichhode M, Nikhil K (2016) Effect of heavy metals on plants: an overview. Int J Appl Innov Eng Manage 5:2319–4847.

Ashfaq AR, Uzma Y, Niaz A, Muhammad AA, Fahad S, Haider S, Tayebeh Z, Subhan D, Süleyman T, Hesham AElE, Pramila T, Jamal MA, Sulaiman AA, Rahul D (2021) Toxicity of Cadmium and nickel in the context of applied activated carbon biochar for improvement in soil fertility. Saudi Soc Agric Sci. https://doi.org/10.1016/j.sjbs.2021.09.035.

Ashraf AM, Akbar A, Askari SH, Iqbal M, Rasheed R, Hussain I (2018a) Recent advances in abiotic stress tolerance of plants through chemical priming: an overview. In: Rakshit A, Singh HB (eds) Advances in seed priming. Springer Singapore, Singapore, pp 51–79.

Ashraf AM, Iqbal M, Rasheed R, Hussain I, Perveen S, Mahmood S (2018b) Chapter 16: Dynamic proline metabolism: importance and regulation in water-limited environments. In: Ahmad P, Ahanger MA, Singh VP, Tripathi DK, Alam P, Alyemeni MN (eds) Plant metabolites and regulation under environmental stress. Academic Press, pp 323–336.

Ashraf M (2010) Inducing drought tolerance in plants: recent advances. Biotechnol Adv 28:169–183.

Ashraf U, Kanu AS, Mo Z, Hussain S, Anjum SA, Khan I, Abbas RN, Tang X (2015) Lead toxicity in rice: effects, mechanisms, and mitigation strategies—a mini review. Environ Sci Pollut Res 22:18318–18332.

Athar M, Masood IA, Sana S, Ahmed M, Xiukang W, Sajid F, Sher AK, Habib A, Faran M, Zafar H, Farhana G, Fahad S (2021) Bio-diesel production of sunflower through sulphur management in a semi-arid subtropical environment. Environ Sci Pollut Res. https://doi.org/10.1007/s11356-021-16688-z.

Atif B, Hesham A, Fahad S (2021) Biochar coupling with phosphorus fertilization modifies antioxidant activity, osmolyte accumulation and reactive oxygen species synthesis in the leaves and xylem sap of rice cultivars under high-temperature stress. Physiol Mol Biol Plants. https://doi.org/10.1007/s12298-021-01062-7.

Atkinson CJ, Fitzgerald JD, Hipps NA (2010) Potential mechanisms for achieving agricultural benefits from biochar application to temperate soils: a review. Plant Soil 337:1–18.

Ayman EL Sabagh, Hossain A, Barutçular C, Iqbal MA, Islam MS, Fahad S, Sytar O, Çig F, Meena RS, Erman M (2020) Consequences of salinity stress on the quality of crops and its mitigation strategies for sustainable crop production: an outlook of arid and semi- arid regions. In: Fahad S, Hasanuzzaman M, Alam M, Ullah H, Saeed M, Khan AK, Adnan M (eds) Environment, climate, plant and vegetation growth. Springer Publ Ltd, Springer Nature Switzerland AG. Part of Springer Nature. pp 503–534. https://doi.org/10.1007/978-3-030-49732-3.

Aziz K, Daniel KYT, Fazal M, Muhammad ZA, Farooq S, Fan W, Fahad S, Ruiyang Z (2017a) Nitrogen nutrition in cotton and control strategies for greenhouse gas emissions: a review. Environ Sci Pollut Res 24:23471–23487. https://doi.org/10.1007/s11356-017-0131-y.

Aziz K, Daniel KYT, Muhammad ZA, Honghai L, Shahbaz AT, Mir A, Fahad S (2017b) Nitrogen fertility and abiotic stresses management in cotton crop: a review. Environ Sci Pollut Res 24:14551–14566. https://doi.org/10.1007/s11356-017-8920-x.

Bailey-Serres J, Voesenek LACJ (2008) Flooding stress: acclimations and genetic diversity. Annu Rev Plant Biol 59:313–339.

Banerjee A, Roychoudhury A (2018) Seed priming technology in the amelioration of salinity stress in plants. In: Rakshit A, Singh HB (eds) Advances in Seed Priming. Springer, pp 81–93. Springer, Singapore.

Bano A, Ullah F, Nosheen A (2012) Role of abscisic acid and drought stress on the activities of antioxidant enzymes in wheat. Plant Soil Environ 58:181–185.

Baseer M, Adnan M, Fazal M, Fahad S, Muhammad S, Fazli W, Muhammad A, Jr. Amanullah, Depeng W, Saud S, Muhammad N, Muhammad Z, Fazli S, Beena S, Mian AR, Ishaq AM (2019) Substituting urea by organic wastes for improving maize yield in alkaline soil. J Plant Nutr. https://doi.org/10.1080/01904167.2019.1659344.

Battisti DS, Naylor RL (2009) Historical warnings of future food insecurity with unprecedented seasonal heat. Science 323:240–244.

Bayram AY, Seher Ö, Nazlican A (2020) Climate change forecasting and modeling for the year of 2050. In: Fahad S, Hasanuzzaman M, Alam M, Ullah H,Saeed M, Khan AK, Adnan M (ed) Environment, climate, plant and vegetation growth. Springer Publ Ltd, Springer Nature Switzerland AG. Part of Springer Nature. pp 109–122. https://doi.org/10.1007/978-3-030-49732-3.

Biederman LA, Harpole WS (2013) Biochar and its effects on plant productivity and nutrient cycling: a meta-analysis. GCB Bioenergy 5:202–214.

Bisbis MB, Gruda N, Blanke M (2018) Potential impacts of climate change on vegetable production and product quality – A review. J Clean Prod 170:1602–1620.

Blanco-Canqui H (2017) Biochar and soil physical properties. Soil Sci Soc Am J 81:687–711.

Bokhorst SF, Bjerke JW, Tømmervik H, Callaghan TV, Phoenix GK (2009) Winter warming events damage sub-Arctic vegetation: consistent evidence from an experimental manipulation and a natural event. Ecology 97:1408–1415.

Bonanomi G, Ippolito F, Scala F (2015) A "black" future for plant pathology? Biochar as a new soil amendment for controlling plant diseases. J Plant Pathol 97:2.

Bracale M, Coraggio I (2003) Chilling and freezing stresses in plants: cellular responses and molecular strategies for adaptation. In: di Toppi LS, Pawlik-Skowrońska B (eds) Abiotic stresses in plants. Springer Nature Switzerland AG. Part of Springer Nature, pp 23–51.

Branzini A, Santos González R, Zubillaga M (2012) Absorption and translocation of copper, zinc and chromium by *Sesbania virgata*. J Environ Manage 102:50–54.

Buchanan BB, Gruissem W, Jones RL (2000) Biochemistry & molecular biology of plants. American Society of Plant Physiologists, Rockville, MD.

Buet A, Galatro A, Ramos-Artuso F, Simontacchi M (2019) Nitric oxide and plant mineral nutrition: current knowledge. J Exp Bot. 70(17):4461-4476

Burke JJ, Orzech Usda-Ars K (1988) The heat-shock response in higher plants: a biochemical model. Plant Cell Environ 11:441–444.

Cakmak I, Kirkby EA (2008) Role of magnesium in carbon partitioning and alleviating photooxidative damage. Physiol Plant 133:692–704.

Cakmak I, Römheld V (1997) Boron deficiency-induced impairments of cellular functions in plants. Plant Soil 193:71–83.

Calderón A, Sevilla F, Jiménez A (2018) Redox protein thioredoxins: function under salinity, drought and extreme temperature conditions. In: Gupta DK, Palma, JM, Corpas FJ (eds) Antioxidants and antioxidant enzymes in higher plants. Springer International Publishing, Cham, pp 123–162.

Camacho-Cristóbal JJ, Navarro-Gochicoa MT, Rexach J, González-Fontes A, Herrera-Rodríguez MB (2018) Chapter 6: Plant response to boron deficiency and boron use efficiency in crop plants. In: Hossain MA, Kamiya T, Burritt DJ, Phan Tran LS, Fujiwara T (eds) Plant micronutrient use efficiency. Academic Press, pp 109–121.

Carter JL, Colmer TD, Veneklaas EJ (2006) Variable tolerance of wetland tree species to combined salinity and waterlogging is related to regulation of ion uptake and production of organic solutes. New Phytologist 169:123–134.

Castorina G, Persico M, Zilio M, Sangiorgio S, Carabelli L, Consonni G (2018) The maize lilliputian1 (lil1) gene, encoding a brassinosteroid cytochrome P450 C-6 oxidase, is involved in plant growth and drought response. Annal Bot 122:227–238.

Chan KY, Van Zwieten L, Meszaros I, Downie A, Joseph S (2008a) Agronomic values of greenwaste biochar as a soil amendment. Soil Res 45:629–634.

Chan KY, Van Zwieten L, Meszaros I, Downie A, Joseph S (2008b) Using poultry litter biochars as soil amendments. Soil Res 46:437–444.

Chan KY, Xu Z (2009) Biochar: nutrient properties and their enhancement. Biochar Environ Manage: Sci Technol 1:67–84.

Chandrakar V, Dubey A, Keshavkant S (2018) Modulation of arsenic-induced oxidative stress and protein metabolism by diphenyleneiodonium, 24-epibrassinolide and proline in Glycine max L. Acta Botanica Croatica 77:51–61.

Chang W, Qiujuan J, Evgenios A, Haitao L, Gezi L, Jingjing Z, Fahad S, Ying J (2021) Hormetic effects of zinc on growth and antioxidant defense system of wheat plants. Sci Total Environ. https://doi.org/10.1016/j.scitotenv.2021.150992.

Chen T, Yuan F, Song J, Wang B (2016) Nitric oxide participates in waterlogging tolerance through enhanced adventitious root formation in the euhalophyte Suaeda salsa. Funct Plant Biol 43:244.

Chen Y, Guo Z, Dong L, Fu Z, Zheng Q, Zhang G, Qin L, Sun X, Shi Z, Fahad S, Xie F, Saud S (2021) Turf performance and physiological responses of native Poa species to summer stress in Northeast China. PeerJ 9:e12252. http://doi.org/10.7717/peerj.12252.

Cheng CH, Lehmann J, Thies JE, Burton SD, Engelhard MH (2006) Oxidation of black carbon by biotic and abiotic processes. Org Geochem 37:1477–1488.

Chinnusamy V, Zhu J, Zhu JK (2007) Cold stress regulation of gene expression in plants. Trends Plant Sci 12:444–451.

Colmer TD, Flowers TJ (2008) Flooding tolerance in halophytes. New Phytol 179:964–974.

Colmer TD, Voesenek L (2009) Flooding tolerance: suites of plant traits in variable environments. Funct Plant Biol 36:665–681.

Cramer GR, Urano K, Delrot S, Pezzotti M, Shinozaki K (2011) Effects of abiotic stress on plants: a systems biology perspective. BMC Plant Biol 11:163.

Dat J, Vandenabeele S, Vranová E, Van Montagu M, Inzé D, Van Breusegem F (2000) Dual action of the active oxygen species during plant stress responses. Cel Mol Life Sci 57:779–795.

Deepranjan S, Ardith S, Siva OD, Sonam S, Shikha, Manoj P, Amitava R, Sayyed RZ., Abdul G, Mohammad JA, Subhan D, Fahad S, Rahul D (2021) Optimizing nutrient use efficiency, productivity, energetics, and economics of red cabbage following mineral fertilization and biopriming with compatible rhizosphere microbes. Sci Rep 11:15680.

Dembitsky VM (2003) Oxidation, epoxidation and sulfoxidation reactions catalysed by haloperoxidases. Tetrahedron 59:4701–4720.

Depeng W, Fahad S, Saud S, Muhammad K, Aziz K, Mohammad NK, Hafiz MH, Wajid N (2018) Morphological acclimation to agronomic manipulation in leaf dispersion and orientation to promote "Ideotype" breeding: evidence from 3D visual modeling of "super" rice (*Oryza sativa* L.). Plant Physiol Biochem 135:499–510. https://doi.org/10.1016/j.plaphy.2018.11.010.

Ding F, Wang M, Liu B, Zhang S (2017) Exogenous melatonin mitigates photoinhibition by accelerating non-photochemical quenching in tomato seedlings exposed to moderate light during chilling. Front Plant Sci 8:244.

Ding Y, Liu Y, Liu S, Li Z, Tan X, Huang X, Zeng G, Zhou L, Zheng B (2016) Biochar to improve soil fertility. A review. Agron Sustain Dev 36:36.

Ding Z, Hu X, Wan Y, Wang S, Gao B (2016) Removal of lead, copper, cadmium, zinc, and nickel from aqueous solutions by alkali-modified biochar: batch and column tests. J Ind Eng Chem 33:239–245.

Drenovsky RE, Khasanova A, James JJ (2012) Trait convergence and plasticity among native and invasive species in resource-poor environments. Am J Bot 99:629–639.

Dwivedi YK, Shareef MA, Simintiras AC, Lal B, Weerakkody V (2016) A generalised adoption model for services: a cross-country comparison of mobile health (m-health). Gov Inf Q 33:174–187.

Edelstein M, Pitchforth E, Asres G, Silverman M, Kulkarni N (2008) Awareness of health effects of cooking smoke among women in the Gondar Region of Ethiopia: a pilot survey. BMC Int Health Human Rights 8:10.

El-Mashad AAA, Mohamed HI (2012) Brassinolide alleviates salt stress and increases antioxidant activity of cowpea plants (Vigna sinensis). Protoplasma 249:625–635.

Elmer WH, Pignatello JJ (2011) Effect of biochar amendments on mycorrhizal associations and Fusarium crown and root rot of asparagus in replant soils. Plant Disease 95:960–966.

El-Naggar A, Lee SS, Rinklebe J, Farooq M, Song H, Sarmah AK, Zimmerman AR, Ahmad M, Shaheen SM, Ok YS (2019) Biochar application to low fertility soils: a review of current status, and future prospects. Geoderma 337:536–554.

El-Ramady H, Alshaal T, Elhawat N, Ghazi A, Elsakhawy T, Omara AED, El-Nahrawy S, Elmahrouk M, Abdalla N, Domokos-Szabolcsy É, Schnug E (2018) Plant nutrients and their roles under saline soil conditions. In: Hasanuzzaman M, Fujita M, Oku H, Nahar K, Hawrylak-Nowak, B (eds) Plant nutrients and abiotic stress tolerance. Springer Singapore, Singapore, pp 297–324.

Emre B, Ömer SU, Martín LB, Andre D, Fahad S, Rahul D, Muhammad Z-ul-H, Ghulam SH, Subhan D (2021) Studying soil erosion by evaluating changes in physico-chemical properties of soils under different land-use types. J Saudi Soc Agric Sci 20:190–197.

Etesami H, Jeong BR (2018) Silicon (Si): review and future prospects on the action mechanisms in alleviating biotic and abiotic stresses in plants. Ecotoxicol Environ Safe 147:881–896.

Ezin V, Pena RDL, Ahanchede A (2010) Flooding tolerance of tomato genotypes during vegetative and reproductive stages. Braz J Plant Physiol 22:131–142.

Faghani E, Gharechahi J, Komatsu S, Mirzaei M, Khavarinejad RA, Najafi F, Farsad LK, Salekdeh GH (2015) Comparative physiology and proteomic analysis of two wheat genotypes contrasting in drought tolerance. J Proteomics 114:1–15.

Fahad S, Abdul B, Adnan M (eds) (2018b) Global wheat production. IntechOpen, United Kingdom. http://dx.doi.org/10.5772/intechopen.72559.

Fahad S, Adnan M, Hassan S, Saud S, Hussain S, Wu C, Wang D, Hakeem KR, Alharby HF, Turan V, Khan MA, Huang J (2019b) Rice responses and tolerance to high temperature. In: Hasanuzzaman M, Fujita M, Nahar K, Biswas JK (eds) Advances in rice research for abiotic stress tolerance. Woodhead Publ Ltd, Abington Hall Abington, Cambridge, UK, pp 201–224.

Fahad S, Bajwa AA, Nazir U, Anjum SA, Farooq A, Zohaib A, Sadia S, NasimW, Adkins S, Saud S, Ihsan MZ, Alharby H, Wu C, Wang D, Huang J (2017) Crop production under drought and heat stress: plant responses and management options. Front Plant Sci 8:1147. https://doi.org/10.3389/fpls.2017.01147.

Fahad S, Bano A (2012) Effect of salicylic acid on physiological and biochemical characterization of maize grown in saline area. Pak J Bot 44:1433–1438.

Fahad S, Chen Y, Saud S, Wang K, Xiong D, Chen C, Wu C, Shah F, Nie L, Huang J (2013) Ultraviolet radiation effect on photosynthetic pigments, biochemical attributes, antioxidant enzyme activity and hormonal contents of wheat. J Food Agric Environ 11(3&4):1635–1641.

Fahad S, Hasanuzzaman M, Alam M, Ullah H, Saeed M, Ali Khan I, Adnan M (eds) (2020) Environment, climate, plant and vegetation growth. Springer Nature Switzerland AG. https://doi.org/10.1007/978-3-030-49732-3.

Fahad S, Hussain S, Bano A, Saud S, Hassan S, Shan D, Khan FA, Khan F, Chen Y, Wu C, Tabassum MA, Chun MX, Afzal M, Jan A, Jan MT, Huang J (2014a) Potential role of phytohormones and plant growth-promoting rhizobacteria in abiotic stresses: consequences for changing environment. Environ Sci Pollut Res 22(7):4907–4921. https://doi.org/10.1007/s11356-014-3754-2.

Fahad S, Hussain S, Matloob A, Khan FA, Khaliq A, Saud S, Hassan S, Shan D, Khan F, Ullah N, Faiq M, Khan MR, Tareen AK, Khan A, Ullah A, Ullah N, Huang J (2014b) Phytohormones and plant responses to salinity stress: a review. Plant Growth Regul 75(2):391–404. https://doi.org/10.1007/s10725-014-0013-y.

Fahad S, Hussain S, Saud S, Hassan S, Chauhan BS, Khan F, et al (2016a) Responses of rapid viscoanalyzer profile and other rice grain qualities to exogenously applied plant growth regulators under high day and high night temperatures. PLoS One 11(7):e0159590. https://doi.org/10.1371/journal.pone.0159590.

Fahad S, Hussain S, Saud S, Hassan S, Ihsan Z, Shah AN, Wu C, Yousaf M, Nasim W, Alharby H, Alghabari F, Huang J (2016c) Exogenously applied plant growth regulators enhance the morphophysiological growth and yield of rice under high temperature. Front Plant Sci 7:1250. https://doi.org/10.3389/fpls.2016.01250.

Fahad S, Hussain S, Saud S, Hassan S, Tanveer M, Ihsan MZ, Shah AN, Ullah A, Nasrullah KF, Ullah S, Alharby H, Wu C, Huang J (2016d) A combined application of biochar and phosphorus alleviates heat-induced adversities on physiological, agronomical and quality attributes of rice. Plant Physiol Biochem 103:191–198.

Fahad S, Hussain S, Saud S, Khan F, Hassan S, Nasim W, Arif M, Wang F, Huang J (2016b) Exogenously applied plant growth regulators affect heat-stressed rice pollens. J Agron Crop Sci 202:139–150.

Fahad S, Hussain S, Saud S, Tanveer M, Bajwa AA, Hassan S, Shah AN, Ullah A, Wu C, Khan FA, Shah F, Ullah S, Chen Y, Huang J (2015a) A biochar application protects rice pollen from high-temperature stress. Plant Physiol Biochem 96:281–287.

Fahad S, Muhammad ZI, Abdul K, Ihsanullah D, Saud S, Saleh A, Wajid N, Muhammad A, Imtiaz AK, Chao W, Depeng W, Jianliang H (2018a) Consequences of high temperature under changing climate optima for rice pollen characteristics-concepts and perspectives, Archives Agron Soil Sci. https://doi.org/10.1080/03650340.2018.1443213.

Fahad S, Nie L, Chen Y, Wu C, Xiong D, Saud S, Hongyan L, Cui K, Huang J (2015b) Crop plant hormones and environmental stress. Sustain Agric Rev 15:371–400.

Fahad S, Rehman A, Shahzad B, Tanveer M, Saud S, Kamran M, Ihtisham M., Khan SU, Turan V, Rahman MHU (2019a) Rice responses and tolerance to metal/metalloid toxicity. In: Hasanuzzaman M, Fujita M, Nahar K, Biswas JK (eds) Advances in rice research for abiotic stress tolerance. Woodhead Publ Ltd, Abington Hall Abington, Cambridge, UK, pp 299–312.

Fahad S, Saud S, Yajun C, Chao W, Depeng W (eds) (2021f) Abiotic stress in plants. IntechOpen, United Kingdom. http://dx.doi.org/10.5772/intechopen.91549.

Fahad S, Sönmez O, Saud S, Wang D, Wu C, Adnan M, Arif M, Amanullah (eds) (2021e) Engineering tolerance in crop plants against abiotic stress, 1st edn. ed, Footprints of climate variability on plant diversity. CRC Press, Boca Raton, FL.

Fahad S, Sönmez O, Saud S, Wang D, Wu C, Adnan M, Turan V (eds) (2021b) Climate change and plants: biodiversity, growth and interactions, 1st edn. ed, Footprints of climate variability on plant diversity. CRC Press, Boca Raton, FL.

Fahad S, Sönmez O, Saud S, Wang D, Wu C, Adnan M, Turan V (eds) (2021c) Developing climate resilient crops: improving global food security and safety, 1st edn. ed, Footprints of climate variability on plant diversity. CRC Press, Boca Raton, FL.

Fahad S, Sönmez O, Saud S, Wang D, Wu C, Adnan M, Turan V (eds) (2021a) Plant growth regulators for climate-smart agriculture, 1st edn. ed, Footprints of climate variability on plant diversity. CRC Press, Boca Raton, FL.

Fahad S, Sönmez O, Turan V, Adnan M, Saud S, Wu C, Wang D (eds) (2021d) Sustainable soil and land management and climate change, 1st edn. ed, Footprints of climate variability on plant diversity. CRC Press, Boca Raton, FL.

Farah R, Muhammad R, Muhammad SA, Tahira Y, Muhammad AA, Maryam A, Shafaqat A, Rashid M, Muhammad R, Qaiser H, Afia Z, Muhammad AA, Muhammad A, Fahad S (2020) Alternative and non-conventional soil and crop management strategies for increasing water use efficiency. In: Fahad S, Hasanuzzaman M, Alam M, Ullah H, Saeed M, Khan AK, Adnan M (eds) Environment, climate, plant and vegetation growth. Springer Publ Ltd, Springer Nature Switzerland AG. Part of Springer Nature. pp 323–338. https://doi.org/10.1007/978-3-030-49732-3.

Farhana G, Ishfaq A, Muhammad A, Dawood J, Fahad S, Xiuling L, Depeng W, Muhammad F, Muhammad F, Syed AS (2020) Use of crop growth model to simulate the impact of climate change on yield of various wheat cultivars under different agro-environmental conditions in Khyber Pakhtunkhwa, Pakistan. Arabian J Geosci 13:112 https://doi.org/10.1007/s12517-020-5118-1.

Farhat A, Hafiz MH, Wajid I, Aitazaz AF, Hafiz FB, Zahida Z, Fahad S, Wajid F, Artemi C (2020) A review of soil carbon dynamics resulting from agricultural practices. J Environ Manage 268:110319.

Fazli W, Muhmmad S, Amjad A, Fahad S, Muhammad A, Muhammad N, Ishaq AM, Imtiaz AK, Mukhtar A, Muhammad S, Muhammad I, Rafi U, Haroon I, Muhammad A (2020) Plant-microbes interactions and functions in changing climate. In: Fahad S, Hasanuzzaman M, Alam M, Ullah H, Saeed M, Khan AK, Adnan M (eds) Environment, climate, plant and vegetation growth. Springer Publ Ltd, Springer Nature Switzerland AG. Part of Springer Nature. pp 397–420. https://doi.org/10.1007/978-3-030-49732-3.

Feng G, Liu G, Xiao J (2015) The Arabidopsis EIN2 restricts organ growth by retarding cell expansion. Plant Signal Behav 10:e1017169.

Forbes MS, Raison RJ, Skjemstad JO (2006) Formation, transformation and transport of black carbon (charcoal) in terrestrial and aquatic ecosystems. Sci Total Environ 37:190–206.

Franklin KA, Wigge PA. (2014) Temperature and plant development. Wiley Online Library.

Galano A, Tan DX, Reiter RJ (2013) On the free radical scavenging activities of melatonin's metabolites, AFMK and AMK. J Pineal Res 54:245–257.

Gao Z, Zhang Y, Song N, Li X (2017) Biomass-derived renewable carbon materials for electrochemical energy storage. Mater Res Lett 5:69–88.

Gascó G, Paz-Ferreiro J, Cely P, Plaza C, Méndez A (2016) Influence of pig manure and its biochar on soil $CO_2$ emissions and soil enzymes. Ecol Eng 95:19–24.

Gaur A, Adholeya A (2004) Prospects of arbuscular mycorrhizal fungi in phytoremediation of heavy metal contaminated soils. Curr Sci 528–534.

Ghulam M, Muhammad AA, Donald LS, Sajid M, Muhammad FQ, Niaz A, Ateeq ur R, Shakeel A, Sajjad H, Muhammad A, Summia M, Aqib HAK, Fahad S, Rahul D, Mazhar I, Timothy DS (2021) Formalin fumigation and steaming of various composts differentially influence the nutrient release, growth and yield of muskmelon (*Cucumis melo* L.). Sci Rep 11:21057.

Gill SS, Tuteja N (2010) Reactive oxygen species and antioxidant machinery in abiotic stress tolerance in crop plants. Plant Physiol Biochem 48:909–930.

Glaser B, Lehmann J, Zech W (2002) Ameliorating physical and chemical properties of highly weathered soils in the tropics with charcoal–a review. Biol Fert Soils 35:219–230.

Gomez JD, Denef K, Stewart CE, Zheng J, Cotrufo MF (2014) Biochar addition rate influences soil microbial abundance and activity in temperate soils: biochar addition stimulates microbial activity proportionally. Eur J Soil Sci 65:28–39.

Gopakumar L, Bernard NO, Donato V (2020) Soil microarthropods and nutrient cycling. In: Fahad S, Hasanuzzaman M, Alam M, Ullah H, Saeed M, Khan AK, Adnan M (eds) Environment, climate, plant and vegetation growth. Springer Publ Ltd, Springer Nature Switzerland AG. Part of Springer Nature. pp 453–472. https://doi.org/10.1007/978-3-030-49732-3.

Gu X, Wang Y, Lai C, Qiu J, Li S, Hou Y, Martens W, Mahmood N, Zhang S (2015) Microporous bamboo biochar for lithium-sulfur batteries. Nano Res 8:129–139.

Guo J, Yue T, Li X, Yuan Y (2016) Heavy metal levels in kiwifruit orchard soils and trees and its potential health risk assessment in Shaanxi, China. Environ Sci Pollut Res 23:14560–14566.

Guofu L, Zhenjian B, Fahad S, Guowen C, Zhixin X, Hao G, Dandan L, Yulong L, Bing L, Guoxu J, Saud S (2021) Compositional and structural changes in soil microbial communities in response to straw mulching and plant revegetation in an abandoned artificial pasture in Northeast China. Global Ecol Conserv 31:e01871.

Gupta P, De B (2017) Metabolomics analysis of rice responses to salinity stress revealed elevation of serotonin, and gentisic acid levels in leaves of tolerant varieties. Plant Signal Behav 12:e1335845.

Guy CL (1990) Cold acclimation and freezing stress tolerance: role of protein metabolism. Annu Rev Plant Biol 41:187–223.

Habib ur R, Ashfaq A, Aftab W, Manzoor H, Fahd R, Wajid I, Md. Aminul I, Vakhtang S, Muhammad A, Asmat U, Abdul W, Syeda RS, Shah S, Shahbaz K, Fahad S, Manzoor H, Saddam H, Wajid N (2017) Application of CSM-CROPGRO-Cotton model for cultivars and optimum planting dates: evaluation in changing semi-arid climate. Field Crops Res. http://dx.doi.org/10.1016/j.fcr.2017.07.007.

Hafeez M, Farman U, Muhammad MK, Xiaowei L, Zhijun Z, Sakhawat S, Muhammad I, Mohammed AA, Mandela F-G, Nicolas D, Muzammal R, Fahad S, Yaobin L (2021) Metabolic-based insecticide resistance mechanism and ecofriendly approaches for controlling of beet armyworm Spodoptera exigua:a review. Environ Sci Pollut Res. https://doi.org/10.1007/s11356-021-16974-w.

Hafiz MH, Abdul K, Farhat A, Wajid F, Fahad S, Muhammad A, Ghulam MS, Wajid N, Muhammad M, Hafiz FB (2020b) Comparative effects of organic and inorganic fertilizers on soil organic carbon and wheat productivity under arid region. Commun Soil Sci Plant Anal. https://doi.org/10.1080/00103624.2020.1763385.

Hafiz MH, Farhat A, Ashfaq A, Hafiz FB, Wajid F, Carol Jo W, Fahad S, Gerrit H (2020a) Predicting kernel growth of maize under controlled water and nitrogen applications. Int J Plant Prod. https://doi.org/10.1007/s42106-020-00110-8.

Hafiz MH, Farhat A, Shafqat S, Fahad S, Artemi C, Wajid F, Chaves CB, Wajid N, Muhammad M, Hafiz FB (2018) Offsetting land degradation through nitrogen and water management during maize cultivation under arid conditions. Land Degrad Dev:1–10. https://doi.org/10.1002/ldr.2933.

Hafiz MH, Muhammad A, Farhat A, Hafiz FB, Saeed AQ, Muhammad M, Fahad S, Muhammad A (2019) Environmental factors affecting the frequency of road traffic accidents: a case study of sub-urban area of Pakistan. Environ Sci Pollut Res. https://doi.org/10. 1007/s11356-019-04752-8.

Hafiz MH, Wajid F, Farhat A, Fahad S, Shafqat S, Wajid N, Hafiz FB (2016) Maize plant nitrogen uptake dynamics at limited irrigation water and nitrogen. Environ Sci Pollut Res 24(3):2549–2557. https://doi.org/10. 1007/s11356-016-8031-0.

Haider SA, Lalarukh I, Amjad SF, Mansoora N, Naz M, Naeem M, Bukhari SA, Shahbaz M, Ali SA, Marfo TD, Subhan D, Rahul D, Fahad S (2021) Drought stress alleviation bypotassium-nitrate-containingchitosan/montmorillonite microparticles confers changes in Spinacia oleracea L. Sustainability 13:9903. https://doi.org/10.3390/su13179903.

Hajar EWI, Sulaiman AZB, Sakinah AMM (2014) Assessment of heavy metals tolerance in leaves, stems and flowers of *Stevia rebaudiana* plant. Procedia Environ Sci 20:386–393.

Hamza SM, Xiukang W, Sajjad A, Sadia Z, Muhammad N, Adnan M, Fahad S, et al (2021) Interactive effects of gibberellic acid and NPK on morpho-physio-biochemical traits and organic acid exudation pattern in coriander (Coriandrum sativum L.) grown in soil artificially spiked with boron. Plant Physiol Biochem 167:884–900.

Hasanuzzaman M, Nahar K, Alam M, Roychowdhury R, Fujita M (2013a) Physiological, biochemical, and molecular mechanisms of heat stress tolerance in plants. Int J Mol 14:9643–9684.

Hasanuzzaman M, Nahar K, Fujita M (2013b) Extreme temperature responses, oxidative stress and antioxidant defense in plants. Abiotic Stress-Plant Responses Appl Agric 13:169–205.

Hasanuzzaman M, Nahar K, Fujita M (2018) Plants under metal and metalloid stress: responses, tolerance and remediation. Springer.

Herzog M, Striker GG, Colmer TD, Pedersen O (2016) Mechanisms of waterlogging tolerance in wheat–a review of root and shoot physiology. Plant Cell Environ 39:1068–1086.

Hesham FA, Fahad S (2020) Melatonin application enhances biochar efficiency for drought tolerance in maize varieties: modifications in physio-biochemical machinery. Agron J 112(4):1–22.

Ho SH, Yang ZK, Nagarajan D, Chang JS, Ren NQ (2017) High-efficiency removal of lead from wastewater by biochar derived from anaerobic digestion sludge. Bioresour Technol 246:142–149.

Hosseini SA, Rad SN, Ali N, Yvin JC (2019) The ameliorative effect of silicon on maize plants grown in mg-deficient conditions. Int J Mol Sci 20:969.

Howarth CJ (2005) Tolerance to high temperature. Abiotic Stresses: Plant Resist Breed Mol Approaches:1920.

Hu H, Xiong L (2014) Genetic engineering and breeding of drought-resistant crops. Annu Rev Plant Biol 65:715–741.

Hua L, Lu Z, Ma H, Jin S (2014) Effect of biochar on carbon dioxide release, organic carbon accumulation, and aggregation of soil. Environ Prog Sustain Energy 33:941–946.

Huang Li Y, Li Xiao X, Zhang Yun B, Fahad S, Wang F (2021) dep1 improves rice grain yield and nitrogen use efficiency simultaneously by enhancing nitrogen and dry matter translocation. J Integr Agric. https://doi.org/10.1016/S2095-3119(21)63795-4.

Hussain MA, Fahad S, Rahat S, Muhammad FJ, Muhammad M, Qasid A, Ali A, Husain A, Nooral A, Babatope SA, Changbao S, Liya G, Ibrar A, Zhanmei J, Juncai H (2020) Multifunctional role of brassinosteroid and its analogues in plants. Plant Growth Regul. https://doi.org/10.1007/s10725-020-00647-8.

Ibrahim EA (2016) Seed priming to alleviate salinity stress in germinating seeds. J Plant Physiol 192:38–46.

Ibrar K, Aneela R, Khola Z, Urooba N, Sana B, Rabia S, Ishtiaq H, Mujaddad Ur Rehman, Salvatore M (2020) Microbes and environment: global warming reverting the frozen zombies. In: Fahad S, Hasanuzzaman M, Alam M, Ullah H, Saeed M, Khan AK, Adnan M (eds) Environment, climate, plant and vegetation growth. Springer Publ Ltd, Springer Nature Switzerland AG. Part of Springer Nature. pp 607–634. https://doi.org/10.1007/978-3-030-49732-3.

Igwe J, Abia AA (2006) A bioseparation process for removing heavy metals from waste water using biosorbents. Afr J Biotechnol 5.

Ikram U, Khadim D, Muhammad T, Muhammad S, Fahad S (2021) Gibberellic acid and urease inhibitor optimize nitrogen uptake and yield of maize at varying nitrogen levels under changing climate. Environ Sci Pollut Res. https://doi.org/10.1007/s11356-021-16049-w.

Ilyas M, Mohammad N, Nadeem K, Ali H, Aamir HK, Kashif H, Fahad S, Aziz K, Abid U (2020) Drought tolerance strategies in plants: a mechanistic approach. J Plant Growth Regul. https://doi.org/10.1007/s00344-020-10174-5.

Inyang M, Gao B, Yao Y, Xue Y, Zimmerman AR, Pullammanappallil P, Cao X (2012) Removal of heavy metals from aqueous solution by biochars derived from anaerobically digested biomass. Bioresour Technol 110:50–56.

Iqra M, Amna B, Shakeel I, Fatima K, Sehrish L, Hamza A, Fahad S (2020) Carbon cycle in response to global warming. In: Fahad S, Hasanuzzaman M, Alam M, Ullah H, Saeed M, Khan AK, Adnan M (eds) Environment, climate, plant and vegetation growth. Springer Publ Ltd, Springer Nature Switzerland AG. Part of Springer Nature. pp 1–16. https://doi.org/10.1007/978-3-030-49732-3.

Irfan M, Muhammad M, Muhammad JK, Khadim MD, Dost M, Ishaq AM, Waqas A, Fahad S, Saud S, et al (2021) Heavy metals immobilization and improvement in maize (Zea mays L.) growth amended with biochar and compost. Sci Rep 11:18416.

Jabborova D, Sulaymanov K, Sayyed RZ, Alotaibi SH, Enakiev Y, Azimov A, Jabbarov Z, Ansari MJ, Fahad S, Danish S, et al (2021) Mineral fertilizers improves the quality of turmeric and soil. Sustainability 13:9437. https://doi.org/10.3390/su13169437.

Jackson MB, Colmer TD (2005) Response and adaptation by plants to flooding stress. Ann Bot 96:501–505.

Jaishankar M, Tseten T, Anbalagan N, Mathew BB, Beeregowda KN (2014) Toxicity, mechanism and health effects of some heavy metals. Interdiscip Toxicol 7:60–72.

Jaleel CA, Riadh K, Gopi R, Manivannan P, Ines J, Al-Juburi HJ, Chang-Xing Z, Hong-Bo S, Panneerselvam R (2009) Antioxidant defense responses: physiological plasticity in higher plants under abiotic constraints. Acta Physiol Plant 31:427–436.

James RA, Blake C, Byrt CS, Munns R (2011) Major genes for Na+ exclusion, Nax1 and Nax2 (wheat HKT1; 4 and HKT1; 5), decrease Na+ accumulation in bread wheat leaves under saline and waterlogged conditions. J Exp Bot 62:2939–2947.

Jan M, Anwar-ul-Haq M, Adnan NS, Muhammad Y, Javaid I, Xiuling L, Depeng W, Fahad S (2019) Modulation in growth, gas exchange, and antioxidant activities of salt-stressed rice (Oryza sativa L.) genotypes by zinc fertilization. Arab J Geosci 12:775. https://doi.org/10.1007/s12517-019-4939-2.

Jiang Y, Huang B (2001) Drought and heat stress injury to two cool-season turf grasses in relation to antioxidant metabolism and lipid peroxidation. Crop Sci 41:436–442.

Jin J. Li S, Peng X, Liu W, Zhang C, Yang Y, Han L, Du Z, Sun K, Wang X (2018) $HNO_3$ modified biochars for uranium (VI) removal from aqueous solution. Bioresour Technol 256:247–253.

Kabata-Pendias A (1993) Behavioural properties of trace metals in soils. J Appl Geochem 8:3–9.

Kabata-Pendias A, Pendia H (1984) Trace elements in plants and soils. Boca Raton, FL, pp 233–237.

Kamran M, Wenwen C, Irshad A, Xiangping M, Xudong Z, Wennan S, Junzhi C, Shakeel A, Fahad S, Qingfang H, Tiening L (2017) Effect of paclobutrazol, a potential growth regulator on stalk mechanical strength, lignin accumulation and its relation with lodging resistance of maize. Plant Growth Regul 84:317–332. https://doi.org/10.1007/ s10725-017-0342-8.

Ke Q, Wang Z, Ji CY, Jeong JC, Lee HS, Li H, Xu B, Deng X, Kwak SS (2016) Transgenic poplar expressing codA exhibits enhanced growth and abiotic stress tolerance. Plant Physiol Biochem 100:75–84.

Ke Q, Ye J, Wang B, Ren J, Yin L, Deng X, Wang S (2018) Melatonin mitigates salt stress in wheat seedlings by modulating polyamine metabolism. Front Plant Sci 9.

Khadim D, Fahad S, Jahangir MMR, Iqbal M, Syed SA, Shah AK, Ishaq AM, Rahul D, et al (2021a) Biochar and urease inhibitor mitigate $NH_3$ and $N_2O$ emissions and improve wheat yield in a urea fertilized alkaline soil. Sci Rep 11:17413.

Khadim D, Saif-ur-R, Fahad S, Syed SA, Shah AK, et al (2021b) Influence of variable biochar concentration on yield-scaled nitrous oxide emissions, Wheat yield and nitrogen use efficiency. Sci Rep 11:16774.

Khan A, Shah N, Muhammad, Khan MS, Ahmad MS, Farooq M, Adnan M, Jawad SM, Ullah H, Yousafzai AM (2016) Quantitative determination of lethal concentration lc50 of atrazine on biochemical parameters; total protein and serum albumin of freshwater fish grass carp (*Ctenopharyngodonidella*). Pol J Environ Stud 25(4):1–7.

Khan MMH, Niaz A, Umber G, Muqarrab A, Muhammad AA, Muhammad I, Shabir H, Shah F, Vibhor A, Shams H. A-H, Reham A, Syed MBA, Nadiyah MA, Ali TKZ, Subhan D, Rahul D (2021) Synchronization of Boron application methods and rates is environmentally friendly approach to improve quality attributes of Mangifera indica L. On sustainable basis. Saudi J Bio Sci. https://doi.org/10.1016/j.sjbs.2021.10.036.

Kidokoro S, Yoneda K, Takasaki H, Takahashi F, Shinozaki K, Yamaguchi-Shinozaki K (2017) Different cold-signaling pathways function in the responses to rapid and gradual decreases in temperature. Plant Cell 29:760–774.

Kim YH, Hwang SJ, Waqas M, Khan AL, Lee JH, Lee JD, Nguyen HT, Lee IJ (2015) Comparative analysis of endogenous hormones level in two soybean (Glycine max L.) lines differing in waterlogging tolerance. Front Plant Sci 6.

Kloke A, Sauerbeck DR, Vetter H (1984) Changing metal cycles and human health. Springier, Berlin, Heidelberg.

Kookana RS (2010) The role of biochar in modifying the environmental fate, bioavailability, and efficacy of pesticides in soils: a review. Soil Res 48:627–637.

Kotak S, Vierling E, Bäumlein H, von Koskull-Döring P (2007) A novel transcriptional cascade regulating expression of heat stress proteins during seed development of Arabidopsis. Plant Cell 19:182–195.

Kozdrój J, van Elsas JD (2000) Response of the bacterial community to root exudates in soil polluted with heavy metals assessed by molecular and cultural approaches. Soil Biol Biochem 32:1405–1417.

Kramer, Boyer (1995) Google Scholar [WWW Document], n.d. https://scholar.google.com/scholar?hl=en&as_sdt=0%2C5&q=%28Kramer+and+Boyer%2C+1995&btnG=. Accessed 8 October 2019.

Kratsch HA, Wise RR (2000) The ultrastructure of chilling stress. Plant Cell Environ 23:337–350.

Lal R (2015) Restoring soil quality to mitigate soil degradation. Sustainability 7:5875–5895.

Lambers H, Chapin III, FS, Pons TL (2008) Plant physiological ecology. Springer Science & Business Media.

Landi M, Margaritopoulou T, Papadakis IE, Araniti F (2019) Boron toxicity in higher plants: an update. Planta:1–22.

Law RD, Crafts-Brandner SJ (2001) High temperature stress increases the expression of wheat leaf ribulose-1, 5-bisphosphate carboxylase/oxygenase activase protein. Arch Biochem Biophy 386:261–267.

Lehmann J (2007) A handful of carbon. Nature 447:143–144.

Lehmann J, da Silva JP, Steiner C, Nehls T, Zech W, Glaser B (2003) Nutrient availability and leaching in an archaeological Anthrosol and a Ferralsol of the Central Amazon basin: fertilizer, manure and charcoal amendments. Plant Soil 249:343–357.

Lehmann J, Joseph S (2015) Biochar for environmental management: science, technology and implementation. Routledge.

Lehmann J, Rillig MC, Thies J, Masiello CA, Hockaday WC, Crowley D (2011) Biochar effects on soil biota–a review. Soil Biol Biochem 43:1812–1836.

Lehmann J, Rondon M (2006) Bio-char soil management on highly weathered soils in the humid tropics. In: Uphoff N (ed) Biological approaches to sustainable soil systems. CRC Press, Boca Raton, FL, pp 517–529.

Lenochová Z, Soukup A, Votrubová O (2009) Aerenchyma formation in maize roots. Biologia Plantarum 53:263–270.

Lesk C, Rowhani P, Ramankutty N (2016) Influence of extreme weather disasters on global crop production. Nature 529:84.

Li B, Fan CH, Xiong ZQ, Li QL, Zhang M (2015) The combined effects of nitrification inhibitor and biochar incorporation on yield-scaled $N_2O$ emissions from an intensively managed vegetable field in southeastern China. Biogeosciences 12:2003–2017.

Li X, Wei JP, Scott E, Liu JW, Guo S, Li Y, Zhang L, Han WY (2018) Exogenous melatonin alleviates cold stress by promoting antioxidant defense and redox homeostasis in Camellia sinensis L. Molecules 23:165.

Li Z, Song Z, Singh BP, Wang H (2019) The impact of crop residue biochars on silicon and nutrient cycles in croplands. Sci Total Environ 659:673–680.

Liang B, Lehmann J, Solomon D, Kinyangi J, Grossman J, O'neill B, Skjemstad JO, Thies J, Luizao FJ, Petersen J (2006) Black carbon increases cation exchange capacity in soils. Soil Sci Soc Am J 70:1719–1730.

Liu C, Dai Z, Xia J, Chang C, Sun H (2018) Combined effect of salt and drought on boron toxicity in *Puccinellia tenuiflora*. Ecotoxicol Environ Saf 157:395–402.

Ma Y, Liu WJ, Zhang N, Li YS, Jiang H, Sheng GP (2014) Polyethylenimine modified biochar adsorbent for hexavalent chromium removal from the aqueous solution. Bioresour Technol 169:403–408.

Maathuis FJ (2009) Physiological functions of mineral macronutrients. Curr Opin Plant Biol 12:250–258.

Machado FJ, Santana FM, Lau D, Del Ponte EM (2017) Quantitative review of the effects of Triazole and Benzimidazole fungicides on fusarium head blight and wheat yield in Brazil. Plant Dis 101:1633–1641.

Mahar A, Amjad A, Altaf HL, Fazli W, Ronghua L, Muhammad A, Fahad S, Muhammad A, Rafiullah, Imtiaz AK, Zengqiang Z (2020) Promising technologies for Cd-contaminated soils: drawbacks and possibilities. In: Fahad S, Hasanuzzaman M, Alam M, Ullah H, Saeed M, Khan AK, Adnan M (eds) Environment, climate, plant and vegetation growth. Springer Publ Ltd, Springer Nature Switzerland AG. Part of Springer Nature. pp 63–92. https://doi.org/10.1007/978-3-030-49732-3.

Mahatma MK, Bhatnagar R, Solanki RK, Mittal GK (2009) Effect of seed soaking treatments on salinity induced antioxidant enzymes activity, lipid peroxidation and free amino acid content in wheat (*Triticum aestivum* L.) leaves. Ind J Agric Biochem 22:108–112.

Mahmood Ul H, Tassaduq R, Chandni I, Adnan A, Muhammad A, Muhammad MA, Muhammad H-ur-R, Mehmood AN, Alam S, Fahad S (2021) Linking plants functioning to adaptive responses under heat stress conditions: a mechanistic review. J Plant Growth Regul. https://doi.org/10.1007/s00344-021-10493-1.

Major J, Rondon M, Molina D, Riha SJ, Lehmann J (2010) Maize yield and nutrition during 4 years after biochar application to a *Colombian savanna* oxisol. Plant Soil 333:117–128.

Manzer HS, Saud A, Soumya M, Abdullah A. Al-A, Qasi DA, Bander MA. Al-M, Hayssam MA, Hazem MK, Fahad S, Vishnu DR, Om PN (2021) Molybdenum and hydrogen sulfide synergistically mitigate arsenic toxicity by modulating defense system, nitrogen and cysteine assimilation in faba bean (*Vicia faba* L.) seedlings. Environ Pollut 290:117953.

Marschner H, Marschner P (eds) (2012) Marschner's mineral nutrition of higher plants, 3rd edn. Elsevier/Academic Press, London; Waltham, MA.

McCauley AM (2011) Nitrogen fixation by annual legume green manures in a semi-arid cropping system. PhD Thesis, Montana State University-Bozeman, College of Agriculture.

McCauley AM, Jones CA, Miller PR, Burgess MH, Zabinski CA (2012) Nitrogen fixation by pea and lentil green manures in a semi-arid agroecoregion: effect of planting and termination timing. Nutr Cycl Agroecosyst 92:305–314.

McCauley E, Downing JA, Watson S (1989) Sigmoid relationships between nutrients and chlorophyll among lakes. Can J Fish Aquat Sci 46:1171–1175.

Md Jakir H, Allah B (2020) Development and applications of transplastomic plants; a way towards eco-friendly agriculture. In: Fahad S, Hasanuzzaman M, Alam M, Ullah H, Saeed M, Khan AK, Adnan M (eds) Environment, climate, plant and vegetation growth. Springer Publ Ltd, Springer Nature Switzerland AG. Part of Springer Nature. pp 285–322. https://doi.org/10.1007/978-3-030-49732-3.

Md. Enamul H, Shoeb AZM, Mallik AH, Fahad S, Kamruzzaman MM, Akib J, Nayyer S, K. M. Mehedi A, Swati AS, Md Yeamin A, Most SS (2020) Measuring vulnerability to environmental hazards: qualitative to quantitative. In: Fahad S, Hasanuzzaman M, Alam M, Ullah H, Saeed M, Khan AK, Adnan M (eds) Environment, climate, plant and vegetation growth. Springer Publ Ltd, Springer Nature Switzerland AG. Part of Springer Nature. pp 421–452. https://doi.org/10.1007/978-3-030-49732-3.

Mechergui K, Mahmoudi H, Khouja ML, Jaouadi W (2017) Factors influencing seed germination of the pastoral plant *Retama raetam* subsp. bovei (Fabaceae): interactive effects of fruit morphology, salinity, and osmotic stress. Biologija 63.

Miller G, Suzuki N, Ciftci-Yilmaz S, Mittler R (2010) Reactive oxygen species homeostasis and signalling during drought and salinity stresses. Plant, Cell Environ 33:453–467.

Milly PCD, Wetherald RT, Dunne KA, Delworth TL (2002) Increasing risk of great floods in a changing climate. Nature 415:514.

Mittler R (2002) Oxidative stress, antioxidants and stress tolerance. Trends Plant Sci 7:405–410.

Mittler R, Vanderauwera S, Gollery M, Van Breusegem F (2004) Reactive oxygen gene network of plants. Trends Plant Sci 9:490–498.

Miura K, Furumoto T (2013) Cold signaling and cold response in plants. Int J Mol Sci 14:5312–5337.

Mizuta K, Matsumoto T, Hatate Y, Nishihara K, Nakanishi T (2004) Removal of nitrate-nitrogen from drinking water using bamboo powder charcoal. Bioresour Technol 95:255–257.

Mohan D, Sarswat A, Ok YS, Pittman CU (2014) Organic and inorganic contaminants removal from water with biochar, a renewable, low cost and sustainable adsorbent – A critical review. Bioresour Technol, Special Issue on Biosorption 160:191–202.

Moraes de Freitas, GP, Basu S, Ramegowda V, Braga EB, Pereira A (2016) Comparative analysis of gene expression in response to cold stress in diverse rice genotypes. Biochem Biophys Res Commun 471:253–259.

Mubeen M, Ashfaq A, Hafiz MH, Muhammad A, Hafiz UF, Mazhar S, Muhammad Sami ul Din, Asad A, Amjed A, Fahad S, Wajid N (2020) Evaluating the climate change impact on water use efficiency of cotton-wheat in semi-arid conditions using DSSAT model. J Water Climate Change. https://doi.org/10.2166/wcc.2019.179/622035/jwc2019179.pdf.

Mueller ND, Gerber JS, Johnston M, Ray DK, Ramankutty N, Foley JA (2012) Closing yield gaps through nutrient and water management. Nature 490:254–257.

Muhammad N, Muqarrab A, Khurram S, Fiaz A, Fahim N, Muhammad A, Shazia A, Omaima N, Sulaiman AA, Fahad S, Subhan D, Rahul D (2021) Kaolin and Jasmonic acid improved cotton productivity under water stress conditions. J Saudi Soc Agric Sci 28:6606–6614. https://doi.org/10.1016/j.sjbs.2021.07.043.

Muhammad Z, Abdul MK, Abdul MS, Kenneth BM, Muhammad S, Shahen S, Ibadullah J, Fahad S (2019) Performance of Aeluropus lagopoides (mangrove grass) ecotypes, a potential turfgrass, under high saline conditions. Environ Sci Pollut Res. https://doi.org/10.1007/s11356-019-04838-3.

Muhammad Tahir ul Q, Amna F, Amna B, Barira Z, Xitong Z, Ling-Ling C (2020) Effectiveness of conventional crop improvement strategies vs. omics. In: Fahad S, Hasanuzzaman M, Alam M, Ullah H, Saeed M, Khan AK, Adnan M (eds) Environment, climate, plant and vegetation growth. Springer Publ Ltd, Springer Nature Switzerland AG. Part of Springer Nature. pp 253–284. https://doi.org/10.1007/978-3-030-49732-3.

Munns R (2005) Genes and salt tolerance: bringing them together. New Phytol 167:645–663.

Munns R, Tester M (2008) Mechanisms of salinity tolerance. Annu Rev Plant Biol 59:651–681.

Mwanamwenge J, Loss SP, Siddique KHM, Cocks PS (1999) Effect of water stress during floral initiation, flowering and podding on the growth and yield of faba bean (*Vicia faba* L.). Eur J Agron 11:1–11.

Namgay T, Singh B, Singh BP (2010) Influence of biochar application to soil on the availability of As, Cd, Cu, Pb, and Zn to maize (*Zea mays* L.). Soil Res 48:638.

Ni BJ, Huang QS, Wang C, Ni TY, Sun J, Wei W (2019) Competitive adsorption of heavy metals in aqueous solution onto biochar derived from anaerobically digested sludge. Chemosphere 219:351–357.

Niu J, Ahmad AS, Wang R, Li J, Liu M, Song J, Zohaib A, Lv J, Wang S, Zong X (2016) Exogenous application of brassinolide can alter morphological and physiological traits of *Leymus chinensis* (Trin.) Tzvelev under room and high temperatures. Chil J Agr Res 76:27–33.

Noor M, Naveed ur R, Ajmal J, Fahad S, Muhammad A, Fazli W, Saud S, Hassan S (2020) Climate change and costal plant lives. In: Fahad S, Hasanuzzaman M, Alam M, Ullah H, Saeed M, Khan AK, Adnan M (eds) Environment, climate, plant and vegetation growth. Springer Publ Ltd, Springer Nature Switzerland AG. Part of Springer Nature. pp 93–108. https://doi.org/10.1007/978-3-030-49732-3.

Ogbomida ET, Nakayama SM, Bortey-Sam N, Oroszlany B, Tongo I, Enuneku AA, Ozekeke O, Ainerua MO, Fasipe IP, Ezemonye LI (2018) Accumulation patterns and risk assessment of metals and metalloid in muscle and offal of free-range chickens, cattle and goat in Benin City, Nigeria. Ecotoxicol Environ Safe 151:98–108.

Oguntunde PG, Abiodun BJ, Ajayi AE, van de Giesen N (2008) Effects of charcoal production on soil physical properties in Ghana. J Plant Nutr Soil Sci 171:591–596.

Oladele SO, Adeyemo AJ, Awodun MA (2019) Influence of rice husk biochar and inorganic fertilizer on soil nutrients availability and rain-fed rice yield in two contrasting soils. Geoderma 336:1–11.

Osakabe Y, Osakabe K, Shinozaki K, Tran LSP (2014) Response of plants to water stress. Front Plant Sci 5:86.

Pachauri RK, Allen MR, Barros VR, Broome J, Cramer W, Christ R, Church JA, Clarke L, Dahe Q, Dasgupta P, Dubash NK, Edenhofer O, Elgizouli I, Field CB, Forster P, Friedlingstein P, Fuglestvedt J, Gomez-Echeverri L, Hallegatte S, Hegerl G, Howden M, Jiang K, Jimenez Cisneroz B, Kattsov V, Lee H, Mach KJ, Marotzke J, Mastrandrea MD, Meyer L, Minx J, Mulugetta Y, O'Brien K, Oppenheimer M, Pereira JJ, Pichs-Madruga R, Plattner GK, Pörtner HO, Power SB, Preston B, Ravindranath NH, Reisinger A, Riahi K, Rusticucci M, Scholes R, Seyboth K, Sokona Y, Stavins R, Stocker TF, Tschakert P, van Vuuren D, Vanpserle JP (2014) Climate change 2014: Synthesis Report. Contribution of Working Groups I, II and III to the Fifth Assessment Report of the Intergovernmental Panel on Climate Change. IPCC, Geneva, Switzerland.

Parent B, Tardieu F (2012) Temperature responses of developmental processes have not been affected by breeding in different ecological areas for 17 crop species. New Phytol 194:760–774.

Park JH, Ok YS, Kim SH, Cho JS, Heo JS, Delaune RD, Seo DC (2016) Competitive adsorption of heavy metals onto sesame straw biochar in aqueous solutions. Chemosphere 142:77–83.

Paul MV, Iyer S, Amerhauser C, Lehmann M, van Dongen, JT, Geigenberger P (2016) Oxygen sensing via the ethylene response transcription factor RAP2.12 affects plant metabolism and performance under both normoxia and hypoxia. Plant Physiol 172:141–153.

Paz-Ferreiro J, Lu H, Fu S, Méndez A, Gascó G (2014) Use of phytoremediation and biochar to remediate heavy metal polluted soils: a review. Solid Earth 5:65–75.

Pedersen O, Perata P, Voesenek LACJ (2017) Flooding and low oxygen responses in plants. Funct Plant Biol 44.

Penfield S (2008) Temperature perception and signal transduction in plants. New Phytol 179:615–628.

Petridis A, Therios I, Samouris G, Koundouras S, Giannakoula A (2012) Effect of water deficit on leaf phenolic composition, gas exchange, oxidative damage and antioxidant activity of four Greek olive (*Olea europaea* L.) cultivars. Plant Physiol Biochem 60:1–11.

Pinheiro C, Passarinho JA, Ricardo CP (2004) Effect of drought and rewatering on the metabolism of *Lupinus albus* organs. J Plant Physiol 161:1203–1210.

Prasad PV, Pisipati SR, Mutava RN, Tuinstra MR (2008) Sensitivity of grain sorghum to high temperature stress during reproductive development. Crop Sci 48:1911–1917.

Prasad PVV, Djanaguiraman M (2011) High night temperature decreases leaf photosynthesis and pollen function in grain sorghum. Funct Plant Biol 38:993–1003.

Prasad PVV, Staggenborg SA, Ristic Z (2008) Impacts of drought and/or heat stress on physiological, developmental, growth, and yield processes of crop plants. Response of crops to limited water: understanding and modeling water stress effects on plant growth processes. Adv Agric 1:301–355.

Pratt K, Moran D (2010) Evaluating the cost-effectiveness of global biochar mitigation potential. Biomass Bioenergy 34:1149–1158.

Probert RJ (2000) The role of temperature in the regulation of seed dormancy and germination. Seeds: Ecol Regener Plant Communities 2:261–292.

Qadir M, Quillérou E, Nangia V, Murtaza G, Singh M, Thomas RJ, Drechsel P, Noble AD (2014) Economics of salt-induced land degradation and restoration. Natural Resources Forum, 38 (4):282–295. https://doi.org/10.1111/1477-8947.12054.

Manzoor Qadir, Emmanuelle Quillérou, Vinay Nangia, Ghulam Murtaza, Murari Singh, et al.. Economics of salt-induced land degradation and restoration. Natural Resources Forum, Blackwell Publishing, 2014, 38 (4), pp. 282–295. ff10.1111/1477-8947.12054ff. ffhal-01954413f.

Qambrani NA, Rahman Md M, Won S, Shim S, Ra C (2017) Biochar properties and eco-friendly applications for climate change mitigation, waste management, and wastewater treatment: a review. Renew Sust Energ Rev 79:255–273.

Qi F, Lamb D, Naidu R, Bolan NS, Yan Y, Ok YS, Rahman MM, Choppala G (2018) Cadmium solubility and bioavailability in soils amended with acidic and neutral biochar. Sci Total Environ 610–611:1457–1466.

Rahnama A, James RA, Poustini K, Munns R (2010) Stomatal conductance as a screen for osmotic stress tolerance in durum wheat growing in saline soil. Funct Plant Biol 37:255–263.

Rajkovich S, Enders A, Hanley K, Hyland C, Zimmerman AR, Lehmann J (2012) Corn growth and nitrogen nutrition after additions of biochars with varying properties to a temperate soil. Biol Fert Soils 48:271–284.

Randolph P, Bansode RR, Hassan OA, Rehrah D, Ravella R, Reddy MR, Watts DW, Novak JM, Ahmedna M (2017) Effect of biochars produced from solid organic municipal waste on soil quality parameters. J Environ Manag 192:271–280.

Rashid M, Qaiser H, Khalid SK, Mohammad I. Al-Wabel, Zhang A, Muhammad A, Shahzada SI, Rukhsanda A, Ghulam AS, Shahzada MM, Sarosh A, Muhammad FQ (2020) Prospects of biochar in alkaline soils to mitigate climate change. In: Fahad S, Hasanuzzaman M, Alam M, Ullah H, Saeed M, Khan AK, Adnan M (eds) Environment, climate, plant and vegetation growth. Springer Publ Ltd, Springer Nature Switzerland AG. Part of Springer Nature. pp 133–150. https://doi.org/10.1007/978-3-030-49732-3.

Ravanbakhsh M, Sasidharan R, Voesenek LA, Kowalchuk GA, Jousset A (2017) ACC deaminase-producing rhizosphere bacteria modulate plant responses to flooding. J Ecol 105:979–986.

Reddy, DHK, Lee SM (2014) Magnetic biochar composite: facile synthesis, characterization, and application for heavy metal removal. Colloid Surf A Physicochem Eng Asp 454:96–103.

Regmi P, Moscoso JLG, Kumar S, Cao X, Mao J, Schafran G (2012) Removal of copper and cadmium from aqueous solution using switchgrass biochar produced via hydrothermal carbonization process. J Environ Manag 109:61–69.

Rehman M, Fahad S, Saleem MH, Hafeez M, Muhammad Habib ur R, Liu F, Deng G (2020) Red light optimized physiological traits and enhanced the growth of ramie (Boehmeria nivea L.). Photosynthetica 58(4):922–931.

Rengel Z (2015) Availability of Mn, Zn and Fe in the rhizosphere. J Soil Sci Plant Nut 15:397–409.

Rengel Z, Marschner P (2005) Nutrient availability and management in the rhizosphere: exploiting genotypic differences: research review. New Phytol 168:305–312.

Revathi S, Venugopal S (2013) Physiological and biochemical mechanisms of heavy metal tolerance. Int J Environ Sci 3:1339.

Ristic Z, Bukovnik U, Prasad PV, West M (2008) A model for prediction of heat stability of photosynthetic membranes. Crop Sci 48:1513–1522.

Sadam M, Muhammad Tahir ul Q, Ghulam M, Muhammad SK, Faiz AJ (2020) Role of biotechnology in climate resilient agriculture. In: Fahad S, Hasanuzzaman M, Alam M, Ullah H, Saeed M, Khan AK, Adnan M (eds) Environment, climate, plant and vegetation growth. Springer Publ Ltd, Springer Nature Switzerland AG. Part of Springer Nature. pp 339–366. https://doi.org/10.1007/978-3-030-49732-3.

Safi UK, Ullah F, Mehmood S, Fahad S, Ahmad Rahi A, Althobaiti F, et al (2021) Antimicrobial, antioxidant and cytotoxic properties of Chenopodium glaucum L.. PLoS One 16(10):e0255502. https://doi.org/10.1371/journal.pone.0255502.

Sahrish N, Shakeel A, Ghulam A, Zartash F, Sajjad H, Mukhtar A, Muhammad AK, Ahmad K, Fahad S, Wajid N, Sezai E, Carol Jo W, Gerrit H (2022) Modeling the impact of climate warming on potato phenology. Eur J AgroN 132:126404.

Saibo NJ, Lourenço T, Oliveira MM (2008) Transcription factors and regulation of photosynthetic and related metabolism under environmental stresses. Ann Bot 103:609–623.

Saifullah, Dahlawi S, Naeem A, Rengel Z, Naidu R (2018) Biochar application for the remediation of salt-affected soils: challenges and opportunities. Sci Total Environ 625:320–335.

Sairam RK, Kumutha D, Ezhilmathi K, Chinnusamy V, Meena RC (2009) Waterlogging induced oxidative stress and antioxidant enzyme activities in pigeon pea. Biol Plant 53:493–504.

Sairam RK, Tyagi A (2004) Physiological and molecular biology of salinity stress tolerance in deficient and cultivated genotypes of chickpea. Plant Growth Regul 57.

Sajid H, Jie H, Jing H, Shakeel A, Satyabrata N, Sumera A, Awais S, Chunquan Z, Lianfeng Z, Xiaochuang C, Qianyu J, Junhua Z (2020) Rice production under climate change: adaptations and mitigating strategies. In: Fahad S, Hasanuzzaman M, Alam M, Ullah H, Saeed M, Khan AK, Adnan M (eds) Environment, Climate, Plant and Vegetation Growth. Springer Publ Ltd, Springer Nature Switzerland AG. Part of Springer Nature. pp 659–686. https://doi.org/10.1007/978-3-030-49732-3.

Sajjad H, Muhammad M, Ashfaq A, Waseem A, Hafiz MH, Mazhar A, Nasir M, Asad A, Hafiz UF, Syeda RS, Fahad S, Depeng W, Wajid N (2019) Using GIS tools to detect the land use/land cover changes during forty years in Lodhran district of Pakistan. Environ Sci Pollut Res. https://doi.org/10.1007/s11356-019-06072-3.

Saleem MH, Fahad S, Adnan M, Mohsin A, Muhammad SR, Muhammad K, Qurban A, Inas AH, Parashuram B, Mubassir A, Reem MH (2020a) Foliar application of gibberellic acid endorsed phytoextraction of copper and alleviates oxidative stress in jute (Corchorus capsularis L.) plant grown in highly copper-contaminated soil of China. Environ Sci Pollution Res. https://doi.org/10.1007/s11356-020-09764–3

Saleem MH, Fahad S, Shahid UK, Mairaj D, Abid U, Ayman ELS, Akbar H, Analía L, Lijun L (2020c) Copper-induced oxidative stress, initiation of antioxidants and phytoremediation potential of flax (Linum usitatissimum L.) seedlings grown under the mixing of two different soils of China. Environ Sci Poll Res. https://doi.org/10.1007/s11356-019-07264-7.

Saleem MH, Rehman M, Fahad S, Tung SA, Iqbal N, Hassan A, Ayub A, Wahid MA, Shaukat S, Liu L, Deng G (2020b) Leaf gas exchange, oxidative stress, and physiological attributes of rapeseed (Brassica napus L.) grown under different light-emitting diodes. Photosynthetica 58(3):836–845.

Saman S, Amna B, Bani A, Muhammad Tahir ul Qamar, Rana MA, Muhammad SK (2020) QTL mapping for abiotic stresses in cereals. In: Fahad S, Hasanuzzaman M, Alam M, Ullah H, Saeed M, Khan AK, Adnan M (eds) Environment, climate, plant and vegetation growth. Springer Publ Ltd, Springer Nature Switzerland AG. Part of Springer Nature. pp 229–252. https://doi.org/10.1007/978-3-030-49732-3.

Sandhu SS, Kumar S (2017) Impact of three types of biochar on the hydrological properties of eroded and depositional landscape positions. Soil Sci Soc Am J 81:878–888.

Saud S, Chen Y, Fahad S, Hussain S, Na L, Xin L, Alhussien SA (2016) Silicate application increases the photosynthesis and its associated metabolic activities in Kentucky bluegrass under drought stress and post-drought recovery. Environ Sci Pollut Res 23(17):17647–17655. https://doi.org/10.1007/s11356-016-6957-x.

Saud S, Chen Y, Long B, Fahad S, Sadiq A (2013) The different impact on the growth of cool season turf grass under the various conditions on salinity and drought stress. Int J Agric Sci Res 3:77–84.

Saud S, Fahad S, Cui G, Chen Y, Anwar S (2020) Determining nitrogen isotopes discrimination under drought stress on enzymatic activities, nitrogen isotope abundance and water contents of Kentucky bluegrass. Sci Rep 10:6415. https://doi.org/10.1038/s41598-020-63548-w.

Saud S, Fahad S, Yajun C, Ihsan MZ, Hammad HM, Nasim W, Amanullah Jr, Arif M, Alharby H (2017) Effects of nitrogen supply on water stress and recovery mechanisms in Kentucky bluegrass plants. Front Plant Sci 8:983. https://doi.org/10.3389/fpls.2017.00983.

Saud S, Li X, Chen Y, Zhang L, Fahad S, Hussain S, Sadiq A, Chen Y (2014) Silicon application increases drought tolerance of Kentucky bluegrass by improving plant water relations and morph physiological functions. SciWorld J 2014:1–10. https://doi.org/10.1155/2014/368694.

Scandalios JG (2002) The rise of ROS. Trends Biochem Sci 27:483–486.

Scandalios JG (2005) Oxidative stress: molecular perception and transduction of signals triggering antioxidant gene defenses. Braz J Med Biol Res 38:995–1014.

Schmuths H, Bachmann K, Weber WE, Horres R, Hoffmann MH (2006) Effects of preconditioning and temperature during germination of 73 natural accessions of Arabidopsis thaliana. Ann Bot 97:623–634.

Senol C (2020) The effects of climate change on human behaviors. In: Fahad S, Hasanuzzaman M, Alam M, Ullah H, Saeed M, Khan AK, Adnan M (eds) Environment, climate, plant and vegetation growth. Springer Publ Ltd, Springer Nature Switzerland AG. Part of Springer Nature. pp 577–590. https://doi.org/10.1007/978-3-030-49732-3

Shafi MI, Adnan M, Fahad S, Fazli W, Ahsan K, Zhen Y, Subhan D, Zafar-ul-Hye M, Martin B, Rahul D (2020) Application of single superphosphate with humic acid improves the growth, yield and phosphorus uptake of wheat (Triticum aestivum L.) in calcareous soil. Agron 10:1224. https://doi.org/10.3390/agronomy10091224.

Shah F, Lixiao N, Kehui C, Tariq S, Wei W, Chang C, Liyang Z, Farhan A, Fahad S, Huang J (2013) Rice grain yield and component responses to near 2°C of warming. Field Crop Res 157:98–110.

Shah SAR, Khan SA, Junaid K, Sattar S, Zaman M, Saleem N, Adnan M (2015) Screening of mustard genotypes for antixenosis and multiplication against mustard aphid, *Lipaphiserysimi*(Kalt) (Aphididae: Homoptera). J Ento Zool Stud 3(6):84–87.

Shao HB, Chu LY, Jaleel CA, Manivannan P, Panneerselvam R, Shao MA (2009) Understanding water deficit stress-induced changes in the basic metabolism of higher plants-biotechnologically and sustainably improving agriculture and the ecoenvironment in arid regions of the globe. Crit Rev Biotechnol 29:131–151.

Sharif R, Xie C, Zhang H, Arnao M, Ali M, Ali Q, Muhammad I, Shalmani A, Nawaz M, Chen P, Li Y (2018) Melatonin and its effects on plant systems. Molecules 23:2352.

Sharma B, Singh S, Siddiqi NJ (2014) Biomedical implications of heavy metals induced imbalances in redox systems. BioMed Res Int 2014:1–26.

Sharma P, Dubey RS (2005) Lead toxicity in plants. Braz J Plant Physiol 17:35–52.

Sharma RK, Agrawal M, Marshall F (2006) Heavy metal contamination in vegetables grown in wastewater irrigated areas of Varanasi, India. B Environ Contam Tox 77:312–318.

Sharma SS, Dietz KJ, Mimura T (2016) Vacuolar compartmentalization as indispensable component of heavy metal detoxification in plants. Plant Cell Env 39:1112–1126.

Shen Z, Hou D, Jin F, Shi J, Fan X, Tsang DC, Alessi DS (2019) Effect of production temperature on lead removal mechanisms by rice straw biochars. Sci Total Environ 655:751–758.

Sial TA, Khan MN, Lan Z, Kumbhar F, Ying Z, Zhang J, Sun D, Li X (2019) Contrasting effects of banana peels waste and its biochar on greenhouse gas emissions and soil biochemical properties. Process Saf Environ 122:366–377.

Sidra K, Javed I, Subhan D, Allah B, Syed IUSB, Fatma B, Khaled DA, Fahad S, Omaima N, Ali TKZ, Rahul D (2021) Physio-chemical characterization of indigenous agricultural waste materials for the development of potting media. J Saudi Soc Agric Sci. https://doi.org/10.1016/j.sjbs.2021.08.058.

Singh B, Singh BP, Cowie AL (2010) Characterisation and evaluation of biochars for their application as a soil amendment. Soil Res 48:516–525.

Skirycz A, Inzé D (2010) More from less: plant growth under limited water. Curr Opin Biotechnol 21:197–203.

Soliman MH, Alayafi AAM, El Kelish AA, Abu-Elsaoud AM (2018) Acetylsalicylic acid enhance tolerance of *Phaseolus vulgaris* L. to chilling stress, improving photosynthesis, antioxidants and expression of cold stress responsive genes. Bot Stud 59.

Son EB, Poo KM, Chang JS, Chae KJ (2018) Heavy metal removal from aqueous solutions using engineered magnetic biochars derived from waste marine macro-algal biomass. Sci Total Environ 615:161–168.

Song J, Shi W, Liu R, Xu Y, Sui N, Zhou J, Feng G (2017) The role of the seed coat in adaptation of dimorphic seeds of the euhalophyte *Suaeda salsa* to salinity. Plant Spec Biol 32:107–114.

Spokas KA, Cantrell KB, Novak JM, Archer DW, Ippolito JA, Collins HP, Boateng AA, Lima, IM, Lamb MC, McAloon AJ (2012) Biochar: a synthesis of its agronomic impact beyond carbon sequestration. J Environ Qual 41:973–989.

Spokas KA, Koskinen WC, Baker JM, Reicosky DC (2009) Impacts of woodchip biochar additions on greenhouse gas production and sorption/degradation of two herbicides in a Minnesota soil. Chemosphere 77:574–581.

Striker GG, Colmer TD (2017) Flooding tolerance of forage legumes. J Exp Bot 68:1851–1872.

Subhan D, Zafar-ul-Hye M, Fahad S, Saud S, Martin B, Tereza H, Rahul D (2020) Drought stress alleviation by ACC deaminase producing *Achromobacter xylosoxidans* and *Enterobacter cloacae*, with and without timber waste biochar in maize. Sustainability 12(6286). https://doi.org/10.3390/su12156286.

Subhashini V, Swamy A (2013) Phytoremediation of Pb and Ni Contaminated Soils Using Catharanthus roseus (L.). Uni J Environ Res Technol 3.

Tan X, Liu Y, Zeng G, Wang X, Hu X, Gu Y, Yang Z (2015) Application of biochar for the removal of pollutants from aqueous solutions. Chemosphere 125:70–85.

Tariq M, Ahmad S, Fahad S, Abbas G, Hussain S, Fatima Z, Nasim W, Mubeen M, ur Rehman MH, Khan MA, Adnan M (2018) The impact of climate warming and crop management on phenology of sunflower-based cropping systems in Punjab, Pakistan. Agric For Meterol 256:270–282.

Terazawa K, Maruyama Y, Morikawa Y (1992) Photosynthetic and stomatal responses of Larix kaempferi seedlings to short-term waterlogging. Ecol Res 7:193–197.

Tong YP (2005) 2-(2-Hydroxyphenyl)-1, 3-benzoxazole. Acta Crystallogr Sect E: Struct Rep Online 61:o3076–o3078.

Tuteja N, Gill SS, Tuteja R (2011) Plant responses to abiotic stresses: shedding light on salt, drought, cold and heavy metal stress. Omics and Plant Abiotic Stress Tolerance 1:39–64.

Unsar NU, Muhammad R, Syed HMB, Asad S, Mirza AQ, Naeem I, Muhammad H ur R, Fahad S, Shafqat S (2020) Insect pests of cotton crop and management under climate change scenarios. In: Fahad S, Hasanuzzaman M, Alam M, Ullah H, Saeed M, Khan AK, Adnan M (eds) Environment, climate, plant and vegetation growth. Springer Publ Ltd, Springer Nature Switzerland AG. Part of Springer Nature. pp 367–396. https://doi.org/10.1007/978-3-030-49732-3.

Valliyodan B, Ye H, Song L, Murphy M, Shannon JG, Nguyen HT (2016) Genetic diversity and genomic strategies for improving drought and waterlogging tolerance in soybeans. J Exp Bot:433.

Van Ha C, Leyva-González MA, Osakabe Y, Tran UT, Nishiyama R, Watanabe Y, Tanaka M, Seki, M, Yamaguchi S, Van Dong N (2014) Positive regulatory role of strigolactone in plant responses to drought and salt stress. Proc Nat Acad Sci 111:851–856.

Wahid F, Fahad S, Subhan D, Adnan M, Zhen Y, Saud S, Manzer HS, Martin B, Tereza H, Rahul D (2020) Sustainable management with mycorrhizae and phosphate solubilizing bacteria for enhanced phosphorus uptake in calcareous soils. Agriculture 10(334). https://doi.org/10.3390/agriculture10080334.

Wajid N, Ashfaq A, Asad A, Muhammad T, Muhammad A, Muhammad S, Khawar J, Ghulam MS, Syeda RS, Hafiz MH, Muhammad IAR, Muhammad ZH, Muhammad Habib ur R, Veysel T, Fahad S, Suad S, Aziz K, Shahzad A (2017) Radiation efficiency and nitrogen fertilizer impacts on sunflower crop in contrasting environments of Punjab. Pak Environ Sci Pollut Res 25:1822–1836. https://doi.org/10.1007/s11356-017-0592-z.

Wang H, Gao B, Wang S, Fang J, Xue Y, Yang K (2015a) Removal of Pb (II), Cu (II), and Cd (II) from aqueous solutions by biochar derived from $KMNO_4$ treated hickory wood. Bioresour Technol 197:356–362.

Wang S, Gao B, Li Y, Mosa A, Zimmerman AR, Ma LQ, Harris WG, Migliaccio KW (2015b) Manganese oxide-modified biochars: preparation, characterization, and sorption of arsenate and lead. Bioresour Technol 181:13–17.

Wani SH, Dutta T, Neelapu NRR, Surekha C (2017) Transgenic approaches to enhance salt and drought tolerance in plants. Plant Gene 11:219–231.

Wani SH, Kumar V, Shriram V, Sah SK (2016) Phytohormones and their metabolic engineering for abiotic stress tolerance in crop plants. Crop J 4:162–176.

Waraich EA, Ahmad R, Ashraf MY, Saifullah, Ahmad M (2011) Improving agricultural water use efficiency by nutrient management in crop plants. Acta Agric Scand Sect B-Soil Plant Sci 61:291–304.

Warnock DD, Lehmann J, Kuyper TW, Rillig MC (2007) Mycorrhizal responses to biochar in soil–concepts and mechanisms. Plant and Soil 300:9–20.

White PJ, Brown PH (2010) Plant nutrition for sustainable development and global health. Ann Bot 105:1073–1080.

Willis WL, Reid L (2008) Investigating the effects of dietary probiotic feeding regimens on broiler chicken production and Campylobacter jejuni presence. Poult Sci 87:606–611.

Wright AJ, Kroon H, Visser EJ, Buchmann T, Ebeling A, Eisenhauer N, Fischer C, Hildebrandt A, Ravenek J, Roscher C (2017) Plants are less negatively affected by flooding when growing in species-rich plant communities. New Phytol 213:645–656.

Wu C, Kehui C, She T, Ganghua L, Shaohua W, Fahad S, Lixiao N, Jianliang H, Shaobing P, Yanfeng D (2020) Intensified pollination and fertilization ameliorate heat injury in rice (Oryza sativa L.) during the flowering stage. Field Crops Res 252:107795.

Wu C, Tang S, Li G, Wang S, Fahad S, Ding Y (2019a) Roles of phytohormone changes in the grain yield of rice plants exposed to heat: a review. PeerJ 7:e7792. https://doi.org/10.7717/peerj.7792.

Wu F, Jia Z, Wang S, Chang SX, Startsev A (2013) Contrasting effects of wheat straw and its biochar on greenhouse gas emissions and enzyme activities in a Chernozemic soil. Biol Fert Soils 49:555–565.

Wu X, Lu X, Riaz M, Yan L, Jiang C (2019b) Boron toxicity induced specific changes of cell ultrastructure and architecture of components in leaf center and tip of trifoliate orange [Poncirus trifoliata (L.) Raf.]. J Environ Manag 246:426–433.

Xu CY, Hosseini-Bai S, Hao Y, Rachaputi RCN, Wang H, Xu Z, Wallace H (2015) Effect of biochar amendment on yield and photosynthesis of peanut on two types of soils. Environ Sci Pollut Res 22:6112–6125.

Xu X, Cao X, Zhao L (2013a) Comparison of rice husk-and dairy manure-derived biochars for simultaneously removing heavy metals from aqueous solutions: role of mineral components in biochars. Chemosphere 92:955–961.

Xu X, Cao X, Zhao L, Wang H, Yu H, Gao B (2013b) Removal of Cu, Zn, and Cd from aqueous solutions by the dairy manure-derived biochar. Environ Sci Pollut Res 20:358–368.

Xue Y, Gao B, Yao Y, Inyang M, Zhang M, Zimmerman AR, Ro KS (2012) Hydrogen peroxide modification enhances the ability of biochar (hydrochar) produced from hydrothermal carbonization of peanut hull to remove aqueous heavy metals: batch and column tests. Chem Eng J 200:673–680.

Yaghoubi P, Reddy KR (2011) Characteristics of biochar-amended soil cover for landfill gas mitigation. In: Pan-Am CGS Geotechnical Conference.

Yamaguchi-Shinozaki K, Shinozaki K (2006) Transcriptional regulatory networks in cellular responses and tolerance to dehydration and cold stresses. Annu Rev Plant Biol 57:781–803.

Yang Z, Zhang Z, Zhang T, Fahad S, Cui K, Nie L, Peng S, Huang J (2017) The effect of season-long temperature increases on rice cultivars grown in the central and southern regions of China. Front Plant Sci 8:1908. https://doi.org/10.3389/fpls.2017.01908.

Ye C, Tenorio FA, Redoña ED, Morales–Cortezano PS, Cabrega GA, Jagadish KS, Gregorio GB (2015) Fine-mapping and validating qHTSF4. 1 to increase spikelet fertility under heat stress at flowering in rice. Theor Appl Genet 128:1507–1517.

Yu XY, Ying GG, Kookana RS (2009) Reduced plant uptake of pesticides with biochar additions to soil. Chemosphere 76:665–671.

Yuan J, Meng J, Liang X, E Y, Yang X, Chen W (2017) Organic molecules from biochar leacheates have a positive effect on rice seedling cold tolerance. Front Plant Sci 8.

Yuan JH, Xu RK, Qian W, Wang RH (2011) Comparison of the ameliorating effects on an acidic ultisol between four crop straws and their biochars. J Soils Sediments 11:741–750.

Zafar-ul-Hye M, Muhammad N, Subhan D, Fahad S, Rahul D, Mazhar A, Ashfaq AR, Martin B, Ji˘rí H, Zahid HT, Muhammad N (2020a) Alleviation of cadmium adverse effects by improving nutrients uptake in bitter gourd through cadmium tolerant rhizobacteria. Environments 7(54). https://doi.org/10.3390/environments7080054.

Zafar-ul-Hye M, Tahzeeb-ul-Hassan M, Muhammad A, Fahad S, Martin B, Tereza D, Rahul D, Subhan D (2020b) Potential role of compost mixed biochar with rhizobacteria in mitigating lead toxicity in spinach. Sci Rep 10:12159. https://doi.org/10.1038/s41598-020-69183-9.

Zahida Z, Hafiz FB, Zulfiqar AS, Ghulam MS, Fahad S, Muhammad RA, Hafiz MH, Wajid N, Muhammad S (2017) Effect of water management and silicon on germination, growth, phosphorus and arsenic uptake in rice. Ecotoxicol Environ Saf 144:11–18.

Zahir SM, Zheng HG, Ala Ud D, Amjad A, Ata Ur R, Kashif J, Shah F, Saud S, Adnan M, Fazli W, Saud A, Manzer HS, Shamsher A, Wajid N, Hafiz MH, Fahad S (2021) Synthesis of silver nanoparticles using Plantago lanceolata extract and assessing their antibacterial and antioxidant activities. Sci Rep 11:20754.

Zandalinas SI, Mittler R, Balfagón D, Arbona V, Gómez-Cadenas A (2018) Plant adaptations to the combination of drought and high temperatures. Physiol Plant 162:2–12.

Zavalloni C, Alberti G, Biasiol S, Delle Vedove G, Fornasier F, Liu J, Peressotti A (2011) Microbial mineralization of biochar and wheat straw mixture in soil: a short-term study. Appl Soil Ecol 50:45–51.

Zhang A, Cui L, Pan G, Li L, Hussain Q, Zhang X, Zheng J, Crowley D (2010) Effect of biochar amendment on yield and methane and nitrous oxide emissions from a rice paddy from Tai Lake plain, China. Agr Ecosyst Environ 139:469–475.

Zhang HJ, Zhang N, Yang RC, Wang L, Sun QQ, Li DB, Cao YY, Weeda S, Zhao B, Ren S, Guo YD (2014) Melatonin promotes seed germination under high salinity by regulating antioxidant systems, ABA and GA 4 interaction in cucumber (*Cucumis sativus* L.). J Pineal Res 57:269–279.

Zhou Y, Xia X, Yu G, Wang J, Wu J, Wang M, Yang Y, Shi K, Yu Y, Chen Z, Gan J, Yu J (2015) Brassinosteroids play a critical role in the regulation of pesticide metabolism in crop plants. Sci Rep 5.

Zhou Y, Zhang Z, Zhang J, Xia S (2016) New insight into adsorption characteristics and mechanisms of the biosorbent from waste activated sludge for heavy metals. J Environ Sci 45:248–256.

Zia-ur-Rehman M (2020) Environment, climate change and biodiversity. In: Fahad S, Hasanuzzaman M, Alam M, Ullah H, Saeed M, Khan AK, Adnan M (eds) Environment, climate, plant and vegetation growth. Springer Publ Ltd, Springer Nature Switzerland AG. Part of Springer Nature. pp 473–502. https://doi.org/10.1007/978-3-030-49732-3.

Zimmerman AR, Gao B, Ahn MY (2011) Positive and negative carbon mineralization priming effects among a variety of biochar-amended soils. Soil Biol Biochem 43:1169–1179.

Zwart DC, Kim SH (2012) Biochar amendment increases resistance to stem lesions caused by Phytophthora spp. in tree seedlings. Hort Sci 47:1736–1740.

# 11

# *Biomolecular Intervention in Understanding Plant's Adaptation to Climate Change*

**Muhammad Ilyas, Amjad-ur-Rahman, Fazli Wahid, and Nazish Huma Khan**
*University of Swabi*

**Tariq Mahmood**
*Quaid-i-Azam University*

**CONTENTS**



## 11.1 Introduction

Climate change impacts the agro-ecosystems of an area by changing the key variable (rising temperature, rainfall, drought, waterlogging) which affects plant growth and development, thereby altering crop productivity, quality and plant–pathogen interaction (Deepranjan et al. 2021; Haider et al. 2021; Huang Li et al. 2021; Ikram et al. 2021; Jabborova et al. 2021; Khadim et al. 2021a, b; Manzer et al. 2021; Muzammal et al. 2021; Abdul et al. 2021a, b; Ashfaq et al. 2021; Amjad et al. 2021; Atif et al. 2021; Athar et al. 2021; Adnan et al. 2018a, b; Adnan et al. 2019; Akram et al. 2018a, b; Aziz et al. 2017a, b; Chang et al. 2021; Chen et al. 2021; Emre et al. 2021; Habib et al. 2017; Hafiz et al. 2016; Hafiz et al. 2019; Ghulam et al. 2021; Guofu et al. 2021; Hafeez et al. 2021; Khan et al. 2021; Kamran et al. 2017; Muhammad et al. 2019; Safi et al. 2021; Sajjad et al. 2019; Saud et al. 2013; Saud et al. 2014; Saud et al. 2017; Saud et al. 2016; Shah et al. 2013; Saud et al. 2020; Qamar et al. 2017; Hamza et al. 2021; Irfan et al. 2021;Wajid et al. 2017; Yang et al. 2017; Zahida et al. 2017; Depeng et al. 2018; Hussain et al. 2020; Hafiz et al. 2020a, b; Shafi et al. 2020; Wahid et al. 2020; Subhan et al. 2020; Zafar-ul-Hye et al. 2020a, b; Zafar et al. 2021; Adnan et al. 2020; Ilyas et al. 2020; Saleem et al. 2020a, b, c; Rehman et al. 2020; Farhat et al. 2020; Wu et al. 2020; Mubeen et al. 2020; Farhana et al. 2020; Jan et al. 2019; Wu et al. 2019; Ahmad et al. 2019; Baseer et al. 2019; Hafiz et al. 2018; Tariq et al. 2018; Fahad and Bano 2012; Fahad et al. 2017; Fahad et al. 2013; Fahad et al. 2014a, b; Fahad et al. 2016a, b, c, d; Fahad et al. 2015a, b;

DOI: 10.1201/9781003286400-11

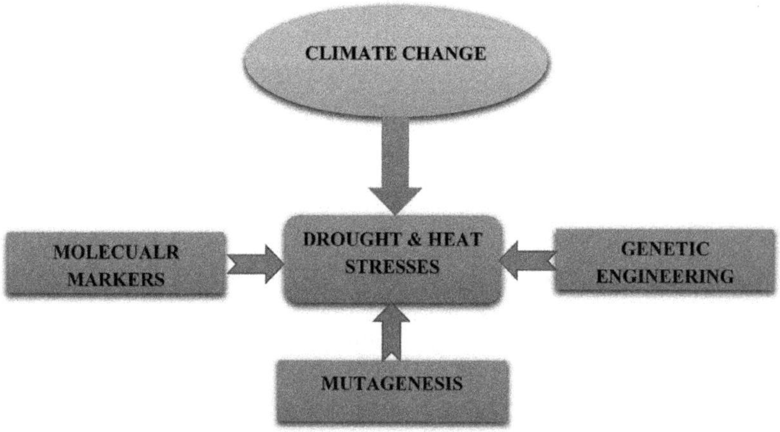


**FIGURE 11.1** Major biomolecular approaches for combating climate change.

Fahad et al. 2018a, b; Fahad et al. 2019a, b; Fahad et al. 2020; Fahad et al. 2021a, b, c, d, e, f; Hesham and Fahad 2020). Climate change disturbs food availability, access, and stability which differ across regions as well as changes the food quality by decreasing protein, micronutrient contents, or modifying lipid composition (Wheeler and Von Braun 2013). According to FAO estimates, by 2050 the global demand for food production will increase by 50%–70% due to an increase in population and consumption pattern. Even without climate change, this would be a significant challenge but recent studies have shown that grains yield could be reduced by ~5% for each degree (C°) rise in global temperature (Lobell et al. 2011). This presents substantial challenges which can be addressed by reducing the yield gap, food wastage, changing consumption patterns, and increasing the effectiveness of food systems. The option of increasing yield potentials through crop improvement is a promising avenue to explore. Ray et al. (2013) analyzed the trends in crop yields over the past 50 years and concluded that the current "business as usual" increase in yield is not enough to double the production in the next 40 years or by 2050 because the annual growth is less than necessary 2.4% (e.g. just 0.9% for wheat). Hence, new approaches should be adopted to tackle the problem. However, the possible solution to the problem may be to develop new cultivars that provide a means to adapt farming to the changing climate (Iqra et al. 2020; Akbar et al. 2020; Mahar et al. 2020; Noor et al. 2020; Bayram et al. 2020; Amanullah and Fahad 2018a, b; Amanullah and Fahad 2017; Amanullah et al. 2020; Amanullah et al. 2021; Rashid et al. 2020; Arif et al. 2020; Amir et al. 2020; Saman et al. 2020; Muhammad Tahir et al. 2020; Md Jakir and Allah 2020; Mahmood et al. 2021; Farah et al. 2020; Sadam et al. 2020; Unsar et al. 2020; Fazli et al. 2020; Md. Enamul et al. 2020; Gopakumar et al. 2020; Zia-ur-Rehman 2020; Ayman et al. 2020; Al-Wabel et al. 2020a, b; Senol 2020; Amjad et al. 2020; Ibrar et al. 2020; Sajid et al. 2020; Muhammad et al. 2021; Sidra et al. 2021; Zahir et al. 2021; Sahrish et al. 2022).

Crops can be adapted to the new environment through two main strategies: either to develop new crops (which is a long-term strategy starting from domestication) or by inserting specific traits into existing varieties using modern biomolecular techniques. However, a deep understanding of the molecular basis of the underlying cellular processes is essential to any successful and safe intervention. Various strategies to adapt plants to changing environment include the development of transgenics with desirable genes, mutagenics development, and the use and development of molecular markers. These not only help to understand the basic mechanism of plant adaptation but also help to change their genetic makeup to make them resilient in changing environment with many such examples discussed below (Figure 11.1).

## 11.2 GMOs and Climate Change

Advances in crop biotechnology are opening up new avenues by incorporating genes of interest in selected varieties against environmental stresses. In the last 20 years, a significant development in this area has considerably enlarged the toolbox for plant breeding which includes insertion or deletion of the target gene, its transfer at inter- or intra-specific levels, and functional modifications via site-directed mutagenesis (Lusser et al. 2012; Ilyas et al. 2016a, b; Ali et al. 2016). The technology has already made a significant impact on cotton, canola (oil rapeseed), soybean, and maize farming on a global scale. Studies have shown that insect-resistant cotton/maize, herbicide-resistant soybean out-yield their conventional counterpart in both economic and agronomic terms thus improving their global production. The net benefit of transgenic crops raised from 19.8 billion dollars in 1996 to 98.2 in 2011 at farm levels in which farmers in the developing world gained a significant portion of 51.2% (Brookes and Barfoot 2013).

Transgenic crops can significantly contribute to the climate change mitigation efforts by providing means to adapt crops to the changing environment especially against, heat, drought, and salinity. However, significant progress in the development of climate-resilient cultivars can be made by integrating conventional plant breeding techniques with modern genetic engineering. In the former, the genetic markers related to morphological traits controlled by a single locus are used while molecular breeding (biochemical/or DNA markers) deals with the activation/regulation of genes involved in the entire mechanism of stress adaptation including signaling, transcriptional control, protection of membrane/protein and free-radical and toxic compound scavenging. Transfer of genes at inter-specific and inter-generic levels is another reproductive barrier for conventional breeding: easily soluble by transgenic approach (Roy et al. 2011). Further, the identification of key genes underlying stress tolerance in plants and their incorporation in susceptible varieties take less time and introduce only the desired gene (avoiding the undesired genes as in conventional breeding). Such technologies have made it possible to engineer greater tolerance in plants against various stresses: specific examples with a particular focus on heat and drought tolerance are discussed below.

## 11.3 Heat Tolerance and Transgenics

Plants respond to high temperature (HT) by reprogramming their genetic machinery: either overexpressing heat-shock proteins (HSPs) genes or altering the expression of heat-shock factors (HSFs) that regulate many HSP and non-HSP genes. HSPs are molecular chaperones involved in protein synthesis, protein/membrane stabilization, and folding, whereas HSFs activation is largely governed by enzymes (calmodulins and kinases). Besides, some trans-acting factors (DREB2A, Bzip28, and WRKY proteins) are also important in combating HT stress. However, equipping plants with strategies like elevating the genetic levels of proteins involved in osmotic adjustment, reactive oxygen species (ROS) removal, modulation of photosynthetic reactions, polyamine production, and saturation of membrane-associated lipids have also been proven useful. Similarly, cyclic nucleotide-gated $Ca^{+2}$ channel protein which regulates the influx of calcium ions across the cellular membrane was proved to be the key player in inducing HT tolerance (Grover et al. 2013).

### 11.3.1 Molecular Chaperone

Heat disrupts the homeostasis of cellular proteins by modifying their optimal structure resulting in the loss of their biological activities triggering the activation of many proteins and molecular chaperones as a survival mechanism to minimize the damage as the accumulation of degraded protein is lethal to cell functionality. HSP synthesis or up-regulation is the hallmark of HT stress response working as a molecular chaperone either protecting cells or stabilizing denatured protein. Their suppression via anti-sense/RNAi mechanism in the trans-host system or overexpression in yeast or bacteria affects immunity against heat. They also have implications in various fundamental plant processes (aging, vigor, longevity and seedling establishment, drought, salinity, etc.) (Jiang et al. 2018). Various studies using HSPs, HSFs, and regulatory elements as a defense mechanism against HT are given in Table 11.1.

### 11.3.2 Osmolytes Regulation

Plant cells accumulate various compatible osmolytes (amino acids, soluble sugar, sugar alcohol, etc.) as an adaptive mechanism against stress to adjust osmotic potential and protect proteins. Because it

**TABLE 11.1**

Transgenic Approaches for Inducing Heat Tolerance Employing Heat-Shock Proteins (HSPs), Heat-Shock Factors (HSFs) and Heat-Shock Regulatory Regions (Promoters) Using Transgenic Approaches

| Gene/Host Plant | Transgenic | Comment | Reference |
|---|---|---|---|
| *AhDnaK/HSP70* | *N. tabacum* | Thermotolerance at 40°C | Ono et al. (2001) |
| *AtHSP100* (*A. thaliana*) | *A. thaliana* | Transformants tolerated sudden shifts to HT; antisense/co-suppression lines had diminished acquired thermotolerance (45°C; 2 hours, with or without a conditioning pretreatment at 38°C; 90 minutes). | Queitsch et al. (2000) |
| *AtHSP100* (*Oryza Sativa*) | *O. sativa* | Better growth in the recovery phase after HT stress (45°C; 3 hours, 47°C; 2 hours and 50°C; 40 minutes) | Katiyar-Agarwal et al. (2003) |
| *SlHSP21* | *Solanum lycopersicum* | Protect PSII from temperature-dependent oxidative stress; early accumulation of carotenoids noted (40°C, 47°C, 50°C; 2 hours and then exposed to high light). | Rhoads et al. (2005) |
| *DcHSP17.7* (*Daucus carota*) | *S. tuberosum* | Transgenic plants showed improved membrane stability and enhanced tuberization | Ahn and Zimmerman (2006) |
| *OsHSP100* | *N. tabacum* | Transgenics survived HT stress (50°C; 50 minutes) relatively better than non-transgenic control plants. | Chang et al. (2007) |
| *AhDnaK/HSP70* | *O. sativa* | Enhancement in ascorbate peroxidase and catalase and Calvin cycle activities/ thermotolerance at 33°C–37°C; 16 hours. | Uchida et al. (2008) |
| *RcHSP17.8* | *A. thaliana* | Higher thermotolerance (45°C; 2 hours)/lower relative electrolyte leakage/more proline content | Jiang et al. (2009) |
| *NtHSP70-1* | *N. tabacum* | seedlings grew better after HT stress/ trans-protein prevented degradation of nDNA during HT stress (45°C; 2 hours) | Cho and Choi (2009) |
| *ScHSP26* | *A. thaliana* | Thermotolerance at 45°C; 16 hours/improved free-proline/enhanced expression of proline biosynthetic pathway genes | Xue et al. (2010) |
| *ThHSP70* | *A. thaliana* | Enhanced tolerance to HT stress (45°C; 5 hours); increased transcript levels of SOS1 and APX1. HSF and HSP genes were down-regulated. | Montero-Barrientos et al. (2010) |
| *OsHSP70* (*mitochondrial*) | | Enhanced tolerance to HT stress (48°C; 15 minutes); programmed cell death was suppressed in overexpression lines. | Qi et al. (2011) |
| *ZmHSP16.9* | *Nicotiana tabacum* | Increased seed germination rate, root length/antioxidant enzyme activity at 40°C; 9 hours. | Sun et al. (2012) |
| *NnHSP17.5* | *A. thaliana* | Better seed germination vigor/ improved basal thermotolerance (44°C; 60 and 70 minutes). | Zhou et al. (2012) |
| *OsHSP26* | *Festuca arundinacea* | Reduction of electrolyte leakage/increase in thiobarbituric acid-reactive substances/higher PSII activity at 42°C; 24 hours | Kim et al. (2012) |
| *LimHSP16.45* | *A. thaliana* | Enhanced cell viability at HT; higher heat-shock granules; increase SOD & catalase activities | Mu et al. (2013) |
| *PfHSP17.1* (*Primula forrestii*) | *A. thaliana* | Thermotolerance (42°C for 2 hours); resistance to salt drought stress | Zhang et al. (2013) |
| *CgHSP70* | *A. thaliana* | Cytoplasm and nucleus-localized expression; tolerance to heat, drought, and salinity; increased peroxidase (POD) activity, higher proline content and inhibited malondialdehyde (MDA) content (45°C for 24 hours) | Song et al. (2014) |

*(Continued)*

**TABLE 11.1 (*Continued*)**

Transgenic Approaches for Inducing Heat Tolerance Employing Heat-Shock Proteins (HSPs), Heat-Shock Factors (HSFs) and Heat-Shock Regulatory Regions (Promoters) Using Transgenic Approaches

| Gene/Host Plant | Transgenic | Comment | Reference |
|---|---|---|---|
| *OsHSP18.6* (*Oryza sativa*) | *Oryza sativa* | Induced by drought, salt and cold, especially under heat; expressed in root, stem, leaf, internode, and spikelet; lower malondialdehyde (MDA) level; enhanced catalase (CAT)/superoxide dismutase (SOD) activities | Wang et al. (2015) |
| *MsHSP17.7* (*Medicago sativa L*) | *A. thaliana* *E. coli* | Induced by thermotolerance, salinity, drought, oxidative stress; increase root length in salt stress; low malondialdehyde (MDA) levels; 37°C for 0, 2, 4, 8, 12, and 24 hours), salt and peroxide tolerance in E. coli | Li et al. (2016) |
| *CsHSP17.2* (*Camellia sinensis*) | *Arabidopsis thaliana* *Escherichia coli*, *Pichia pastoris* | Up-regulated by heat stress (45°C for 4 hours); greater proline content; higher germination rate; higher hypocotyl elongation length than wild type plant; expression levels of several HS-responsive genes increased; overexpressed in prokaryotes; expressed in the nucleus and epidermal cells | Wang et al. (2017) |
| *ZmHsf04* (*Zea mays*) | *A. thaliana* | Heat/salt-stress response; up-regulated the expression of heat-specific HSPs (*AtHsp25.3-P*, *AtHsp18.2-CI*, and *AtHsp70B*) and stress-related genes (*AtAPX2* and *AtGolS1*) | Jiang et al. (2018) |
| *PlHSP70* | *A. thaliana* | Constitutive expression induce tolerance; high chlorophyll content; lower $H_2O_2$, superoxide free radicle, REC; more intact cell membrane, chloroplast, and starch grain (40°C on a 14 hours) | Zhao et al. (2019) |
| *CaHSP16.4* (*Capsicum annuum L.*) | *A. thaliana* | Thermotolerance; drought stress; expressed in cytoplasm and nucleus; silencing increase malonaldehyde content & reduce total chlorophyll content; higher SOD/catalase/ascorbic acid peroxidase and glutathione peroxidase activities | Huang et al. (2019) |
| **Engineering HSF level** | | | |
| *AtHsfA1b* (*A. thaliana*) | *Lycopersicon esculentum* | Thermal and chilling tolerance; the higher activity of soluble isoforms of ascorbate peroxidase (40°C for 2 hours) | Li et al. (2003) |
| *HSFA1* (*Glycine max*) | *G. max* | HT stress tolerance (45°C; 3 hours) tolerance | Zhu et al. (2006) |
| *AtHSFA2* (*A. thaliana*) | *A. thaliana* | Thermotolerance (45°C; 90 minutes) and acquired (37°C; 1 hour: 22°C; 3 hours: 45°C; 90 minutes); salt & osmotic stresses; role in cell proliferation; dominant-negative mutants displayed | Ogawa et al. (2007) |
| *AtHSFA3* (*A. thaliana*) | *A. thaliana* | HT tolerance (46°C; 45 minutes); induction of numerous HT stress-linked genes | Yoshida et al. (2008) |
| *OsHsfA2e* *O. Sativa* | *A. thaliana* | Thermotolerance; localized in nucleus; tolerance to high salinity | Yokotani et al. (2008) |
| *HSF7* (*O. sativa*) | *A. thaliana* | Enhanced thermotolerance (42°C; 16 hours); transcription of some HSF genes was heightened | Liu et al. (2009) |
| *HSF1* (*Boea hygrometrica*) | *A. thaliana* *N. tabacum* | Thermotolerance increased to 48°C; 2 hours 30 minutes for tobacco and 45°C; 1 hour for *Arabidopsis* and acquired (40°C; 3 hours: 50°C; 2 hours for tobacco and 37°C; 1 hour: 22°C; 3 hour: 45°C; 90 minutes for *Arabidopsis*) thermotolerance via regulation of genes involved in stress protection and mitotic cell cycle | Zhu et al. (2009) |

*(Continued)*

**TABLE 11.1 (*Continued*)**

Transgenic Approaches for Inducing Heat Tolerance Employing Heat-Shock Proteins (HSPs), Heat-Shock Factors (HSFs) and Heat-Shock Regulatory Regions (Promoters) Using Transgenic Approaches

| Gene/Host Plant | Transgenic | Comment | Reference |
|---|---|---|---|
| *HSFA2* (*Lilium longiflorum*) | *A. thaliana* | Overexpression activates genes like *HSP101, HSP70, HSP25.3,* and *APX2*, so improving HT stress (45°C; 1 hour) tolerance in transgenic plants | Xin et al. (2010) |
| *NnHSP17.5* (*Nelumbo nucifera*) | *A. thaliana* | Specifically expressed in seeds under normal conditions; strongly up-regulated in germinating seeds upon heat and oxidative stresses; enhanced seed germination vigor; increased SOD activity; thermotolerance | Zhou et al. (2012) |
| *SlHsfA3* (*Solanum lycopersicum, Sl*) | *A. thaliana* | Highly induced by HT, moderately salinity, and to some extent by drought, but was not induced by ABA; late flowering; induced *SlHsp26.1-P* and *SlHsp21.5-ER* | Li et al. (2013) |
| *HSFA1b* (*A. thaliana*) | *A. thaliana* | HSFA1b directly regulates 55 HSE1b-containing genes; which control the remaining 454 genes, together accounting for the stress defense and developmental phenotypes; thermotolerance (36°C for 15 minutes) | Bechtold et al. (2013) |
| *HsfA1a* (*A. thaliana*) | *A. thaliana* | A 94-fold increase in mRNA level which subsequently induces the expression of *Hsp18.2* and *Hsp70* genes in response to heat stress, low/high pH changes, and $H_2O_2$ | Qian et al. (2014) |
| *HSFA1s* (*A. thaliana*) | *A. thaliana* | The HSFA1s level increased; melatonin-mediated thermotolerance; HSFA1s-activated transcripts of heat-responsive genes (HSFA2, heat stress-associated 32 (HSA32), heat-shock protein 90 (HSP90), and 101 (HSP101)) | Shi et al. (2015) |
| *HsfA3* (*A. thaliana*) | *A. thaliana* | HS tolerance; oxidative stress tolerance; improved fresh weight, chlorophyll fluorescence, and ion leakage under light conditions; induce galactinol biosynthesis | Song et al. (2016) |
| *LlHsfA2b* (*Lilium longiflorum*) | *A. thaliana* | enhanced tolerance to heat and oxidative ($H_2O_2$) stresses; expressed in nucleus & mesophyll cells; regulate downstream heat-related genes | Xin et al. (2017) |
| *TaHsfC2a* (*Triticum aestivum*) | *Triticum aestivum* | Heat stress tolerance; up-regulated by ABA, drought; up-regulate heat protection genes (*TaHSP70d* and *TaGalSyn*) | Hu et al. (2018) |
| *HSFA1b* (*A. thaliana*) | *A. thaliana* | Thermotolerance; smaller rosettes in seedlings and mature plants; longer inflorescence; early flowering | Obomighie (2019) |
| **Engineering HSP Promoter** | | | |
| ***sHSP26*** (*Triticum aestivum*) | *A. thaliana* | Heat (37°C), drought, cold, salt stresses inducibility; the probable role of imperfect CCAAT-box element or some novel *cis*-element in heat stress | Khurana et al. (2013) |
| ***TasHSP26*** (*Triticum aestivum*) | *A. thaliana* | HT expression regulation; (35°C for 10 minutes) drought stress regulation | Reddy et al. (2014) |
| ***CsHSP17.2*** (*Camellia sinensis*) | *A. thaliana* | Highly responsive to HT stress (45°C for 4 hours) | Wang et al. (2017) |

*HT:* high temperature, nDNA: nuclear DNA.

makes the cell environment reductive, maintains higher energy status, and reduces protein damage. Engineering its biosynthesis in transformants provides tolerance at supra-optimal temperature (Grover et al. 2013). *Oryza sativa* expressing barley peroxisomal betaine aldehyde dehydrogenase gene (*HvBADH*) which catalyzes the conversion of choline into betaine aldehyde was tolerant to HT, cold, and salt stresses (Kishitani et al. 2000). Spinach *BADH* gene protects RUBICSO activity and $CO_2$ assimilation in transgenic tobacco under HT (Yang et al. 2005). Wheat transformed with *Atriplex hortensis BADH* gene showed greater thermotolerance via stabilization of lipid composition in thylakoid membrane (Wang et al. 2010a, b). Rice expressing spinach monooxygenase protein (catalyzes the conversion of betaine aldehyde dehydrogenase into glycinebetaine) was more thermotolerant (Shirasawa et al. 2006). *Arthrobacter globiformis* choline oxidase gene (catalyzes choline into glycinebetaine) in *Arabidopsis* and *Brassica campestris* increase glycinebetaine production causing greater thermotolerance (Wang et al. 2010a, b). Increase in sugar and its alcohols increases thermotolerance as revealed by trehalose-6-phosphate synthase and trehalose-6-phosphate phosphatase genes transformation from *Saccharomyces cerevisiae* into *Arabidopsis* (Miranda et al. 2007). *Medicago*, overexpressing yeast TPS1 and TPP2 genes controlled by rd29A promoter (water stress-induced promoter) enhanced growth, plant biomass, and HT tolerance (Suárez et al. 2009).

### 11.3.3 Antioxidation and Detoxification Pathways

HT increases the level of $H_2O_2$ and other ROS (causing oxidative stress) related via overlapping gene expression signifying that antioxidant producing or ROS removing genes are the potential candidate against HT. Thermotolerant *Arabidopsis*/tobacco was developed using barley peroxisomal ascorbate peroxidase and tomato glutathione peroxidases genes respectively (Chen et al. 2004). Red algae (*Cyanidioschyzon merolae*) peroxidase gave similar results in *Arabidopsis* via chloroplast structure stability (Hirooka et al. 2009). Likewise, peroxidase of tomato gave a similar result in tomato and tobacco (Sun et al. 2010). Besides, Cu/Zn superoxide dismutase is also important against HT as observed in *Arabidopsis* transformed with superoxide reductase gene from archeal hyper-thermophile (*Pyrococcus furiosus*) (Im et al. 2009). Other proteins that affect ROS activity include glutaredoxin from the fern pteris conferring thermotolerance in *Arabidopsis* (Sundaram and Rathinasabapathi 2010). Rice aldo-keto reductase exhibited improved HT tolerance in tobacco (Turóczy et al. 2011). Tomato expressing monodehydroascorbate reductase resulted in resistance to heat and ROS (Li et al. 2010a, b). Overexpression of *Arabidopsis* 2-cysperoxiredoxin in *Festuca arundinacea* and potato caused HT tolerance (Kim et al. 2012).

### 11.3.4 Membrane and Other Metabolic Processes

Membrane incites heat-shock response by changing its properties such as accumulation of saturated and unsaturated fatty acids. Silencing of chloroplast-localized fatty acid desaturase gene increase thermotolerance in tobacco by reducing trienoic fatty acids accumulation (Murakami et al. 2000). Similar results were obtained in rice with increased dienoic FA content by co-suppressing fatty acid desaturase genes (Sohn and Back 2007). Grover et al (2013) reviewed many such studies of manipulating metabolic protein linked to increasing thermotolerance can be found.

### 11.3.5 Transcriptional Factors (TFs)

HSFs regulate the expression of many HSPs or heat-related non-HSPs genes by binding to heat-shock elements in their promoter. HSFs induce thermotolerance through the formation of HSF–HSP regulatory network or undergo changes in the homo/hetero-oligomeric state causing high thermo-expression or protein–protein interaction causing their altered expression. Many HSFs have been characterized for their potential role against various stresses in rice and Arabidopsis among which overexpression seems a promising approach.

Dehydration response element-binding proteins (DREB): *DREB2A* (knock out) and *DREB2CA* (overexpression) induce thermotolerance in *Arabidopsis* via HSE in their promoters (Sakuma et al. 2006).

Similar result was also noted for *ZmDREB2A* in *Arabidopsis* (Qin et al. 2007). *AtDREB1A* increases HT stress in chrysanthemum with increased photosynthetic, RUBISCO, and sucrose-phosphate synthase activities (Hong et al. 2009). Overexpression of *AtDREB2C* and *OsDREB2B* led to an increase in thermotolerance (Matsukura et al. 2010). Similarly, activation of *DDF1* gene coding for AP2TF of the CBF/DREB1 subfamily conferred HT and drought tolerance in *Arabidopsis* (Kang et al. 2011).

WRKY TF: Overexpression of *WRKY40* from pepper induces pathogens and heat tolerance in *Nicotiana tabacum* (Dang et al. 2013). Similarly, *OsWRKY11* conferred heat and drought tolerance in rice (Wu et al. 2009). Overexpression of *AtWRKY25* elevates the level of HSF-A2, HSF-B1, HSF-B2A, and HSF-1000 leading to HT tolerance in rice, and the same is true for *AtWRKY39* in *Arabidopsis* (Li et al. 2010). Moreover, overexpression of *OsZFP177* (Zn-finger protein) induces thermotolerance in tobacco (Huang et al. 2008). *ABP9* (ABA-responsive element) enhanced photosynthetic capacity, C-use efficiency, HT, and water stress tolerance efficiency in *Arabidopsis* (Zhang et al. 2008). *MBF1C* is the key regulator of heat tolerance in *Arabidopsis* which works independently of HSF–HSP regulon (Suzuki et al. 2011; Khan et al. 2017). Likewise, the Zn-finger domain containing protein *AtSAP10* enhanced tolerance to HT and heavy metals in *Arabidopsis* (Dixit and Dhankher 2011).

### 11.3.6 Signaling Components

Heat stress signaling is regulated by various secondary messengers among which $Ca^{+2}$ is important controlled by plasma membrane via calmodulins and kinases (CaMs) pathways ($Ca^{+2}$-CaM-kinase) and involved in the activation of many HSFs. HT tolerance in *Arabidopsis* and *Physcomitrella patens* is controlled by cyclic nucleotide-gated $Ca^{+2}$ channel genes which regulate the influx of $Ca^{+2}$ across the plasma membrane (Finka et al. 2012). Plant overexpressing *CaM3* gene showed enhanced thermotolerance (Zhang et al. 2009). Similar results were obtained in *Arabidopsis* using *AtCBK3* (CAM-binding protein kinase) (Liu et al. 2008). Overexpression of *OsCaM1-1* and knocking out of *GSK3* (Glycogen synthase kinase 3-like) conferred enhanced thermotolerance in rice (Koh et al. 2007; Wu et al. 2012). Besides, nitric oxide signaling is also important against HT tolerance. Overexpression of phosphoinositide-specific phospholipase (*AtPLC9*) gene in *Arabidopsis* and tobacco conferred heat tolerance (Zheng et al. 2012). Similarly, *HOT5* null mutants lacking functional S-nitroso glutathione reductase are sensitive to HT stress. HT stress response is also associated with specific hormone signals. Ethylene receptor mutants are HT sensitive (Clarke et al. 2009). Similarly, $\beta$-aminobutyric acid (*BABA*) gene enhanced HT tolerance in *Arabidopsis* through ABA-pathway (Zimmerli et al. 2008). *Agrostis stolonifera* overexpressing adenine isopentyl transferase showed enhanced thermotolerance (Xu et al. 2009). Overexpression of *GASA4* promotes gibberellic acid-mediated responses provide immunity against HT stress (Ko et al. 2007). Similarly, auxin-mediated signal transduction during HT stress has also been reported (Carranco et al. 2010).

## 11.4 Conclusion

Climate change is one of the biggest modern-day challenges with drought and heat stresses being the most common consequences. Molecular techniques carry the potential to make crops adaptable to changing climatic conditions. It is obvious that the transgenic approach is one of the most widely used techniques for inducing tolerance against drought and heat stresses followed by mutagenic and molecular markers approaches. However, very few studies are available regarding the use of modern-day technology (CRISPER/cas9) which needs much more work. Overall, most of the studies are related to rice, *Arabidopsis*, tobacco, and wheat which suggest that a much more holistic approach is needed to cover other important plant species. Molecular marker studies are mostly conducted in wheat, while few studies are available regarding other plant species. Mutagenic studies are mostly comprised of traditional methods of mutagenesis while no serious effort has been made using modern techniques. In the future, the combination of these approaches may provide a more suitable option for inducing tolerance in diverse plant species. The present effort will provide a roadmap for future breeding programs by using the current molecular approaches.

## REFERENCES


Abdul S, Muhammad AA, Shabir H, Hesham A. El E, Sajjad H, Niaz A, Abdul G, Sayyed RZ, Fahad S, Subhan D, Rahul D (2021a) Zinc nutrition and arbuscular mycorrhizal symbiosis effects on maize (Zea mays L.) growth and productivity. J Saudi Soc Agric Sci. https://doi.org/10.1016/j.sjbs.2021.06.096.

Abdul S, Muhammad AA, Subhan D, Niaz A, Fahad S, Rahul D, Mohammad JA, Omaima N, Muhammad Habib ur R, Bernard RG (2021b) Effect of arbuscular mycorrhizal fungi on the physiological functioning of maizeunder zinc-deficient soils. Sci Rep 11:18468.

Adnan M, Fahad S, Khan IA, Saeed M, Ihsan MZ, Saud S, Riaz M, Wang D, Wu C (2019) Integration of poultry manure and phosphate solubilizing bacteria improved availability of Ca bound P in calcareous soils. 3 Biotech 9(10):368.

Adnan M, Fahad S, Muhammad Z, Shahen S, Ishaq AM, Subhan D, Zafar-ul-Hye M, Martin LB, Raja MMN, Beena S, Saud S, Imran A, Zhen Y, Martin B, Jiri H, Rahul D (2020) Coupling phosphate-solubilizing bacteria with phosphorus supplements improve maize phosphorus acquisition and growth under lime induced salinity stress. Plants 9(900). https://doi.org/10.3390/plants9070900.

Adnan M, Shah Z, Sharif M, Rahman H (2018b) Liming induces carbon dioxide ($CO_2$) emission in PSB inoculated alkaline soil supplemented with different phosphorus sources. Environ Sci Pollut Res 25(10):9501–9509.

Adnan M, Zahir S, Fahad S, Arif M, Mukhtar A, Imtiaz AK, Ishaq AM, Abdul B, Hidayat U, Muhammad A, Inayat-Ur R, Saud S, Muhammad ZI, Yousaf J, Amanullah, Hafiz MH, Wajid N (2018a) Phosphate-solubilizing bacteria nullify the antagonistic effect of soil calcification on bioavailability of phosphorus in alkaline soils. Sci Rep 8:4339. https://doi.org/10.1038/s41598-018-22653-7.

Ahmad S, Kamran M, Ding R, Meng X, Wang H, Ahmad I, Fahad S, Han Q (2019) Exogenous melatonin confers drought stress by promoting plant growth, photosynthetic capacity and antioxidant defense system of maize seedlings. PeerJ 7:e7793. http://doi.org/10.7717/peerj.7793.

Ahn YJ, Zimmerman JL (2006) Introduction of the carrot *HSP17.7* into potato (*Solanum tuberosum* L.) enhances cellular membrane stability and tuberization in vitro. Plant Cell Environ 29:95–104.

Akbar H, Timothy JK, Jagadish T, M. Golam M, Apurbo KC, Muhammad F, Rajan B, Fahad S, Hasanuzzaman M (2020) Agricultural land degradation: Processes and problems undermining future food security. In: Fahad S, Hasanuzzaman M, Alam M, Ullah H, Saeed M, Khan AK, Adnan M (eds) Environment, climate, plant and vegetation growth. Springer Publ Ltd, Springer Nature Switzerland AG. Part of Springer Nature, pp 17–62. https://doi.org/10.1007/978-3-030-49732-3.

Akram R, Turan V, Hammad HM, Ahmad S, Hussain S, Hasnain A, Maqbool MM, Rehmani MIA, Rasool A, Masood N, Mahmood F, Mubeen M, Sultana SR, Fahad S, Amanet K, Saleem M, Abbas Y, Akhtar HM, Waseem F, Murtaza R, Amin A, Zahoor SA, ul Din MS, Nasim W (2018a) Fate of organic and inorganic pollutants in paddy soils. In: Hashmi MZ, Varma A (eds) Environmental pollution of paddy soils, soil biology. Springer International Publishing AG, Switzerland, pp 197–214.

Akram R, Turan V, Wahid A, Ijaz M, Shahid MA, Kaleem S, Hafeez A, Maqbool MM, Chaudhary HJ, Munis, MFH, Mubeen M, Sadiq N, Murtaza R, Kazmi DH, Ali S, Khan N, Sultana SR, Fahad S, Amin A, Nasim W (2018b) Paddy land pollutants and their role in climate change. In: Hashmi MZ, Varma A (eds) Environmental pollution of paddy soils, soil biology. Springer International Publishing AG, Switzerland, pp 113–124.

Al-Wabel MI, Abdelazeem S, Munir A, Khalid E, Adel RAU (2020b) Extent of climate change in Saudi Arabia and its impacts on agriculture: A case study from Qassim Region. In: Fahad S, Hasanuzzaman M, Alam M, Ullah H, Saeed M, Khan AK, Adnan M (eds) Environment, climate, plant and vegetation growth. Springer Publ Ltd, Springer Nature Switzerland AG. Part of Springer Nature. pp 635–658. https://doi.org/10.1007/978-3-030-49732-3.

Al-Wabel MI, Ahmad M, Usman AR, Akanji M, Rafique MI (2020a) Advances in pyrolytic technologies with improved carbon capture and storage to combat climate change. In: Fahad S, Hasanuzzaman M, Alam M, Ullah H, Saeed M, Khan AK, Adnan M (eds) Environment, climate, plant and vegetation growth. Springer Publ Ltd, Springer Nature Switzerland AG. Part of Springer Nature. pp 535–576. https://doi.org/10.1007/978-3-030-49732-3.

Ali M, Ullah Z, Mian IA, Khan M, Khan N, Adnan M, Saeed M (2016) Response of maize to nitrogen levels and seed priming. Pure Appl Biol 5(3):578–587.

Amanullah, Fahad S (eds) (2017) Rice - technology and production. IntechOpen, Croatia. http://dx.doi.org/10.5772/64480.

Amanullah, Fahad S (eds) (2018a) Corn - Production and human Health in changing climate. IntechOpen, United Kingdom. http://dx.doi.org/10.5772/intechopen.74074.

Amanullah, Fahad S (eds) (2018b) Nitrogen in agriculture – updates. IntechOpen, Croatia. http://dx.doi.org/10.5772/65846.

Amanullah, Muhammad I, Haider N, Shah K, Manzoor A, Asim M, Saif U, Izhar A, Fahad S, Adnan M, et al. (2021) Integrated foliar nutrients application improve wheat (Triticum Aestivum L.) productivity under calcareous soils in drylands. Commun Soil Sci Plant Anal. https://doi.org/10.1080/00103624.2021.1956521.

Amanullah, Shah K, Imran, Hamdan AK, Muhammad A, Abdel RA, Muhammad A, Fahad S, Azizullah S, Brajendra P (2020) Effects of climate change on irrigation water quality. In: Fahad S, Hasanuzzaman M, Alam M, Ullah H, Saeed M, Khan AK, Adnan M (eds) Environment, climate, plant and vegetation growth. Springer Publ Ltd, Springer Nature Switzerland AG. Part of Springer Nature. pp 123–132. https://doi.org/10.1007/978-3-030-49732-3.

Amir M, Muhammad A, Allah B, Sevgi Ç, Haroon ZK, Muhammad A, Emre A (2020) Bio fortification under climate change: The fight between quality and quantity. In: Fahad S, Hasanuzzaman M, Alam M, Ullah H, Saeed M, Khan AK, Adnan M (eds) Environment, climate, plant and vegetation growth. Springer Publ Ltd, Springer Nature Switzerland AG. Part of Springer Nature. pp 173–228. https://doi.org/10.1007/978-3-030-49732-3.

Amjad I, Muhammad H, Farooq S, Anwar H (2020) Role of plant bioactives in sustainable agriculture. In: Fahad S, Hasanuzzaman M, Alam M, Ullah H, Saeed M, Khan AK, Adnan M (eds) Environment, climate, plant and vegetation growth. Springer Publ Ltd, Springer Nature Switzerland AG. Part of Springer Nature. pp 591–606. https://doi.org/10.1007/978-3-030-49732-3.

Amjad SF, Mansoora N, Din IU, Khalid IR, Jatoi GH, Murtaza G, Yaseen S, Naz M, Danish S, Fahad S, et al. (2021) Application of zinc fertilizer and mycorrhizal inoculation on physio- biochemical parameters of wheat grown under water-stressed environment. Sustainability 13:11007. https://doi.org/10. 3390/su131911007.

Arif M, Talha J, Muhammad R, Fahad S, Muhammad A, Amanullah, Kawsar A, Ishaq AM, Bushra K, Fahd R (2020) Biochar; a remedy for climate change. In: Fahad S, Hasanuzzaman M, Alam M, Ullah H, Saeed M, Khan AK, Adnan M (eds) Environment, climate, plant and vegetation growth. Springer Publ Ltd, Springer Nature Switzerland AG. Part of Springer Nature. pp 151–172. https://doi.org/10.1007/978-3-030-49732-3.

Ashfaq AR, Uzma Y, Niaz A, Muhammad AA, Fahad S, Haider S, Tayebeh Z, Subhan D, Süleyman T, Hesham AElE, Pramila T, Jamal MA, Sulaiman AA, Rahul D (2021) Toxicity of Cadmium and nickel in the context of applied activated carbon biochar for improvement in soil fertility. Saudi Soc Agric Sci. https://doi.org/10.1016/j.sjbs.2021.09.035.

Athar M, Masood IA, Sana S, Ahmed M, Xiukang W, Sajid F, Sher AK, Habib A, Faran M, Zafar H, Farhana G, Fahad S (2021) Bio-diesel production of sunflower through sulphur management in a semi-arid subtropical environment. Environ Sci Pollut Res. https://doi.org/10.1007/s11356-021-16688-z.

Atif B, Hesham A, Fahad S (2021) Biochar coupling with phosphorus fertilization modifies antioxidant activity, osmolyte accumulation and reactive oxygen species synthesis in the leaves and xylem sap of rice cultivars under high-temperature stress. Physiol Mol Biol Plants. https://doi.org/10.1007/s12298-021-01062-7.

Ayman EL Sabagh, Hossain A, Barutçular C, Iqbal MA, Islam MS, Fahad S, Sytar O, Çig F, Meena RS, Erman M (2020) Consequences of salinity stress on the quality of crops and its mitigation strategies for sustainable crop production: An outlook of arid and semi- arid regions. In: Fahad S, Hasanuzzaman M, Alam M, Ullah H, Saeed M, Khan AK, Adnan M (eds) Environment, climate, plant and vegetation growth. Springer Publ Ltd, Springer Nature Switzerland AG. Part of Springer Nature. pp 503–534. https://doi.org/10.1007/978-3-030-49732-3.

Aziz K, Daniel KYT, Fazal M, Muhammad ZA, Farooq S, Fan W, Fahad S, Ruiyang Z (2017a) Nitrogen nutrition in cotton and control strategies for greenhouse gas emissions: A review. Environ Sci Pollut Res 24:23471–23487. https://doi.org/10.1007/s11356-017-0131-y.

Aziz K, Daniel KYT, Muhammad ZA, Honghai L, Shahbaz AT, Mir A, Fahad S (2017b) Nitrogen fertility and abiotic stresses management in cotton crop: A review. Environ Sci Pollut Res 24:14551–14566. https://doi.org/10.1007/s11356-017-8920-x.

Baseer M, Adnan M, Fazal M, Fahad S, Muhammad S, Fazli W, Muhammad A, Jr. Amanullah, Depeng W, Saud S, Muhammad N, Muhammad Z, Fazli S, Beena S, Mian AR, Ishaq AM (2019) Substituting urea by organic wastes for improving maize yield in alkaline soil. J Plant Nutr. https://doi.org/10.1080/01904167.2019.1659344.

Bayram AY, Seher Ö, Nazlican A (2020) Climate change forecasting and modeling for the year of 2050. In: Fahad S, Hasanuzzaman M, Alam M, Ullah H, Saeed M, Khan AK, Adnan M (eds) Environment, climate, plant and vegetation growth. Springer Publ Ltd, Springer Nature Switzerland AG. Part of Springer Nature. pp 109–122. https://doi.org/10.1007/978-3-030-49732-3.

Bechtold U, Albihlal WS, Lawson T, Fryer MJ, Sparrow PA, Richard F, Persad R, Bowden L, Hickman R, Martin C (2013) *Arabidopsis* Heat shock transcription factora1b overexpression enhances water productivity, resistance to drought, and infection. J Exp Bot 64:3467–3481.

Brookes G, Barfoot P (2013) The global income and production effects of genetically modified (GM) crops 1996–2011. GM Crops Food 4:74–83.

Carranco R, Espinosa JM, Prieto-Dapena P, Almoguera C, Jordano J (2010) Repression by an auxin/indole acetic acid protein connects auxin signaling with heat shock factor-mediated seed longevity. Proc Natl Acad Sci USA 107:21908–21913.

Chang CC, Huang P-S, Lin H-R, Lu C-H (2007) Transactivation of protein expression by rice *HSP101* in planta and using *Hsp101* as a selection marker for transformation. Plant Cell Physiol 48:1098–1107.

Chang W, Qiujuan J, Evgenios A, Haitao L, Gezi L, Jingjing Z, Fahad S, Ying J (2021) Hormetic effects of zinc on growth and antioxidant defense system of wheat plants. Sci Total Environ. https://doi.org/10.1016/j.scitotenv.2021.150992.

Chen S, Vaghchhipawala Z, Li W, Asard H, Dickman MB (2004) Tomato phospholipid hydroperoxide glutathione peroxidase inhibits cell death induced by Bax and oxidative stresses in yeast and plants. Plant Physiol 135:1630–1641.

Chen Y, Guo Z, Dong L, Fu Z, Zheng Q, Zhang G, Qin L, Sun X, Shi Z, Fahad S, Xie F, Saud S. (2021) Turf performance and physiological responses of native Poa species to summer stress in Northeast China. PeerJ 9:e12252. http://doi.org/10.7717/peerj.12252.

Cho EK, Choi YJ (2009) A nuclear-localized HSP70 confers thermoprotective activity and drought-stress tolerance on plants. Biotechnol Lett 31:597–606.

Clarke SM, Cristescu SM, Miersch O, Harren FJ, Wasternack C, Mur LA (2009) Jasmonates act with Salicylic acid to confer basal thermotolerance in *Arabidopsis thaliana*. New Phytol 182:175–187.

Dang FF, Wang YN, Yu L, Eulgem T, Lai Y, Liu ZQ, Wang X, Qiu AL, Zhang TX, Lin J (2013) CaWRKY40, a WRKY protein of pepper, plays an important role in the regulation of tolerance to heat stress and resistance to *Ralstonia solanacearum* infection. Plant Cell Environ 36:757–774.

Deepranjan S, Ardith S, Siva OD, Sonam S, Shikha, Manoj P, Amitava R, Sayyed RZ, Abdul G, Mohammad JA, Subhan D, Fahad S, Rahul D (2021) Optimizing nutrient use efficiency, productivity, energetics, and economics of red cabbage following mineral fertilization and biopriming with compatible rhizosphere microbes. Sci Rep 11:15680.

Depeng W, Fahad S, Saud S, Muhammad K, Aziz K, Mohammad NK, Hafiz MH, Wajid N (2018) Morphological acclimation to agronomic manipulation in leaf dispersion and orientation to promote "Ideotype" breeding: Evidence from 3D visual modeling of "super" rice (*Oryza sativa* L.). Plant Physiol Biochem 135:499–510. https://doi.org/10.1016/j.plaphy.2018.11.010.

Dixit AR, Dhankher OP (2011) A novel stress-associated protein 'AtSAP10'from *Arabidopsis thaliana* confers tolerance to nickel, manganese, zinc, and high temperature stress. PLoS One 6:e20921.

El-Rawy ME, Youssef M (2014) Evaluation of drought and heat tolerance in wheat based on seedling traits and molecular analysis. J Crop Sci Biotech 17:183–189.

Emre B, Ömer SU, Martín LB, Andre D, Fahad S, Rahul D, Muhammad Z-ul-H, Ghulam SH, Subhan D (2021) Studying soil erosion by evaluating changes in physico-chemical properties of soils under different land-use types. J Saudi Soc Agric Sci 20:190–197.

Fahad S, Abdul B, Adnan M (eds) (2018b) Global wheat production. IntechOpen, United Kingdom. http://dx.doi.org/10.5772/intechopen.72559.

Fahad S, Adnan M, Hassan S, Saud S, Hussain S, Wu C, Wang D, Hakeem KR, Alharby HF, Turan V, Khan MA, Huang J (2019b) Rice responses and tolerance to high temperature. In: Hasanuzzaman M, Fujita M, Nahar K, Biswas JK (eds) Advances in rice research for abiotic stress tolerance. Woodhead Publ Ltd, Abington Hall Abington, Cambridge, UK, pp 201–224.

Fahad S, Bajwa AA, Nazir U, Anjum SA, Farooq A, Zohaib A, Sadia S, NasimW, Adkins S, Saud S, Ihsan MZ, Alharby H, Wu C, Wang D, Huang J (2017) Crop production under drought and heat stress: Plant responses and management options. Front Plant Sci 8:1147. https://doi.org/10.3389/fpls.2017.01147.

Fahad S, Bano A (2012) Effect of salicylic acid on physiological and biochemical characterization of maize grown in saline area. Pak J Bot 44:1433–1438.

Fahad S, Chen Y, Saud S, Wang K, Xiong D, Chen C, Wu C, Shah F, Nie L, Huang J (2013) Ultraviolet radiation effect on photosynthetic pigments, biochemical attributes, antioxidant enzyme activity and hormonal contents of wheat. J Food Agric Environ 11(3&4):1635–1641.

Fahad S, Hasanuzzaman M, Alam M, Ullah H, Saeed M, Ali Khan I, Adnan M (eds) (2020) Environment, climate, plant and vegetation growth. Springer Nature Switzerland AG. https://doi.org/10.1007/978-3-030-49732-3.

Fahad S, Hussain S, Bano A, Saud S, Hassan S, Shan D, Khan FA, Khan F, Chen Y, Wu C, Tabassum MA, Chun MX, Afzal M, Jan A, Jan MT, Huang J (2014a) Potential role of phytohormones and plant growth-promoting rhizobacteria in abiotic stresses: Consequences for changing environment. Environ Sci Pollut Res 22(7):4907–4921. https://doi.org/10.1007/s11356-014-3754-2.

Fahad S, Hussain S, Matloob A, Khan FA, Khaliq A, Saud S, Hassan S, Shan D, Khan F, Ullah N, Faiq M, Khan MR, Tareen AK, Khan A, Ullah A, Ullah N, Huang J (2014b) Phytohormones and plant responses to salinity stress: A review. Plant Growth Regul 75(2):391–404. https://doi.org/10.1007/s10725-014-0013-y.

Fahad S, Hussain S, Saud S, Hassan S, Chauhan BS, Khan F, et al. (2016a) Responses of rapid viscoanalyzer profile and other rice grain qualities to exogenously applied plant growth regulators under high day and high night temperatures. PLoS One 11(7):e0159590. https://doi.org/10.1371/journal.pone.0159590.

Fahad S, Hussain S, Saud S, Hassan S, Ihsan Z, Shah AN, Wu C, Yousaf M, Nasim W, Alharby H, Alghabari F, Huang J (2016c) Exogenously applied plant growth regulators enhance the morphophysiological growth and yield of rice under high temperature. Front Plant Sci 7:1250. https://doi.org/10.3389/fpls.2016.01250.

Fahad S, Hussain S, Saud S, Hassan S, Tanveer M, Ihsan MZ, Shah AN, Ullah A, Nasrullah KF, Ullah S, Alharby H, Wu C, Huang J (2016d) A combined application of biochar and phosphorus alleviates heat-induced adversities on physiological, agronomical and quality attributes of rice. Plant Physiol Biochem 103:191–198.

Fahad S, Hussain S, Saud S, Khan F, Hassan S, Nasim W, Arif M, Wang F, Huang J (2016b) Exogenously applied plant growth regulators affect heat-stressed rice pollens. J Agron Crop Sci 202:139–150.

Fahad S, Hussain S, Saud S, Tanveer M, Bajwa AA, Hassan S, Shah AN, Ullah A, Wu C, Khan FA, Shah F, Ullah S, Chen Y, Huang J (2015a) A biochar application protects rice pollen from high-temperature stress. Plant Physiol Biochem 96:281–287

Fahad S, Muhammad ZI, Abdul K, Ihsanullah D, Saud S, Saleh A, Wajid N, Muhammad A, Imtiaz AK, Chao W, Depeng W, Jianliang H (2018a) Consequences of high temperature under changing climate optima for rice pollen characteristics-concepts and perspectives. Archives Agron Soil Sci. https://doi.org/10.1080/03650340.2018.1443213.

Fahad S, Nie L, Chen Y, Wu C, Xiong D, Saud S, Hongyan L, Cui K, Huang J (2015b) Crop plant hormones and environmental stress. Sustain Agric Rev 15:371–400.

Fahad S, Saud S, Yajun C, Chao W, Depeng W (eds) (2021f) Abiotic stress in plants. IntechOpen, United Kingdom. http://dx.doi.org/10.5772/intechopen.91549.

Fahad S, Rehman A, Shahzad B, Tanveer M, Saud S, Kamran M, Ihtisham M., Khan SU, Turan V, Rahman MHU (2019a) Rice responses and tolerance to metal/metalloid toxicity. In: Hasanuzzaman M, Fujita M, Nahar K, Biswas JK (eds) Advances in rice research for abiotic stress tolerance. Woodhead Publ Ltd, Abington Hall Abington, Cambridge, UK, pp 299–312.

Fahad S, Sönmez O, Saud S, Wang D, Wu C, Adnan M, Arif M, Amanullah (eds) (2021e) Engineering tolerance in crop plants against abiotic stress, 1st edn. ed, Footprints of climate variability on plant diversity. CRC Press, Boca Raton, FL.

Fahad S, Sönmez O, Saud S, Wang D, Wu C, Adnan M, Turan V (eds) (2021b) Climate change and plants: biodiversity, growth and interactions, 1st edn. ed, Footprints of climate variability on plant diversity. CRC Press, Boca Raton, FL.

Fahad S, Sönmez O, Saud S, Wang D, Wu C, Adnan M, Turan V (eds) (2021c) Developing climate resilient crops: improving global food security and safety, 1st edn. ed, Footprints of climate variability on plant diversity. CRC Press, Boca Raton, FL.

Fahad S, Sönmez O, Saud S, Wang D, Wu C, Adnan M, Turan V (eds) (2021a) Plant growth regulators for climate-smart agriculture, 1st edn. ed, Footprints of climate variability on plant diversity. CRC Press, Boca Raton, FL.

Fahad S, Sönmez O, Turan V, Adnan M, Saud S, Wu C, Wang D (eds) (2021d) Sustainable soil and land management and climate change, 1st edn. ed, Footprints of climate variability on plant diversity. CRC Press, Boca Raton, FL.

Farah R, Muhammad R, Muhammad SA, Tahira Y, Muhammad AA, Maryam A, Shafaqat A, Rashid M, Muhammad R, Qaiser H, Afia Z, Muhammad AA, Muhammad A, Fahad S (2020) Alternative and non-conventional soil and crop management strategies for increasing water use efficiency. In: Fahad S, Hasanuzzaman M, Alam M, Ullah H, Saeed M, Khan AK, Adnan M (eds) Environment, climate, plant and vegetation growth. Springer Publ Ltd, Springer Nature Switzerland AG. Part of Springer Nature. pp 323–338. https://doi.org/10.1007/978-3-030-49732-3.

Farhana G, Ishfaq A, Muhammad A, Dawood J, Fahad S, Xiuling L, Depeng W, Muhammad F, Muhammad F, Syed AS (2020) Use of crop growth model to simulate the impact of climate change on yield of various wheat cultivars under different agro-environmental conditions in Khyber Pakhtunkhwa, Pakistan. Arabian J Geosci 13:112. https://doi.org/10.1007/s12517-020-5118-1.

Farhat A, Hafiz MH, Wajid I, Aitazaz AF, Hafiz FB, Zahida Z, Fahad S, Wajid F, Artemi C (2020) A review of soil carbon dynamics resulting from agricultural practices. J Environ Manage 268:110319.

Fazli W, Muhmmad S, Amjad A, Fahad S, Muhammad A, Muhammad N, Ishaq AM, Imtiaz AK, Mukhtar A, Muhammad S, Muhammad I, Rafi U, Haroon I, Muhammad A (2020) Plant-microbes interactions and functions in changing climate. In: Fahad S, Hasanuzzaman M, Alam M, Ullah H, Saeed M, Khan AK, Adnan M (eds) Environment, climate, plant and vegetation growth. Springer Publ Ltd, Springer Nature Switzerland AG. Part of Springer Nature. pp 397–420. https://doi.org/10.1007/978-3-030-49732-3.

Finka A, Cuendet AFH, Maathuis FJ, Saidi Y, Goloubinoff P (2012) Plasma membrane cyclic nucleotide gated calcium channels control land plant thermal sensing and acquired thermotolerance. Plant Cell 24:3333–3348.

Ghulam M, Muhammad AA, Donald LS, Sajid M, Muhammad FQ, Niaz A, Ateeq ur R, Shakeel A, Sajjad H, Muhammad A, Summia M, Aqib HAK, Fahad S, Rahul D, Mazhar I, Timothy DS (2021) Formalin fumigation and steaming of various composts differentially influence the nutrient release, growth and yield of muskmelon (*Cucumis melo* L.). Sci Rep 11:21057.

Gopakumar L, Bernard NO, Donato V (2020) Soil microarthropods and nutrient cycling. In: Fahad S, Hasanuzzaman M, Alam M, Ullah H, Saeed M, Khan AK, Adnan M (eds) Environment, climate, plant and vegetation growth. Springer Publ Ltd, Springer Nature Switzerland AG. Part of Springer Nature. pp 453–472. https://doi.org/10.1007/978-3-030-49732-3.

Grover A, Mittal D, Negi M, Lavania D (2013) Generating high temperature tolerant transgenic plants: Achievements and challenges. Plant Sci 205–206:38–47.

Guofu L, Zhenjian B, Fahad S, Guowen C, Zhixin X, Hao G, Dandan L, Yulong L, Bing L, Guoxu J, Saud S (2021) Compositional and structural changes in soil microbial communities in response to straw mulching and plant revegetation in an abandoned artificial pasture in Northeast China. Global Ecol Conserv 31:e01871.

Habib ur R, Ashfaq A, Aftab W, Manzoor H, Fahd R, Wajid I, Md. Aminul I, Vakhtang S, Muhammad A, Asmat U, Abdul W, Syeda RS, Shah S, Shahbaz K, Fahad S, Manzoor H, Saddam H, Wajid N (2017) Application of CSM-CROPGRO-Cotton model for cultivars and optimum planting dates: Evaluation in changing semi-arid climate. Field Crops Res. http://dx.doi.org/10.1016/j.fcr.2017.07.007.

Hafeez M, Farman U, Muhammad MK, Xiaowei L, Zhijun Z, Sakhawat S, Muhammad I, Mohammed AA, Mandela F-G, Nicolas D, Muzammal R, Fahad S, Yaobin L (2021) Metabolic-based insecticide resistance mechanism and ecofriendly approaches for controlling of beet armyworm Spodoptera exigua:a review. Environ Sci Pollut Res. https://doi.org/10.1007/s11356-021-16974-w.

Hafiz MH, Abdul K, Farhat A, Wajid F, Fahad S, Muhammad A, Ghulam MS, Wajid N, Muhammad M, Hafiz FB (2020b) Comparative effects of organic and inorganic fertilizers on soil organic carbon and wheat productivity under arid region. Commun Soil Sci Plant Anal. https://doi.org/10.1080/00103624.2020.1763385.

Hafiz MH, Farhat A, Ashfaq A, Hafiz FB, Wajid F, Carol Jo W, Fahad S, Gerrit H (2020a) Predicting kernel growth of maize under controlled water and nitrogen applications. Int J Plant Prod. https://doi.org/10.1007/s42106-020-00110-8.

Hafiz MH, Farhat A, Shafqat S, Fahad S, Artemi C, Wajid F, Chaves CB, Wajid N, Muhammad M, Hafiz FB (2018) Offsetting land degradation through nitrogen and water management during maize cultivation under arid conditions. Land Degrad Dev:1–10. https://doi.org/10.1002/ldr.2933.

Hafiz MH, Muhammad A, Farhat A, Hafiz FB, Saeed AQ, Muhammad M, Fahad S, Muhammad A (2019) Environmental factors affecting the frequency of road traffic accidents: A case study of sub-urban area of Pakistan. Environ Sci Pollut Res. https://doi.org/10. 1007/s11356-019-04752-8.

Hafiz MH, Wajid F, Farhat A, Fahad S, Shafqat S, Wajid N, Hafiz FB (2016) Maize plant nitrogen uptake dynamics at limited irrigation water and nitrogen. Environ Sci Pollut Res 24(3):2549–2557. https://doi.org/10. 1007/s11356-016-8031-0.

Haider SA, Lalarukh I, Amjad SF, Mansoora N, Naz M, Naeem M, Bukhari SA, Shahbaz M, Ali SA, Marfo TD, Subhan D, Rahul D, Fahad S (2021) Drought stress alleviation by potassium-nitrate-containing chitosan/montmorillonite microparticles confers changes in Spinacia oleracea L. Sustainability 13:9903. https://doi.org/10.3390/su13179903.

Hamza SM, Xiukang W, Sajjad A, Sadia Z, Muhammad N, Adnan M, Fahad S, et al. (2021) Interactive effects of gibberellic acid and NPK on morpho-physio-biochemical traits and organic acid exudation pattern in coriander (Coriandrum sativum L.) grown in soil artificially spiked with boron. Plant Physiol Biochem 167:884–900.

Hesham FA, Fahad S (2020) Melatonin application enhances biochar efficiency for drought tolerance in maize varieties:Modifications in physio-biochemical machinery. Agron J 112(4):1–22.

Hirooka S, Misumi O, Yoshida M, Mori T, Nishida K, Yagisawa F, Yoshida Y, Fujiwara T, Kuroiwa H, Kuroiwa T (2009) Expression of the Cyanidioschyzon merolae stromal ascorbate peroxidase in *Arabidopsis thaliana* enhances thermotolerance. Plant Cell Rep 28:1881–1893.

Hong B, Ma C, Yang Y, Wang T, Yamaguchi-Shinozaki K, Gao J (2009) Over-expression of *AtDREB1A* in *chrysanthemum* enhances tolerance to heat stress. Plant Mol Biol 70:231–240.

Hu XJ, Chen D, Lynne Mclntyre C, Fernanda Dreccer M, Zhang ZB, Drenth J, Kalaipandian S, Chang H, Xue GP (2018) Heat shock factor *C2a* serves as a proactive mechanism for heat protection in developing grains in wheat via an ABA-mediated regulatory pathway. Plant Cell Environ 41:79–98.

Huang J, Wang MM, Jiang Y, Bao YM, Huang X, Sun H, Xu DQ, Lan HX, Zhang HS (2008) Expression analysis of rice *A20/AN1*-type zinc finger genes and characterization of *ZFP177* that contributes to temperature stress tolerance. Gene 420:135–144.

Huang LJ, Cheng GX, Khan A, Wei AM, Yu QH, Yang SB, Luo DX, Gong ZH (2019) *CaHSP16. 4*, a small heat shock protein gene in pepper, is involved in heat and drought tolerance. Protoplasma 256:39–51.

Huang Li Y, Li Xiao X, Zhang Yun B, Fahad S, Wang F (2021) dep1 improves rice grain yield and nitrogen use efficiency simultaneously by enhancing nitrogen and dry matter translocation. J Integrative Agric. https://doi.org/10.1016/S2095-3119(21)63795-4.

Hussain MA, Fahad S, Rahat S, Muhammad FJ, Muhammad M, Qasid A, Ali A, Husain A, Nooral A, Babatope SA, Changbao S, Liya G, Ibrar A, Zhanmei J, Juncai H (2020) Multifunctional role of brassinosteroid and its analogues in plants. Plant Growth Regul. https://doi.org/10.1007/s10725-020-00647-8.

Ibrar K, Aneela R, Khola Z, Urooba N, Sana B, Rabia S, Ishtiaq H, Mujaddad Ur Rehman, Salvatore M (2020) Microbes and environment: Global warming reverting the frozen zombies. In: Fahad S, Hasanuzzaman M, Alam M, Ullah H, Saeed M, Khan AK, Adnan M (eds) Environment, climate, plant and vegetation growth. Springer Publ Ltd, Springer Nature Switzerland AG. Part of Springer Nature. pp 607–634. https://doi.org/10.1007/978-3-030-49732-3.

Ikram U, Khadim D, Muhammad T, Muhammad S, Fahad S (2021) Gibberellic acid and urease inhibitor optimize nitrogen uptake and yield of maize at varying nitrogen levels under changing climate. Environ Sci Pollut Res. https://doi.org/10.1007/s11356-021-16049-w.

Ilyas H, Alam SS, Khan I, Shah B, Naeem A, Khan N, Ullah W, Adnan M, Shah SRA, Junaid K, Ahmed N, Iqbal M (2016a) Medicinal plants rhizosphere exploration for the presence of potential biocontrol fungi. J Ento Zool Stud 4(3):108–113.

Ilyas M, Mohammad N, Nadeem K, Ali H, Aamir HK, Kashif H, Fahad S, Aziz K, Abid U (2020) Drought tolerance strategies in plants: A mechanistic approach. J Plant Growth Regul. https://doi.org/10.1007/s00344-020-10174-5.

Ilyas M, Rasheed A, Mahmood T (2016b) Functional characterization of germin and germin-like protein genes in various plant species using transgenic approaches. Biotechnol Lett 38:1405–1421.

Im YJ, Ji M, Lee A, Killens R, Grunden AM, Boss WF (2009) Expression of *Pyrococcus furiosus* superoxide reductase in *Arabidopsis* enhances heat tolerance. Plant Physiol 151:893–904.

Iqra M, Amna B, Shakeel I, Fatima K, Sehrish L, Hamza A, Fahad S (2020) Carbon cycle in response to global warming. In: Fahad S, Hasanuzzaman M, Alam M, Ullah H, Saeed M, Khan AK, Adnan M (eds) Environment, climate, plant and vegetation growth. Springer Publ Ltd, Springer Nature Switzerland AG. Part of Springer Nature. pp 1–16. https://doi.org/10.1007/978-3-030-49732-3.

Irfan M, Muhammad M, Muhammad JK, Khadim MD, Dost M, Ishaq AM, Waqas A, Fahad S, Saud S, et al. (2021) Heavy metals immobilization and improvement in maize (Zea mays L.) growth amended with biochar and compost. Sci Rep 11:18416.

Jabborova D, Sulaymanov K, Sayyed RZ, Alotaibi SH, Enakiev Y, Azimov A, Jabbarov Z, Ansari MJ, Fahad S, Danish S, et al. (2021) Mineral fertilizers improves the quality of turmeric and soil. Sustainability 13:9437. https://doi.org/10.3390/su13169437.

Jan M, Anwar-ul-Haq M, Adnan NS, Muhammad Y, Javaid I, Xiuling L, Depeng W, Fahad S (2019) Modulation in growth, gas exchange, and antioxidant activities of salt-stressed rice (Oryza sativa L.) genotypes by zinc fertilization. Arab J Geosci 12:775. https://doi.org/10.1007/s12517-019-4939-2.

Jiang C, Xu J, Zhang H, Zhang X, Shi J, Li M, Ming F (2009) A cytosolic class I small heat shock protein, *RcHSP17. 8*, of *Rosa chinensis* confers resistance to a variety of stresses to *Escherichia coli*, yeast and *Arabidopsis thaliana*. Plant Cell Environ 32:1046–1059.

Jiang Y, Zheng Q, Chen L, Liang Y, Wu J (2018) Ectopic overexpression of maize heat shock transcription factor gene *ZmHsf04* confers increased thermo and salt-stress tolerance in transgenic *Arabidopsis*. Acta Physiol Plant 40:9.

Kamran M, Wenwen C, Irshad A, Xiangping M, Xudong Z, Wennan S, Junzhi C, Shakeel A, Fahad S, Qingfang H, Tiening L (2017) Effect of paclobutrazol, a potential growth regulator on stalk mechanical strength, lignin accumulation and its relation with lodging resistance of maize. Plant Growth Regul 84:317–332. https://doi.org/10.1007/s10725-017-0342-8.

Kang HG, Kim J, Kim B, Jeong H, Choi SH, Kim EK, Lee HY, Lim PO (2011) Overexpression of *FTL1/DDF1*, an *AP2* transcription factor, enhances tolerance to cold, drought, and heat stresses in *Arabidopsis thaliana*. Plant Sci 180:634–641.

Katiyar-Agarwal S, Agarwal M, Grover A (2003) Heat-tolerant basmati rice engineered by over-expression of *hsp101*. Plant Mol Biol 51:677–686.

Khadim D, Fahad S, Jahangir MMR, Iqbal M, Syed SA, Shah AK, Ishaq AM, Rahul D, et al. (2021a) Biochar and urease inhibitor mitigate $NH_3$ and $N_2O$ emissions and improve wheat yield in a urea fertilized alkaline soil. Sci Rep 11:17413.

Khadim D, Saif-ur-R, Fahad S, Syed SA, Shah AK, et al. (2021b) Influence of variable biochar concentration on yield-scaled nitrous oxide emissions, Wheat yield and nitrogen use efficiency. Sci Rep 11:16774.

Khan MA, Basir A, Adnan M, Shah AS, Noor M, Khan A, Shah JA, Ali Z, Rehman A (2017) Wheat phenology and density and fresh and dry weight of weeds as affected by potassium sources levels and tillage practices. Pak J Weed Sci Res 23(4):451–462.

Khan MMH, Niaz A, Umber G, Muqarrab A, Muhammad AA, Muhammad I, Shabir H, Shah F, Vibhor A, Shams H. A-H, Reham A, Syed MBA, Nadiyah MA, Ali TKZ, Subhan D, Rahul D (2021) Synchronization of Boron application methods and rates is environmentally friendly approach to improve quality attributes of Mangifera indica L. On sustainable basis. Saudi J Bio Sci. https://doi.org/10.1016/j.sjbs.2021.10.036.

Khurana N, Chauhan H, Khurana P (2013) Wheat chloroplast targeted *sHSP26* promoter confers heat and abiotic stress inducible expression in transgenic *Arabidopsis* Plants. PLoS One 8:e54418.

Kim KH, Alam I, Kim YG, Sharmin SA, Lee KW, Lee SH, Lee BH (2012) Overexpression of a chloroplast-localized small heat shock protein *OsHSP26* confers enhanced tolerance against oxidative and heat stresses in tall fescue. Biotechnol Lett 34:371–377.

Kishitani S, Takanami T, Suzuki M, Oikawa M, Yokoi S, Ishitani M, Alvarez-Nakase A, Takabe T, Takabe T (2000) Compatibility of glycinebetaine in rice plants: evaluation using transgenic rice plants with a gene for peroxisomal betaine aldehyde dehydrogenase from barley. Plant Cell Environ 23:107–114.

Ko CB, Woo YM, Lee DJ, Lee MC, Kim CS (2007) Enhanced tolerance to heat stress in transgenic plants expressing the *GASA4* gene. Plant Physiol Biochem 45:722–728.

Koh S, Lee SC, Kim MK, Koh JH, Lee S, An G, Choe S, Kim SR (2007) T-DNA tagged knockout mutation of rice *OsGSK1*, an orthologue of *Arabidopsis* BIN2, with enhanced tolerance to various abiotic stresses. Plant Mol Biol 65:453–466.

Li F, Wu QY, Sun YL, Wang LY, Yang XH, Meng QW (2010a) Overexpression of chloroplastic monodehydroascorbate reductase enhanced tolerance to temperature and Methyl viologen-mediated oxidative stresses. Physiol Plant 139:421–434.

Li H-Y, Chang CS, Lu L-S, Liu CA, Chan MT, Charng YY (2003) Overexpression of *Arabidopsis thaliana* heat shock factor gene (*AtHsfA1b*) enhances chilling tolerance in transgenic tomato. Bot Bull Acad Sin 44:129–140.

Li S, Zhou X, Chen L, Huang W, Yu D (2010b) Functional characterization of *Arabidopsis thaliana WRKY39* in heat stress. Mol Cells 29:475–483.

Li Z, Zhang L, Wang A, Xu X, Li J (2013) Ectopic overexpression of *SlHsfA3*, a heat stress transcription factor from tomato, confers increased thermotolerance and salt hypersensitivity in germination in transgenic *Arabidopsis*. PLoS One 8:e54880.

Li ZY, Long RC, Zhang TJ, Yang QC, Kang JM (2016) Molecular cloning and characterization of the *MsHSP17. 7* gene from *Medicago sativa* L. Mol Biol Rep 43:815–826.

Liu HT, Gao F, Li GL, Han JL, Liu DL, Sun DY, Zhou RG (2008) The calmodulin-binding protein kinase 3 is part of heat-shock signal transduction in *Arabidopsis thaliana*. Plant J 55:760–773.

Liu JG, Qin QL, Zhang Z, Peng RH, Xiong AS, Chen JM, Yao QH (2009) *OsHSF7* gene in rice, *Oryza sativa* L., encodes a transcription factor that functions as a high temperature receptive and responsive factor. BMB Rep 42:16–21.

Lobell DB, Schlenker W, Costa-Roberts J (2011) Climate trends and global crop production since 1980. Science 333:616–620.

Lusser M, Parisi C, Plan D, Rodríguez-Cerezo E (2012) Deployment of new biotechnologies in plant breeding. Nat Biotechnol 30:231–239.

Mahar A, Amjad A, Altaf HL, Fazli W, Ronghua L, Muhammad A, Fahad S, Muhammad A, Rafiullah, Imtiaz AK, Zengqiang Z (2020) Promising technologies for Cd-contaminated soils: Drawbacks and possibilities. In: Fahad S, Hasanuzzaman M, Alam M, Ullah H, Saeed M, Khan AK, Adnan M (eds) Environment, climate, plant and vegetation growth. Springer Publ Ltd, Springer Nature Switzerland AG. Part of Springer Nature. pp 63–92. https://doi.org/10.1007/978-3-030-49732-3.

Mahmood Ul H, Tassaduq R, Chandni I, Adnan A, Muhammad A, Muhammad MA, Muhammad H-ur-R, Mehmood AN, Alam S, Fahad S (2021) Linking plants functioning to adaptive responses under heat stress conditions: A mechanistic review. J Plant Growth Regul. https://doi.org/10.1007/s00344-021-10493-1.

Manzer HS, Saud A, Soumya M, Abdullah A. Al-A, Qasi DA, Bander MA. Al-M, Hayssam MA, Hazem MK, Fahad S, Vishnu DR, Om PN (2021) Molybdenum and hydrogen sulfide synergistically mitigate arsenic toxicity by modulating defense system, nitrogen and cysteine assimilation in faba bean (*Vicia faba* L.) seedlings. Environ Pollut 290:117953.

Matsukura S, Mizoi J, Yoshida T, Todaka D, Ito Y, Maruyama K, Shinozaki K, Yamaguchi-Shinozaki K (2010) Comprehensive analysis of rice *DREB2*-type genes that encode transcription factors involved in the expression of abiotic stress-responsive genes. Mol Genet Genomics 283: 185–196.

Md Jakir H, Allah B (2020) Development and applications of transplastomic plants; a way towards eco-friendly agriculture. In: Fahad S, Hasanuzzaman M, Alam M, Ullah H, Saeed M, Khan AK, Adnan M (eds) Environment, climate, plant and vegetation growth. Springer Publ Ltd, Springer Nature Switzerland AG. Part of Springer Nature. pp 285–322. https://doi.org/10.1007/978-3-030-49732-3.

Md. Enamul H, Shoeb AZM, Mallik AH, Fahad S, Kamruzzaman MM, Akib J, Nayyer S, K. M. Mehedi A, Swati AS, Md Yeamin A, Most SS (2020) Measuring vulnerability to environmental hazards: Qualitative to quantitative. in: Fahad S, Hasanuzzaman M, Alam M, Ullah H, Saeed M, Khan AK, Adnan M (eds) Environment, climate, plant and vegetation growth. Springer Publ Ltd, Springer Nature Switzerland AG. Part of Springer Nature. pp 421–452. https://doi.org/10.1007/978-3-030-49732-3.

Miranda JA, Avonce N, Suárez R, Thevelein JM, Van Dijck P, Iturriaga G (2007) A bifunctional TPS–TPP enzyme from yeast confers tolerance to multiple and extreme abiotic-stress conditions in transgenic *Arabidopsis*. Planta 226:1411–1421.

Montero-Barrientos M, Hermosa R, Cardoza RE, Gutierrez S, Nicolas C, Monte E (2010) Transgenic expression of the *Trichoderma harzianum hsp70* gene increases *Arabidopsis* resistance to heat and other abiotic stresses. J Pl Physiol 167:659–665.

Mu C, Zhang S, Yu G, Chen N, Li X, Liu H (2013) Overexpression of small heat shock protein *LimHSP16.45* in *Arabidopsis* enhances tolerance to abiotic stresses. PLoS One 8:e82264.

Mubeen M, Ashfaq A, Hafiz MH, Muhammad A, Hafiz UF, Mazhar S, Muhammad Sami ul Din, Asad A, Amjed A, Fahad S, Wajid N (2020) Evaluating the climate change impact on water use efficiency of cotton-wheat in semi-arid conditions using DSSAT model. J Water Climate Change. https://doi.org/10.2166/wcc.2019.179/622035/jwc2019179.pdf.

Muhammad N, Muqarrab A, Khurram S, Fiaz A, Fahim N, Muhammad A, Shazia A, Omaima N, Sulaiman AA, Fahad S, Subhan D, Rahul D (2021) Kaolin and Jasmonic acid improved cotton productivity under water stress conditions. J Saudi Soc Agric Sci 28:6606–6614. https://doi.org/10.1016/j.sjbs.2021.07.043.

Muhammad Z, Abdul MK, Abdul MS, Kenneth BM, Muhammad S, Shahen S, Ibadullah J, Fahad S (2019) Performance of Aeluropus lagopoides (mangrove grass) ecotypes, a potential turfgrass, under high saline conditions. Environ Sci Pollut Res. https://doi.org/10.1007/s11356-019-04838-3.

Muhammad Tahir ul Q, Amna F, Amna B, Barira Z, Xitong Z, Ling-Ling C (2020) Effectiveness of conventional crop improvement strategies vs. omics. In: Fahad S, Hasanuzzaman M, Alam M, Ullah H, Saeed M, Khan AK, Adnan M (eds) Environment, climate, plant and vegetation growth. Springer Publ Ltd, Springer Nature Switzerland AG. Part of Springer Nature. pp 253–284. https://doi.org/10.1007/978-3-030-49732-3.

Murakami Y, Tsuyama M, Kobayashi Y, Kodama H, Iba K (2000) Trienoic fatty acids and plant tolerance of high temperature. Science 287:476–479.

Muzammal R, Fahad S, Guanghui D, Xia C, Yang Y, Kailei T, Lijun L, Fei-Hu L, Gang D (2021) Evaluation of hemp (Cannabis sativa L.) as an industrial crop: A review. Environ Sci Pollut Res. https://doi.org/10.1007/s11356-021-16264-5.

Noor M, Naveed ur R, Ajmal J, Fahad S, Muhammad A, Fazli W, Saud S, Hassan S (2020) Climate change and costal plant lives. In: Fahad S, Hasanuzzaman M, Alam M, Ullah H, Saeed M, Khan AK, Adnan M (eds) Environment, climate, plant and vegetation growth. Springer Publ Ltd, Springer Nature Switzerland AG. Part of Springer Nature. pp 93–108. https://doi.org/10.1007/978-3-030-49732-3.

Obomighie I (2019) The role of Arabidopsis thaliana Clade A1 heat shock transcription factors in growth and development. University of Essex, UK.

Ogawa D, Yamaguchi K, Nishiuchi T (2007) High-level overexpression of the *Arabidopsis HsfA2* gene confers not only increased themotolerance but also salt/osmotic stress tolerance and enhanced callus growth. J Exp Bot 58:3373–3383.

Ono K, Hibino T, Kohinata T, Suzuki S, Tanaka Y, Nakamura T, Takabe T (2001) Overexpression of DnaK from a halotolerant *cyanobacterium Aphanothece* halophytica enhances the high-temperature tolerance of tobacco during germination and early growth. Plant Sci 160:455–461.

Qamar uz Z, Zubair A, Muhammad Y, Muhammad ZI, Abdul K, Fahad S, Safder B, Ramzani PMA, Muhammad N (2017) Zinc biofortification in rice: Leveraging agriculture to moderate hidden hunger in developing countries. Arch Agron Soil Sci 64:147–161. https://doi.org/10.1080/03650340.2017.1338343.

Qi Y, Wang H, Zou Y, Liu C, Liu Y, Wang Y, Zhang W (2011) Over-expression of mitochondrial heat shock protein 70 suppresses programmed cell death in rice. FEBS Lett 585:231–239.

Qian J, Chen J, Liu YF, Yang LL, Li WP, Zhang LM (2014) Overexpression of *Arabidopsis HsfA1a* enhances diverse stress tolerance by promoting stress-induced *Hsp* expression. GMR 13:1233–1243.

Qin F, Kakimoto M, Sakuma Y, Maruyama K, Osakabe Y, Tran LSP, Shinozaki K, Yamaguchi-Shinozaki K (2007) Regulation and functional analysis of *ZmDREB2A* in response to drought and heat stresses in *Zea mays* L. Plant J 50:54–69.

Queitsch C, Hong SW, Vierling E, Lindquist S (2000) Heat shock protein 101 plays a crucial role in thermotolerance in *Arabidopsis*. Plant Cell 12:479–492.

Rashid M, Qaiser H, Khalid SK, Al-Wabel MI, Zhang A, Muhammad A, Shahzada SI, Rukhsanda A, Ghulam AS, Shahzada MM, Sarosh A, Muhammad FQ (2020) Prospects of biochar in alkaline soils to mitigate climate change. In: Fahad S, Hasanuzzaman M, Alam M, Ullah H, Saeed M, Khan AK, Adnan M (eds) Environment, climate, plant and vegetation growth. Springer Publ Ltd, Springer Nature Switzerland AG. Part of Springer Nature. pp 133–150. https://doi.org/10.1007/978-3-030-49732-3.

Ray DK, Mueller ND, West PC, Foley JA (2013) Yield trends are insufficient to double global crop production by 2050. PLoS One 8: e66428.

Reddy PS, Kishor PBK, Seiler C, Kuhlmann M, Eschen-Lippold L, Lee J, Reddy MK, Sreenivasulu N (2014) Unraveling regulation of the small heat shock proteins by the heat shock factor *HvHsfB2c* in barley: its implications in drought stress response and seed development. PLoS One 9:e89125.

Rehman M, Fahad S, Saleem MH, Hafeez M, Muhammad Habib ur R, Liu F, Deng G (2020) Red light optimized physiological traits and enhanced the growth of ramie (Boehmeria nivea L.). Photosynthetica 58(4):922–931.

Rhoads DM, White SJ, Zhou Y, Muralidharan M, Elthon TE (2005) Altered gene expression in plants with constitutive expression of a mitochondrial small heat shock protein suggests the involvement of retrograde regulation in the heat stress response. Physiol Plant 123:435–444.

Roy B, Noren S, Mandal AB, Basu AK (2011) Genetic engineering for abiotic stress tolerance in agricultural crops. Biotechnology 10:1–22.

Sadam M, Muhammad Tahir ul Q, Ghulam M, Muhammad SK, Faiz AJ (2020) Role of biotechnology in climate resilient agriculture. In: Fahad S, Hasanuzzaman M, Alam M, Ullah H, Saeed M, Khan AK, Adnan M (eds) Environment, climate, plant and vegetation growth. Springer Publ Ltd, Springer Nature Switzerland AG. Part of Springer Nature. pp 339–366. https://doi.org/10.1007/978-3-030-49732-3.

Safi UK, Ullah F, Mehmood S, Fahad S, Ahmad Rahi A, Althobaiti F, et al. (2021) Antimicrobial, antioxidant and cytotoxic properties of Chenopodium glaucum L.. PLoS One 16(10):e0255502. https://doi.org/10.1371/journal.pone.0255502.

Sahrish N, Shakeel A, Ghulam A, Zartash F, Sajjad H, Mukhtar A, Muhammad AK, Ahmad K, Fahad S, Wajid N, Sezai E, Carol Jo W, Gerrit H (2022) Modeling the impact of climate warming on potato phenology. Eur J Agron 132:126404.

Sajid H, Jie H, Jing H, Shakeel A, Satyabrata N, Sumera A, Awais S, Chunquan Z, Lianfeng Z, Xiaochuang C, Qianyu J, Junhua Z (2020) Rice production under climate change: Adaptations and mitigating strategies. In: Fahad S, Hasanuzzaman M, Alam M, Ullah H, Saeed M, Khan AK, Adnan M (eds) Environment, Climate, Plant and Vegetation Growth. Springer Publ Ltd, Springer Nature Switzerland AG. Part of Springer Nature. pp 659–686. https://doi.org/10.1007/978-3-030-49732-3.

Sajjad H, Muhammad M, Ashfaq A, Waseem A, Hafiz MH, Mazhar A, Nasir M, Asad A, Hafiz UF, Syeda RS, Fahad S, Depeng W, Wajid N (2019) Using GIS tools to detect the land use/land cover changes during forty years in Lodhran district of Pakistan. Environ Sci Pollut Res. https://doi.org/10.1007/s11356-019-06072-3.

Sakuma Y, Maruyama K, Osakabe Y, Qin F, Seki M, Shinozaki K, Yamaguchi-Shinozaki K (2006) Functional analysis of an *Arabidopsis* transcription factor, *DREB2A*, involved in drought-responsive gene expression. Plant Cell 18:1292–1309.

Saleem MH, Fahad S, Adnan M, Mohsin A, Muhammad SR, Muhammad K, Qurban A, Inas AH, Parashuram B, Mubassir A, Reem MH (2020a) Foliar application of gibberellic acid endorsed phytoextraction of copper and alleviates oxidative stress in jute (Corchorus capsularis L.) plant grown in highly copper-contaminated soil of China. Environ Sci Pollut Res. https://doi.org/10.1007/s11356-020-09764-3.

Saleem MH, Fahad S, Shahid UK, Mairaj D, Abid U, Ayman ELS, Akbar H, Analía L, Lijun L (2020c) Copper-induced oxidative stress, initiation of antioxidants and phytoremediation potential of flax (Linum usitatissimum L.) seedlings grown under the mixing of two different soils of China. Environ Sci Poll Res. https://doi.org/10.1007/s11356-019-07264-7.

Saleem MH, Rehman M, Fahad S, Tung SA, Iqbal N, Hassan A, Ayub A, Wahid MA, Shaukat S, Liu L, Deng G (2020b) Leaf gas exchange, oxidative stress, and physiological attributes of rapeseed (Brassica napus L.) grown under different light-emitting diodes. Photosynthetica 58(3):836–845.

Saman S, Amna B, Bani A, Muhammad Tahir ul Qamar, Rana MA, Muhammad SK (2020) QTL mapping for abiotic stresses in cereals. In: Fahad S, Hasanuzzaman M, Alam M, Ullah H, Saeed M, Khan AK, Adnan M (eds) Environment, climate, plant and vegetation growth. Springer Publ Ltd, Springer Nature. Switzerland AG. Part of Springer Nature. pp 229–252. https://doi.org/10.1007/978-3-030-49732-3.

Saud S, Chen Y, Fahad S, Hussain S, Na L, Xin L, Alhussien SA (2016) Silicate application increases the photosynthesis and its associated metabolic activities in Kentucky bluegrass under drought stress and post-drought recovery. Environ Sci Pollut Res 23(17):17647–17655. https://doi.org/10.1007/s11356-016-6957-x.

Saud S, Chen Y, Long B, Fahad S, Sadiq A (2013) The different impact on the growth of cool season turf grass under the various conditions on salinity and drought stress. Int J Agric Sci Res 3:77–84.

Saud S, Fahad S, Cui G, Chen Y, Anwar S (2020) Determining nitrogen isotopes discrimination under drought stress on enzymatic activities, nitrogen isotope abundance and water contents of Kentucky bluegrass. Sci Rep 10:6415. https://doi.org/10.1038/s41598-020-63548-w.

Saud S, Fahad S, Yajun C, Ihsan MZ, Hammad HM, Nasim W, Amanullah Jr, Arif M, Alharby H (2017) Effects of nitrogen supply on water stress and recovery mechanisms in Kentucky bluegrass plants. Front Plant Sci 8:983. https://doi.org/10.3389/fpls.2017.00983.

Saud S, Li X, Chen Y, Zhang L, Fahad S, Hussain S, Sadiq A, Chen Y (2014) Silicon application increases drought tolerance of Kentucky bluegrass by improving plant water relations and morph physiological functions. SciWorld J 2014:1–10. https://doi.org/10.1155/2014/368694.

Senol C (2020) The effects of climate change on human behaviors. In: Fahad S, Hasanuzzaman M, Alam M, Ullah H, Saeed M, Khan AK, Adnan M (eds) Environment, climate, plant and vegetation growth. Springer Publ Ltd, Springer Nature Switzerland AG. Part of Springer Nature. pp 577–590. https://doi.org/10.1007/978-3-030-49732-3.

Shafi MI, Adnan M, Fahad S, Fazli W, Ahsan K, Zhen Y, Subhan D, Zafar-ul-Hye M, Martin B, Rahul D (2020) Application of single superphosphate with humic acid improves the growth, yield and phosphorus uptake of wheat (Triticum aestivum L.) in calcareous soil. Agron 10:1224. https://doi.org/10.3390/agronomy10091224.

Shah F, Lixiao N, Kehui C, Tariq S, Wei W, Chang C, Liyang Z, Farhan A, Fahad S, Huang J (2013) Rice grain yield and component responses to near 2°C of warming. Field Crop Res 157:98–110.

Shi H, Tan DX, Reiter RJ, Ye T, Yang F, Chan Z (2015) Melatonin induces class *A1* heat-shock factors (*HSFA 1s*) and their possible involvement of thermotolerance in *Arabidopsis*. J Pineal Res 58:335–342.

Shirasawa K, Takabe T, Takabe T, Kishitani S (2006) Accumulation of glycinebetaine in rice plants that overexpress choline monooxygenase from spinach and evaluation of their tolerance to abiotic stress. Ann Bot 98:565–571.

Sidra K, Javed I, Subhan D, Allah B, Syed IUSB, Fatma B, Khaled DA, Fahad S, Omaima N, Ali TKZ, Rahul D (2021) Physio-chemical characterization of indigenous agricultural waste materials for the development of potting media. J Saudi Soc Agric Sci. https://doi.org/10.1016/j.sjbs.2021.08.058.

Sohn S, Back K (2007) Transgenic rice tolerant to high temperature with elevated contents of dienoic fatty acids. Biol Plant 51: 340–342.

Song A, Zhu X, Chen F, Gao H, Jiang J, Chen S (2014) A chrysanthemum heat shock protein confers tolerance to abiotic stress. Int J Mol Sci 15: 5063–5078.

Song C, Chung WS, Lim CO (2016) Overexpression of heat shock factor gene *HsfA3* increases galactinol levels and oxidative stress tolerance in *Arabidopsis*. Mol Cells 39:477.

Suárez R, Calderón C, Iturriaga G (2009) Enhanced tolerance to multiple abiotic stresses in transgenic alfalfa accumulating trehalose. Crop Sci 49:1791–1799.

Subhan D, Zafar-ul-Hye M, Fahad S, Saud S, Martin B, Tereza H, Rahul D (2020) Drought stress alleviation by ACC deaminase producing *Achromobacter xylosoxidans* and *Enterobacter cloacae*, with and without timber waste biochar in maize. Sustainability 12(6286). https://doi.org/10.3390/su12156286.

Sun L, Liu Y, Kong X, Zhang D, Pan J, Zhou Y, Wang L, Li D, Yang X (2012) *ZmHSP16.9*, a cytosolic class I small heat shock protein in maize (*Zea mays*), confers heat tolerance in transgenic tobacco. Plant Cell Rep 31:1473–1484.

Sun WH, Duan M, Li F, Shu DF, Yang S, Meng QW (2010) Overexpression of tomato *tAPX* gene in tobacco improves tolerance to high or low temperature stress. Biol Plant 54:614–620.

Sundaram S, Rathinasabapathi B (2010) Transgenic expression of fern *Pteris vittata* glutaredoxin *PvGrx5* in *Arabidopsis thaliana* increases plant tolerance to high temperature stress and reduces oxidative damage to proteins. Planta 231:361.

Suzuki N, Sejima H, Tam R, Schlauch K, Mittler R (2011) Identification of the *MBF1* heat-response regulon of *Arabidopsis thaliana*. Plant J 66:844–851.

Tariq M, Ahmad S, Fahad S, Abbas G, Hussain S, Fatima Z, Nasim W, Mubeen M, ur Rehman MH, Khan MA, Adnan M (2018) The impact of climate warming and crop management on phenology of sunflower-based cropping systems in Punjab, Pakistan. Agric For Meterol 256:270–282.

Turóczy Z, Kis P, Török K, Cserháti M, Lendvai Á, Dudits D, Horváth GV (2011) Overproduction of a rice aldo–keto reductase increases oxidative and heat stress tolerance by malondialdehyde and methylglyoxal detoxification. Plant Mol Biol 75:399–412.

Uchida A, Hibino T, Shimada T, Saigusa M, Takabe T, Araki E, Kajita H, Takabe T (2008) Overexpression of *DnaK* chaperone from a halotolerant cyanobacterium *Aphanothece halophytica* increases seed yield in rice and tobacco. Plant Biotechnol 25:141–150.

Unsar NU, Muhammad R, Syed HMB, Asad S, Mirza AQ, Naeem I, Muhammad H ur R, Fahad S, Shafqat S (2020) Insect pests of cotton crop and management under climate change scenarios. In: Fahad S, Hasanuzzaman M, Alam M, Ullah H, Saeed M, Khan AK, Adnan M (eds) Environment, climate, plant and vegetation growth. Springer Publ Ltd, Springer Nature Switzerland AG. Part of Springer Nature. pp 367–396. https://doi.org/10.1007/978-3-030-49732-3.

Wahid F, Fahad S, Subhan D, Adnan M, Zhen Y, Saud S, Manzer HS, Martin B, Tereza H, Rahul D (2020) Sustainable management with mycorrhizae and phosphate solubilizing bacteria for enhanced phosphorus uptake in calcareous soils. Agriculture 10(334). https://doi.org/10.3390/agriculture10080334.

Wajid N, Ashfaq A, Asad A, Muhammad T, Muhammad A, Muhammad S, Khawar J, Ghulam MS, Syeda RS, Hafiz MH, Muhammad IAR, Muhammad ZH, Muhammad Habib ur R, Veysel T, Fahad S, Suad S, Aziz K, Shahzad A (2017) Radiation efficiency and nitrogen fertilizer impacts on sunflower crop in contrasting environments of Punjab. Pak Environ Sci Pollut Res 25:1822–1836. https://doi.org/10.1007/s11356-017-0592-z.

Wang A, Yu X, Mao Y, Liu Y, Liu G, Liu Y, Niu X (2015) Overexpression of a small heat-shock-protein gene enhances tolerance to abiotic stresses in rice. Plant Breed 134:384–393.

Wang G, Zhang X, Li F, Luo Y, Wang W (2010a) Overaccumulation of glycine betaine enhances tolerance to drought and heat stress in wheat leaves in the protection of photosynthesis. Photosynthetica 48:117–126.

Wang M, Zou Z, Li Q, Sun K, Chen X, Li X (2017) The *CsHSP17.2* molecular chaperone is essential for thermotolerance in *Camellia sinensis*. Sci Rep 7:1237.

Wang QB, Xu W, Xue QZ, Su WA (2010b) Transgenic *Brassica chinensis* plants expressing a bacterial *codA* gene exhibit enhanced tolerance to extreme temperature and high salinity. J Zhejiang Univ Sci B 11:851–861.

Wheeler T, Von Braun J (2013) Climate change impacts on global food security. Science 341:508–513.

Wu C, Kehui C, She T, Ganghua L, Shaohua W, Fahad S, Lixiao N, Jianliang H, Shaobing P, Yanfeng D (2020) Intensified pollination and fertilization ameliorate heat injury in rice (Oryza sativa L.) during the flowering stage. Field Crops Res 252:107795.

Wu C, Tang S, Li G, Wang S, Fahad S, Ding Y (2019) Roles of phytohormone changes in the grain yield of rice plants exposed to heat: a review. PeerJ 7:e7792. https://doi.org/10.7717/peerj.7792.

Wu HC, Luo DL, Vignols F, Jinn TL (2012) Heat shock-induced biphasic $Ca^{2+}$ signature and *OsCaM1-1* nuclear localization mediate downstream signalling in acquisition of thermotolerance in rice (*Oryza sativa* L.). Plant Cell Environ 35:1543–1557.

Wu X, Shiroto Y, Kishitani S, Ito Y, Toriyama K (2009) Enhanced heat and drought tolerance in transgenic rice seedlings overexpressing *OsWRKY11* under the control of *HSP101* promoter. Plant Cell Rep 28:21–30.

Xin H, Zhang H, Chen L, Li X, Lian Q, Yuan X, Hu X, Cao L, He X, Yi M (2010) Cloning and characterization of *HsfA2* from Lily (*Lilium longiflorum*). Plant Cell Rep 29:875–885.

Xin H, Zhang H, Zhong X, Lian Q, Dong A, Cao L, Yi M, Cong R (2017) Over-expression of *LlHsfA2b,* a lily heat shock transcription factor lacking trans-activation activity in yeast, can enhance tolerance to heat and oxidative stress in transgenic *Arabidopsis* seedlings. Plant Cell Tiss Organ Cult 130:617–629.

Xu Y, Tian J, Gianfagna T, Huang B (2009) Effects of *SAG12-ipt* expression on cytokinin production, growth and senescence of creeping bentgrass (*Agrostis stolonifera* L.) under heat stress. Plant Growth Regul 57:281.

Xue Y, Peng R, Xiong A, Li X, Zha D, Yao Q (2010) Over-expression of heat shock protein gene *hsp26* in *Arabidopsis thaliana* enhances heat tolerance. Biol Plant 54:105–111.

Yang X, Liang Z, Lu C (2005) Genetic engineering of the biosynthesis of glycinebetaine enhances photosynthesis against high temperature stress in transgenic tobacco plants. Plant Physiol 138:2299–2309.

Yang Z, Zhang Z, Zhang T, Fahad S, Cui K, Nie L, Peng S, Huang J (2017) The effect of season-long temperature increases on rice cultivars grown in the central and southern regions of China. Front Plant Sci 8:1908. https://doi.org/10.3389/fpls.2017.01908.

Yokotani N, Ichikawa T, Kondou Y, Matsui M, Hirochika H, Iwabuchi M, Oda K (2008) Expression of rice heat stress transcription factor *OsHsfA2e* enhances tolerance to environmental stresses in transgenic *Arabidopsis*. Planta 227:957–967.

Yoshida T, Sakuma Y, Todaka D, Maruyama K, Qin F, Mizoi J, Kidokoro S, Fujita Y, Shinozaki K, Yamaguchi-Shinozaki K (2008) Functional analysis of an *Arabidopsis* heat-shock transcription factor *HsfA3* in the transcriptional cascade downstream of the *DREB2A* stress-regulatory system. Biochem Biophys Res Commun 368:515–521.

Zafar-ul-Hye M, Muhammad N, Subhan D, Fahad S, Rahul D, Mazhar A, Ashfaq AR, Martin B, Jiˇrí H, Zahid HT, Muhammad N (2020a) Alleviation of cadmium adverse effects by improving nutrients uptake in bitter gourd through cadmium tolerant rhizobacteria. Environments 7(54). https://doi.org/10.3390/environments7080054.

Zafar-ul-Hye M, Tahzeeb-ul-Hassan M, Muhammad A, Fahad S, Martin B, Tereza D, Rahul D, Subhan D (2020b) Potential role of compost mixed biochar with rhizobacteria in mitigating lead toxicity in spinach. Sci Rep 10:12159. https://doi.org/10.1038/s41598-020-69183-9.

Zahida Z, Hafiz FB, Zulfiqar AS, Ghulam MS, Fahad S, Muhammad RA, Hafiz MH, Wajid N, Muhammad S (2017) Effect of water management and silicon on germination, growth, phosphorus and arsenic uptake in rice. Ecotoxicol Environ Saf 144:11–18.

Zahir SM, Zheng HG, Ala Ud D, Amjad A, Ata Ur R, Kashif J, Shah F, Saud S, Adnan M, Fazli W, Saud A, Manzer HS, Shamsher A, Wajid N, Hafiz MH, Fahad S (2021) Synthesis of silver nanoparticles using Plantago lanceolata extract and assessing their antibacterial and antioxidant activities. Sci Rep 11:20754.

Zhang GL, Chen LY, Xiao GY, Xiao YH, Chen XB, Zhang ST (2009) Bulked segregant analysis to setect QTL related to heat tolerance in rice (*Oryza sativa* L.) using SSR markers. Agric Sci China 8:482–487.

Zhang L, Gao Y, Pan H, Hu W, Zhang Q (2013) Cloning and characterisation of a *Primula* heat shock protein gene, *PfHSP17. 1*, which confers heat, salt and drought tolerance in transgenic *Arabidopsis thaliana*. Acta Physiol Plant 35:3191–3200.

Zhang SZ, Yang BP, Feng CL, Tang HL (2005) Genetic transformation of tobacco with the trehalose synthase gene from G*rifola frondosa* Fr. enhances the resistance to drought and salt in tobacco. J Int Pl Biol 47:579–587.

Zhang W, Zhou RG, Gao YJ, Zheng SZ, Xu P, Zhang SQ, Sun DY (2009) Molecular and genetic evidence for the key role of *AtCaM3* in heat-shock signal transduction in *Arabidopsis*. Plant Physiol 149:1773–1784.

Zhang X, Wollenweber B, Jiang D, Liu F, Zhao J (2008) Water deficits and heat shock effects on photosynthesis of a transgenic *Arabidopsis thaliana* constitutively expressing *ABP9*, a *bZIP* transcription factor. J Exp Bot 59:839–848.

Zhao D, Xia X, Su J, Wei M, Wu Y, Tao J (2019) Overexpression of herbaceous peony *HSP70* confers high temperature tolerance. BMC Genomics 20:70.

Zheng SZ, Liu YL, Li B, Shang ZL, Zhou RG, Sun DY (2012) Phosphoinositide-specific phospholipase *C9* is involved in the thermotolerance of *Arabidopsis*. Plant J 69:689–700

Zhou Y, Chen H, Chu P, Li Y, Tan B, Ding Y, Tsang EW, Jiang L, Wu K, Huang S (2012) *NnHSP17.5*, a cytosolic class II small heat shock protein gene from *Nelumbo nucifera*, contributes to seed germination vigor and seedling thermotolerance in transgenic *Arabidopsis*. Plant Cell Rep 31:379–389.

Zhu B, Ye C, Lü H, Chen X, Chai G, Chen J, Wang C (2006) Identification and characterization of a novel heat shock transcription factor gene, *GmHsfA1*, in soybeans (*Glycine max*). J Plant Res 119:247–256.

Zhu Y, Wang Z, Jing Y, Wang L, Liu X, Liu Y, Deng X (2009) Ectopic over-expression of *BhHsf1*, a heat shock factor from the resurrection plant *Boea hygrometrica*, leads to increased thermotolerance and retarded growth in transgenic *Arabidopsis* and tobacco. Plant Mol Biol 71:451.

Zia-ur-Rehman M (2020) Environment, Climate Change and Biodiversity. In: Fahad S, Hasanuzzaman M, Alam M, Ullah H, Saeed M, Khan AK, Adnan M (eds) Environment, climate, plant and vegetation growth. Springer Publ Ltd, Springer Nature Switzerland AG. Part of Springer Nature. pp 473–502. https://doi.org/10.1007/978-3-030-49732-3.

Zimmerli L, Hou BH, Tsai CH, Jakab G, Mauch-Mani B, Somerville S (2008) The xenobiotic $\beta$-aminobutyric acid enhances *Arabidopsis* thermotolerance. Plant J 53:144–156.

# 12

# *Halophytes Diversity as an Indicator to Saline Environment in Pakistan*

**Moona Nazish, Muhammad Zafar, Mushtaq Ahmad, and Shazia Sultana**
*Quaid-i-Azam University*

**CONTENTS**



## 12.1 Halophytes

Halophytes represent a heterogeneous and complex ecological plant group including varying plant species in terms of adaptive characters, taxonomic diversity, and habitat (Grigore and Toma 2010). Halophytes are the group of plants which can complete their growth and life cycle in a habitat containing high salt content. Halophytes grow in diverse environmental conditions depending on their ability to respond or tolerate various salts present in the substrate and may be categorized in several ways as in coastal, inland, and near-coastal plant species depending on the type of their occupied habitats such as desert or salt marsh (Song et al. 2008). Halophytes can reproduce in salt concentration of 200 mM NaCl or more (Flowers and Colmer 2008). Halophytes were recognized since Goethe's time (Flowers et al. 1986), but have been taken into scientific consideration through the research papers of Warming (1897; 1909) and Schimper (1903).

Halophytes have been classified in a number of ways. Waisel (1972) classified the halophytes on the basis of ecological distribution, and salt intake quantity. Popp et al. (1993) classified halophytes based on the presence and absence of salt glands. Halophytes can be classified depending on the response to salt content internally as marsh halophytes can be divided into salt accumulating and salt regulating types (Steiner 1934). Iversen (1936) divided the saline habitats on the basis of salt contents. According to him, halophytes can be oligohalophytes, mesohalophytes, and euhalophytes if the salt content is 0.01%–0.1%, 0.1%–1.0%, and 1% or more, respectively. Halophytes have been also classified on the basis of response to saline habitat by Van Eijk (1939). Chapman (1942) divided halophytes into miohalophytes (grow in low saline habitats) and euhalophytes (grow in highly saline habitats). In the last century, few general halophytes classifications were proposed by Prodan (1939), Gusuleac (1933), Iversen (1936), Van Eijk (1939), Topa (1939), Bucur et al. (1957), Patrut et al. (2005), and Grigore (2008b). Cushman (2001) differentiated between facultative, obligate, and habitat in different halophytes using ecophysiological aspects. Sabovljevic and Sabovljevic (2007) classified halophytes into obligate and facultative on the basis of sodium tolerance and demands.

According to Ashraf et al. (2006) halophytes can be of includes and excluders depending on the plant internal salt content. Morphologically, halophytes are excretive capable of releasing excess salts from

DOI: 10.1201/9781003286400-12

the plant body and succulents which store a significant amount of water in the salt bladder to reduce salt toxicity (Weber 2008). According to Youssef (2009), halophytes can be categorized into hydro and xero-halophytes on the basis of geographical distribution. Grigore and Toma (2010) classified the halophytes into three categories (extreme halophytes, mesohalophytes, and glycophytes) on the basis of integrative anatomy which correlates the plant structure with functional, phylogenetic, ecological implications. Mucina et al. (2016) classified halophytic vegetation in 19 classes excluding less saline coastal dunes. The best definition for our purpose is "the plants which can complete their life cycle in salts containing habitats" (Khan and Qaiser 2006), where the concentration of salt in a solution of soil is 5 g/l of the whole number of solids dissolved (7–8 dS/m).

## 12.2 Diversity of Halophytes at Global Perspectives

Saline environments are one of the great contributors to global biodiversity (Flowers and Muscolo 2015). In saline environments, quickly changing gradients necessitate highly specialized plants such as halophytes. Halophytes constitute a large diversity of various plant forms but constitute only 2% of terrestrial plants. About half of families of higher plants contain salt-tolerant plant species with polyphyletic origin (Aslam et al. 2011). About 2200 halophytic species were reported by Menzel and Leith (1999). Aronson (1989) listed about 1550 salt-tolerant plant species depending on their capacity toward salt tolerance and reported that the largest number of salt-tolerant plants belong to the family Chenopodiaceae, while less than 5% of plant species in Fabaceae, Poaceae, and Asteraceae are salt tolerant.

Halophytes present immense diversity in growing habitat and behavior toward salinity tolerance (Flowers et al. 2010). Menzel and Leith (2003) provided the list of 2600 halophytes. A total of 350 halophytes distributed in 20 orders belonging to 256 families have been classified by Flowers et al. (2010). It has been suggested that salt tolerance in angiosperms has a polyphyletic origin.

About 6.5 million ha of world land is salt affected in which about 2 million ha are affected moderately. The land of good quality needs to be used for food growing for the growing population. However, the potential does not lie in the reclamation of salt-affected lands by improving the fertility of the soil but the potential is in the better use of these resources for cultivating cash crop halophytes. Salinity is one of the general constraints in irrigated agriculture, affecting 1/3 of the irrigated land and more than 7 % of the total area of land (Flowers and Colmer 2008). Salt tolerance is a significant trait which is getting attention among scientists globally. The conservation and ecology of halophyte *Limonium multiflorum* have been investigated to determine its conservation status and habitat demands (Caperta et al. 2014). The halophyte ecosystem conservation value has been focused in "Red list of habitats", the EU project which classified habitats in parts such as near-threatened (Black Sea and Arctic coastal salt marsh), vulnerable (Atlantic coastal salt marsh), and endangered (Baltic coastal meadow) (Janssen et al. 2016).

Boestfleisch et al. (2014) documented that the total amount of flavonoids, phenols, ascorbate, ROS scavenging enzymes, and glutathione concentration in salt-tolerant plants of different families such as Amaranthaceae, Rhizophoraceae, Plantaginaceae, and Brassicaceae can be altered by altering the level of salinity in the environment in which they grow. Qasem (2015) conducted a study on the wild industrial and medicinal plants prospects in the saline environment in Jordan Valley and listed the highest number of halophytes belonging to Asteraceae (48 species), Poaceae (18 species), and Chenopodiaceae (18 species). A study on several salt-tolerant plants seeds as *Suaeda fruticosa*, *Bassia scoparia*, *Halogeton glomeratus*, *Salicornia bigelovii*, and *Haloxylon stocksii* recorded 70–80% of edible oil. Nedjimi et al. (2012) studied halophytes biodiversity of Chott Zehrez lake of Djelfa in Algeria and recorded 112 species and 39 genera. Salvi et al. (2017) observed halophytes diversity in the Gulf of Kutch, Gujarat, India and enlisted 27 halophytes.

## 12.3 Diversity of Halophytes in Pakistan

Soil salinity is becoming the main problem in Pakistan. In the Indus basin of Pakistan, salinity is primarily inherited in the arid environment through soil-forming processes, inefficient farm practices and irrigation systems, and usage of brackish groundwater. Increasing salinization and water scarcity with the increasing human population are the major factors threatening food security in the country. Plants with high salt tolerance are one of the probable options for the economical use of saline soil. The national development programs are in absolute need of proper institutional procedures for the economic utilization of saline lands and waters in the rural areas of Pakistan. The potential areas in Pakistan are dry mountains, deserts, wastelands, coastal areas, and waters. In arid regions of Pakistan, saline lands are abundant because the high evaporation from the surface of soil brings up most of the water via capillary rise, which results in salt accumulation on the surface of the soil. Soil and Reclamation Directorate of WAPDA conducted a survey on salinity during 2001–2003 and reported strong saline (7%), moderate saline (4%), and slight saline (10%) areas.

In Pakistan, about 400–600 plants with medicinal uses have been reported to have a usage in traditional health care practices (Ahmad et al. 2014; Shinwari and Qaisar 2011). Khan and Qaiser (2006) documented about 410 salt-tolerant plants in Pakistan and represented 19% flora of Pakistan is halophytic. Majority of the salt-tolerant plant species are found in plains of Balochistan (182), coastal areas (163), and wet mountains (149), while other salt-tolerant plants are present in mountains of arid and semi-arid regions (93), Pothwar plateau (76), and deserts (65). According to them, most of the halophytes of arid regions are perennial herbs or shrubs. About 274 of the reported total number of salt-tolerant plants have potential economical utilization. Kokab and Ahmad (2010) characterized halophytes using potential economic utilization and ecosystem in Pakistan. In the Indus plains of Pakistan, *Salvadora oleoides*, *Artemisia scoparia*, *Chenopodium ficifolium*, *Suaeda monoica*, *Kochia indica*, *Phragmites australis*, *Portulaca quadrifolia*, and *Atriplex pamirica* are commonly found halophytes. Some parts of arid mountains are under scrub forest and used for growing wheat and maize. *Atriplex griffithii*, *Bassia prostrata*, *Haloxylon salicornicum*, *Salsola tomentosa*, *Salsola tragus*, *Suaeda aegyptiaca*, *Tamarix baluchistanica*, and *Chenopodium botrys* are some important halophytes of arid mountains. *Atriplex halimus*, *Aeluropus lagopoides*, *Prosopis cineraria*, and *Tamarix alii* are common deserts halophytes in Pakistan.

The Indus delta in Sindh is the world's sixth largest contagious mangrove forest and supports the growth of mangroves. *Ziziphus nummularia*, *Rhizophora mucronata*, *Thespesia populnea*, *Acacia nilotica*, *Sporobolus ioclados*, and *Urochondra setulosa* are dominant halophytes of this area. *Dysphania ambrosioides*, *Chenopodium glaucum*, *Atriplex lasiantha*, *Triglochin palustris*, *Arundo donax*, *Suaeda heterophylla*, *Halogeton glomeratus*, and *Typha latifolia* are common and important halophytes of wet mountains. The most common halophytes of Pothwar plateau are *Arundo donax*, *Kochia indica*, *Prosopis cineraria*, *Trianthema triquetra*, and *Sporobolus tremulus*. *Atriplex canescens*, *Aeluropus littoralis*, *Haloxylon griffithii*, *Poa pratensis*, *Suaeda arcuata*, *Chloris virgata*, *Salsola imbricata*, *Kochia iranica*, *Puccinellia gigantea*, and *Tamarix karelini* are some represented halophytes of Balochistan plains.

Khan et al. (2011) reported the ethnobotanical and ethnomedicinal significance of halophytes from Noshpho salt mine located in Karak district and documented 33 halophytes belonging to 18 families. The reported dominant families were Asteraceae, Chenopodiaceae, Poaceae, and Asclepediaceae. The traditional ethnobotanical study of coastal halophytes from Hub, Balochistan was carried out by Qasim et al. (2010). They reported 48 wild halophytes belonging to 26 families. Amaranthaceae and Poaceae have been found as dominant families of halophytes.

## 12.4 Diversity of Halophytes in Punjab

The salt range of Punjab ranges over the administrative districts of Mianwali, Chakwal, Attock, Khushab, Jhelum, and Sargodha. The salt range contains two distinctive hill tracts present in the east to south-west direction. The mountains of the salt range running parallel to each other are intervened by spurs, ridges, and hills with various lakes and valleys of the various phytogeographic and agroclimatic authorities. Uchali lake and Soon valley are famous for the salt range of Punjab. Salt range is occupied by people with diverse cultures. The areas under the salt range are underdeveloped and the rural communities mainly depend on the plants of the area and traditional systems of their ancestors for their health care.

The rich diversity of salt-tolerant plants and valued natural resources are present in the salt range. *Acacia modesta*, *Olea ferruginea*, *Dodonaea viscosa*, *Salvadora oleoides*, *Prosopis juliflora*, and

*Reptonia buxifolia* characterize the subtropical dry mountainous climate. The semi-evergreen subtropical type, dry thorny tropical semi-deciduous forests, and deteriorated scrubs type of vegetation are present in the salt range. The semi-evergreen subtropical type of forests occupies the hill tract more than 750 m. *Acacia modesta, Solanum incanum, Chenopodium ficifolium, Capparis decidua, Malcolmia africana, Calotropis procera, Withania somnifera*, and *Alhagi maurorum* are the dominant halophytes of this type of forest. The thorny dry tropical semi-deciduous forest dominantly comprises of halophytes such as *Acacia nilotica, Suaeda fruticosa, Tamarix aphylla, Tamarix dioica, Oxystelma esculentum, Cyperus niveus, Cressa cretica, Cenchrus ciliaris, Bassia indica, Periploca aphylla, Peganum harmala*, and *Caroxylon aphyllum.* Degraded scrubs are the plants under grazing pressure and with restricted growth (Ahmad et al. 2010). *Acacia nilotica, Prosopis juliflora*, and *Olea ferruginea* are the distinguishing salt-tolerant trees of the salt range.

The salt-tolerant members of Chenopodiaceae dominate the arid or semi-arid environments (Kadereit et al. 2003), their autumnal fruiting and flowering habits may lead to uncertainty in identification may be due to the common sample units in early summers, which is also timely for other plants. In salt ranges of Punjab, halophytic vegetation is heterogeneous phonologically with the matrix species of Amaranthaceae. The late fruiting and flowering of the perennial halophytic species of this family are common and these are difficult to identify at an immature stage. Therefore, multiple sampling in different seasons is recommended and historical plot data in salt ranges should be reassessed critically.

Iqbal et al. (2011) carried out a study on the medicinally used plants including many salt-tolerant plants from the salt range of Pind Dadan Khan, District Jhelum, Punjab. The ethnomedicinal survey was carried out by Ahmad and Husain (2008) in Kallar Kahar salt range of Punjab. Their study also includes the medicinal properties of halophytes such as *Chenopodium album, Dysphania ambrosioides* and *Polygonum plebeium.* Shaheen et al. (2014) documented the medicinal properties of many halophytes also in their ethnobotanical study conducted in the Thal desert, Punjab. Biodiversity of medicinal plants among the rural communities of arid regions of Northern Punjab was studied by Ashfaq et al. (2019). They documented also medicinal halophytes such as *Acacia nilotica, Aerva javanica, Alhagi maurorum, Capparis decidua, Chenopodium album, Fagonia indica, Periploca aphylla, Phyla nodiflora, Prosopis juliflora, Rhazya stricta, Solanum incanum, Withania coagulans*, and *Ziziphus nummularia.*

## 12.5 Systematic Exploration of Halophytes

Halophytes adaptations are xeromorphic in nature and strongly related to their vegetative habitats. Therefore, halophytes and xerophytes have many similar adaptations. Succulence is the striking feature in halophytes. The leaves of halophytes are fleshy and thick (Grigore 2008a). Among such halophytes are *Suaeda fruticosa, Salsola imbricata, Caroxylon aphyllum, Salsola kali*, and *Batis maritima.* The mangrove plants possess mucilage cells such as *Sonneratia* sp. Lesage (1890) proved experimentally that cells in halophytes divided transversely and increase salt content decrease chlorophyll content in the plant. The palisade tissues of halophytes are larger and can be massive. The cell sap of halophytes contains large quantities of salts which make the solution concentrated (Grigore et al. 2011). Halophytes of sand habitat possess hairs on the leaf and stem, i.e., *Atriplex* sp. Salt glands are mostly distributed in salt-tolerant plants (Grigore and Toma 2010) which eliminate the excessive salts concentration from the aerial plant parts. The presence of Krantz anatomy structures and the arrangement of the chlorenchymatic tissue in leaves are considered as the distinctive adaptability characters in salt-tolerant plants (Grigore et al. 2012).

A palynological exploration is a useful application in the phylogenetic and taxonomic identification of many plants families (Khan et al. 2018). Nazish et al. (2019) investigated pollen morphological characteristics of 11 salt-tolerant plants of Amaranthaceae using Scanning electron microscopy from Northern Punjab salt range, Pakistan and also reported apolar-type pollens with spheroidal shape, pantoporate apertures, and sunken pore ornamentation as a characteristic feature of the halophytes of Amaranthaceae. Grigore and Toma (2008) observed four halophytes of Chenopodiaceae and reported some adaptive features such as secretory hairs, tracheids, and successive cambia.

Leaf anatomy of some xerophytic and halophytic sedges of Punjab has been examined by Hameed et al. (2012) to observe their adaptations to withstand the environmental stresses. These adaptations include extensive sclerification, and high parenchyma proportion. Halophytes adaptation in the salt-effected soil and plant parts morphology are directly correlated. The examples of these morphological adaptations are red-colored leaves of *Suaeda fruticosa* on high salt concentration, presence of vesiculated-type trichomes on the leaf surface in *Chenopodium* spp. which rupture and eliminate the salt back in the environment, and multicellular salt glands and salt hairs on the stem and leaf surface in *Atriplex* spp. (Hasanuzzaman et al. 2014; Cacador and Duarte 2015).

## 12.6 Halophytes as an Indicator of Saline Environment

Halophytes usually grow in water and soil with high salinity such as salt marshes, mangrove swamps, saline semi-deserts, and seashores. They tolerate high soil salinity using various adaptive methods such as tolerance, avoidance, and resistance. In saline environments, halophytes have less competition.

Halophytes are classified into terresto-halines, aero-halines, and aqua-halines based on the saline environment.

- **Terresto-halines**: Plants that grow on swamps are Hygrohalphytes. Plants that grow on non-drylands are Mesohalophytes. Plants mostly grow on drylands are Xerohalophytes.
- **Aero-halines**: Plants that grow in saline soil with 0.01%–0.1% NaCl are Oligohalophytes. Plants that grow in saline soil with 0.1%–1% concentration of NaCl are Mesohalophytes. Plants that grow in greater than 1% NaCl saline soil are Euhalophytes.
- **Aqua-halines**: If the stem of the plant remains above water level, such plants are Emerged halophytes. If the whole plant grows underwater, these plants are called Hydrohalophytes.

According to Schimper (1903), in addition to the structural characteristics of halophytes, environmental characteristics such as soil and water have also great significance in the adaptation of halophytes in the habitat. Some halophytes grow in colonies and places under periodic inundation such as *Suaeda monica* is halophyte which grows in degraded areas gregariously as compared to other halophytes. Land colonization requires adaptation to free low water energy in the atmosphere which occurred at probably freshwater pools margins about 470 million years ago (Flowers et al. 2010). The small number of halophytic flora in the world suggests salt tolerance seemed later in their evolution, and it is not a basic trait in terrestrial plant species (Flowers and Colmer 2015).

Saline habitats have been taken as comprehensible model ecosystems showing the different affinities of plant communities to soil salinity as a dominant environmental factor. The anatomical, morphological, and ecophysiological adaptations of plant species to the high level of salinity and ecological distribution between terrestrial and marine ecosystems have fascinated various ecologists and botanists (Ranwell 1972; Beeftink 1985).

Halophytes respond to salinity at the level of cellular, tissue, and whole plant. Halophytes used the mechanism to control the accumulation and inorganic ions sequestration to adjust their osmotic balance internally to salinity present externally (Flowers and Yeo 1995). Some halophytes grow on sand dunes. Their succulent leaves make them capable to thrive in semi-desert regions by storing water in adverse environmental conditions. Halophytes play a significant role in maintaining ecological stability and protect coastal habitats as they function as sand dune binders and prevent erosion. Coastal habitats were considered more stressful as compared to inland for plant species due to the greater intensity of light, high salinity of soil, and more periodic seasonal and diurnal environmental conditions (Gulzar and Khan 1998).

The 434 and 397 million hectares of area in the world is sodic and saline affected, respectively. The reasons for salinity include anthropogenic activities such as irrigation with saline water, poor drainage, and excessive irrigation using underground water results in a rise in the water table and natural such as sea salt deposition carried in rain and wind, coastal land inundation by tidal water, and parent material weathering. Saline soil inhibits the plant growth in the form of ion toxicity, osmotic stress, oxidative stress, genotoxicity, nutritional imbalance, membrane disorganization, and cell division and expansion reduction. Halophytes adopt different strategies such as salt accumulation (Sequestration of excessive salt ions into vacuoles, i.e., *Sesuvium portulacastrum*), salt excretion (salt from roots are excluded with salt hairs and salt glands from leaves), and salt exclusion (toxic salt ions are excluded selectively, i.e., *Rhizophora* spp.) (Shah et al. 2016; Nikalje et al. 2017).

Land use practices and climate changes lead to an increase in land surface salinization in various regions (Rozema and Flowers 2008). Lutts and Lefèvre (2015) demonstrated that some halophytes in substrates not only can tolerate high salinization but can also cope with heavy metals. Some mine tailings and salt marshes can have heavy metals at elevated levels. They highlighted the halophytes role in sequestration of metal ions in a few soil remediation conditions. Halophytes as compared to glycophytes are well adapted to metal-affected land due to which they are eco-friendly and can clean up the contaminated coastal environments (Hasanuzzaman et al. 2014). *Limonium* is one of the represented biodiverse genera of Mediterranean halophytic flora with more than 300 lianas, shrubs, and herbs and includes many threatened taxa and local endemics (Brullo and Erben 2016; Al Hassan et al. 2017). Cortinhas et al. (2015) analyzed morphometrically the taxonomically problematic genus, *Limonium*. They suggested that *L. narbonese* and *L. maritimum* are new and evolving plant species something common among salt-tolerant plants.

## 12.7 Saline Soil

Recent estimates and predictions about increase in ambient temperature have created apprehension among scientists (Ali et al. 2016; Fahad et al. 2017; Fahad et al. 2014a,b; Fahad et al. 2016a,b,c,d; Fahad et al. 2015a,b; Fahad et al. 2018; Fahad et al. 2019a,b). This level of rising in temperature can affect all living organisms either directly or by modifying their ambient environment and hence affecting it indirectly (Adnan et al. 2018; Akram et al. 2018a,b; Aziz et al. 2017a,b; Fahad et al. 2013; Habib et al. 2017; Hafiz et al. 2016; Hafiz et al. 2019; Kamran et al. 2017; Muhammad et al. 2019; Sajjad et al. 2019; Saud et al. 2013; Saud et al. 2014; Saud et al. 2017; Saud et al. 2016; Shah et al. 2013;Qamar et al. 2017; Wajid et al. 2017; Hussain et al. 2016; Yang et al. 2017; Zahida et al. 2017; Depeng et al. 2018). Salinity is one of the general phenomena and fundamental characteristics of arid and semi-arid regions of the world (Fahad and Bano 2012). It is a common form of land degradation and results in significant social, economic, and environmental consequences in arid and semi-arid zones (Wang et al. 2018). There is a close relationship between soil salinity and soil aridity. Rafiq (1975) classified salt-affected soil on the basis of physical, chemical, and morphological characteristics.

- Porous saline-sodic soil
- Dense saline-sodic soil
- Porous saline soil with small saline-sodic patches
- Porous saline soil with saline-sodic upper surfaces
- Saline soil comprising gypsum

Sandu (1984) defined saline soil as its fertility is affected by soluble salts in high content in the soil profile, by the exchangeable sodium in colloidal complex, and by the located mineralized water table in upper layers. Salinization intensity is evaluated by alkalization and salinity levels at different locations and depth horizons (Grigore and Toma 2017). Out of the 7 billion ha arable land in the world, 340 million ha land is salt affected. Salinized zones are increasing 10% annually due to several reasons, these reasons include native rocks weathering, low precipitation, evaporation of the upper surface, poor traditional practices, and saline water irrigation. According to an estimation, 50% arable land of the world will be more saline by 2050 (Shrivastava and Kumar 2015; Sana et al. 2015). In the last century, the processes of biological, chemical, and physical land degradation have resulted in severe consequences to natural resources globally. Salinized regions are continuing to increase over time due to the irrigation in new regions or places (Patel et al. 2011). Soil salinity results in major threats to human health and environmental resources in various countries (Yensen 2008).

The salinity of soil is taken as a constraint for the growth of plants as about 6% of the total area of world land is salt affected (Khan et al. 2011). In Pakistan, 6 million ha of soil is salt affected and 2.7 million ha of this salt-affected soil lies in Punjab. Soil salinity is extending in various continents such as Asia, Australia, America, and Africa (Rengasamy 2006). The large areas in several countries are facing salinization including Pakistan, India, Egypt, Mexico, and South America (Khan and Qaiser 2006). Pakistan ranks eighth in terms of saline-affected areas (Corbishley and Pearce 2007). The saline soil of Khyber Pakhtunkhwa, Pakistan in a laboratory has been analyzed and results showed calcareous soil with low organic matter and high exchangeable sodium percentage (ESP), electrical conductivity (EC), and pH (Niazi et al. 1990).

In the salt range and deserted places of Punjab, soils of certain places are salty due to the presence of calcium sulfate, sodium chloride, potassium chloride, sodium bicarbonate, etc. In these salty and dry places, *Chenopodium album*, *Salsola foetida*, *Suaeda fruticosa*, *Haloxylon salicorneum*, and *Tamarix aphylla* grow successfully and make micro-edaphic arrangements. The soil with high concentrations of chloride and sodium ion are lethal for plant species because salt stress causes stomatal closure, hyperosmotic shock, inhibition of growth affecting cell expansion and division, photosynthesis inhibition, and imbalance in nutrients.

## REFERENCES

Adnan M, Zahir S, Fahad S, Arif M, Mukhtar A, Imtiaz AK, Ishaq AM, Abdul B, Hidayat U, Muhammad A, Inayat-Ur R, Saud S, Muhammad ZI, Yousaf J, Amanullah Hafiz MH, Wajid N (2018) Phosphate-solubilizing bacteria nullify the antagonistic effect of soil calcification on bioavailability of phosphorus in alkaline soils. Sci Rep 8:4339. https://doi.org/10.1038/s41598-018-22653-7.

Ahmad I, Ahmad MSA, Hussain M, et al. (2010) Spatiotemporal aspects of plant community structure in open scrub rangelands of sub-mountainous Himalayan plateaus. Pak J Bot 42(5):3431–3440.

Ahmad M, Sultana S, Fazl-i-Hadi S, et al. (2014) An Ethnobotanical study of Medicinal Plants in high mountainous region of Chail valley (District Swat-Pakistan). J Ethnobiol Ethnomed 10:36.

Ahmad SS, Husain SZ (2008) Ethnomedicinal survey of plants from salt range (Kallar Kahar) of Pakistan. Pak J Bot 40(3):1005–1011.

Akram R, Turan V, Hammad HM, Ahmad S, Hussain S, Hasnain A, Maqbool MM, Rehmani MIA, Rasool A, Masood N, Mahmood F, Mubeen M, Sultana SR, Fahad S, Amanet K, Saleem M, Abbas Y, Akhtar HM, Waseem F, Murtaza R, Amin A, Zahoor SA, ul Din MS, Nasim W (2018a) Fate of organic and inorganic pollutants in paddy soils. In: Hashmi MZ, Varma A (eds) Environmental pollution of paddy soils, soil biology. Springer International Publishing AG, Switzerland, pp 197–214.

Akram R, Turan V, Wahid A, Ijaz M, Shahid MA, Kaleem S, Hafeez A, Maqbool MM, Chaudhary HJ, Munis, MFH, Mubeen M, Sadiq N, Murtaza R, Kazmi DH, Ali S, Khan N, Sultana SR, Fahad S, Amin A, Nasim W (2018b) Paddy land pollutants and their role in climate change. In: Hashmi MZ, Varma A (eds) Environmental pollution of paddy soils, soil biology. Springer International Publishing AG, Switzerland, pp 113–124.

Ali A, Imtiyaz M, Adnan M, Arshad M, Rahman IU, Jamal Y, Muhammad H, Saleem N, Rahman Z (2016) Effect of zinc activities on shoot, root biomass and phosphorus uptake in wheat gynotypes. Am Eurasian J Agric Environ Sci 16(1):204–208.

Al Hassan M, Estrelles E, Soriano P, et al (2017) Unraveling salt tolerance mechanisms in halophytes: A comparative study on four Mediterranean *Limonium* species with different geographic distribution patterns. Front Plant Sci 8:1438.

Aronson J (1989) Halophytes: Salt tolerant plants for the world- A computerized global database of halophytes with emphasis on their economic uses. University of Arizona Press, Tucson, AZ.

Ashfaq S, Ahmad M, Zafar M, et al (2019) Medicinal plant biodiversity used among the rural communities of Arid regions of Northern Punjab, Pakistan. Ind J Tradit Knowl 18(2):226–241.

Ashraf M, Hameed M, Arshad M, et al (2006) Salt potential of some forage grasses from Cholistan desert of Pakistan. In: Khan MA, Weber DG (eds) Ecophysiol high salinity tolerance plants. Springer, Netherlands, pp 31–54.

Aslam R, Bostan N, Amen N, et al (2011) A critical review on halophytes: Salt tolerant plants. J Med Plants Res 5(33):7108–7118.

Aziz K, Daniel KYT, Fazal M, Muhammad ZA, Farooq S, Fan W, Fahad S, Ruiyang Z (2017a) Nitrogen nutrition in cotton and control strategies for greenhouse gas emissions: A review. Environ Sci Pollut Res 24:23471–23487. https://doi.org/10.1007/s11356-017-0131-y.

Aziz K, Daniel KYT, Muhammad ZA, Honghai L, Shahbaz AT, Mir A, Fahad S (2017b) Nitrogen fertility and abiotic stresses management in cotton crop: A review. Environ Sci Pollut Res 24:14551–14566. https://doi.org/10.1007/s11356-017-8920-x.

Beeftink WG (1985) Population dynamics of annual *Salicornia* species in the tidal salt marshes of the Oosterschelde, the Netherlands, Vegetatio 61:127–136.

Boestfleisch C, Wagenseil NB, Buhmann AK, et al (2014) Manipulating the antioxidant capacity of halophytes to increase their cultural and economic value through saline cultivation. AoB Plants 6:46.

Brullo S, Erben M (2016) The genus *Limonium* (Plumbaginaceae) in Greece. Phytotaxa 240:1–212.

Bucur N, Dobrescu C, Turcu Gh (1957) ContribuŃii la studiul halofiliei plantelor din păsuni si fâneŃe de sărătură din Depresiunea Jijia-Bahlui (partea a I-a). Stud. si Cerc. (Biol. si St. Agr.), Acad. R.P.Române., filiala Iasi 8(2):277–317.

Cacador I, Duarte B (2015) Mechanisms of salt stress tolerance in halophytes: Biophysical and Biochemical adaptations. In Wani SH, Hossain MA (eds) Managing salt tolerance in plants: Molecular and genomic perspectives. CRC Press, Boca Raton, FL, pp 19–34.

Caperta AD, Espirito-Santo MD, Silva V, et al (2014) Habitat specificity of a threatened and endemic, cliff-dwelling halophyte. AoB Plants 6:32.

Chapman VJ (1942) The new perspective in the halophytes. Quart Rev Biol 17(4):291–311.

Corbishley J, Pearce D (2007) Growing trees on salt-effected land. ACIAR Impact Assessment Series Report No. 51, ACIAR July 2007 Clarendon Press, Oxford.

Cortinhas A, Erben M, Paes AP, et al (2015) Taxonomic complexity in the halophyte *Limonium vulgare* and related taxa (Plumbaginaceae): Insights from analysis of morphological, reproductive and karyological data. Ann Bot 115(3):369–383.

Cushman JC (2001) Osmoregulation on plants: Implications for agriculture. Am Zool 414:758–769.

Depeng W, Fahad S, Saud S, Muhammad K, Aziz K, Mohammad NK, Hafiz MH, Wajid N (2018) Morphological acclimation to agronomic manipulation in leaf dispersion and orientation to promote "Ideotype" breeding: Evidence from 3D visual modeling of "super" rice (*Oryza sativa* L.). Plant Physiol Biochem. https://doi.org/10.1016/j.plaphy.2018.11.010.

Fahad S, Adnan M, Hassan S, Saud S, Hussain S, Wu C, Wang D, Hakeem KR, Alharby HF, Turan V, Khan MA, Huang J (2019b) Rice responses and tolerance to high temperature. In: Hasanuzzaman M, Fujita M, Nahar K, Biswas JK (eds) Advances in rice research for abiotic stress tolerance. Woodhead Publ Ltd, Abington Hall Abington, Cambridge, UK, pp 201–224.

Fahad S, Bajwa AA, Nazir U, Anjum SA, Farooq A, Zohaib A, Sadia S, Nasim W, Adkins S, Saud S, Ihsan MZ, Alharby H, Wu C,Wang D, Huang J (2017) Crop production under drought and heat stress: Plant responses and management options. Front Plant Sci 8:1147. https://doi.org/10.3389/fpls.2017.01147.

Fahad S, Bano A (2012) Effect of salicylic acid on physiological and biochemical characterization of maize grown in saline area. Pak J Bot 44:1433–1438.

Fahad S, Chen Y, Saud S, Wang K, Xiong D, Chen C, Wu C, Shah F, Nie L, Huang J (2013) Ultraviolet radiation effect on photosynthetic pigments, biochemical attributes, antioxidant enzyme activity and hormonal contents of wheat. J Food Agric Environ 11(3&4):1635–1641.

Fahad S, Hussain S, Bano A, Saud S, Hassan S, Shan D, Khan FA, Khan F, Chen Y, Wu C, Tabassum MA, Chun MX, Afzal M, Jan A, Jan MT, Huang J (2014a) Potential role of phytohormones and plant growth-promoting rhizobacteria in abiotic stresses: Consequences for changing environment. Environ Sci Pollut Res 22(7):4907–4921. https://doi.org/10.1007/s11356-014-3754-2.

Fahad S, Hussain S, Matloob A, Khan FA, Khaliq A, Saud S, Hassan S, Shan D, Khan F, Ullah N, Faiq M, Khan MR, Tareen AK, Khan A, Ullah A, Ullah N, Huang J (2014b) Phytohormones and plant responses to salinity stress: A review. Plant Growth Regul 75(2):391–404. https://doi.org/10.1007/s10725-014-0013-y.

Fahad S, Hussain S, Saud S, Hassan S, Chauhan BS, Khan F, et al (2016a) Responses of rapid viscoanalyzer profile and other rice grain qualities to exogenously applied plant growth regulators under high day and high night temperatures. PLoS One 11(7):e0159590. https://doi.org/10.1371/journal.pone.0159590.

Fahad S, Hussain S, Saud S, Hassan S, Ihsan Z, Shah AN, Wu C, Yousaf M, Nasim W, Alharby H, Alghabari F, Huang J (2016c) Exogenously applied plant growth regulators enhance the morphophysiological growth and yield of rice under high temperature. Front Plant Sci 7:1250. https://doi.org/10.3389/fpls. 2016.01250.

Fahad S, Hussain S, Saud S, Hassan S, Tanveer M, Ihsan MZ, Shah AN, Ullah A, Nasrullah KF, Ullah S, Alharby H, Wu C, Huang J (2016d) A combined application of biochar and phosphorus alleviates heat-induced adversities on physiological, agronomical and quality attributes of rice. Plant Physiol Biochem 103:191–198.

Fahad S, Hussain S, Saud S, Khan F, Hassan S, Nasim W, Arif M, Wang F, Huang J (2016b) Exogenously applied plant growth regulators affect heat-stressed rice pollens. J Agron Crop Sci 202:139–150.

Fahad S, Hussain S, Saud S, Tanveer M, Bajwa AA, Hassan S, Shah AN, Ullah A,Wu C, Khan FA, Shah F, Ullah S, Chen Y, Huang J (2015a) A biochar application protects rice pollen from high-temperature stress. Plant Physiol Biochem 96:281–287.

Fahad S, Muhammad ZI, Abdul K, Ihsanullah D, Saud S, Saleh A, Wajid N, Muhammad A, Imtiaz AK, Chao W, Depeng W, Jianliang H (2018) Consequences of high temperature under changing climate optima for rice pollen characteristics-concepts and perspectives, Archives Agron Soil Sci. https://doi.org/10.1080/03650340.2018.1443213.

Fahad S, Nie L, Chen Y, Wu C, Xiong D, Saud S, Hongyan L, Cui K, Huang J (2015b) Crop plant hormones and environmental stress. Sustain Agric Rev 15:371–400.

Fahad S, Rehman A, Shahzad B, Tanveer M, Saud S, Kamran M, Ihtisham M., Khan SU, Turan V, Rahman MHU (2019a) Rice responses and tolerance to metal/metalloid toxicity. In: Hasanuzzaman M, Fujita M, Nahar K, Biswas JK (eds) Advances in rice research for abiotic stress tolerance. Woodhead Publ Ltd, Abington Hall Abington, Cambridge, UK, pp 299–312.

Flowers TJ, Colmer TD (2008) Salinity tolerance in halophytes. New Phytologist 179:945–963.

Flowers TJ, Colmer TD (2015) Plant salt tolerance: Adaptations in halophytes. Ann Bot 115(3):327–331.

Flowers TJ, Galal HK, Bromham L (2010) Evolution of halophytes: Multiple origins of salt tolerance. Funct Plant Biol 37:604–612.

Flowers TJ, Hajibagheri MA, Clipson NJW (1986) Halophytes. Q Rev Biol 6:313–337.

Flowers TJ, Muscolo A (2015) Introduction to the special issue: Halophytes in a changing world. AOB Plants 7:plv020.

Flowers TJ, Yeo AR (1995) Breeding for salinity resistance in crop plants: Where next. Aust J Plant Physiol 22:875–884.

Grigore M-N (2008a) Introducere ın Halofitologie. Elemente de Anatomie Integrativaˇ. Ed. PIM, Iasi.

Grigore MN (2008b) Halofitotaxonomia. Lista plantelor de sărătură din România. Edit. Pim Iasi.

Grigore MN, Boscaiu M, Vicente O (2011) Assessment of the relevance of osmolyte biosynthesis for salt tolerance of halophytes under natural conditions. Eur J Plant Sci Biotechnol 5(2):12–19.

Grigore MN, Toma C (2008) Ecological anatomy of halophyte species from the Chenopodiaceae family. Advanced topics on mathematical biology and ecology. In: Proceedings of the 4th WSEAS International Conference on Mathematical Biology and Ecology – MABE '08, Acapulco, Mexico, January 25–27, 2008, pp 62–67.

Grigore MN, Toma C (2010) Halofitele. Aspecte de anatomie ecologica Edit Univ "Al. I. Cuza" Iasi.

Grigore MN, Toma C (2017) Definition and classification of halophytes. In: Grigore M-N, Toma C (eds) Anatomical adaptation of halophytes. Switzerland, pp 3–28.

Grigore MN, Toma C, Zamfirache MM, et al (2012) Ecological anatomy in halophytes with C4 photosynthesis: Discussing adaptative features in endangered ecosystems. Carp J Earth Environ Sci 7(2):13–21.

Gulzar S, Khan MA (1998) Diurnal water relations of inland and coastal halophytic populations from Pakistan. J Arid Environ 40(3):295–305.

Gusuleac M (1933) Urme de vegetaŃie halofilă în Bucovina. Bul Fac St CernăuŃi 7:329–339.

Habib ur R, Ashfaq A, Aftab W, Manzoor H, Fahd R, Wajid I, Md. Aminul I, Vakhtang S, Muhammad A, Asmat U, Abdul W, Syeda RS, Shah S, Shahbaz K, Fahad S, Manzoor H, Saddam H, Wajid N (2017) Application of CSM-CROPGRO-Cotton model for cultivars and optimum planting dates: Evaluation in changing semi-arid climate. Field Crops Res. http://dx.doi.org/10.1016/j.fcr.2017.07.007.

Hafiz MH, Muhammad A, Farhat A, Hafiz FB, Saeed AQ, Muhammad M, Fahad S, Muhammad A (2019) Environmental factors affecting the frequency of road traffic accidents: A case study of sub-urban area of Pakistan. Environ Sci Pollut Res. https://doi.org/10. 1007/s11356-019-04752-8.

Hafiz MH, Wajid F, Farhat A, Fahad S, Shafqat S, Wajid N, Hafiz FB (2016) Maize plant nitrogen uptake dynamics at limited irrigation water and nitrogen. Environ Sci Pollut Res 24(3):2549–2557. https://doi.org/10. 1007/s11356-016-8031-0.

Hameed M, Nawaz T, Asharf M, et al (2012) Leaf anatomical adaptations of some halophytic and xerophytic sedges of the Punjab. Pak J Bot 44:159–164.

Hasanuzzaman M, Nahar K, Alam M, et al (2014) Potential use of halophytes to remediate saline soils. BioMed Res Int 12.

Hussain H, Raziq F, Khan I, Shah B, Altaf M, Attaullah, Ullah W, Naeem A, Adnan M, Junaid K, Shah SRA, Iqbal M (2016) Effect of *Bipolarismaydis*(Y. Nisik& C. Miyake) shoemaker at various growth stages of different maize cultivars. J Ento Zool Stud 4(2):439–444.

Iqbal H, Sher Z, Khan ZU (2011) Medicinal plants from salt range Pind Dadan Khan, district Jhelum, Punjab, Pakistan. J Med Plants Res 5(11):2157–2168.

Iversen J (1936) Biologische Pflanzentypen als Hilfsmittel in der Vegetations for schung. Dissertation. Medd. fra Skalling Laboratoriet, Copenhagen.

Janssen JAM, Rodwell JS, Garcia Criado M, et al (2016) European Red list of habitats. Part 2: Terrestrial and freshwater habitats. European Commission, Brussels, BE.

Kadereit GT, Borsch T, Weising K (2003) Phylogeny of Amaranthaceae and Chenopodiaceae and the evolution of C4 photosynthesis. Int J Plant Sci 164:959–986.

Kamran M, Wenwen C, Irshad A, Xiangping M, Xudong Z, Wennan S, Junzhi C, Shakeel A, Fahad S, Qingfang H, Tiening L (2017) Effect of paclobutrazol, a potential growth regulator on stalk mechanical strength, lignin accumulation and its relation with lodging resistance of maize. Plant Growth Regul 84:317–332. https://doi.org/10.1007/s10725-017-0342-8.

Khan M, Musharaf S, Shinwari ZK (2011) Ethnobotanical importance of halophytes of Noshpho salt mine, District Karak, Pakistan. Res Pharm Biotech 3:46–52.

Khan MA, Qaiser M (2006) Halophytes of Pakistan: Characteristics, distribution and potential economic usages. Sabkha Ecosyst 129–153.

Khan R, Abidin ZU, Ahmad M, et al (2018) Palyno-morphological characteristics of gymnosperm flora of Pakistan and its taxonomic implications with LM and SEM methods. Microsc Res Tech 81(1):74–87.

Kokab S, Ahmad S (2010) Characterizing salt tolerant plants using ecosystem and economic utilization potentials for Pakistan. Manag Nat Resour Sustain Future Agric 2(12):1–20.

Lesage PM (1890) Recherches experimentales sur les modifications des feuilles chez les plantes maritimes. Rév Gén Bot 2:55–65.

Lutts S, Lefèvre I (2015) How can we take advantage of halophyte properties to cope with heavy metal toxicity in salt-affected areas? Ann Bot 115(3):509–528.

Menzel U, Leith H (1999) Halophyte uses in different climates I: Ecological and Ecophysiological studies. In: Leith H, Moschenko M, Lohman M, et al (eds) Progress in biometeriology. Backhuys Publishers, The Netherlands, pp 77–88.

Menzel U, Leith H (2003) Halophyte Database V. 2.0 update. In: Leith II, Mochtehenko M (eds) Cash crop halophytes. Kluwer, Dordrecht.

Mucina L, Bultmann H, Dierben K, et al (2016) Vegetation of Europe: Hierarchical floristic classification system of vascular plant, bryophyte, lichen, and algal communities. App Veg Sci 19:3–264.

Muhammad Z, Abdul MK, Abdul MS, Kenneth BM, Muhammad S, Shahen S, Ibadullah J, Fahad S (2019) Performance of Aeluropus lagopoides (mangrove grass) ecotypes, a potential turfgrass, under high saline conditions. Environ Sci Pollut Res. https://doi.org/10.1007/s11356-019-04838-3.

Nazish M, Zafar M, Ahmad M, et al (2019) Palyno-morphological investigations of halophytic taxa of Amaranthaceae through SEM from salt range of Northern Punjab, Pakistan. Microsc Res Tech 2018(1):1–13.

Nedjimi B, Beladel B, Guit B (2012) Biodiversity of halophytic vegetation in Chott Zehrez lake of Djelfa (Algeria). Am J Plant Sci 3:1527–1534.

Niazi MHK, Hussain N, Dogar MS, et al (1990) A review of soil salinity research in Pakistan and future horizons. In: Conference: Proceedings of the INDO-PAK Workshop on Soil Salinity and water Management, IWASRI, UNDP and PARC Islamabad, Pakistan 1, pp 119–134.

Nikalje GC, Nikam TD, Suprasanna P (2017) Looking at halophytic adaptation to high salinity through genomics landscape. Curr Genom 18(6):542–552.

Patel BB, Patel BB, Dave RS (2011) Studies on infiltration of saline–alkali soils of several parts of Mehsana and Patan districts of north Gujarat. J Appl Technol Environ Sanitation 1(1):87–92.

Patrut DI, Adelina P, Ioan C (2005) Biodiversitatea halofitelor din Campia Banatului. Edit, Eurobit, Timişoara.

Popp M, Polania J, Weiper M (1993) Physiological adaptations to different salinity levels in mangrove. In: Towards the rational use of high salinity tolerant plants. Kluwer, Dordrecht, 1, pp 217–224.

Prodan I (1939) Flora pentru detrminarea şi descrierea plantelor ce cresc în România. II. (ediţia aII-a). Edit. Cartea Românească, Cluj-Napoca 253–305.

Qamar uz Z, Zubair A, Muhammad Y, Muhammad ZI, Abdul K, Fahad S, Safder B, Ramzani PMA, Muhammad N (2017) Zinc biofortification in rice: Leveraging agriculture to moderate hidden hunger in developing countries. Arch Agron Soil Sci 64:147–161. https://doi.org/10.1080/03650340.2017.1338343.

Qasem JR (2015) Prospects of wild medicinal and industrial plants of saline habitats in the Jordan Valley. Pak J Bot 47(2):551–570.

Qasim M, Gulzar S, Shinwari ZK, et al (2010) Traditional ethnobotanical uses of halophytes from Hub, Balochistan. Pak J Bot 42(3):1543–1551.

Rafiq M (1975) Saline, saline-alkali and water logged soil of Indus plain, their characteristics, causes of formation and measures needed for reclamation. In: Proceedings of International Conference in Waterlogging and Salinity, Oct 13–17, Univ Engg Tech, Lahore.

Ranwell DS (1972) Ecology of salt marshes and sand dunes. Chapman and Hall, London.

Rengasamy P (2006) World salinization with emphasis on Australia. J Exp Bot 57(5):1017–1023.

Rozema J, Flowers T (2008) Crops for a salinized world. Science 322:1478–1480.

Sabovljevic M, Sabovljevic A (2007) Contribution to the coastal bryophytes to the Northern Mediterranean: Are these halophytes among bryophytes? Phytol Balcanica 13:131–135.

Sajjad H, Muhammad M, Ashfaq A, Waseem A, Hafiz MH, Mazhar A, Nasir M, Asad A, Hafiz UF, Syeda RS, Fahad S, Depeng W, Wajid N (2019) Using GIS tools to detect the land use/land cover changes during forty years in Lodhran district of Pakistan. Environ Sci Pollut Res. https://doi.org/10.1007/s11356-019-06072-3.

Salvi H, Das L, Brahmbhatt B, et al (2017) Diversity of halophytes in Gulf of Kachchh, Gujarat. Int J Life Sci Scienti Res 3(3):995–1002.

Sana A, Ahmed N, Khan IA, Shah B, Khan A, Adnan M, Junaid K, Rasheed MT, Zaki AB, Huma Z, Ahmed S (2015) Evaluating larvicidal action of *Coriandrumsativum* (Dhania) and Mentha (mint) plant extracts against *Aedesaegyptiand Aedesalbopictuslarvae* under laboratory conditions. J Ento Zool Stud 3(6):156–159.

Sandu G (1984) Solurile saline si alcalice din R.S.R. Ameliorarea lor. Ed. Ceres, Bucuresti.

Saud S, Chen Y, Fahad S, Hussain S, Na L, Xin L, Alhussien SA (2016) Silicate application increases the photosynthesis and its associated metabolic activities in Kentucky bluegrass under drought stress and post-drought recovery. Environ Sci Pollut Res 23(17):17647–17655. https://doi.org/10.1007/s11356-016-6957-x.

Saud S, Chen Y, Long B, Fahad S, Sadiq A (2013) The different impact on the growth of cool season turf grass under the various conditions on salinity and drought stress. Int J Agric Sci Res 3:77–84.

Saud S, Fahad S, Yajun C, Ihsan MZ, Hammad HM, Nasim W, Amanullah Jr, Arif M, Alharby H (2017) Effects of nitrogen supply on water stress and recovery mechanisms in Kentucky bluegrass plants. Front Plant Sci 8:983. https://doi.org/10.3389/fpls.2017.00983.

Saud S, Li X, Chen Y, Zhang L, Fahad S, Hussain S, Sadiq A, Chen Y (2014) Silicon application increases drought tolerance of Kentucky bluegrass by improving plant water relations and morph physiological functions. SciWorld J 2014:1–10. https://doi.org/10.1155/2014/368694.

Schimper AFW (1903) Plant geography upon a physiological basis. Clarendon Press, Oxford.

Shah B, Khan IA, Khan A, Din MMU, Adnan M, Junaid K, Shah SRA, Zaman M, Ahmad N, Akbar R, Fayaz W, Rahman IU (2016) Determination of physio-morphic basis of resistance in different maize cultivars against insect pests. J Ento Zool Stud 4(1):317–321.

Shah F, Lixiao N, Kehui C, Tariq S, Wei W, Chang C, Liyang Z, Farhan A, Fahad S, Huang J (2013) Rice grain yield and component responses to near 2°C of warming. Field Crop Res 157:98–110.

Shaheen H, Qureshi R, Akram A, Gulfraz M (2014) Inventory of medicinal flora from Thal desert, Punjab, Pakistan. Afr J Tradit Complement Med 11(3):282–290.

Shinwari ZK, Qaisar M (2011) Efforts on conservation and sustainable use of medicinal plants of Pakistan. Pak J Bot 43:5–10.

Shrivastava P, Kumar R (2015) Oil salinity: A serious environmental issue and plant growth promoting bacteria as one of the tools for its alleviation. Saudi J Biol Sci 22(2):123–131.

Song J, Fana H, Zhaoa Y, et al (2008) Effect of salinity on germination, seedling emergence, seedling growth and ion accumulation of a euhalophyte *Suaeda salsa* in an intertidal zone and on saline inland. Aqu Bot 88:331–337.

Steiner M (1934) To the ecology of salt marsh of the Nordostilchen united countries of Nordamerika. Jahrb Know Offered 81:94.

Topa E (1939) VegetaŃia halofitelor din nordul României în legătură cu cea din restul Ńării. Bul Fac St Cernăuti 13:1–78.

Van Eijk M (1939) Analyse der Wirkung des NaCl auf die Entwicklung Sukkulenze und Transpiration bei *Salicornia herbacea*, sowie Untersuchungen über den Einfluss der Salzaufnahme auf die Wurzelatmung bei *Aster tripolium*. Rec Trav Bot Neerl 36:559–657.

Waisel Y (1972) Biology of halophytes. Academic press, New York and London.

Wajid N, Ashfaq A, Asad A, Muhammad T, Muhammad A,Muhammad S, Khawar J, Ghulam MS, Syeda RS, Hafiz MH, Muhammad IAR, Muhammad ZH, Muhammad Habib ur R, Veysel T, Fahad S, Suad S, Aziz K, Shahzad A (2017) Radiation efficiency and nitrogen fertilizer impacts on sunflower crop in contrasting environments of Punjab. Pak Environ Sci Pollut Res 25:1822–1836. https://doi.org/10.1007/s11356-017-0592-z.

Wang J, Ding J, Abulimiti A, et al (2018) Quantitative estimation of soil salinity by means of different modeling methods and visible-near infrared (VIS_NIR) spectroscopy, Ebinur Lake Wetland, Northwest China. Peer J:1–24.

Warming E (1897) Halophyt-studier. D. Kgl. Danske Vidensk. Selsk. Skr., 6, Raekke, naturvidenskabeling og mathematisk Afd. VIII 4:173–272.

Warming E (1909) Oecology of Plants. An introduction to the study of plant communities. Clarendon Press, Oxford.

Weber DJ (2008) Adaptive mechanisms of halophytes in desert regions. In: Ashraf M, Ozturk M, Athar HR (eds) Salinity and water stress. Springer Netherlands, Neterhlands, 44, pp 179–185.

Yang Z, Zhang Z, Zhang T, Fahad S, Cui K, Nie L, Peng S, Huang J (2017) The effect of season-long temperature increases on rice cultivars grown in the central and southern regions of China. Front Plant Sci 8:1908. https://doi.org/10.3389/fpls.2017.01908.

Yensen NP (2008) Halophyte uses for the twenty-first century. In: Khan MA, Weber DJ (eds) Ecophysiology of high salinity tolerant plants. Springer, Dordrecht, pp 367–396.

Youssef AM (2009) Salt tolerance mechanisms in some halophytes from Saudi Arabia and Egypt. Res J Agric Biol Sci 5:191–206.

Zahida Z, Hafiz FB, Zulfiqar AS, Ghulam MS, Fahad S, Muhammad RA, Hafiz MH,Wajid N, Muhammad S (2017) Effect of water management and silicon on germination, growth, phosphorus and arsenic uptake in rice. Ecotoxicol Environ Saf 144:11–18.

# 13

# *Bacterial Contamination in Food Chain: Sources, Impact and Control*

**Muhammad Shahid, Mohsin Khurshid, Muhammad Aamer Mehmood, Sana Malik, Irfan Manzoor, and Sher Muhammad**
*Government College University Faisalabad*

**CONTENTS**



## 13.1 Introduction

### 13.1.1 Food Contamination: A Major Future Challenge

Food products are under continuous threat of microbial, chemical and physical contamination. Bacteria and other parasites are the main contaminants of biological food contamination. Contaminants can enter food chain through various routes, thus making food products unfit for use. Food contamination has been recognized as a huge global risk worldwide by the World Health Organization (Keiji Fukuda 2015; WHO 2015). It is obvious that food contamination spreads around and may affect the people living in other places. People commonly experience the food-borne and water-borne diseases throughout the world at different stages of their lives. Mortality rate due to the consumption of contaminated food is high. Food may be contaminated by the presence of antibiotics, fresh food contamination and intentional food contamination, but contamination of fresh produce is a hot issue (Scallan et al. 2011). The bacterial contaminations of food include an epidemic due to *Escherichia coli*-contaminated cucumber in Germany and other European Union countries in 2011. Serious complications have been reported in the United Kingdom in 2012 due to the *Cryptosporidium*-infected bagged salads. In 2016, *Listeria monocytogenes*-contaminated pre-packed salad products resulted in serious complications among the users, whereas an outbreak has been reported in Australia in 2016 due to *Salmonella*-contaminated

DOI: 10.1201/9781003286400-13

lettuce in canned salads. Another serious future threat is the emergence of antibiotic-resistant bacteria in food products that constitute a serious threat in the future (Pruden et al. 2006). Many reports are available online now to describe the role of antibiotic resistance bacteria in food spoilage. Despite the clinical arena, the food products treated with antibiotics made the situation much worse. Food products are being continuously contaminated with antibiotics which is responsible for antibiotic resistance development in bacteria. Sometimes, a criminal activity of deliberate addition of a poisonous material in food product is also responsible for food contamination and this is called intentional contamination. Hence, fraudulent food contamination is a serious challenge for food safety stakeholders. Many global and regional issues related to food contamination need to be addressed through scientific research and technological advancement. Therefore, the factors such as types of food contaminants, their sources, associated hazards for the consumers and measures to prevent/eliminate contamination should be in view before designing food safety plans.

## 13.2 Biofilm Formation in Food Processes

Microorganisms have the ability to grow on wet places of food processing units into micro-colonies forming biofilms. The presence of undesirable biofilms in food processing industries is very common and may lead to disease dissemination, food spoilage and cleaning issues (Cappitelli et al. 2014). The process of formation of biofilms is most common among bacteria due to their ubiquitous nature and engagement in multistep processes required for biofilm formation (Van Houdt et al. 2010). The following factors affect the formation of bacterial biofilms.

### 13.2.1 Bacterial Cell Surface

Bacterial cells depend upon many physical and chemical processes for the attachment and formation of biofilms. They adhere to the surface with the help of numerous bonding forces like Van der Waal forces, electrostatic and steric forces. Thus, the bacterial surface possessing the ability to develop above mentioned attractive forces is much more likely to contaminate the food chain.

### 13.2.2 Flagella

Many bacteria carry peritrichous or polar flagella and are motile. The colonization process is depending on the bacterial motility and flagella helped the bacteria to move toward and away from the stimulus. The flagellar motility helped in developing preliminary cell to cell adhesion and biofilm formation in several bacterial species like *L. monocytogenes*, *E. coli* (Pratt and Kolter 1998) and *Yersinia enterocolitica* (Kim et al. 2008). It has been found that there is a decrease in cell attachment and biofilm formation in the absence of flagella. In the case of *Bacillus cereus*, it is concluded that flagellar motility is necessary for biofilm formation to glass surfaces (Houry et al. 2010).

### 13.2.3 Surface Appendages

Fimbriae and other fine thread-like appendages come under the definition of surface appendages. These structures, protruding from cell surface, perform a key role in the attachment of bacteria to various surfaces (McClelland et al. 2001). Type 1 fimbriae are found in the members of *Enterobacteriaceae* (Vizcarra et al. 2016), in bacterial strains that produce Shiga toxins (Cookson et al. 2002) and *S. enterica* serovar *Enteritidis* (Austin et al. 1998). Type 1 and some other types of fimbriae are specifically involved in biofilm formation, whereas type 3 and type 4 fimbriae promote adherence to the polymers (Di Martino et al. 2003) and biofilm formation at a rapid pace, respectively (Bechet et al. 2003; Shah et al. 2015). Type 4 fimbriae are commonly present in water- and food-borne pathogens called *Aeromonas caviae* (Neyts et al. 2000) and *Pseudomonas aeruginosa* (Giltner et al. 2006). Another structure resembles with fimbriae called pilli which is principally involved in the conjugation process, thus helping in horizontal gene transfer.

### 13.2.4 Outer Layer Polysaccharides

The lipopolysaccharides, forming the outer surface of Gram-negative bacteria, play a vital role in biofilm formation (Barak et al. 2007). Changes in peptidoglycan structure in many Gram-positive bacteria can also facilitate bacterial attachment to various surfaces and ultimately on biofilm formation. Many bacteria have the ability to produce extracellular polysaccharides layer which is considered to be very important for cell attachment and biofilm formation.

## 13.3 Bacteria Involved in Food Contamination

Food contamination in general refers to the foods either contains contain microorganisms such as bacteria or parasites, or the presence of toxic substances, therefore the food becomes inappropriate for consumption. The food contaminant may be physical, chemical or biological in nature, although the biological contaminations especially with bacteria are more common. The bacterial contaminants follow several routes all the way through the food supply chain to enter. The most commonly involved bacterial species in food contamination include *Salmonella* species, *Campylobacter jejuni*, *Bacillus cereus*, *Clostridium botulinum*, *C. perfrigens*, *L. monocytogenes*, pathogenic *E. coli* strains, *Staphylococcus aureus*, *Shigella* spp., *Vibrio parahaemolyticus*, *V. cholera*, *V. vulnificus* and *Y. enterocolitica*. However, heavy metals and pesticides are the chemicals that contaminate the food chain frequently (Fukuda et al. 2015; Hussain, 2016; Karch et al. 2012; Kirezieva et al. 2013; McKerr et al. 2015; Poimenidou et al. 2016) (Figures 13.1 and 13.2).

Emergence of antimicrobial-resistant bacteria is now recognized as an impending threat to public health. The studies have widely described the prevalence of antimicrobial-resistant bacteria in food products. Therefore, the World Health Organization has suggested a global policy to address the incipient threat (Pruden et al. 2006). The clinical field was considered a major contributor; however, the current situation is more complicated due to the misuse of antibiotics in food and agriculture production. Consequently, the foods contaminated with antibiotic-resistant bacteria will be a foremost food safety issue in near future (Lammie and Hughes 2016; Manning and Soon 2016).

Food-borne diseases are broadly classified into two major groups known as intoxication and infection. The presence of preformed toxins in the food produced by certain pathogens results in intoxication, while consumption of food containing viable pathogenic microorganisms leads to a variety of infections. The food may contain the toxin even if the producer microorganism is not present or eating animals that have ingested toxin-producing microbes may cause intoxication. The onset of sign and symptoms is rapid

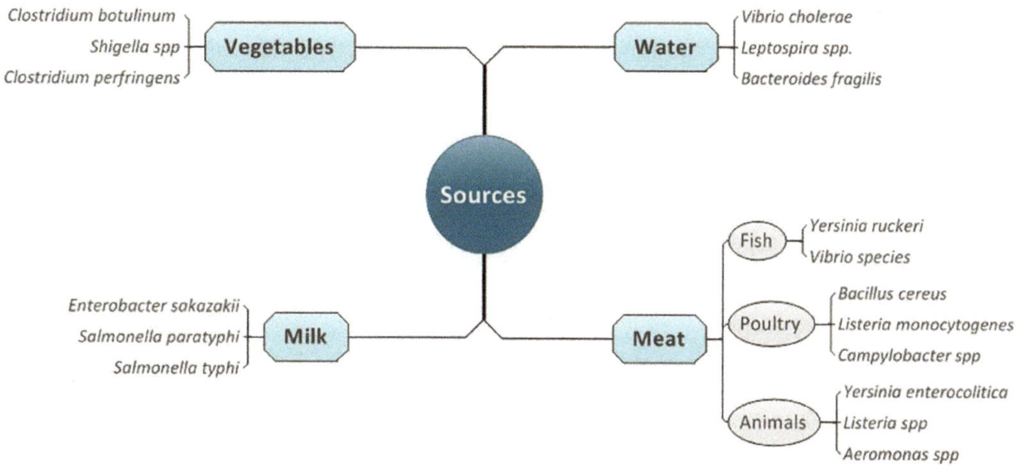


**FIGURE 13.1** Nature of bacterial contamination with respect to the sources.

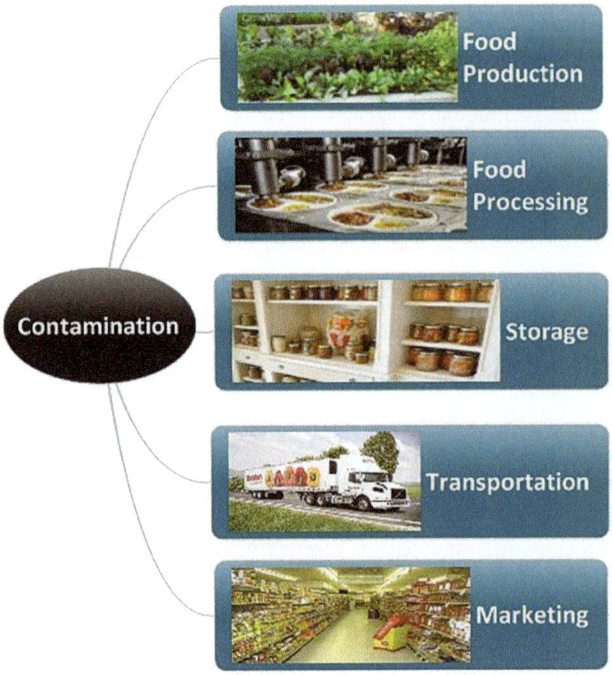


**FIGURE 13.2** Possible routes of bacterial contamination in the food chain.

in case of food intoxication and the person needs to be hospitalized (Addis and Sisay 2015). About 66% of food-associated problems are due to the harmful bacteria. The bacteria present in food products grow when the conditions are favorable and produce toxins that are absorbed following ingestion through the gastrointestinal epithelium and cause cellular damage. The bacterially produced toxins are transported through blood and neurons within the body, and affect the functioning of various tissues, organs or systems like the kidney, liver, central or peripheral nervous system. Diarrhea, abdominal cramps, vomiting, nausea and headache are the common symptoms of food-borne illness. The diagnosis is usually made based on a history of food intake and the clinical symptoms. The prevention of food-borne disorders merely depends upon the extent of food safety measures from the food production to the processing and distribution of food. Furthermore, the hygiene of the cooking, storage and distribution area, separation of raw and cooked foods, proper cooking and placement temperature of food and usage of safe water and raw materials are some of the essential steps to maintain the food safety (Grass et al. 2013; Hussain 2016; Switaj et al. 2015). Therefore, the implementation of strict control measures throughout the food chain is suggested as a prerequisite to developing the hygienic environment during the manufacturing, storage and marketing of foods (Gil et al. 2015).

### 13.3.1 *Salmonella*

*Salmonella* is a Gram-negative, facultative anaerobe bacterium that doesn't form spores. Most of the Salmonella strains grow best at a temperature between 35°C and 37°C; therefore, they are easily killed during the process of pasteurization (Ray and Bhunia 2007). *Salmonella* species has been recognized as a major cause of contamination of food-related items in the European Union (EFSA 2010). The pathogenic species has been isolated and identified from various sources including poultry, sheep, cows, pigs and birds (Control and Prevention 2010).

Salmonella infections known as salmonellosis is an important food-borne disease and considered a foremost menace to public health both in developing and developed countries (Carrasco et al. 2012). The subtleties of Salmonella infection are variable among the humans affected mainly by lifestyle, industry, technology and travel. Several serotypes are widely spread in nature and are commonly existed in the gut of almost all wild and domestic animals resulting as a source of a variety of infections (Carrasco et al. 2012; Foley et al. 2008; Silva et al. 2014). The *Salmonella* species represents the second most common zoonotic disease reported to cause human diseases following *Campylobacter* infection and a serious food-borne disease risk worldwide associated with 131,468 confirmed cases of human infections in the European Union in 2008. Although the reported cases of human salmonellosis in 2008 showed a decrease (13.5%) from 2007, still various European countries show an increasing trend (EFSA 2010).

Various reports highlighted the occurrence of infectious *Salmonella* spp. in meat and/or meat products (Williams et al. 2014). Generally, the poultry, eggs and meat are recognized as a constant source of Salmonella serotypes and are often associated with infectious diseases. In addition to meat and meat products, another reservoir of *Salmonella* infections includes a variety of foods like fruits, vegetables, milk, dairy and fishery products (Mani-Lopez et al. 2012; Tafida et al. 2013; Todd 1997). Various reports confirmed poultry meat contamination with Salmonella in most of the nations, such as Australia, the United Kingdom, Spain, Greece, Italy and Malaysia. The prevalence rates in these countries were significantly higher; however, the isolated serotypes showed disparities within countries with predominance of *S. Typhimurium*, *S. Enteritidis*, *S. Newport*, *S. Hadar*, *S. Heidelberg* and *S. Virchow*. These studies nominate the poultry meat as the major reservoir of *Salmonella* pathogen, and thus the baked poultry products are at the higher risk of re-contamination with this pathogen (Reij et al. 2004).

The *Salmonella* contamination in eggs is another important health concern. The recent consumer preferences have influenced the egg production industry, with a preference for egg production methods in cage-free environment. Furthermore, the increased desire for raw and untreated foods possibly increases the risk of salmonella infections among customers. The poultry products like eggs are more likely to be contaminated with *S. enteritidis* as this is the major pathogen for such products (Gast et al.2014). The *S. enteritidis* have remarkable capability to survive within the harsh environment of egg white. Lysozyme and ovotransferrin are the egg proteins that play an essential role in defense against bacterial attacks. A number of other proteins and peptides have been recognized for their protective potential against bacterial invaders. Although these proteins are extensively studied, their antibacterial abilities in egg white environment are not properly understood. Temperature, nutrient restriction, alkalinity and viscosity are the key factors which might influence the activity of antibacterial proteins within the eggs (Baron et al. 2016; Silva et al. 2014; Whiley and Ross 2015).

### 13.3.2 *Staphylococcus aureus*

*Staphylococcus aureus* is ubiquitous to various environments considered as commensal as well as an infectious microbe that can cause serious complications of skin and can also damage various vital organs. *S aureus* is considered a pathogen of public health importance due to the combination of toxin-mediated diseases, capacity to invade and acquisition of antibiotic resistance (Lowy 1998). *S. aureus* food-borne disease (SFD) is the most important concern in worldwide public health programs and the frequently reported cause of food-associated diseases in the United States (Hennekinne et al. 2012; Kadariya et al. 2014). SFD is the commonest food-borne disease worldwide (Hennekinne et al. 2012). The symptoms of *S. aureus*-mediated FDB involve the rapid onset of infection generally 3–5 hours following consumption of *S. aureus*-contaminated food that is mainly attributed to the production of the toxins during its growth (Le Loir et al. 2003). The incubation period of food intoxication caused by *S. aureus* is mainly dependent on the amount of toxin ingested (Murray et al. 2005). SFD can be caused by a very small dose of toxin as a large outbreak was reported with consumption of chocolate milk contaminated with 0.5 ng/mL of *Staphylococcal enterotoxin* (Kadariya et al. 2014; Murray et al. 2005).

The onset of staphylococcal food-borne diseases is abrupt and the symptoms mainly include nausea, vomiting, hypersalivation and abdominal cramping. In the case of diarrhea, the symptoms involve the signs of water balance disturbances in consequence of significant fluid loss. The SFD usually resolves itself with time and symptoms generally disappear within 1 or 2 days, however can be fetal, specifically in immune-compromised persons, newborns and aged patients. The antibiotics therapy is generally not recommended as the symptoms are mainly due to toxins. The microbiological examination of food does not detect *S. aureus* especially when cooked food is contaminated with the preformed toxin. Nearly 10% of affected SFD cases are presented to a hospital (Kadariya et al. 2014; Murray et al. 2005).

Different types of food items provide an optimal medium for the growth of *S. aureus* although eggs, salads, poultry, milk, dairy, meat and bakery products have been commonly associated with SFD outbreaks. The foods involved with SFD may vary within countries, principally due to disparities in food habits (Murray 2005). SFD outbreaks can lead to serious consequences affecting thousands of people in case the food items are widely distributed. In Japan, more than 13,000 cases of SFD were reported in 2000 because of milk contamination at a dairy food production plant (Kadariya et al. 2014; Murray et al. 2005).

The most common foods that are associated with staphylococcal food poisoning vary broadly between the counties. 53% of the reported cases of staphylococcal food poisonings in the United Kingdom (between 1969 and 1990) occurred because of consumption of meat products, especially ham, 22% cases were due to poultry products, 8% were due to dairy products, 7% were due to fish and 3.5% were due to eggs (Le Loir et al. 2003; Adnan et al. 2016). The food poisonings due to *Staphylococcus* spp. reported during 1999–2000 in France are as follows: 32% of the cases were due to dairy products, 22% due to meat, 15% due to sausages and pies, 11% due to seafood, 11% due to egg and egg products and 9.5% due to poultry (Mohammed et al. 2014). In the United States, the majority of reported cases of staphylococcal food poisoning during 1975–1982 were due to meat and salads whereas only 1.4% cases were due to milk products and seafood. Therefore, the source of food poisoning caused by *S. aureus* differs greatly from country to country attributed to the different food habits among the nations (Hennekinne et al. 2012). For instance, the unpasteurized milk and cheese consumption is more common in France in comparison with other European countries that can expound on the involvement of milk products in staphylococcal food-borne disease in France (Le Loir et al. 2003).

A wide array of heat-stable *Staphylococcus* enterotoxins (SEs) are produced by *Staphylococcus aureus* that belongs to nine major serological types (SEA, SEB, SEC, SED, SEE, SEG, SEH, SEI and SEJ) of a large family of pyrogenic toxin superantigens (Kadariya et al., 2014). These pyrogenic toxins are nonspecific T-cell proliferation and immunosuppression (Hennekinne et al. 2012; Kadariya et al. 2014). SEA is more tolerant to proteolytic enzymes and is most commonly implicated in SFD. In the United States, the SEA was identified from 77.8% of SFD cases followed by SED (37.5%). The most common enterotoxin involved in SFD outbreaks in Japan, the United Kingdom and France was SEA (Argudin et al. 2010). In the United States, a food handler was the source of SEC-producing methicillin-resistant *Staphylococcus aureus* (MRSA) via contaminated coleslaw that caused the outbreak of food poisoning (Jones et al. 2002). SEC was involved in SFD epidemic in Canada in 1980, in Taiwan during 2001–2003 and in Japan (2009). *S. aureus* causes mastitis in goats, sheep and cattle and a study reported that the milk of animals suffering from mastitis was found to contain *S. aureus* oriented SEC toxin. Various reports confirmed the existence of SEC synthesizing *S. aureus* strains in udder and teats skin of goats (Bardiau et al. 2013; Kadariya et al. 2014; Kant et al. 2015; Ostyn et al. 2010).

MRSA associated with livestock, has been isolated and identified as an emerging pathogen during the past few years causing an array of human infections (Leedom Larson et al. 2011; Ostyn et al. 2010). The MRSA was firstly detected in pigs and farmers of France and the Netherlands in 2005. Afterward, the livestock-associated MRSA has been isolated from various countries in Asia, Europe and North America. The studies revealed that ST398 is responsible for 20% of MRSA infections in humans and comprises of 42% of newly reported cases of MRSA in the Netherlands, signifying the role of animals as a reservoir for MRSA infections in humans (Geenen et al. 2013; Kadariya et al. 2014; Kock et al. 2013; Wendlandt et al. 2013).

MRSA has been increasingly reported in Europe and the United States from commercially available meats and so it becomes a serious health threat for both meat consumers and sellers (Alba et al. 2015; Frana et al. 2013; Hanson et al. 2011; Jackson et al. 2013; Kadariya et al. 2014; O'Brien et al. 2012). The frequently implicated species of meat-producing animals include pigs, cattle and poultry (Fessler et al. 2010; Kadariya et al. 2014; Mulders et al. 2010). Various studies have reported the MRSA existence in uncooked meat products, with a low occurrence in Asia (about 1%) and 11.9% prevalence in the Netherlands. A study from Georgia, United States reported the chance of *S. aureus* to exist is 45% in pork meat products and 63% in beef meat products, whereas, the MRSA occurrence found by this study was 3% in pork meat and 4% in beef meat (Jackson et al. 2013; Kadariya et al. 2014). Another study from

Louisiana reported a 39.2% prevalence of *S. aureus* in retail meat samples in the United States and 5% were found to contain MRSA. A relatively higher frequency of *S. aureus* (64.8%) was reported in pork meat from Minnesota, New Jersey and Iowa with 6.6% MRSA strains. These studies draw attention to the contamination of commercially available meat with *S. aureus* that might serve as a possible way to transmit this pathogen to the human population (Kadariya et al. 2014).

### 13.3.3 *Bacillus cereus*

*Bacillus cereus* is well-known bacteria that can cause food-borne illnesses. Various Bacillus species have been problematic for food producers because of their capacity to form endospores. The presence of heat and their resistant spores in raw and cooked foods enable their survival at higher cooking temperatures therefore require foods to be served hot after cooking or quickly cool down to prevent their growth. The wide range of protein toxins produced by *Bacillus cereus* is related to diarrhea and other emetic illness (Logan et al. 2012; Newell et al. 2010; Tewari and Abdullah 2015). The vegetative cells under transmission electron microscopy reveal cytoplasmic membrane surrounding the cellular material. Few strains also contain an additional outer surface protein (S layer). These spore-forming bacteria are resistant to harsh environmental conditions including heat, freezing and drying (Bottone et al. 2010; Tewari and Abdullah 2015).

*Bacillus cereus* is the aerobic endospore-forming bacteria that cause a wide range of opportunistic infections which results in distinct food-borne illness syndromes. During the last few years, *B cereus* has been truly appreciated as a pathogen, though the diarrheal disease was described as early as in the 1950s (Bottone et al. 2010; Logan et al. 2012). The true burden of illness caused by *B. cereus* is unknown because these bacteria are not always included in epidemiological studies of food-borne diseases and because of relatively intermittent cases rather than any major outbreaks. The diarrheal syndrome and emetic syndromes arise from the fact that the spores of *B. cereus* are capable of survival at normal cooking temperatures. The spore starts germination and multiplies under improper storage conditions. The infective or intoxicating dose depends on the capacity of the strain to produce toxins and influence both types of illness. Therefore, the cases of diarrheal and emetic syndromes might be either due to both strains that produce diarrheal and emetic toxins present in the food or a single strain that produces both diarrheal and emetic toxins. The toxins produced by *B. cereus* strains and their associations for food safety are not much clear; however, the temperature, food components and other environmental factors influence the toxin production within the food and intestinal environments (Ceuppens et al. 2011; Logan et al. 2012).

### 13.3.4 *Clostridium botulinum*

*Clostridium botulinum* is a Gram-positive, anaerobic bacteria capable of forming spores and is commonly distributed in soil, and decaying vegetations. The various strains (A–G) of *C. botulinum* are classified on the basis of their serologic specificity. The majority of epidemics in the human population are due to the consumption of fish and other seafood products. Two strains (i.e. C and D) are principally involved in the infections in animals (Abdel-Moein et al. 2016; Addis and Sisay 2015).

Botulism is widely spread with no geographical limitations and the sporadic cases occur in many countries; however, the source of toxin and comparative risk for the disease progression differs significantly between different counties because of differences in food storage practices. The outbreaks linked with the consumption of toxin-containing feeds are frequent in Europe and the United States, while the outbreaks among the animal's dependent on pasture are majorly reported from Australia, the United States, the Gulf Coast and South Africa (Smith et al. 2015). Ingestion of botulism type C toxin-contaminated feed from cat carcass resulted in an incident resulting in the death of 427 out of 444 dairy cattle. Although the *C. botulinum* spores are distributed all over the world, the recorded outbreaks were reported mostly from the north of the tropic of cancer. The global spread of *C. botulinum* spores and reported cases from various parts of the world requires significant attention toward the canning of fruits and vegetables at the domestic level in most of the countries (Addis and Sisay 2015; Radostits et al. 2006).

*C. botulinum* strains produce highly potent neurotoxin during their growth that results in botulism; a neuroparalytic disorder of both humans and animals. The improperly treated botulism causes respiratory muscle paralysis that ultimately leads to death. The absorption of botulinum toxin occurs from wound, stomach or small intestine that ultimately enters the nerve cell by endocytosis after binging with receptor. The heavy chain of toxin helps to bind the toxin with the receptor and translocation to the cell whereas light chain plays its role in limiting the discharge of acetylcholine at myoneural connection. The flaccid paralysis because of lack of acetylcholine affects the muscle including the diaphragm leading to death due to respiratory failure (Abdel-Moein and Hamza 2016; Rossetto et al. 2014). The incubation period of *C. botulinum* is 12–36 hours and the common symptoms include increased thirst, vomiting, dryness of mouth, constipation, blurred vision and difficulty in breathing, swallowing and speaking (Anniballi et al. 2013; Kalb et al. 2015). A temperature of 80°C for 30 minutes or boiling at 100°C for 10 minutes can completely destroy the preformed toxin in food. Therefore, the low acid canned foods need to be boiled before consumption. Moreover, the food should not be tasted if its odor is changed or shows signs of gas formation. The preventive measures of food-borne botulism include the effective control of commercially and homemade canned foods. The cooking at 121°C or higher temperature is required to ensure the killing of all the spores. Home-canned vegetables have to boil properly for no less than 3 minutes before serving to denature the potential botulism toxins (Addis and Sisay 2015; Jay et al. 2012).

### 13.3.5 *Clostridium perfringens*

*Clostridium perfringens* is another spore-forming Gram-positive anaerobic bacteria abundantly distributed in vegetation, environment, animal feces and sewage. Enterotoxin A that produces a *C. perfringens* pare is commonly associated with food-borne illness, while other enterotoxins (B to G) do not cause food-borne disease (Hayes et al. 2013). The spores of *C. perfringens* may survive in the food during cooking germinate when the food products are inadequately stored. The vegetative cells release enterotoxins in the intestine. At least 12 different toxins are produced by *C. perfringens* type A strains, though the food poisoning is caused by alpha and theta toxins. The *C. perfringens* enterotoxin doses not act as superantigen and symptoms begin when toxin binds to the receptors on epithelial cells of the gastrointestinal tract resulting in excess fluid accumulation in the intestine. The toxin is localized in the plasma membrane complex of the intestinal cells and induces alterations in membrane permeability that may cause cell death (Stiles et al. 2013; Uzal et al. 2014).

The food-borne illness caused by *Clostridium perfringens* type A have been considered as one of the most widespread food-borne diseases. There were 121 confirmed outbreaks reported in the United States between 1983 and 1997 with a total of 9316 cases and 13 deaths. A study reported the epidemiology of food-borne outbreaks linked with *C. perfringens* in the United States during 1998–2010. A total of 823 outbreaks causing 28,543 (10%) illnesses, were suspected to be caused by *C. perfringens*. Out of these outbreaks, 289 (35%) were confirm caused by *C. perfringens*, which caused disease in 15,208 patients and also 8 deaths occured. Therefore, *C. perfringens* is designated as the second most common bacterial pathogen associated with food-borne disease outbreaks, following *Salmonella* (Grass et al. 2013).

The majority of *C. perfringens* outbreaks were attributed to meat and poultry products. The organism is ubiquitous in the environment, is associated with animals and can grow easily under pH ranges of most poultry and meat products, i.e. between 6.0 and 7.0. *C. perfringens* type A associated with human disease was also isolated from raw poultry and meat and in food services. Contamination of meats, poultry and other food items has been known to occur through cross-contamination by other foods, contact of carcasses with fecal material and contaminated surfaces during food processing (Juneja and Sofos 2010). Most of the toxin-producing *C. perfringens* strains appear to contaminate the surface of meats, and low levels of *C. perfringens* spores have been isolated from deeper muscle tissues (Grass et al. 2013).

### 13.3.6 *Escherichia coli*

*Escherichia coli* is a member of a bacterial family known as enterobactereriae and are Gram-negative rods. The *E. coli* O157:H7 serotype is reported as a major contaminant of food products and the causative agent of food-associated infections. *E. coli* have more than 60 verotoxin-producing serotypes that cause various types of gastrointestinal infections in humans including hemorrhagic colitis, diarrhea and hemolytic uremic syndrome (Graeter et al. 2014; Koochakzadeh et al. 2014). Entero-hemorrhagic *E. coli* and other *E. coli* strains secreting numerous verotoxins are reported as a major causative agent of food-borne epidemics in humans and animals. One of the important serotypes is *E. coli* O157:H7 that seems to be prevalent in most parts of the world. The verotoxin produced by the strains is similar to Shiga toxin, which is responsible for causing diarrhea in humans and animals. The main reservoir of *E. coli* O157:H7 includes cattle and some other ruminants as these strains transiently reside in their gastrointestinal tract. The human and animal's fecal contamination is the main source of food-borne infection. The role of cattle as a reservoir of the *E. coli* O157: H7 was evident from multiple studies that reported the isolation of these pathogens from raw meat, ground beef and cow milk (Kundu et al. 2014; Miszczycha et al. 2012; Surendran Nair et al. 2016). Entero-hemorrhagic *E. coli* strains may be able to produce cytotoxins known as Shiga-like toxins that were functionally and antigenically similar to Shiga toxin previously known as verotoxin produced by *Shigella dysenteriae*. The Shiga-like toxins incite cell secretion and damage human colon and ileum epithelial cells. The toxin causes localized fluid accumulation and lesions in the colon and crypt epithelial cells in animals (Kunsmann et al. 2015; Tran et al. 2014).

The fresh vegetables are often contaminated with enteric pathogens and in recent years frequent outbreaks were reported from different parts of the world. In Canada, the United States, Europe and Asia, these outbreaks are mainly associated with the intake of contaminated spinach, lettuce and different types of sprouts. In Japan, approximately 10,000 cases were reported in one of the largest outbreaks from 14 different locations that were linked with the intake of white radish sprouts contaminated with *E. coli* O157:H7 (Martinez-Vaz et al. 2014). More than 200 individuals were encountered serious illnesses due to the consumption of bagged spinach contaminated with *E. coli* O157:H7 in various states of Canada in 2006. In 2011, another multistate outbreak with 58 cases of food poisoning was reported and linked to the contamination of romaine lettuce with *E. coli* O157:H7 in five cities of the United States (Martinez-Vaz et al. 2014; Slayton et al. 2013). Recently, around 3,800 individuals in Germany and France were infected after consuming fenugreek seeds contaminated with *E. coli* O104:H4. The reported outbreaks associated with the consumption of vegetables demonstrate the risks as well as the challenges for food safety both for consumers and producers (Frank et al. 2011; Gault et al. 2011; Martinez-Vaz et al. 2014; Slayton et al. 2013).

## 13.4 Conclusion

Beliefs in the public relative to food security and infectivity are frequently embedded in customs and traditions. The choice of diet is primarily dependent upon the financial state and social factors of the consumer; however, the policies relative to food safety are formed and organized by the regulatory bodies. These bodies can establish laboratories for contaminants testing and impose legislation strictly. For the people's universal health and everyday life, social stability and the government's and country's image and economic development, the consistent harmless supply of foodstuff that is free from injurious contaminants is essential. Unluckily, however, many prospects exist for foodstuff to be contaminated from farm to table, which leads to massive destructive impacts. However, the good thing is that with positive safety measures, food microbiological infectivity and the significant impacts of food spoilage can be alleviated. Microbiological food protection is actually a farm-to-table matter that could handle food microbial infectivity and the subsequent impacts have to happen from the primary production to the dining table and food contaminants or pathogens or antimicrobial resistance in the food chain signify an arrangement of numerous features. These preventive procedures come under good manufacturing practices, good agricultural practices, good transportation practices, good hygienic practices and good storage practices to reduce microbial food protection risks. Education of society regarding livestock rearing and safer crop cultivation and safer cooking routine and choice of a balanced diet should also be encouraged. The control and elimination of the predictable menaces through the above-named practices could assist to build up safe foodstuff globally and minimize the socioeconomic burden of diseases that spread through food contamination. Amazingly, recent reports show that food pathogens or contaminants still signify significant causes of unintended diseases, injury and demise. In addition to demise and diseases, the utilization of pathogen-contaminated foodstuff has also created psychological, community and financial impacts that were quite destructive on the consumers, food dealers, companies and nations.

## REFERENCES

Abdel-Moein KA, Hamza DA (2016) Occurrence of human pathogenic *Clostridium botulinum* among healthy dairy animals: an emerging public health hazard. Pathog Glob Health 110(1):25–29. https://doi.org/10.1080/20477724.2015.1133107.

Addis M, Sisay D (2015) A review on major food borne bacterial illnesses. J Trop Dis Public Health 3:176

Adnan M, Shah Z, Arif M, Khan MJ, Mian IA, Sharif M, Alam M, Basir A, Ullah H, Rahman I, Saleem N (2016) Impact of rhizobial inoculum and inorganic fertilizers on nutrients (NPK) availability and uptake in wheat crop. Can J Soil Sci 96:169–176.

Alba P, Feltrin F, Cordaro G, et al. (2015) Livestock-associated methicillin resistant and methicillin susceptible *Staphylococcus aureus* sequence type (CC)1 in European farmed animals: high genetic relatedness of isolates from Italian cattle herds and humans. PLoS One 10(8):e0137143. https://doi.org/10.1371/journal.pone.0137143.

Anniballi F, Fiore A, Lofstrom C, Skarin H, et al (2013) Management of animal botulism outbreaks: from clinical suspicion to practical countermeasures to prevent or minimize outbreaks. Biosecur Bioterror 11 Suppl 1:S191–S199. https://doi.org/10.1089/bsp.2012.0089.

Argudin MA, Mendoza MC, et al (2010) Food poisoning and *Staphylococcus aureus* enterotoxins. Toxins (Basel) 2(7):1751–1773. https://doi.org/10.3390/toxins2071751.

Austin JW, Sanders G, Kay WW, et al (1998) Thin aggregative fimbriae enhance *Salmonella enteritidis* biofilm formation. FEMS Microbiol Lett 162(2):295–301.

Barak JD, Jahn CE, Gibson DL, et al (2007) The role of cellulose and O-antigen capsule in the colonization of plants by *Salmonella enterica*. Mol Plant Microbe Interact 20(9):1083–1091.

Bardiau M, Yamazaki K, Duprez JN, et al (2013) Genotypic and phenotypic characterization of methicillin-resistant *Staphylococcus aureus* (MRSA) isolated from milk of bovine mastitis. Lett Appl Microbiol 57(3):181–186. https://doi.org/10.1111/lam.12099.

Baron F, Nau F, Guerin-Dubiard C, et al (2016) Egg white versus *Salmonella Enteritidis*! A harsh medium meets a resilient pathogen. Food Microbiol 53(Pt B):82–93. https://doi.org/10.1016/j.fm.2015.09.009.

Bechet M, et al (2003) Factors associated with the adherence and biofilm formation by Aeromonas caviae on glass surfaces. J Appl Microbiol 94(6):1072–1078.

Bottone EJ, et al (2010) *Bacillus cereus*, a volatile human pathogen. Clin Microbiol Rev 23(2):382–398. https://doi.org/10.1128/CMR.00073-09.

Cappitelli F, et al (2014) Biofilm formation in food processing environments is still poorly understood and controlled. Food Engi Rev 6(1–2):29–42.

Carrasco E, Morales-Rueda A, García-Gimeno RM, et al (2012) Cross-contamination and recontamination by *Salmonella* in foods: a review. Food Res Int 45(2):545–556.

Ceuppens S, Rajkovic A, Heyndrickx M, Tsilia V, et al (2011) Regulation of toxin production by Bacillus cereus and its food safety implications. Crit Rev Microbiol 37(3):188–213. https://doi.org/10.3109/1040841X.2011.558832.

Control & Prevention (2010) Surveillance for foodborne disease outbreaks—United States, 2007. MMWR. Morbidity and mortality weekly report 59(31):973.

Cookson AL, et al (2002) The role of type 1 and curli fimbriae of Shiga toxin-producing *Escherichia coli* in adherence to abiotic surfaces. Int J Med Microbiol 292(3):195–205.

Di Martino P, Cafferini N, Joly B, et al (2003) *Klebsiella pneumoniae* type 3 pili facilitate adherence and biofilm formation on abiotic surfaces. Res Microbiol 154(1):9–16.

EFSA (2010) Trends and sources of zoonoses and zoonotic agents and food-borne outbreaks in the European Union in 2008. EFSA 8:1–370.

Fessler A, Scott C, Kadlec K, et al (2010) Characterization of methicillin-resistant *Staphylococcus aureus* ST398 from cases of bovine mastitis. J Antimicrob Chemother 65(4):619–625. https://doi.org/10.1093/jac/dkq021.

Foley SL, et al (2008) *Salmonella* challenges: prevalence in swine and poultry and potential pathogenicity of such isolates. J Anim Sci 86(14 Suppl):E149–E162. https://doi.org/10.2527/jas.2007-0464.

Frana TS, Beahm AR, Hanson BM, et al (2013) Isolation and characterization of methicillin-resistant *Staphylococcus aureus* from pork farms and visiting veterinary students. PLoS One 8(1):e53738. https://doi.org/10.1371/journal.pone.0053738.

Frank C, Faber MS, Askar M, et al (2011) Large and ongoing outbreak of haemolytic uraemic syndrome, Germany, May 2011. Euro Surveill 16(21):19878.

Fukuda K (2015) Food safety in a globalized world. Bull World Health Organ 93(4):212. https://doi.org/10.2471/BLT.15.154831.

Gast RK, Guraya R, Jones DR, et al (2014) Contamination of eggs by *Salmonella Enteritidis* in experimentally infected laying hens housed in conventional or enriched cages. Poult Sci 93(3):728–733. https://doi.org/10.3382/ps.2013-03641.

Gault G, Weill FX, Mariani-Kurkdjian P, et al (2011) Outbreak of haemolytic uraemic syndrome and bloody diarrhoea due to Escherichia coli O104:H4, south-west France, June 2011. Euro Surveill 16(26):19905.

Geenen PL, Graat EA, Haenen A, et al (2013) Prevalence of livestock-associated MRSA on Dutch broiler farms and in people living and/or working on these farms. Epidemiol Infect 141(5):1099–1108. https://doi.org/10.1017/S0950268812001616.

Gil MI, Selma MV, Suslow T, et al (2015) Pre- and postharvest preventive measures and intervention strategies to control microbial food safety hazards of fresh leafy vegetables. Crit Rev Food Sci Nutr 55(4):453–468. https://doi.org/10.1080/10408398.2012.657808.

Giltner CL, Van Schaik EJ, Audette GF, et al (2006) The *Pseudomonas aeruginosa* type IV pilin receptor binding domain functions as an adhesin for both biotic and abiotic surfaces. Mol Microbiol 59(4):1083–1096.

Graeter L, Hertenstein E, Accurso C, et al (2014) Elsevier's medical laboratory science examination review. Elsevier Health Sciences.

Grass JE, et al (2013) Epidemiology of foodborne disease outbreaks caused by *Clostridium perfringens* United States, 1998–2010. Foodborne Pathog Dis 10(2):131–136. https://doi.org/10.1089/fpd.2012.1316.

Hanson BM, Dressler AE, Harper AL (2011) Prevalence of *Staphylococcus aureus* and methicillin-resistant *Staphylococcus aureus* (MRSA) on retail meat in Iowa. J Infect Public Health 4(4):169–174. https://doi.org/10.1016/j.jiph.2011.06.001.

Hayes PR, et al (2013) Food hygiene microbiology and HACCP. Springer Science & Business Media.

Hennekinne JA, et al (2012) *Staphylococcus aureus* and its food poisoning toxins: characterization and outbreak investigation. FEMS Microbiol Rev 36(4):815–836. https://doi.org/10.1111/j.1574-6976.2011.00311.x.

Houry A, Briandet R, Aymerich S (2010) Involvement of motility and flagella in *Bacillus cereus* biofilm formation. Microbiology 156(4):1009–1018.

Hussain MA (2016) Food contamination: major challenges of the future. Foods 5(2):21.

Jackson CR, et al (2013). Prevalence and characterization of methicillin-resistant *Staphylococcus aureus* isolates from retail meat and humans in Georgia. J Clin Microbiol 51(4):1199–1207. https://doi.org/10.1128/JCM.03166-12.

Jay JM, et al (2012) Modern food microbiology. Springer Science & Business Media.

Jones TF, Kellum ME, Porter SS, et al (2002). An outbreak of community-acquired foodborne illness caused by methicillin-resistant *Staphylococcus aureus*. Emerg Infect Dis 8(1):82–84.

Juneja VK Sofos JN (2010) Pathogens and toxins in foods: challenges and interventions. ASM Press.

Kadariya J, et al (2014) *Staphylococcus aureus* and staphylococcal food-borne disease: an ongoing challenge in public health. Biomed Res Int 2014:827965. https://doi.org/10.1155/2014/827965.

Kalb SR, Krilich JC, Dykes JK (2015) Detection of Botulinum Toxins A, B, E, and F in Foods by Endopep-MS. J Agric Food Chem. https://doi.org/10.1021/jf505482b.

Kant R, Taponen S, Koort J, et al (2015) Genome sequences of four *Staphylococcus aureus* strains isolated from bovine mastitis. Genome Announc 3(2). https://doi.org/10.1128/genomeA.00334-15.

Karch H, Denamur E, Dobrindt U, et al (2012) The enemy within us: lessons from the 2011 European *Escherichia coli* O104:H4 outbreak. EMBO Mol Med 4(9):841–848. https://doi.org/10.1002/emmm.201201662.

Kim TJ, et al (2008) Effect of flagellar mutations on *Yersinia enterocolitica* biofilm formation. Appl Environ Microbiol 74(17):5466–5474.

Kirezieva K, Jacxsens L, Uyttendaele M (2013) Assessment of food safety management systems in the global fresh produce chain. Food Res International 52(1):230–242.

Kock R, Schaumburg F, Mellmann A (2013) Livestock-associated methicillin-resistant *Staphylococcus aureus* (MRSA) as causes of human infection and colonization in Germany. PLoS One 8(2):e55040. https://doi.org/10.1371/journal.pone.0055040.

Koochakzadeh A, Askari Badouei M, Mazandarani E (2014) Survey on O157:H7 enterohemorrhagic *Escherichia coli* (EHEC) in cattle in Golestan province, Iran. Iran J Microbiol 6(4):276–280.

Kundu D, Gill A, Lui C (2014) Use of low dose e-beam irradiation to reduce *E. coli* O157:H7, non-O157 (VTEC) *E. coli* and *Salmonella* viability on meat surfaces. Meat Sci 96(1):413–418. https://doi.org/10.1016/j.meatsci.2013.07.034.

Kunsmann L, Ruter C, Bauwens A (2015) Virulence from vesicles: novel mechanisms of host cell injury by *Escherichia coli* O104:H4 outbreak strain. Sci Rep 5:13252. https://doi.org/10.1038/srep13252.

Lammie SL, Hughes JM (2016) Antimicrobial resistance, food safety, and one health: the need for convergence. Annu Rev Food Sci Technol 7:287–312. https://doi.org/10.1146/annurev-food-041715-033251.

Le Loir Y, et al (2003) *Staphylococcus aureus* and food poisoning. Genet Mol Res 2(1):63–76.

Leedom Larson KR, Harper AL, Hanson BM, et al (2011) Methicillin-resistant *Staphylococcus aureus* in pork production shower facilities. Appl Environ Microbiol 77(2):696–698. https://doi.org/10.1128/AEM.01128-10.

Logan NA, et al (2012) *Bacillus* and relatives in foodborne illness. J Appl Microbiol 112(3):417–429. https://doi.org/10.1111/j.1365-2672.2011.05204.x.

Lowy FD (1998) *Staphylococcus aureus* infections. New England J Med 339(8):520–532.

Mani-Lopez E, et al (2012) Organic acids as antimicrobials to control Salmonella in meat and poultry products. Food Res Int 45(2):713–721.

Manning L, Soon JM (2016) Food safety, food fraud, and food defense: a fast evolving literature. J Food Sci 81(4):R823–R834. https://doi.org/10.1111/1750-3841.13256.

Martinez-Vaz BM, Fink RC, Diez-Gonzalez F (2014) Enteric pathogen-plant interactions: molecular connections leading to colonization and growth and implications for food safety. Microbes Environ 29(2):123–135.

McClelland M, Sanderson KE, Spieth J, et al (2001) Complete genome sequence of *Salmonella* enterica serovar Typhimurium LT2. Nature 413(6858):852–856.

McKerr C, Adak GK, Nichols G (2015) An Outbreak of Cryptosporidium parvum across England & Scotland Associated with Consumption of Fresh Pre-Cut Salad Leaves, May 2012. PLoS One 10(5):e0125955. https://doi.org/10.1371/journal.pone.0125955.

Miszczycha SD, Ganet S, Duniere L (2012) Novel real-time PCR method to detect *Escherichia coli* O157:H7 in raw milk cheese and raw ground meat. J Food Prot 75(8):1373–1381. https://doi.org/10.4315/0362-028X.JFP-11-498.

Mohammed TA, et al (2014) Isolation of *Staphylococcus aureus* from some semi preserved sudanese. Food Res J Microbiol 9(5):246.

Mulders MN, Haenen AP, Geenen PL (2010) Prevalence of livestock-associated MRSA in broiler flocks and risk factors for slaughterhouse personnel in The Netherlands. Epidemiol Infect 138(5):743–755. https://doi.org/10.1017/S0950268810000075.

Murray R, et al (2005) Recognition and management of *Staphylococcus aureus* toxin-mediated disease. Int Med J 35(s2):S106–S119. https://doi.org/10.1111/j.1444-0903.2005.00984.x.

Newell DG, Koopmans M, Verhoef L (2010) Food-borne diseases - the challenges of 20 years ago still persist while new ones continue to emerge. Int J Food Microbiol 139 Suppl 1:S3–S15. https://doi.org/10.1016/j.ijfoodmicro.2010.01.021.

Neyts K, Huys G, Uyttendaele M (2000) Incidence and identification of mesophilic *Aeromonas* spp. from retail foods. Lett Appl Microbiol 31(5): 359–363.

O'Brien AM, Hanson BM, Farina SA (2012) MRSA in conventional and alternative retail pork products. PLoS One 7(1):e30092. https://doi.org/10.1371/journal.pone.0030092.

Ostyn A, De Buyser ML, Guillier F (2010) First evidence of a food poisoning outbreak due to staphylococcal enterotoxin type E, France, 2009. Euro Surveill 15(13):19528.

Poimenidou SV, Bikouli VC, Gardeli C (2016) Effect of single or combined chemical and natural antimicrobial interventions on *Escherichia coli* O157: H7, total microbiota and color of packaged spinach and lettuce. Int J Food Microbiol 220:6–18.

Pratt LA, Kolter R (1998) Genetic analysis of *Escherichia coli* biofilm formation: roles of flagella, motility, chemotaxis and type I pili. Mol Microbiol 30(2):285–293.

Pruden A, et al (2006) Antibiotic resistance genes as emerging contaminants: studies in northern Colorado. Environ Sci Tech 40(23):7445–7450.

Radostits OM, Gay CC, Hinchcliff KW (2006) Veterinary medicine: a textbook of the diseases of cattle, sheep, goats, pigs and horses. Saunders Co, London.

Ray B, Bhunia A (2007) Fundamental food microbiology. CRC press, Boca Raton, FL.

Reij MW, et al (2004) Recontamination as a source of pathogens in processed foods. Int J Food Microbiol 91(1):1–11. https://doi.org/10.1016/S0168-1605(03)00295-2.

Rossetto O, et al (2014) Botulinum neurotoxins: genetic, structural and mechanistic insights. Nat Rev Microbiol 12(8):535–549. https://doi.org/10.1038/nrmicro3295.

Scallan E, Hoekstra RM, Angulo FJ, et al (2011) Foodborne illness acquired in the United States—major pathogens. Emerg Infect Dis 17(1):7–15.

Shah SRA, Khan SA, Saljoqi AUR, Junaid K, Khan I, Zaman M, Said F, Adnan M, Saleem N (2015) Assessment of the varietal preference and resistance against LipaphisErysimi (K.) in segregating mustard genotypes under agro-ecological conditions of Peshawar, Pakistan. J Ento Zool Stud 3(5):100–103.

Silva C, et al (2014) One health and food-borne disease: *Salmonella* transmission between humans, animals, and plants. Microbiol Spectr 2(1): OH-0020-2013. https://doi.org/10.1128/microbiolspec.OH-0020-2013.

Slayton RB, Turabelidze G, Bennett SD, et al (2013) Outbreak of Shiga toxin-producing *Escherichia coli* (STEC) O157:H7 associated with romaine lettuce consumption, 2011. PLoS One 8(2):e55300. https://doi.org/10.1371/journal.pone.0055300.

Smith TJ, et al (2015) Historical and current perspectives on Clostridium botulinum diversity. Res Microbiol 166(4):290–302. https://doi.org/10.1016/j.resmic.2014.09.007.

Stiles BG, Barth G, Barth H, et al (2013) *Clostridium perfringens* epsilon toxin: a malevolent molecule for animals and man? Toxins (Basel) 5(11):2138–2160. https://doi.org/10.3390/toxins5112138.

Surendran Nair M, Lau P, Belskie K, et al (2016) Potentiating the heat inactivation of Escherichia coli O157:H7 in ground beef patties by natural antimicrobials. Front Microbiol 7:15. https://doi.org/10.3389/fmicb.2016.00015.

Switaj TL, et al (2015) Diagnosis and management of foodborne illness. Am Fam Physician 92(5):358–365.

Tafida SY, Kabir J, Kwaga JKP, et al (2013) Occurrence of Salmonella in retail beef and related meat products in Zaria, Nigeria. Food Control 32(1):119–124.

Tewari A, Abdullah S (2015) *Bacillus cereus* food poisoning: international and Indian perspective. J Food Sci Technol 52(5):2500–2511. https://doi.org/10.1007/s13197-014-1344-4.

Todd EC (1997) Epidemiology of foodborne diseases: a worldwide review. World Health Stat Q 50(1–2):30–50.

Tran SL, Billoud L, Lewis SB, et al (2014) Shiga toxin production and translocation during microaerobic human colonic infection with Shiga toxin-producing *E. coli* O157:H7 and O104:H4. Cell Microbiol 16(8):1255–1266. https://doi.org/10.1111/cmi.12281.

Uzal FA, Freedman JC, Shrestha A, et al (2014) Towards an understanding of the role of *Clostridium perfringens* toxins in human and animal disease. Future Microbiol 9(3):361–377. https://doi.org/10.2217/fmb.13.168.

Van Houdt R, et al (2010) Biofilm formation and the food industry, a focus on the bacterial outer surface. J Appl Microbiol 109(4):1117–1131.

Vizcarra IA, Hosseini V, Kollmannsberger P, et al (2016) How type 1 fimbriae help Escherichia coli to evade extracellular antibiotics. Sci Rep 6:18109

Wendlandt S, et al (2013) Methicillin-resistant *Staphylococcus aureus*: a food-borne pathogen? Annu Rev Food Sci Technol 4:117–139. https://doi.org/10.1146/annurev-food-030212-182653.

Whiley H, Ross K (2015) *Salmonella* and eggs: from production to plate. Int J Environ Res Public Health 12(3):2543–2556. https://doi.org/10.3390/ijerph120302543.

Williams MS, Ebel ED, Golden NJ (2014) Temporal patterns in the occurrence of *Salmonella* in raw meat and poultry products and their relationship to human illnesses in the United States. Food Control 35(1):267–273.

WHO (2015) WHO estimates of the global burden of foodborne diseases: Oodborne disease burden epiemiology reference group 2007–2015. World Health Organization.

# 14

# *Molecular and Ecological Mechanisms of Grasshopper Food Selection and Habitat Adaptation*

**Xinghu Qin**
*University of St Andrews*

**Huihui Wu, Weizhong Zhang, Xunbing Huang, and Zehua Zhang**
*Chinese of Academy of Agricultural Sciences*

**CONTENTS**



## 14.1 Introduction

Except for Antarctica and Eurasia, north of the 55° north latitude, grasshoppers are widely distributed in almost all places where plants exist (Guang et al. 2002), indicating that grasshoppers have an extremely wide range of ecological adaptability to the natural environment and ecological conditions. *Oedaleus asiaticus* B. Bienko, an important pest widely distributed across the northern China grassland (Kang and Chen 1994b), has been reported repeatedly outbreak during five continuous years, causing great damage to grassland and crops. Other dominant grasshoppers, such as *Dasyhippus barbipes*, *Chorthippus albonemus*, and *Calliptamus abbreviatus* recurrently outbreak in heavy grazing areas, posing a big threat to Inner Mongolia grass products and husbandry.

At present, related research on *O. asiaticus* mainly includes food quantity measurement (Qin 2016; Qin et al. 2017a), olfactory response (Chen and Kang 2000; Li et al. 2018), occurrence and dynamics (Shen et al. 2015; Zhang et al. 2015), behavior (Cease et al. 2010), growth and development (Cease et al. 2017), biological characteristics (Zhang et al. 2013a), outbreak prediction (Guo and Bao-Ping 2012; Huang et al. 2017a), yield loss and economic thresholds (Guang et al. 2007; Lu et al. 2008; Lu and Zhang 2009), photoperiod (Hao and Kang 2004; Chen et al. 2009), prevention and control (LIU et al. 2006; Cease et al. 2015; Qin et al. 2017a), genetic diversity (Gao et al. 2011; Han et al. 2013), and response to climate warming (Liu et al. 2006), recent on food preference and food utilization (Huang et al. 2017b), and phenotypic plasticity (Qin et al. 2017a).

However, a mechanism for grasshopper plague under global climate change, long-term grazing, and human inference is still complicated and remains unclear. Studies related to the ecological and molecular mechanism of grasshopper environment adaptation are still under study. It is a great need to develop new tools to reach an effective and proper grasshopper control and management.

DOI: 10.1201/9781003286400-14

## 14.2 Grasshopper Food Preference and Selection

The grasshopper performance related to its outbreaks is affected by biological and abiotic factors. Among them, food plants are the most important factors affecting the viability of grasshoppers (Zhang et al. 2013a). According to the degree of selectivity of insects to host plants, herbivorous insects can be divided into three categories: (1) Monophagy, that is, insects that only feed on one species or feed on their closely related plants; (2) Oligophagy, that is, insects that feed on plants of one or several genera (or families); and (3) Polyphagy, that is, insects that can feed on multiple families. Strictly speaking, there is no clear boundary between the three (Hengeveld 1980; Biondi 1996). Both herbivorous grasshopper and adult grasshopper often feed on the same species of host plants. In addition, the host plant community can provide food for grasshoppers and is the sheltering site for biological activities, such as resting, mating, avoiding predators, and oviposition. Many people have studied the selective relationship between grasshoppers and host plants. Williams (1954) believes that different subfamilies of grasshopper have different feeding habits, and subfamily is more diverse and prefers gramineous plants (Williams 1954), while some species prefer broadleaf plants. "Soviet fauna" described 71 host plants grasshoppers, wherein 17 grasshoppers, including Oxya genus *Oxya*, sugarcane Fieber *Hieroglyphus* and *Hieroglyphus* and nude *Conophyma Conophyma mirabile*, *Mirabile et al* prefer grasses such as rice, sugar cane, and corn, while others prefer dicotyledonous host plants. Uvarov (1977) believes that Italian cockroaches do not prefer monocots, but they were found to prefer Cruciferae, Compositae, Labiatae, and Convolvulus host plants (Uvarov 1977). Musain classified grasshopper into three criteria: frequently-eating, occasional-eating, and anti-feeding, based on the preference of grasshopper for feeding on host plants. Li Hongchang et al divided the typical grassland grasshoppers in Inner Mongolia into four levels: hyperphagic, preferred, favorable, and non-preferred (Li et al. 1987; Yonglin 1987). Huang Chunmei (1995) reported the relationship between the dominant species of grasshoppers and the classification system in Xinjiang, China, and found that the subfamily, Locustinae, Catantopinae, and Oedipodinae prefers the dicotyledon, monocotyledon and both of them Oedipodinae prefer the dicotyledon, respectively (Chunmei 1995).

It is known that food is the most important factor affecting the viability of grasshoppers (Zhang et al. 2013a). Grasshoppers have certain feeding selectivity and the same species have different feeding intakes at a different developmental stage. However, feeding by grasshoppers is not always harmful to plants. Lu Hui et al. studied the effects of *O. asiaticus* on the forage yield at different ages. It was concluded that intake by *O. asiaticus* in the larval stage had an obvious overcompensation effect on pasture, with the amount of grass production increasing significantly. However, with the age of grasshopper increasing, the pasture yield decreased because the food intake increased rapidly (Lu and Zhang 2009). Yang Qunfang et al. studied the feeding habits of Tibetan migratory grasshoppers and found that the feeding capacity of 1–3-year-old Tibetan grasshoppers is relatively small. After the fourth instar, the food intake of Tibetan grasshoppers has increased dramatically. The fifth instar nymphs and adults can be called the gluttony period. The study found that the grasshopper species differed in food consumption in Tibet (Qing and Chunxian 2008; Zaki et al. 2015). For example, first instar Tibetan grasshopper consumed about two times the amount of wormwood than that of barley. However, with the age increasing, the amount of intake increased rapidly, and the proportion of turfgrass decreased. Feng Guanghan studied the *Myrmeleotettix palpalis*, *Chorthippus dubius*, *Euchorthippus cheui*, and *Angaracris rhodopa* under outdoor conditions, and their results showed that the grasshopper had a big difference in food intake, and food intake in the female is larger than that of the male (Guanghan 2000). At the same temperature, male adults have higher transformation efficiency than female adults, and the digestive efficiency and utilization rate of female adults in the ovarian development stage is higher than that in the somatic growth stage (Mordue et al. 1970; Singhal 1979).

## 14.3 Effects of Plant Community on the Performance of Grasshopper

Population density, growth, and reproductive capacity of grasshoppers are significantly affected by food vegetation (Andersen et al. 2001). The growth, development, and reproduction of herbivorous insects

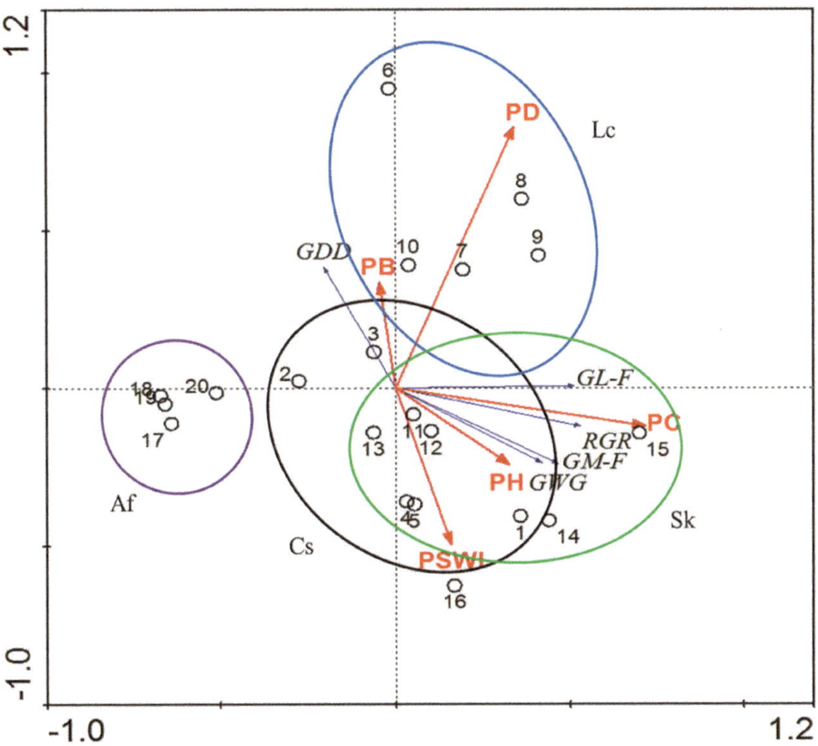

**FIGURE 14.1** The effect of plant community on grasshopper performance. GM-F, grasshopper female mass; RGR, grasshopper relative growth rate; GL-F, grasshopper female length; GDD, grasshopper development time; GWG, grasshoppers' mass gain; GSR, grasshopper survival rate; PSWI, plant Shannon-Wiener index; PH, plant height; PD, plant density; PB, plant biomass; PC, plant coverage; Cs, *C. squarrosa* habitat type; Lc, *L. chinensis* habitat type; Sk, *S. krylovii* habitat type; Af, *A. frigida* habitat type. (Data are taken from Qin 2016.)

depend on feeding on the host plant having enough nutrients for supporting insects' requirements (Facknath 2005; Facknath and Lalljee 2005). The population growth of herbivorous insects, in terms of survival rate, life span, and egg production, are also affected by nutrition required by grasshopper species and living stages. Miura and Ohsaki (2004) found that grasshoppers had higher survival rates when they were fed with four poor plant mixes than those fed with a single plant (Miura and Ohsaki 2004). Arianne found that high-nitrogen foods could decrease the body size and viability of *O. asiaticus. O. asiaticus* prefers low-nitrogen plants (*S. grandis*) which are conducive to the growth and development of grasshopper species (Cease et al. 2012). However, other experiments have shown that the Locusta migratoria laid the highest number of eggs when they feed on gramineous plants, followed by sedge plants, and the lowest number of eggs were observed when fed with dicotyledonous plants. Feeding on different food plants during the adult stage in northern Hebei has a great influence on the fecundity of grasshoppers. The number of eggs in the Yuebei grasshopper with mulberry and maple is two times higher than that fed with white orchids, and the average anomaly rate of eggs fed with white orchids reached 45%, about two times higher than that fed with mulberry and maple. This indicates that fertility can be influenced by the nutritional status of feeding with different plants at the adult stage.

Besides, plant distribution and composition in the community can also have large effects on grasshopper performance. A study found that community structure, e.g., plant density, biomass, height, and ground cover, can significantly influence the relative growth rate, body size, and overall performance of grasshoppers (Qin 2016; Qin et al. 2017a) (Figure 14.1). Characteristics of different plant communities differ in their actions on grasshopper phenotypes. Therefore, plant management practice could be a tool to control the grasshopper plague.

## 14.4 Effects of Plant Communities on Grasshopper Community Structure

Although the grasshopper has a strong survival instinct, its plague still depends on external environmental factors. The climatic and meteorological factors affecting the outbreaks of grasshoppers mainly include temperature, light, precipitation, drought and flood, and global warming. The relationship between plant community diversity, community structure, and grasshopper outbreak indicates that plant community not only provides food, suitable microclimate, and habitat resources for grasshoppers but also shapes the structure of grasshopper community (Joern 1979; Torrusio et al. 2002; Chang et al. 2006).

Zhao Chengzhang et al. show that the density of *O. asiaticus* is significantly and positively correlated with the coverage of the *Stipa krylovii* (Zhang et al. 2012). Within a certain range, the number of occurred *O. asiaticus* is negatively correlated with plant cover and negatively correlated with plant biomass (Yong-Lin 1997; Zhang et al. 2013b; Tariq et al. 2013). Vegetation coverage and height can directly affect the survival of grasshoppers. Studies have shown that grasshopper prefers to lay eggs on bare land in grasslands with a vegetation cover of less than 50% (Zhang et al. 2013b). Due to the unreasonable use of grassland, overgrazing, and climatic factors, grassland degradation and desertification result in more sparse vegetation grasslands that provide suitable places for grasshoppers to lay eggs. With the increase of degradation, such sites have increased significantly. At the same time, this degraded environment is suitable for the growth and development of grasshopper individuals, while a large number of grasshopper individuals further deteriorate this environment, thus forming a vicious circle (Qin et al. 2015; Qin 2016; Qin et al. 2017b).

The distribution and population structure of grasshopper are closely related to plant community types and structures (Qin 2016; Qin et al. 2017b). Different grasshopper species have different sensitivities and adaptabilities to food and habitat (Holmes et al. 1979). There are great differences in the structure of grasshopper communities formed in different habitats (Zhongcheng and Yonglin 1998; Sun et al. 2010). Studies have shown that intensity of grazing affects the habitat selection of *O. asiaticus* (Rasheed et al. 2015; Qin et al. 2017b). *O. asiaticus* prefers to inhabit habitats with high grazing intensity and low nitrogen content in plants (Cease et al. 2012). The study found that grasshopper food selection and environment selection are not consistent, and certain mechanisms related to grasshopper aggregation may be responsible for grasshopper habitat selection in Inner Mongolia grassland (Hui-Hui et al. 2012).

## 14.5 Habitat Selection and Habitat Adaptation of Grassland Grasshoppers

Habitat selection, the selection or preference of living space, evolved in the evolutionary process of animals' long-term interaction with natural environmental factors and ecological factors, has become one of the important ecological characteristics of animals. There are many factors to determine the habitat selection of animals, such as habitat characteristics, the characteristics of the animal itself, the validity of food, predation, and competition. An evolutionary survival species would evaluate pros and cons and select those habitats that will maximize reproductive success (Kang and Chen 1994a). The composition and changes of habitat structure have an important impact on the population of animals (Kang and Chen 1994a; Qin et al. 2015). Meanwhile, in a favorable habitat to animals, habitat quality will decline with the increase of population density and affect the habitat selection of animals (Wiens 1976; Southwood 1977). The habitat selection of animals is an important research field in the current international ecological research. Since MacArthur et al. (MacArthur and Pianka 1966) first proposed the theory of optimal habitat selection, other theories such as the individual's non-density-constrained habitat selection theory (Rosenzweig 1974), habitat selection theory under species competition (Fretwell and Lucas 1970), two species or population density constrained habitat selection theory (Morris 1989) soon were proposed to explain the mechanism of habitat selection.

Grasshopper habitat study with remote sensing in China started in the 1990s. The research team represented by experts such as Jiang Jianjun, Ni Shaoxiang, Wei Yuchun, and others used the TM image data, grassland-type map, geographic aid data, and GPS field survey data to study the grasshopper distribution and monitoring. They obtained the remote sensing image and established a geographic database and expert knowledge system. They did a systematic and complete analysis of the habitat factors, climatic factors, and topographic factors with grasshopper dynamics in the Qinghai Lake area. Great achievements have been made from those projects that provide reliable and valuable data and necessary guidelines for the prevention and management of grasshoppers in the area.

The adaptation of habitat is to enable animals to live in a specific and specific environment, which is conducive to the accumulation of life experience and better adapt to their living environment. Ecological and geographical conditions of grassland, including climate, land, and vegetation, can be classified into different ecosystem types. They inevitably produce a very profound even decisive impact on many aspects of the geographical distribution of grasshoppers. That is to say, the species composition, population characteristics, and dynamics of population changes of existing grassland insects in different types of zones (or regions) occur and are developed under certain ecological and geographical conditions (material and energy). The complexity and diversity of grassland climate and topography have led to the diversity of ecological conditions in different environments, which determines or affects the species and geographical distribution of grasshoppers. In the grassland, due to differences in topography, soil, and microclimate conditions, habitat types with different vegetation compositions are developed to be different sub-types. The difference in habitat types has an important impact on the dynamics of the important primary consumers and grassland grasshoppers in the grassland ecosystem. The adaptability of grasshoppers to typical grasslands is determined according to the suitability of grasshoppers (Kang and Chen 1994b). Kang Le analyzed the ecological factors influencing the grasshopper composition and structure as well as the grasshopper habitat selection, explained and discussed the internal competition relations between grasshopper community and plant community, grasshoppers and habitat, as well as the grasshopper diversity and resource utilization from the perspective of energy, nutrition niche, niche and multidimensional resource theory (Kang and Chen 1995; Ahmad et al. 2021). They also suggested studying the ecological effects of the grasshopper community from the environmental factors themselves (Kang et al. 2007).

The grasshopper habitat selection is related to habitat horizontal structure, habitat vertical structure, habitat biodiversity, soil water content, and dispersal ability (Kang et al. 2007; Qin et al. 2015; Qin 2016). Zhang Weizhong (2013) showed that the main factors affecting the viability of grasshoppers were vegetation coverage, insect density, and food intake (Zhang et al. 2013b). Grasshopper preferred to inhabit habitats dominated by *Stipa grandis* rather than *Leymus chinensis*. Wuhui Hui discovered that *O. asiaticus* food consumption decreased with the increase in the plant total coverage measured by PCR quantitative method, which explains an important reason that how plant characteristics influence grasshopper growth and development (Wu 2012).

## 14.6 Molecular and Ecological Basis of Habitat Adaptation

Different grasshopper species differed in their nutritional niche widths (Qin et al. 2015). In general, if the grasshopper lives in a harsh environment, the nutrient niche is relatively wide, because the organism extends its ability to feed and ingest food resources to achieve an adaptation to the resources in the environment. When the environment is not suitable for survival, or when there is a relative shortage of food resources due to outbreaks of grasshoppers, they will compete and maximize the utilization of resources. If the environment is more suitable and the food resources are relatively sufficient, the grasshoppers will feed on the food resources that are most suitable for their growth and development.

A large number of studies have shown that locust's adaptation to habitat is the result of long-term geographical, environmental, and climatic evolution. Locust's adaptation to habitat is the result of locust's continuous response to environmental changes and genetic variation to its offspring. Based on migratory locust genome sequencing, the cuticle protein gene plays an important role in the adaptability of locusts. In addition, genes in regulating neural plasticity, DNA methylation, and alternative splicing are responsible for rapid adaptation to the environment, which are helpful to improve the grasshopper survival rate, revealing that the locusts can adapt to the environment from the responses of immunity, hypoxia adaptation, nutrition metabolism and social behavior (Wang et al. 2014).

Researchers have found that rapid adaptation (tolerance to stress acclimation) to a harsh environment results in high expression of responsive genes, e.g., heat shock protein Hsp70, Hsp90, and AK. Even after stress acclimation, when the expression level of Hsp70 returned to normal, the effect of acclimation on the survival rate of insects under temperature stress was still significant. However, acclimation time and temperature should not be too drastic. Obviously, insects that develop at non-optimal acclimation temperatures pay a considerable reproductive or developmental cost (Garland and Adolph 1994). Many other researchers have identified that the Hsp family is involved in adaptation to thermal stress in habitat selection (Edgerly et al. 2005; Li et al. 2019).

Besides Hsp, AK (Arginine kinase), AMPK (Adenosine Monophosphate Activated Protein kinase), ALDH, and Vg (Vitellogenin) are related to energy metabolism and environmental stress tolerance in insects. A large number of results showed that when organisms were not suitable for the environment or under nutritional stress, AK expression increased, which could enhance their resistance to the outside world, including enhancing their feeding ability, improving their life vitality, and resisting adverse environment (Pereira et al. 2003; Zhao et al. 2008). AMPK gene is also the regulatory hub of biological energy (Gwinn et al. 2008; Hardie et al. 2012), studies have confirmed that AMPK activation ACTS as an energy receptor to initiate metabolic adaptation during hypoxia, and is an important factor regulating the hypoxia response genes of cells (Hardie 2008). When insects are under stress conditions such as hypoxia, exercise, and energy consumption, ATP consumption increases and AMP concentration increases, which increases the AMP:ATP ratio and activates the AMPK kinase system. When AMPK is activated, ATP is increased and decreased by increasing feeding, glycolysis, and inhibiting protein synthesis (Yang et al. 2013). ALDH is an enzyme that has an oxidative effect on aldehydes in the cytoplasm and mitochondria. Studies have confirmed that disturbance of ALDH expression in drosophila resulted in weaker growth and decreased vitality (Chakraborty and Fry 2011).

AMPK regulates the balance of energy storage substances such as glucose and fat (Gowans and Hardie 2014). The AMPK activity of fasting rats was significantly higher than that of feeding rats (Cohen and Hardie 1991). Under fasting conditions, AMPK activity increased in the liver and fat of small-tailed cold sheep (Xu et al. 2015). High-fat diet components in rats can affect the decreased expression of AMPK in liver tissues (Xiaojie et al. 2009). In addition, AMPK plays an important role in the regulation of hypothalamus feeding, with activating AMPK promoting feeding and inhibiting AMPK causing anorexia (Andersson et al. 2004). When hypoxia occurs, AMPK can be activated to act as an energy receptor to initiate metabolic adaptation, which is an important factor regulating the response of genes to hypoxia (Gowans and Hardie 2014). With the continuous improvement of biotechnology, new genetic technologies have been used to reveal the mechanism of locust adaptation to environmental changes. Currently, genome (Wang et al. 2014), transcriptome (Chen et al. 2010; Qin et al. 2017a), SNP (Wang et al. 2014; Wang and Kang 2014; Berdan et al. 2015), and SSR (Guo et al. 2017) have been adopted to reveal the locusts' response to environmental changes and heritable mechanism.

Vitellogenin is synthesized in a fat body. Usually, after the synthesis of vitellogenin in body fat, it is transferred to the ovary through hemolymph, then absorbed by oocytes through endocytosis, and then deposited in the egg yolk proteins, providing amino acids, fat, carbohydrates, vitamins, and other developmental functional substances for the embryo (Byrne et al. 1989; Falchuk and Montorzi 2001). The study of vitellogenin began in 1954. The vitellogenin was found through studying the silkworm moth. In the 1970s, vitellogenin has been studied in more than 20 insects in the family of Orthoptera (desert grasshoppers) (Tetlak et al. 2015), Lepidoptera (bombyx mori, silkworm moth) (Chino et al. 1977; Sorge et al. 2000), Diptera (fruit fly) (Huybrechts and DeLoof 1982), etc. Research has focused on synthetic sites, synthetic hormone regulation, and physicochemical properties. With respect to the special role of vitellogenin in supporting insect reproduction, it has been reported to be one of the key genes in regulating grasshopper's performance and viability (Zhang et al. 2013a; Qin 2016) in relation to food adaptation and habitat adaption.

Recent research indicated that LCT, peritrophin-1 and cuticle protein (ICP), peritrophin-1 and lactase-phlorizin hydrolase (LCT) interactively regulated feeding and body size exposing to the habitats in O. asiaticus (Qin et al. 2017a). The temperature and humidity in the habitat stimulate the grasshopper body cuticle. Meanwhile, the plants in the habitats affect the expression of LCT. The LCT can affect the synthesis of chitin, which comprises the cuticle of grasshopper regulated by ICP (Figure 14.2). This environment-food circle is a typical mechanism that grasshopper adapts to habitat changes. Food and environmental conditions in the habitats can shape the grasshopper body size and affect grasshopper survival. One important phenomenon is that habitat community structure can mediate the tradeoff between body and survival, therefore regulating the grasshopper dynamic in the community.

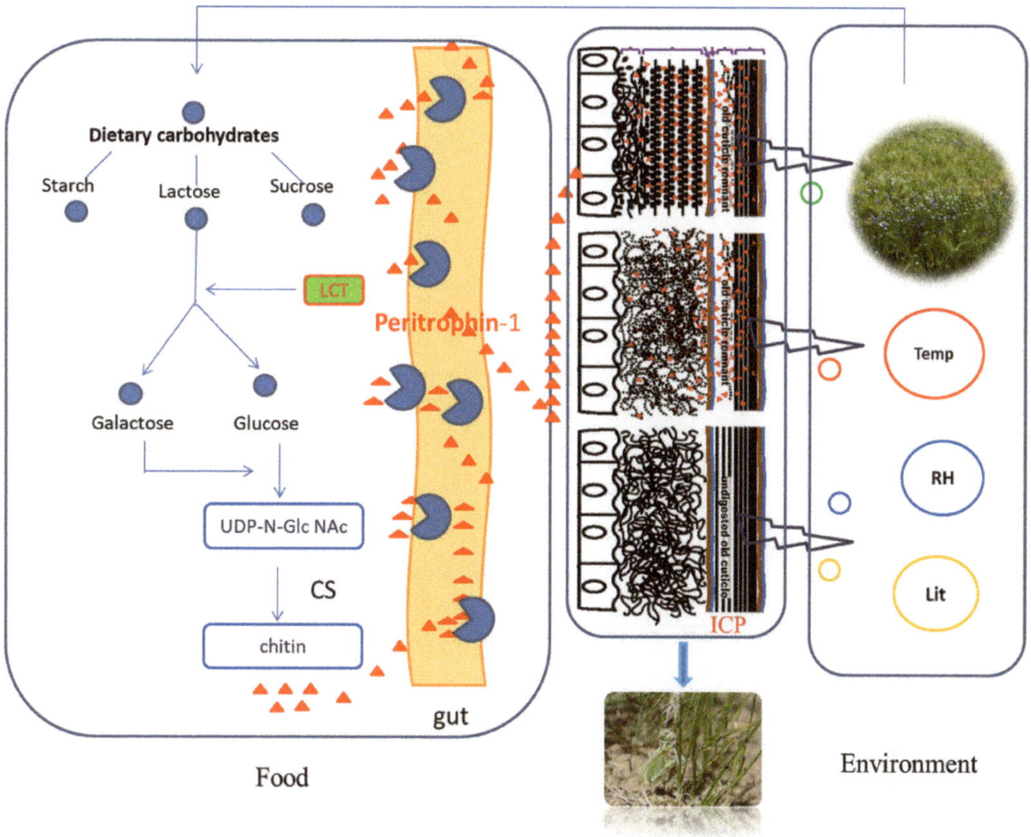


**FIGURE 14.2** The stimuli-response of grasshopper (*O. asiaticus*) in environment-feeding circulation to shape the phenotype of grasshopper. Peritrophin-1; ICP, insect cuticle protein; LCT, lactase-phlorizin hydrolase.

## 14.7 Conclusions

Molecular ecological evidence provides us implications that grasshoppers can be affected by external environmental factors, such as food composition, plant community, and microclimate. These factors stimulate the inner regulating system (e.g., Hsp, LCT) to respond to the food and environment. Food quality meets the diet requirement of grasshopper development and the environment provides the essential climate conditions for grasshoppers. Grasshopper can respond to adapt to the plant community changes and can also be regulated by these factors. Thus, either controlling the target genes or the target ecological factors can make regulations on grasshopper dynamics.

In summary, molecular and ecological factors interact with grasshopper performance, finding molecular and ecological approaches that can affect the grasshopper performance will be the potential tools to control grasshoppers.

## REFERENCES

Ahmad K, Aslam M, Saleem MH, Ijaz M, Ul-Allah SAMI, El-Sheikh AH, Adnan M, Ali S (2021) Genetic diversity and characterization of salt stress tolerance traits in maize (Zea Mays L.) under normal and saline conditions. Pak J Bot 54:3.

Andersen AN, Ludwig JA, Lowe LM, Rentz, D (2001) Grasshopper biodiversity and bioindicators in Australian tropical savannas: responses to disturbance in Kakadu National Park. Austral Ecol 26:213–222.

Andersson U, Filipsson K, Abbott CR, Woods A, Smith K, Bloom SR, et al. (2004) AMP-activated protein kinase plays a role in the control of food intake. J Biol Chem 279:12005–12008.
Berdan EL, Mazzoni CJ, Waurick I, Roehr JT, Mayer F (2015) A population genomic scan in Chorthippus grasshoppers unveils previously unknown phenotypic divergence. Mol Ecol 24:3918–3930.
Biondi M (1996) Proposal for an ecological and zoogeographical categorization of the Mediterranean species of the flea beetle genus Longitarsus Berthold Chrysomelidae. Biology 3:13–35.
Byrne BM, De Jong H, Fouchier RA, Williams DL, Gruber M, Ab G (1989) Rudimentary phosvitin domains in a minor chicken vitellogenin gene. Biochemistry 28:2572–2577.
Cease AJ, Elser JJ, Fenichel EP, Hadrich JC, Harrison JF, Robinson BE (2015) Living with locusts: connecting soil nitrogen, locust outbreaks, livelihoods, and livestock markets. Bioscience 65:551–558.
Cease AJ, Elser JJ, Ford CF, Hao S, Kang L, Harrison, JF (2012) Heavy livestock grazing promotes locust outbreaks by lowering plant nitrogen content. Science 335:467–469.
Cease AJ, Hao, S, Kang, L, Elser JJ, Harrison JF (2010) Are color or high rearing density related to migratory polyphenism in the band-winged grasshopper, Oedaleus asiaticus? J Insect Physiol 56:926–936.
Cease AJ, Harrison JF, Hao S, Niren DC, Zhang G, Kang L, et al (2017) Nutritional imbalance suppresses migratory phenotypes of the Mongolian locust (Oedaleus asiaticus). R Soc Open Sci 4:161039.
Chakraborty M, Fry JD (2011) Drosophila lacking a homologue of mammalian ALDH2 have multiple fitness defects. Chem-Biol Interact 191:296–302.
Chang G, Lian Z-M, Jiang G-F, Wang Y-Q, Liu H (2006) Ordination and environmental factors analysis of grasshopper community in Luo River District of Loess Plateau. Chinese Bulletin Entomol 8.
Chen G, Hao S, Pang B, Kang L (2009) Effect of photoperiod on the development, survival, eclosion and reproduction of 4th instar nymph of three grasshopper species in Inner Mongolia. Chin Bull Entomology 46:51–56.
Chen H, Kang L (2000) Olfactory responses of two species of grasshoppers to plant odours. Entomol Exp et Appl 95:129–134.
Chen S, Yang P, Jiang, F, Wei, Y, Ma, Z, Kang, L (2010) De novo analysis of transcriptome dynamics in the migratory locust during the development of phase traits. PLoS One 5:e15633.
Chino H, Yamagata M, Sato S (1977) Further characterization of lepidopteran vitellogenin from haemolymph and mature eggs. Insect Biochem 7:125–131.
Chunmei H (1995) Feeding habits and subfamily systematics of acrididae in barkol grassland of xinjiang.J Entomotaxonomia 17:128–134.
Cohen P, Hardie DG (1991) The actions of cyclic AMP on biosynthetic processes are mediated indirectly by cyclic AMP-dependent protein kinase. Biochimica et Biophysica Acta (BBA)-Mol Cell Res 1094:292–299.
Edgerly J, Tadimalla A, Dahlhoff E (2005) Adaptation to thermal stress in lichen-eating webspinners (Embioptera): habitat choice, domicile construction and the potential role of heat shock proteins. Funct Ecol 19:255–262.
Facknath S (2005) Leaf age and life history variables of a leafminer: the case of Liriomyza trifolii on potato leaves. Entomol Exp et Appl 115:79–87.
Facknath S, Lalljee B (2005) Effect of soil-applied complex fertiliser on an insect–host plant relationship: Liriomyza trifolii on Solanum tuberosum. Entomol Exp et Appl 115:67–77.
Falchuk KH, Montorzi M (2001) Zinc physiology and biochemistry in oocytes and embryos. In: Maret W (ed) Zinc Biochemistry, Physiology, and Homeostasis. Springer Nature, Switzerland, pp 199–209.
Fretwell S, Lucas H (1970) On territorial behaviour and other factors affecting habitat distribution in birds II. Acta Biotheo 19:1–36
Gao S, Han J, Liu A, Yan Z, Chang X (2011) RAPD Analysis on genetic divergence among populations of Oedaleus asiaticus B Bienko and Oedaleus infernalis Saussure. Acta Agric Bor Sin 26:94–100.
Garland Jr T, Adolph, SC (1994) Why not to do two-species comparative studies: limitations on inferring adaptation. Physiol Zool 67:797–828.
Gowans GJ, Hardie DG (2014) AMPK: a cellular energy sensor primarily regulated. *Biochem Soc Trans.* 42(1):71–75. doi: 10.1042/BST20130244
Guang HS, Lian QQ, Jun WZ, Le K, Chang LH, Lin CY, et al (2002) Management strategies and control techniques for locust and grasshopper plagues around the world: status and perspectives. J Acta Entomol Sinica 4.
Guang L, Zehua Z, Lisheng Z (2007) Control index of Oedaleus asiaticus in Keerqin grasslands. Plant Prot-Beijing 33:63.

Guanghan LCF (2000) Investigations of the species and biology of grasshoppers in high mountain grassland in Gansu province. J Plant Prot 1.

Guo Q, Wang J-X, Su L-Z, Lv W, Sun Y-H, Li Y (2017) Development and evaluation of a novel set of EST-SSR markers based on transcriptome sequences of black locust (Robinia pseudoacacia L). Genes 8:177.

Guo ZX-RCY, Bao-Ping Y-HP (2012) Population dynamics of Oedaleus asiaticus on desert grasslands in Inner Mongolia. J Chinese J Appl Entomol 6.

Gwinn DM, Shackelford DB, Egan DF, Mihaylova MM, Mery A, Vasquez DS, et al (2008) AMPK phosphorylation of raptor mediates a metabolic checkpoint. Mol cell 30:214–226.

Han H, Zhou X, Pang B, Zhang M, Li H (2013) Microsatellite marker analysis of the genetic diversity of Oedaleus asiaticus (Orthoptera: Acrididae) populations in Inner Mongolia, northern China. J Acta Entomol Sinica 56:79–87.

Hao S-G, Kang L (2004) Postdiapause development and hatching rate of three grasshopper species (Orthoptera: Acrididae) in Inner Mongolia. Environ Entomol 33:1528–1534.

Hardie D (2008) AMPK: a key regulator of energy balance in the single cell and the whole organism. Int J obesity 32:S7.

Hardie DG, Ross FA, Hawley SA (2012) AMPK: a nutrient and energy sensor that maintains energy homeostasis. Nat Rev Mol Cell Biol 13:251.

Hengeveld R (1980) Polyphagy, oligophagy and food specialization in ground beetles (Coleoptera, Carabidae). Neth J Zool 30:564–584.

Holmes N, Smith D, Johnston A (1979) Effect of grazing by cattle on the abundance of grasshoppers on fescue grassland. Rangeland Ecol Manag/ J Range Manag Arch 32:310–311.

Huang X, McNeill M, Ma J, Qin X, Tu X, Cao G, et al (2017a) Biological and ecological evidences suggest Stipa krylovii (Pooideae), contributes to optimal growth performance and population distribution of the grasshopper Oedaleus asiaticus. Bull Entomol Res 107:401–409.

Huang X, Whitman DW, Ma J, McNeill MR, Zhang Z (2017b) Diet alters performance and transcription patterns in Oedaleus asiaticus (Orthoptera: Acrididae) grasshoppers. PLoS One 12:e0186397.

Hui-Hui WU, Yun-Hu XU, Cao GC, Gexigeduren, Liu ZY, Bing HE, et al (2012) Ecological effects of typical grassland types in inner mongolia on grasshopper community. Sci Agricul Sinica 20:4178–4186. https://doi.org/10.3864/j.issn.0578-1752.2012.20.007.

Huybrechts R, DeLoof A (1982) Similarities in vitellogenin and control of vitellogenin synthesis within the genera Sarcophaga, Calliphora, Phormia and Lucilia (Diptera). Comparative Biochem Physiol-B: Comparative Biochem 72:339–344.

Joern A (1979) Feeding patterns in grasshoppers (Orthoptera: Acrididae): factors influencing diet specialization. Oecologia 38:325–347.

Kang L, Chen Y (1994a) Multidimensional analysis of resource utilization in assemblages of rangeland grasshoppers (Orthoptera: Acrididae). Insect Sci 1:264–282.

Kang L, Chen Y (1994b) Trophic niche of grasshoppers within steppe ecosystem in Inner Mongolia. Kun chong xue bao J Acta Entomol Sinica 37:178–189.

Kang L, Chen Y (1995) Dynamics of grasshopper communities under different grazing intensities in Inner Mongolian steppes. Insect Sci 2:265–281.

Kang L, Han X, Zhang Z, Sun OJ (2007) Grassland ecosystems in China: review of current knowledge and research advancement. Philos Trans R Soc B: Biol Sci 362:997–1008.

Li H, Wang P, Zhang L, Xu X, Cao Z, Zhang L (2018) Expressions of olfactory proteins in locust olfactory organs and a palp odorant receptor involved in plant aldehydes detection. Front Physiol 9.

Li H, Wang Z, Chen Y (1987) Food consumption and utilization by three species of Acridoids (adult stage) in typical steppe. Acta Ecol Sin 7:331–338.

Li R, Wang Y-T, Jiang G-F (2019) The transcriptome analysis of the bamboo grasshopper provides insights into hypothermic stress acclimation. Int J Biol macromol 134:237–246.

Liu G-s, Zhuang L-w, Guo A-h (2006) Primary study on climatic prediction on nymph stages of ODA (Oedalens asiaticus) in rangeland of Inner Mongolia. J Pratacultural Sci 1.

Lu H, Han J, Zhang Z (2008) Impacts of damage caused by Oedaleus asiaticus on the compensatory growth of plants in steppe. Pratacultural Sci 25:112–116.

Lu H, Zhang L-d (2009) Study on hyper-spectral remote sensing models for monitoring damage of Oedaleus asiaticus (Orthoptera: Acrididae). Spectrosc Spect Anal 29:745–748.

MacArthur RH, Pianka ER (1966) On optimal use of a patchy environment. Am Nat 100:603–609.

Miura K, Ohsaki N (2004) Diet mixing and its effect on polyphagous grasshopper nymphs. Ecol Res 19:269–274.

Mordue W, Highnam K, Hill L, Luntz A (1970) Environmental effects upon endocrine-mediated processes in locusts. Mem Soc Endocrinol 18:111–136.

Morris DW (1989) Density-dependent habitat selection: testing the theory with fitness data *Evolutionary*. Ecol 3:80–94.

Pereira CA, Alonso GD, Ivaldi S, Silber AM, Alves MJM, Torres HN, et al (2003) Arginine kinase overexpression improves Trypanosoma cruzi survival capability. FEBS Lett 554:201–205.

Qin X (2016) Study on the adaptation of Oedaleus asiaticus to typical steppe habitats in Inner Mongolia. Chinese Academy of Agricultural Sciences, Beijing.

Qin X, Hao K, Ma J, Huang X, Tu X, Ali M, et al (2017a) Molecular ecological basis of grasshopper (Oedaleus asiaticus) phenotypic plasticity under environmental selection. Front Physiol 8:770.

Qin X, Hui-Hui WU, Huang X, Wang GJ, Cao GC, Nong XQ, et al (2015) Community structure and ecological niche of grasshopper in typical steppes in Inner Mongolia. Plant Prot 41:17.

Qin X, Ma J, Huang X, Kallenbach RL, Lock TR, Ali M, et al (2017b) Population dynamics and transcriptomic responses of Chorthippus albonemus (Orthoptera: Acrididae) to herbivore grazing intensity. Front Ecol Evol 5:136.

Qing YQLZL, Chunxian YGFCJ (2008) Feeding habits and economic threshold of Locusta migratoria tibetensis. J Acta Phytophylacica Sinica 5.

Rasheed MT, Inayatullah M, Shah B, Ahmed N, Khan A, Ali M, Ahmed S, Junaid K, Adnan M, Huma Z (2015) Relative abundance of insect pollinators on two cultivars of sunflower in Islamabad. J Ento Zool Stud 3(6):164–165.

Rosenzweig ML (1974) On the optimal aboveground activity of bannertail kangaroo rats. J Mammal 55:193–199.

Shen J, Zhang N, He B, Liu C-Y, Li Y, Zhang H-Y, et al (2015) Construction of a GeogDetector-based model system to indicate the potential occurrence of grasshoppers in Inner Mongolia steppe habitats. Bull Entomol Res 105:335–346.

Singhal R (1979) Effect of temperature on consumption, digestion and utilization of food inPoecilocerus pictus Fabr(Orthoptera: Acrididae). Proc: Animal Sci 88:229–239.

Sorge D, Nauen R, Range S, Hoffmann KH (2000) Regulation of vitellogenesis in the fall armyworm, Spodoptera frugiperda (Lepidoptera: Noctuidae). J Insect Physiol 46:969–976.

Southwood TR (1977) Habitat, the templet for ecological strategies? J Animal Ecol 46:337–365.

Sun T, Long R, Liu Z (2010) Community composition, temporal dynamics and biological characteristics of grasshoppers on alpine grassland of the Qilian Mountains China. Chin J Appl Environ Biol 16:550–554.

Tariq M, Khan Z, Arif M, Ali K, Waqas M, Naveed K, Ali M, Khan MA, Shafi B, Adnan M (2013) Effect of nitrogen application timings on the seed yield of Brassica cultivars and associated weeds. Pak J Weed Sci Res 19(4):493–502.

Tetlak AG, Burnett JB, Hahn DA, Hatle JD (2015) Vitellogenin-RNAi and ovariectomy each increase lifespan, increase protein storage, and decrease feeding, but are not additive in grasshoppers. Biogerontol 16:761–774.

Torrusio S, Cigliano MM, De Wysiecki ML (2002) Grasshopper (Orthoptera: Acridoidea) and plant community relationships in the Argentine pampas. J Biogeogr 29:221–229.

Uvarov B (1977) Grasshoppers and locusts A handbook of general acridology Vol 2 Behaviour, ecology, biogeography, population dynamics. Centre for Overseas Pest Research.

Wang X, Fang X, Yang P, Jiang X, Jiang F, Zhao D, et al (2014) The locust genome provides insight into swarm formation and long-distance flight. Nat Comm 5:2957.

Wang X, Kang L (2014) Molecular mechanisms of phase change in locusts. Annu Rev Entomol 59:225–244.

Wiens JA (1976) Population responses to patchy environments. Ann Rev Ecol Syst 7:81–120.

Williams L (1954) The feeding habits and food preferences of Acrididae and the factors which determine them. Trans R Entomol Soc London 105:423–454.

Wu H (2012) Study on the food adaptability of dominant grasshoppers in Inner Mongolia Steppe. In: Chinese Academy of Agricultural Sciences, Beijing.

Xiaojie D, Youmin W, Liping W (2009) Effects of high-fat diet on expression and activity of AMPK in liver tissues of rats. J Acta Universitatis Medicinalis Anhui 6.

Xu J, Chen L, Tang L, Chang L, Liu S, Tan J, et al (2015) Electroacupuncture inhibits weight gain in diet-induced obese rats by activating hypothalamicLKB1-AMPK signaling. BMC Complemen Altern Med 15:147.

Yang L, Sha H, Davisson RL, Qi L (2013) Phenformin activates the unfolded protein response in an AMP-activated protein kinase (AMPK)-dependent manner. J Biol Chem 288:13631–13638.

Yong-Lin YZ-C (1997) Observation of plant height selection in 3 species of grasshoppers in grasslands of the Xilin Gol River Basin, Inner Mongolia. J Entomol Knowledge 4.

Yonglin LHWZC (1987) Food consumption and utilization by three species of acridoids (adult stage) in typical steppe. J Acta Ecologica Sinica 4:005.

Zaki AB, Ahmad N, Khan IA, Shah B, Khan A, Rasheed MT, Adnan M, Junaid K, Huma Z, Ahmed S (2015) Adulticidal efficacy of *Azadirachtaindica* (neem tree), *Sesamumindicum*(til) and *Pinussabinaena*(pine tree) extracts against *Aedesaegyptiunder* laboratory conditions. J Ento Zool Stud 3(6):112–116.

Zhang J, Zhao C, Yin C, Li L, Hou Z, Zhang J (2012) Relationship between the spatial pattern of nymph and adult abundance of Oedaleus asiaticus (Orthoptera: Acrididae) and topography in the alpine grassland in the upper reaches of Heihe River analyzed with the GAM model. J Acta Entomol Sinica 55:1368–1375.

Zhang N, Zhang H-Y, He B, Xin Z-Y, Lin H (2015) Spatiotemporal heterogeneity of the potential occurrence of Oedaleus decorus asiaticus in Inner Mongolia steppe habitats. J Arid Environ 116:33–43.

Zhang W-z, He B, Cao G-c, Zhang Z-h, Wu Y-h, Liu S-c, et al (2013a) Quantitative analysis of the effects of Stipa krylovii and Leymus chinensis on the factors of vitiality of Oedaleus decorus asiaticus. Acta Prataculture Sinica 22:302–309.

Zhang WZ, Bing HE, Cao GC, Zhang ZH, Ya-Han WU, Liu SC, et al (2013b) Quantitative analysis of the effects of Stipa krylovii and Leymus chinensis on the factors of vitiality of Oedaleus decorus asiaticus. Acta Prataculturae Sinica 22:302–309.

Zhao Y, Yang G, Wang-Pruski G, You M (2008) Phyllotreta striolata (Coleoptera: Chrysomelidae): arginine kinase cloning and RNAi-based pest control. Eur J Entomol 105.

Zhongcheng Y, Yonglin C (1998) Habitat selection in animals. Chin J Ecol 2.

# 15

## Climate Change: A Global Perspective

**Imran, Amanullah, Shah Khalid, and Muhammad Arif**
*The University of Agriculture*

**Abdel Rahman Altawaha**
*Al Hussein Bin Talal University*

**Shah Fahad**
*Hainan University*
*The University of Haripur*

**Muhammad Adnan**
*University of Swabi*

**CONTENTS**



### 15.1 Introduction

Carbon pollution and climate change mean both more floods and droughts making it harder and harder for farmers to grow good-quality, nutritious food. Farmers around the world depend on a stable climate to grow their crops and put food on our plates. But as climate change leads to more droughts, floods, and extreme weather, we see harvests wither or wash away. In 2011, during Mexico's worst-ever drought on record, more than 2 million acres of crops were lost. Climate change also makes soil less suitable for farming, as nutrients for plant growth are stripped out by droughts and floods.

Some deniers will argue that carbon dioxide is "plant food" and that more of it will help crops flourish. But this is a huge oversimplification of science. Plants also need water, but does that mean a flood is a good way to get it to them? Absolutely not. The same holds true for carbon. Experiments show that some plants may do better in a world with more carbon dioxide, but many actually show signs of damage. Not convinced? Recent research has revealed that elevated levels of carbon dioxide in our atmosphere can actually have a "junk food effect" on typically nutritious foods. As increasing carbon dioxide speeds up photosynthesis, plants produce more carbohydrates to keep up with their rapid growth at the expense of the minerals, vitamins, and proteins that take longer to build.

Climate refers to the average weather conditions in a place over many years (usually at least 30 years, to account for the range of natural variations from one year to another). For example, the climate in Swat Pakistan (Especially upper swat) is cold and snowy in the winter, while Peshawar Pakistan climate is hot and humid (Imran et al. 2015a,b,c,d,e; Imran et al. 2016a,b) the climate in one area, like Swat or Peshawar, is called a regional climate. The regional climate has great importance in respect to

DOI: 10.1201/9781003286400-15

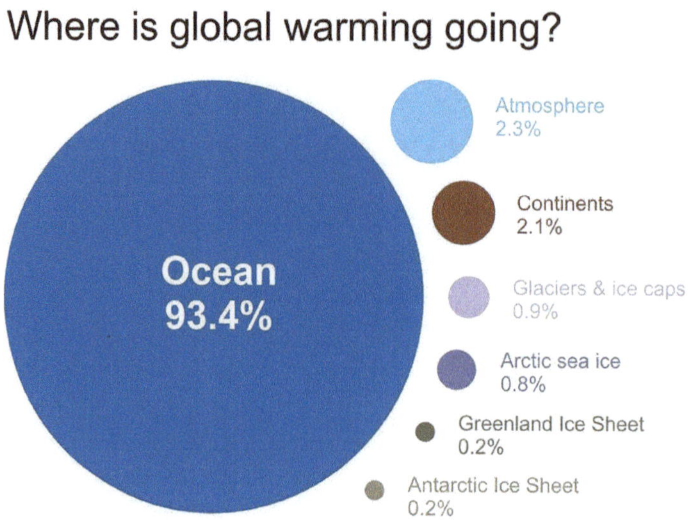


**FIGURE 15.1** A trend of global warming and its negative effects on the ocean, atmosphere, continents, glacier & ice caps, arctic sea ice, green land ice sheet, and Antarctic ice sheet.

agricultural productivity, food security, and livelihood (Figure 15.1). Recent studies have shown that plants and crops respond positively to regional climate and show an ameliorating effect in terms of grain yield, plant height, thousand-grain weight, dry matter portioning, biological yield, oil yield, and quality of the crops (Imran and Asad 2015; Naveed et al. 2016; Imran et al. 2015a,b,c,d,e; Fatima et al. 2016). A significant climate change occurs slowly and gradually and influences life on the earth. Climate can be defined as "expected weather." When changes in the expected weather occur, it is known as climate change. They can be defined by the differences between average weather conditions at two separate times. Climate may change in different ways over different time scales and at different geographical scales. In recent times, scientists have become interested in global warming, due to mankind's impact on the climate system, through the enhancement of the natural greenhouse effect (Iqbal et al. 2016; Samreen et al. 2016; Anwar et al. 2016; Khan et al. 2016a,b; Imran et al. 2016a,b; Deepranjan et al. 2021; Haider et al. 2021; Huang Li et al. 2021; Ikram et al. 2021; Jabborova et al. 2021; Khadim et al. 2021a,b; Manzer et al. 2021; Muzammal et al. 2021; Abdul et al. 2021a,b; Ashfaq et al. 2021; Amjad et al. 2021; Atif et al. 2021; Athar et al. 2021; Adnan et al. 2018a,b; Adnan et al. 2019; Akram et al. 2018a,b; Aziz et al. 2017a,b; Chang et al. 2021; Chen et al. 2021; Emre et al. 2021; Habib et al. 2017; Hafiz et al. 2016; Hafiz et al. 2019; Ghulam et al. 2021; Guofu et al. 2021; Hafeez et al. 2021; Khan et al. 2021; Kamran et al. 2017; Muhammad et al. 2019; Safi et al. 2021; Sajjad et al. 2019; Saud et al. 2013; Saud et al. 2014; Saud et al. 2017; Saud et al. 2016; Shah et al. 2013; Saud et al. 2020; Qamar et al. 2017; Hamza et al. 2021; Irfan et al. 2021; Wajid et al. 2017; Yang et al. 2017; Zahida et al. 2017; Depeng et al. 2018; Hussain et al. 2020; Hafiz et al. 2020a,b; Shafi et al. 2020; Wahid et al. 2020; Subhan et al. 2020; Zafar-ul-Hye et al. 2020a,b; Zafar et al. 2021; Adnan et al. 2020; Ilyas et al. 2020; Saleem et al 2020a,b,c; Rehman et al. 2020; Farhat et al. 2020; Wu et al. 2020; Mubeen et al. 2020; Farhana et al. 2020; Jan et al. 2019; Wu et al. 2019; Ahmad et al. 2019; Baseer et al. 2019; Hafiz et al. 2018; Tariq et al. 2018). The earth is currently getting warmer because people are adding heat-trapping greenhouse gases to the atmosphere. The term "global warming" refers to warmer temperatures, while "climate change" refers to the broader set of changes that go along with warmer temperatures, including changes in weather patterns, the oceans, ice and snow, and ecosystems around the world (Imran et al. 2015a,b,c,d,e; Iqbal et al. 2016; Samreen et al. 2016; Anwar et al. 2016; Khan et al. 2016a,b and Imran et al. 2016a,b). The average climate around the world is called global climate. When scientists talk about global climate change, they're talking about a pattern of changes happening around the world over many years. One of the most important trends that scientists look for it is the average temperature of the earth, which has been increasing for many years

(Figure 15.1). The increase in temperature adversely affected all sectors of life but the most vulnerable is agriculture. Water shortage is enhancing day by day due to climate change at arid and semi-arid regions, which leads to demolish crop production practices (Imran and Asad 2015; Naveed et al. 2016; Ali et al. 2013; Imran et al. 2015a,b,c,d,e; Khan et al. 2016a,b; Fahad and Bano 2012; Fahad et al. 2017; Fahad et al. 2013; Fahad et al. 2014a,b; Fahad et al. 2016a,b,c,d; Fahad et al. 2015a,b; Fahad et al. 2018a,b; Fahad et al. 2019a,b; Fahad et al. 2020; Fahad et al. 2021a,b,c,d,e,f; Hesham and Fahad 2020).

## 15.2 Climate Change and Health Problems

Climate change impacts human health in countless ways, but the following four are worth emphasizing here.

**Rising temperatures**: As temperatures climb around the globe, we expect to see more heatwaves and ever-more intense onesfrom the previous years. Extreme heat can "overpower the human body" and cause dehydration, heatstroke, and major organ damage (Figure 15.2). And certain populations are more at risk from the impacts of heatwaves than others, including the elderly, children, and the poor.

**Air quality**: Pollution from burning fossil fuels is bad enough for the air we breathe, but many impacts of climate change also impact air quality. For instance, climate change has been linked to more wildfires. Wildfire smoke carries fine particles that "can penetrate deep into your lungs." Exposure has been linked to burning eyes, heart and lung diseases, and even death.

**Vector-borne diseases**: Vector-borne diseases are illnesses spread by insects or arachnids such as mosquitoes, fleas, mites, and ticks. As our climate becomes warmer, some insects will see their geographic ranges grow – bringing the Lyme disease and West Nile or Zika viruses they carry along with them to new regions.

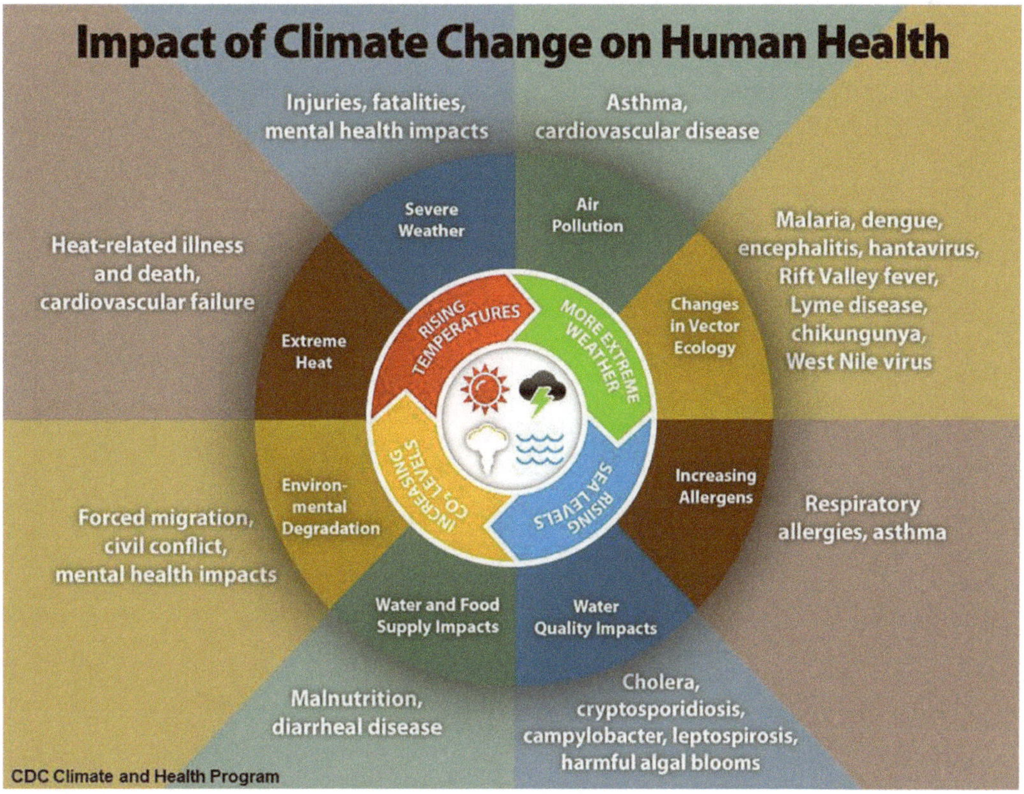


**FIGURE 15.2** How climate change will affect human health in various ways.

**Extreme weather**: While we go into more detail, climate change has been linked to many types of extreme weather, including hurricanes and floods. Not only can these extreme weather events have immediate fatal consequences, but they can lead to major injuries and the spread of waterborne illnesses such as wound infections, dermatitis, conjunctivitis, and ear, nose, and throat infections (Figure 15.2).

## 15.3 Climate Change and Water Security

Everyone deserves access to water for their survival. Climate change is making our water supply much less secure (Figure 15.3). Climate change poses a huge threat to something humans need above all else water (Iqra et al. 2020; Akbar et al. 2020; Mahar et al. 2020; Noor et al. 2020; Bayram et al. 2020; Amanullah and Fahad 2018a,b; Amanullah and Fahad 2017; Amanullah et al. 2020; Amanullah et al. 2021; Rashid et al. 2020; Arif et al. 2020; Amir et al. 2020; Saman et al. 2020; Muhammad Tahir et al. 2020; Md Jakir and Allah 2020; Mahmood et al. 2021; Farah et al. 2020; Sadam et al. 2020; Unsar et al. 2020; Fazli et al. 2020; Md. Enamul et al. 2020; Gopakumar et al. 2020; Zia-ur-Rehman 2020; Ayman et al. 2020; Al-Wabel et al. 2020a,b; Senol 2020; Amjad et al. 2020; Ibrar et al. 2020; Sajid et al. 2020; Muhammad et al. 2021; Sidra et al. 2021; Zahir et al. 2021; Sahrish et al. 2022).

We need it for drinking, of course, but also for growing food. We're seeing the dire consequences climate change can have on our water supplies right now. While many factors have led to the water crisis, climate change has undoubtedly made the situation much worse. We know that increasing global temperatures can lead to droughts. We have been experiencing record drought periods in several regions and also half of its average annual rainfall since 2015 (Figure 15.3). Many cities around the world could face a similar fate. We know that the largest reservoir in the United States – Lake Mead – contained less than half the amount of water in 2015 as it did in 2000.

Climate change significantly increases water demand, limiting crop productivity and by reducing water availability in areas where irrigation is most needed or has a comparative advantage. Global atmospheric

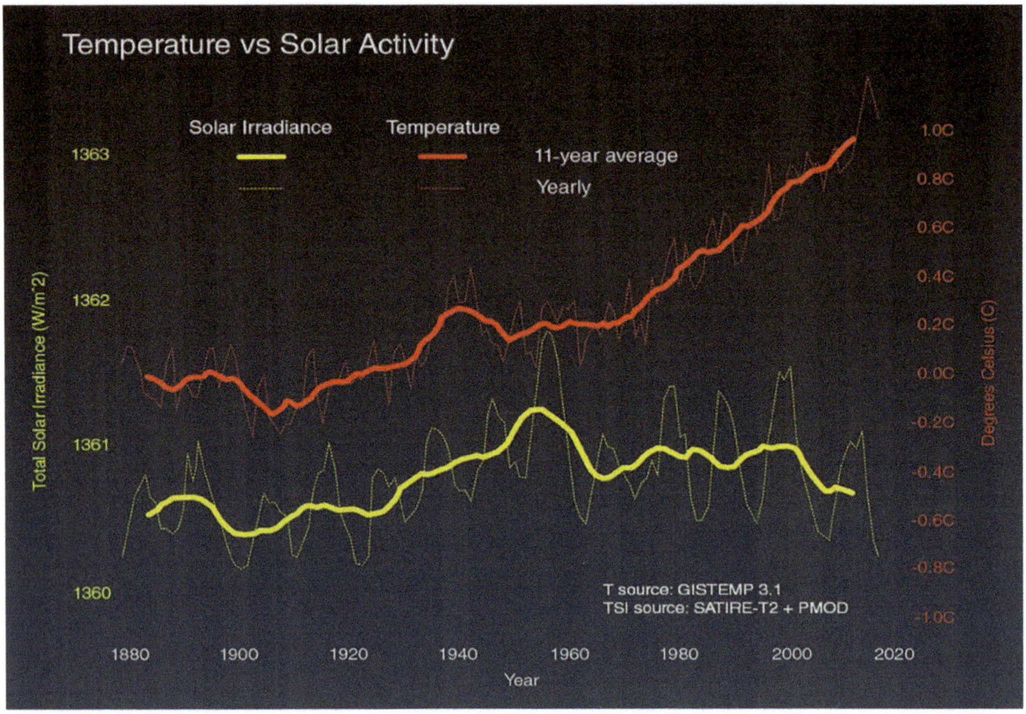


**FIGURE 15.3** A trend of global warming, rising temperature, and solar irradiance.

temperature is predicted to rise by approximately 4°C by 2080, consistent with a doubling of atmospheric carbon dioxide concentration. Mean temperatures are expected to rise at a faster rate in the upper latitudes, with slower rates in equatorial regions. Mean temperature rise at altitude is expected to be higher than at sea level, resulting in an intensification of convective precipitation and acceleration of snowmelt and glacier retreat. In response to global warming, the hydrological cycle is expected to accelerate as rising temperatures increase the rate of evaporation from land and sea. Thus rainfall is predicted to rise in the tropics and higher latitudes but decreases in the already dry semi-arid to arid mid-latitudes and the interior of large continents. Water-scarce areas of the world will generally become drier and hotter. Both rainfall and temperatures are predicted to become more variable, with a consequent higher incidence of droughts and floods, sometimes in the same place. Runoff patterns are harder to predict as they are governed by land use as well as uncertain changes in rainfall amounts and patterns. Substantial reductions (up to −40%) in regional runoff have been modeled in southeastern Australia and in other areas where annual potential evapotranspiration exceeds rainfall. Relatively small reductions in rainfall will translate into much larger reductions in runoff. For example, a 5% fall precipitation in Morocco will result in a 25% reduction in runoff. In glacier-fed river systems, the timing of flows will change, although mean annual runoff may be less affected. As temperature rises, the efficiency of photosynthesis increases.

Artificial irrigation provides approximately 40% of the world's food, including most of its agronomic and horticultural output, from an estimated 20% of agricultural land, or about 300 million ha worldwide. The Green Revolution technology of high inputs of nitrogen fertilizer, applied to responsive short-season varieties of rice and wheat, often required irrigation to realize its potential in Pakistan. Private investment was done in same era in groundwater to make the availability of cheap pumps, power and well construction methods not only did irrigated areas continue to grow, and access to groundwater services expanded. Consequently, aquifers are depleted in many parts of the world, such as China, Pakistan, and the United States, and sometimes facilitated by perverse incentives of subsidized energy and support prices for irrigated products. As the global population heads for more than nine billion people by 2050 (under medium growth projections), the world is rapidly becoming urbanized and wealthier. Food preferences are changing to reflect this, with declining trends in the consumption of staple carbohydrates, and an increase in demand for luxury products – milk, meat, fruits, and vegetables – that are heavily reliant on irrigation in many parts of the world. Future global food demand is expected to increase by some 70% by 2050, but will approximately double for developing countries. All other things being equal (that is a world without climate change), the amount of water withdrawn by irrigated agriculture will need to increase by 11% to match the demand for biomass production. Prices for rice doubled and although commodity prices have fallen back since, the fundamentals (oil price, biofuel development and continued rising food demand) are now expected to drive a period of high volatility in food prices. In the wake of this market turmoil, food security and agricultural livelihoods have regained importance in development planning, although some countries such as China seem ever more likely to balance further agricultural development and investment with imports. Pakistan has a large stock of underperforming canal irrigation infrastructure and a vibrant groundwater sector that is competitively depleting its own lifeblood. Both create significant environmental externalities, which need to be managed. Not only that, there are calls for water to be reserved to maintain environmental flows in rapidly developing river basins and restored to ecosystems in over-allocated ones.

Water management in crop production tends to be concentrated on food crops where the timing and reliability of supply are critical. Water management (irrigation, drainage, and water conservation and control) achieves stability of crop production by maintaining soil conditions close to optimum for crop growth. Irrigation allows the cultivation of crops when rainfall is insufficient, insures high-value, high-risk horticulture from failure and has played a major role in achieving national and regional food security in Asia, as well as improving individual livelihoods. The extent and area of irrigation have grown massively in the 20th century but have depleted surface and groundwater flows, often with severe consequences for aquatic ecosystems and those dependent on them. It is increasingly recognized, although rarely common practice, that greater net socio-economic benefit can be obtained from maintaining the integrity of managed ecosystems. In the future, food security strategies will be more complex. Higher temperatures will increase water demand, and as rainfall declines, many will seek more irrigation to ensure food security and maintain livelihoods. At the same time, water supplies available for irrigation will become more variable and will decline in many parts of the world. New agricultural demands will be further tempered by the need to achieve better equity in access to reliable food supplies than in the past. As irrigation has been practiced on only 20% of the world's cultivated land, there have been many, often the poorest, who have missed out on its benefits. The need to maintain viable aquatic ecosystems will place further stress on water resources, especially where the poorest are dependent on them for their livelihoods. Water allocations to agriculture may fall in many parts of the world owing to the combined impacts of climate change, environmental needs, and competition from higher-value economic sectors. There will be strong pressure to produce more with less water and spread the benefits of all water use more widely and wisely. This task will be even more challenging because higher temperatures will reduce potential land and water productivity. These are not academic considerations. Climatic variability in southeastern Australia has had more profound impacts on water allocations and associated livelihoods in agriculture than even the most prudent farmers had anticipated and big changes lie ahead.

Rainfed lands account for more than 80% of global crop area and 60% of global food output but are especially susceptible to the impacts of climate change, more so in the arid and semi-arid regions. In general, the productivity of rainfed agriculture in Pakistan is considerably lower than in more secure irrigated conditions, where the use of other factor inputs is generally higher. Many sophisticated adaptations have been developed by farmers to allow cropping in precarious arid and semi-arid conditions, including mixed and companion cropping, floodwater spreading, and runoff harvesting. Substantial literature exists on the likely impacts of climate change on rainfed agriculture, with only a skimpy outline given here. Rainfed and pastoral agriculture dominate land use in many countries and are therefore a major determinant of hydrology and runoff in a river basin. The agriculture in Pakistan, especially Khyber Pakhtunkhwa, is predominantly rainfed, with low rainfalls by small subsistence farms. There is a growing realization that the subsistence model of small-holder development cannot power rapid economic growth in Pakistan and that it is unlikely to make the required difference to the lives of the poorest. Various forms of foreign direct investment in agriculture (including land purchase) have generated much debate and apprehension over the past 5–10 years, but the development of an indigenous commercially based agricultural sector is limited. The transaction costs of raising the intensity of farming are lower in commercial farming than in subsistence production, indicating a window of opportunity for production growth. One of the main debates on how to feed the world's growing population is the future balance between irrigated and rainfed agriculture. Higher productivity and relative security of irrigated agriculture are offset by the undesirable environmental consequences of diverting, storing, and consuming streamflow and groundwater. This debate will be further complicated by the impacts of climate change on agriculture. Rainfed agriculture is reliable and productive at higher latitudes and in the humid tropics. However, in the mid-latitudes (sub-humid tropics, seasonally arid, semi-arid, and arid tropics), rainfed agriculture generally has low productivity and is prone to drought and crop failures. Rainfed staples (such as wheat, maize, sorghum, and millet) are generally grown in conditions of deficient or sufficient rainfall. However, rice is naturally adapted to wet environments with excessive rainfall, requiring some measure of water control.

## 15.4 Conclusion

The effect of climate change is not only limited to water availability and uneven rainfall distribution, but it also intensifies floods leading to the damp soil, causes storms, and also increases the air temperature which increases the risk of drought and affects various agronomic crop yields and thus the food security worsens the human living environments. Therefore, a well-coordinated program is necessary to create awareness among different sections of the society including the policymakers, general public, organizations, industrialists, and farmers to sustain the food security and the environment in the days to come where climate change has to be experienced around the world.

## REFERENCES

Abdul S, Muhammad AA, Shabir H, Hesham A. El E, Sajjad H, Niaz A, Abdul G, Sayyed RZ, Fahad S, Subhan D, Rahul D (2021a) Zinc nutrition and arbuscular mycorrhizal symbiosis effects on maize (Zea mays L.) growth and productivity. J Saudi Soc Agric Sci. https://doi.org/10.1016/j.sjbs.2021.06.096.

Abdul S, Muhammad AA, Subhan D, Niaz A, Fahad S, Rahul D, Mohammad JA, Omaima N, Muhammad Habib ur R, Bernard RG (2021b) Effect of arbuscular mycorrhizal fungi on the physiological functioning of maizeunder zinc-deficient soils. Sci Rep 11:18468.

Adnan M, Fahad S, Khan IA, Saeed M, Ihsan MZ, Saud S, Riaz M, Wang D, Wu C (2019) Integration of poultry manure and phosphate solubilizing bacteria improved availability of Ca bound P in calcareous soils. 3 Biotech 9(10):368.

Adnan M, Fahad S, Muhammad Z, Shahen S, Ishaq AM, Subhan D, Zafar-ul-Hye M, Martin LB, Raja MMN, Beena S, Saud S, Imran A, Zhen Y, Martin B, Jiri H, Rahul D (2020) Coupling phosphate-solubilizing bacteria with phosphorus supplements improve maize phosphorus acquisition and growth under lime induced salinity stress. Plants 9(900). https://doi.org/10.3390/plants9070900.

Adnan M, Shah Z, Sharif M, Rahman H (2018b) Liming induces carbon dioxide ($CO_2$) emission in PSB inoculated alkaline soil supplemented with different phosphorus sources. Environ Sci Pollut Res 25(10):9501–9509.

Adnan M, Zahir S, Fahad S, Arif M, Mukhtar A, Imtiaz AK, Ishaq AM, Abdul B, Hidayat U, Muhammad A, Inayat-Ur R, Saud S, Muhammad ZI, Yousaf J, Amanullah, Hafiz MH, Wajid N (2018a) Phosphate-solubilizing bacteria nullify the antagonistic effect of soil calcification on bioavailability of phosphorus in alkaline soils. Sci Rep 8:4339. https://doi.org/10.1038/s41598-018-22653-7.

Ahmad S, Kamran M, Ding R, Meng X, Wang H, Ahmad I, Fahad S, Han Q (2019) Exogenous melatonin confers drought stress by promoting plant growth, photosynthetic capacity and antioxidant defense system of maize seedlings. PeerJ 7:e7793. http://doi.org/10.7717/peerj.7793.

Akbar H, Timothy JK, Jagadish T, M. Golam M, Apurbo KC, Muhammad F, Rajan B, Fahad S, Hasanuzzaman M (2020) Agricultural land degradation: Processes and problems undermining future food security. In: Fahad S, Hasanuzzaman M, Alam M, Ullah H, Saeed M, Khan AK, Adnan M (eds) Environment, climate, plant and vegetation growth. Springer Publ Ltd, Springer Nature Switzerland AG. Part of Springer Nature, pp 17–62. https://doi.org/10.1007/978-3-030-49732-3.

Akram R, Turan V, Hammad HM, Ahmad S, Hussain S, Hasnain A, Maqbool MM, Rehmani MIA, Rasool A, Masood N, Mahmood F, Mubeen M, Sultana SR, Fahad S, Amanet K, Saleem M, Abbas Y, Akhtar HM, Waseem F, Murtaza R, Amin A, Zahoor SA, ul Din MS, Nasim W (2018a) Fate of organic and inorganic pollutants in paddy soils. In: Hashmi MZ, Varma A (eds) Environmental pollution of paddy soils, soil biology. Springer International Publishing AG, Switzerland, pp 197–214.

Akram R, Turan V, Wahid A, Ijaz M, Shahid MA, Kaleem S, Hafeez A, Maqbool MM, Chaudhary HJ, Munis, MFH, Mubeen M, Sadiq N, Murtaza R, Kazmi DH, Ali S, Khan N, Sultana SR, Fahad S, Amin A, Nasim W (2018b) Paddy land pollutants and their role in climate change. In: Hashmi MZ, Varma A (eds) Environmental pollution of paddy soils, soil biology. Springer International Publishing AG, Switzerland, pp 113–124.

Ali K, Arif M, Khan Z, Tariq M, Waqas M, Gul B, Bibi S, Zia-ud-Din, Ali M, Shafi B, Adnan M (2013) Effect of cutting on productivity and associated weeds of canola. Pak J Weed Sci Res 19(4):393–401.

Al-Wabel MI, Abdelazeem S, Munir A, Khalid E, Adel RAU (2020b) Extent of climate change in Saudi Arabia and its impacts on agriculture: A case study from Qassim Region. In: Fahad S, Hasanuzzaman M, Alam M, Ullah H, Saeed M, Khan AK, Adnan M (eds) Environment, climate, plant and vegetation growth. Springer Publ Ltd, Springer Nature Switzerland AG. Part of Springer Nature. pp 635–658. https://doi.org/10.1007/978-3-030-49732-3.

Al-Wabel MI, Ahmad M, Usman AR, Akanji M, Rafique MI (2020a) Advances in pyrolytic technologies with improved carbon capture and storage to combat climate change. In: Fahad S, Hasanuzzaman M, Alam M, Ullah H, Saeed M, Khan AK, Adnan M (eds) Environment, climate, plant and vegetation growth. Springer Publ Ltd, Springer Nature Switzerland AG. Part of Springer Nature. pp 535–576. https://doi.org/10.1007/978-3-030-49732-3.

Amanullah, Fahad S (eds) (2017) Rice - technology and production. IntechOpen, Croatia. http://dx.doi.org/10.5772/64480.

Amanullah, Fahad S (eds) (2018a) Corn - Production and human Health in changing climate. IntechOpen, United Kingdom. http://dx.doi.org/10.5772/intechopen.74074.

Amanullah, Fahad S (eds) (2018b) Nitrogen in agriculture – updates. IntechOpen, Croatia. http://dx.doi.org/10.5772/65846.

Amanullah, Muhammad I, Haider N, Shah K, Manzoor A, Asim M, Saif U, Izhar A, Fahad S, Adnan M, et al. (2021) Integrated foliar nutrients application improve wheat (Triticum Aestivum L.) productivity under calcareous soils in drylands. Commun Soil Sci Plant Anal. https://doi.org/10.1080/00103624.2021.1956521.

Amanullah, Shah K, Imran, Hamdan AK, Muhammad A, Abdel RA, Muhammad A, Fahad S, Azizullah S, Brajendra P (2020) Effects of climate change on irrigation water quality. In: Fahad S, Hasanuzzaman M, Alam M, Ullah H, Saeed M, Khan AK, Adnan M (eds) Environment, climate, plant and vegetation growth. Springer Publ Ltd, Springer Nature Switzerland AG. Part of Springer Nature. pp 123–132. https://doi.org/10.1007/978-3-030-49732-3.

Amir M, Muhammad A, Allah B, Sevgi Ç, Haroon ZK, Muhammad A, Emre A (2020) Bio fortification under climate change: The fight between quality and quantity. In: Fahad S, Hasanuzzaman M, Alam M, Ullah H, Saeed M, Khan AK, Adnan M (eds) Environment, climate, plant and vegetation growth. Springer Publ Ltd, Springer Nature Switzerland AG. Part of Springer Nature. pp 173–228. https://doi.org/10.1007/978-3-030-49732-3.

Amjad I, Muhammad H, Farooq S, Anwar H (2020) Role of plant bioactives in sustainable agriculture. In: Fahad S, Hasanuzzaman M, Alam M, Ullah H, Saeed M, Khan AK, Adnan M (eds) Environment, climate, plant and vegetation growth. Springer Publ Ltd, Springer Nature Switzerland AG. Part of Springer Nature. pp 591–606. https://doi.org/10.1007/978-3-030-49732-3.

Amjad SF, Mansoora N, Din IU, Khalid IR, Jatoi GH, Murtaza G, Yaseen S, Naz M, Danish S, Fahad S, et al. (2021) Application of zinc fertilizer and mycorrhizal inoculation on physio- biochemical parameters of wheat grown under water-stressed environment. Sustainability 13:11007. https://doi.org/10. 3390/su131911007.

Anwar, Israeel BI, Asad AK, Imran, Wajid AS, Mohammad I, Wajid AK, Ikramullah, Wasim A (2016) Nitrogen and phosphorus fertilization of improved varieties for enhancing phenological traits of wheat. Pure Appl Biol 5(3):511–519. http://dx.doi.org/10.19045/bspab.2016.50065.

Arif M, Talha J, Muhammad R, Fahad S, Muhammad A, Amanullah, Kawsar A, Ishaq AM, Bushra K, Fahd R (2020) Biochar; a remedy for climate change. In: Fahad S, Hasanuzzaman M, Alam M, Ullah H, Saeed M, Khan AK, Adnan M (eds) Environment, climate, plant and vegetation growth. Springer Publ Ltd, Springer Nature Switzerland AG. Part of Springer Nature. pp 151–172. https://doi.org/10.1007/978-3-030-49732-3.

Ashfaq AR, Uzma Y, Niaz A, Muhammad AA, Fahad S, Haider S, Tayebeh Z, Subhan D, Süleyman T, Hesham AElE, Pramila T, Jamal MA, Sulaiman AA, Rahul D (2021) Toxicity of Cadmium and nickel in the context of applied activated carbon biochar for improvement in soil fertility. Saudi Soc Agric Sci. https://doi.org/10.1016/j.sjbs.2021.09.035.

Athar M, Masood IA, Sana S, Ahmed M, Xiukang W, Sajid F, Sher AK, Habib A, Faran M, Zafar H, Farhana G, Fahad S (2021) Bio-diesel production of sunflower through sulphur management in a semi-arid subtropical environment. Environ Sci Pollut Res. https://doi.org/10.1007/s11356-021-16688-z.

Atif B, Hesham A, Fahad S (2021) Biochar coupling with phosphorus fertilization modifies antioxidant activity, osmolyte accumulation and reactive oxygen species synthesis in the leaves and xylem sap of rice cultivars under high-temperature stress. Physiol Mol Biol Plants. https://doi.org/10.1007/s12298-021-01062-7.

Ayman EL Sabagh, Hossain A, Barutçular C, Iqbal MA, Islam MS, Fahad S, Sytar O, Çig F, Meena RS, Erman M (2020) Consequences of salinity stress on the quality of crops and its mitigation strategies for sustainable crop production: An outlook of arid and semi- arid regions. In: Fahad S, Hasanuzzaman M, Alam M, Ullah H, Saeed M, Khan AK, Adnan M (eds) Environment, climate, plant and vegetation growth. Springer Publ Ltd, Springer Nature Switzerland AG. Part of Springer Nature. pp 503–534. https://doi.org/10.1007/978-3-030-49732-3.

Aziz K, Daniel KYT, Fazal M, Muhammad ZA, Farooq S, Fan W, Fahad S, Ruiyang Z (2017a) Nitrogen nutrition in cotton and control strategies for greenhouse gas emissions: A review. Environ Sci Pollut Res 24:23471–23487. https://doi.org/10.1007/s11356-017-0131-y.

Aziz K, Daniel KYT, Muhammad ZA, Honghai L, Shahbaz AT, Mir A, Fahad S (2017b) Nitrogen fertility and abiotic stresses management in cotton crop: A review. Environ Sci Pollut Res 24:14551–14566. https://doi.org/10.1007/s11356-017-8920-x.

Baseer M, Adnan M, Fazal M, Fahad S, Muhammad S, Fazli W, Muhammad A, Jr. Amanullah, Depeng W, Saud S, Muhammad N, Muhammad Z, Fazli S, Beena S, Mian AR, Ishaq AM (2019) Substituting urea by organic wastes for improving maize yield in alkaline soil. J Plant Nutr. https://doi.org/10.1080/01904167.2019.1659344.

Bayram AY, Seher Ö, Nazlican A (2020) Climate change forecasting and modeling for the year of 2050. In: Fahad S, Hasanuzzaman M, Alam M, Ullah H, Saeed M, Khan AK, Adnan M (eds) Environment, climate, plant and vegetation growth. Springer Publ Ltd, Springer Nature Switzerland AG. Part of Springer Nature. pp 109–122. https://doi.org/10.1007/978-3-030-49732-3.

Bender M (1996) Variability in the O-2/N-2 ratio of southernhemisphere air, 1991–1994: Implications for the carbon cycle. GlobalBiogeochem. Cycles, 1, 9–21.

Chang W, Qiujuan J, Evgenios A, Haitao L, Gezi L, Jingjing Z, Fahad S, Ying J (2021) Hormetic effects of zinc on growth and antioxidant defense system of wheat plants. Sci Total Environ. https://doi.org/10.1016/j.scitotenv.2021.150992.

Chen Y, Guo Z, Dong L, Fu Z, Zheng Q, Zhang G, Qin L, Sun X, Shi Z, Fahad S, Xie F, Saud S. (2021) Turf performance and physiological responses of native Poa species to summer stress in Northeast China. PeerJ 9:e12252. http://doi.org/10.7717/peerj.12252.

Deepranjan S, Ardith S, Siva OD, Sonam S, Shikha, Manoj P, Amitava R, Sayyed RZ., Abdul G, Mohammad JA, Subhan D, Fahad S, Rahul D (2021) Optimizing nutrient use efficiency, productivity, energetics, and economics of red cabbage following mineral fertilization and biopriming with compatible rhizosphere microbes. Sci Rep 11:15680.

Depeng W, Fahad S, Saud S, Muhammad K, Aziz K, Mohammad NK, Hafiz MH, Wajid N (2018) Morphological acclimation to agronomic manipulation in leaf dispersion and orientation to promote "Ideotype" breeding: Evidence from 3D visual modeling of "super" rice (*Oryza sativa* L.). Plant Physiol Biochem 135:499–510. https://doi.org/10.1016/j.plaphy.2018.11.010.

Emre B, Ömer SU, Martín LB, Andre D, Fahad S, Rahul D, Muhammad Z-ul-H, Ghulam SH, Subhan D (2021) Studying soil erosion by evaluating changes in physico-chemical properties of soils under different land-use types. J Saudi Soc Agric Sci 20:190–197.

Fahad S, Abdul B, Adnan M (eds) (2018b) Global wheat production. IntechOpen, United Kingdom. http://dx.doi.org/10.5772/intechopen.72559.

Fahad S, Adnan M, Hassan S, Saud S, Hussain S, Wu C, Wang D, Hakeem KR, Alharby HF, Turan V, Khan MA, Huang J (2019b) Rice responses and tolerance to high temperature. In: Hasanuzzaman M, Fujita M, Nahar K, Biswas JK (eds) Advances in rice research for abiotic stress tolerance. Woodhead Publ Ltd, Abington Hall Abington, Cambridge, UK, pp 201–224.

Fahad S, Bajwa AA, Nazir U, Anjum SA, Farooq A, Zohaib A, Sadia S, Nasim W, Adkins S, Saud S, Ihsan MZ, Alharby H, Wu C, Wang D, Huang J (2017) Crop production under drought and heat stress: Plant responses and management options. Front Plant Sci 8:1147. https://doi.org/10.3389/fpls.2017.01147.

Fahad S, Bano A (2012) Effect of salicylic acid on physiological and biochemical characterization of maize grown in saline area. Pak J Bot 44:1433–1438.

Fahad S, Chen Y, Saud S, Wang K, Xiong D, Chen C, Wu C, Shah F, Nie L, Huang J (2013) Ultraviolet radiation effect on photosynthetic pigments, biochemical attributes, antioxidant enzyme activity and hormonal contents of wheat. J Food Agric Environ 11(3&4):1635–1641.

Fahad S, Hasanuzzaman M, Alam M, Ullah H, Saeed M, Ali Khan I, Adnan M (eds) (2020) Environment, climate, plant and vegetation growth. Springer Nature Switzerland AG. https://doi.org/10.1007/978-3-030-49732-3.

Fahad S, Hussain S, Bano A, Saud S, Hassan S, Shan D, Khan FA, Khan F, Chen Y, Wu C, Tabassum MA, Chun MX, Afzal M, Jan A, Jan MT, Huang J (2014a) Potential role of phytohormones and plant growth-promoting rhizobacteria in abiotic stresses: Consequences for changing environment. Environ Sci Pollut Res 22(7):4907–4921. https://doi.org/10.1007/s11356-014-3754-2.

Fahad S, Hussain S, Matloob A, Khan FA, Khaliq A, Saud S, Hassan S, Shan D, Khan F, Ullah N, Faiq M, Khan MR, Tareen AK, Khan A, Ullah A, Ullah N, Huang J (2014b) Phytohormones and plant responses to salinity stress: A review. Plant Growth Regul 75(2):391–404. https://doi.org/10.1007/s10725-014-0013-y.

Fahad S, Hussain S, Saud S, Hassan S, Chauhan BS, Khan F, et al (2016a) Responses of rapid viscoanalyzer profile and other rice grain qualities to exogenously applied plant growth regulators under high day and high night temperatures. PLoS One 11(7):e0159590. https://doi.org/10.1371/journal.pone.0159590.

Fahad S, Hussain S, Saud S, Hassan S, Ihsan Z, Shah AN, Wu C, Yousaf M, Nasim W, Alharby H, Alghabari F, Huang J (2016c) Exogenously applied plant growth regulators enhance the morphophysiological growth and yield of rice under high temperature. Front Plant Sci 7:1250. https://doi.org/10.3389/fpls. 2016.01250.

Fahad S, Hussain S, Saud S, Hassan S, Tanveer M, Ihsan MZ, Shah AN, Ullah A, Nasrullah KF, Ullah S, Alharby H, Wu C, Huang J (2016d) A combined application of biochar and phosphorus alleviates heat-induced adversities on physiological, agronomical and quality attributes of rice. Plant Physiol Biochem 103:191–198.

Fahad S, Hussain S, Saud S, Khan F, Hassan S, Nasim W, Arif M, Wang F, Huang J (2016b) Exogenously applied plant growth regulators affect heat-stressed rice pollens. J Agron Crop Sci 202:139–150.

Fahad S, Hussain S, Saud S, Tanveer M, Bajwa AA, Hassan S, Shah AN, Ullah A, Wu C, Khan FA, Shah F, Ullah S, Chen Y, Huang J (2015a) A biochar application protects rice pollen from high-temperature stress. Plant Physiol Biochem 96:281–287.

Fahad S, Muhammad ZI, Abdul K, Ihsanullah D, Saud S, Saleh A, Wajid N, Muhammad A, Imtiaz AK, Chao W, Depeng W, Jianliang H (2018a) Consequences of high temperature under changing climate optima for rice pollen characteristics-concepts and perspectives, Archives Agron Soil Sci. https://doi.org/10.1080/03650340.2018.1443213.

Fahad S, Nie L, Chen Y, Wu C, Xiong D, Saud S, Hongyan L, Cui K, Huang J (2015b) Crop plant hormones and environmental stress. Sustain Agric Rev 15:371–400.

Fahad S, Rehman A, Shahzad B, Tanveer M, Saud S, Kamran M, Ihtisham M., Khan SU, Turan V, Rahman MHU (2019a) Rice responses and tolerance to metal/metalloid toxicity. In: Hasanuzzaman M, Fujita M, Nahar K, Biswas JK (eds) Advances in rice research for abiotic stress tolerance. Woodhead Publ Ltd, Abington Hall Abington, Cambridge, UK, pp 299–312.

Fahad S, Saud S, Yajun C, Chao W, Depeng W (eds) (2021f) Abiotic stress in plants. IntechOpen, United Kingdom. http://dx.doi.org/10.5772/intechopen.91549.

Fahad S, Sönmez O, Saud S, Wang D, Wu C, Adnan M, Arif M, Amanullah (eds) (2021e) Engineering tolerance in crop plants against abiotic stress, 1st edn. ed, Footprints of climate variability on plant diversity. CRC Press, Boca Raton, FL.

Fahad S, Sönmez O, Saud S, Wang D, Wu C, Adnan M, Turan V (eds) (2021b) Climate change and plants: biodiversity, growth and interactions, 1st edn. ed, Footprints of climate variability on plant diversity. CRC Press, Boca Raton, FL.

Fahad S, Sönmez O, Saud S, Wang D, Wu C, Adnan M, Turan V (eds) (2021c) Developing climate resilient crops: improving global food security and safety, 1st edn. ed, Footprints of climate variability on plant diversity. CRC Press, Boca Raton, FL.

Fahad S, Sönmez O, Saud S, Wang D, Wu C, Adnan M, Turan V (eds) (2021a) Plant growth regulators for climate-smart agriculture, 1st edn. ed, Footprints of climate variability on plant diversity. CRC Press, Boca Raton, FL.

Fahad S, Sönmez O, Turan V, Adnan M, Saud S, Wu C, Wang D (eds) (2021d) Sustainable soil and land management and climate change, 1st edn. ed, Footprints of climate variability on plant diversity. CRC Press, Boca Raton, FL.

Farah R, Muhammad R, Muhammad SA, Tahira Y, Muhammad AA, Maryam A, Shafaqat A, Rashid M, Muhammad R, Qaiser H, Afia Z, Muhammad AA, Muhammad A, Fahad S (2020) Alternative and non-conventional soil and crop management strategies for increasing water use efficiency. In: Fahad S, Hasanuzzaman M, Alam M, Ullah H, Saeed M, Khan AK, Adnan M (eds) Environment, climate, plant and vegetation growth. Springer Publ Ltd, Springer Nature Switzerland AG. Part of Springer Nature. pp 323–338. https://doi.org/10.1007/978-3-030-49732-3.

Farhana G, Ishfaq A, Muhammad A, Dawood J, Fahad S, Xiuling L, Depeng W, Muhammad F, Muhammad F, Syed AS (2020) Use of crop growth model to simulate the impact of climate change on yield of various wheat cultivars under different agro-environmental conditions in Khyber Pakhtunkhwa, Pakistan. Arabian J Geosci 13:112 https://doi.org/10.1007/s12517-020-5118-1.

Farhat A, Hafiz MH, Wajid I, Aitazaz AF, Hafiz FB, Zahida Z, Fahad S, Wajid F, Artemi C (2020) A review of soil carbon dynamics resulting from agricultural practices. J Environ Manage 268:110319.

Fatima, Shah M, Usman A, Sohail K, Afzaal M, Shah B, Adnan M, Ahmed N, Junaid K, Shah SRA, Rahman LU (2016) Rearing and identification of *Callosobruchusmaculatus* (Bruchidae: Coleoptera) in Chickpea. J Ento and Zoo Stud 4(2):264–266.

Fazli W, Muhmmad S, Amjad A, Fahad S, Muhammad A, Muhammad N, Ishaq AM, Imtiaz AK, Mukhtar A, Muhammad S, Muhammad I, Rafi U, Haroon I, Muhammad A (2020) Plant-microbes interactions and functions in changing climate. In: Fahad S, Hasanuzzaman M, Alam M, Ullah H, Saeed M, Khan AK, Adnan M (eds) Environment, climate, plant and vegetation growth. Springer Publ Ltd, Springer Nature Switzerland AG. Part of Springer Nature. pp 397–420. https://doi.org/10.1007/978-3-030-49732-3.

Ghulam M, Muhammad AA, Donald LS, Sajid M, Muhammad FQ, Niaz A, Ateeq ur R, Shakeel A, Sajjad H, Muhammad A, Summia M, Aqib HAK, Fahad S, Rahul D, Mazhar I, Timothy DS (2021) Formalin fumigation and steaming of various composts differentially influence the nutrient release, growth and yield of muskmelon (*Cucumis melo* L.). Sci Rep 11:21057.

Gopakumar L, Bernard NO, Donato V (2020) Soil microarthropods and nutrient cycling. In: Fahad S, Hasanuzzaman M, Alam M, Ullah H, Saeed M, Khan AK, Adnan M (eds) Environment, climate, plant and vegetation growth. Springer Publ Ltd, Springer Nature Switzerland AG. Part of Springer Nature. pp 453–472. https://doi.org/10.1007/978-3-030-49732-3.

Guofu L, Zhenjian B, Fahad S, Guowen C, Zhixin X, Hao G, Dandan L, Yulong L, Bing L, Guoxu J, Saud S (2021) Compositional and structural changes in soil microbial communities in response to straw mulching and plant revegetation in an abandoned artificial pasture in Northeast China. Global Ecol Conserv 31:e01871.

Habib ur R, Ashfaq A, Aftab W, Manzoor H, Fahd R, Wajid I, Md. Aminul I, Vakhtang S, Muhammad A, Asmat U, Abdul W, Syeda RS, Shah S, Shahbaz K, Fahad S, Manzoor H, Saddam H, Wajid N (2017) Application of CSM-CROPGRO-Cotton model for cultivars and optimum planting dates: Evaluation in changing semi-arid climate. Field Crops Res. http://dx.doi.org/10.1016/j.fcr.2017.07.007.

Hafeez M, Farman U, Muhammad MK, Xiaowei L, Zhijun Z, Sakhawat S, Muhammad I, Mohammed AA, Mandela F-G, Nicolas D, Muzammal R, Fahad S, Yaobin L (2021) Metabolic-based insecticide resistance mechanism and ecofriendly approaches for controlling of beet armyworm Spodoptera exigua: A review. Environ Sci Pollut Res. https://doi.org/10.1007/s11356-021-16974-w.

Hafiz MH, Abdul K, Farhat A, Wajid F, Fahad S, Muhammad A, Ghulam MS, Wajid N, Muhammad M, Hafiz FB (2020b) Comparative effects of organic and inorganic fertilizers on soil organic carbon and wheat productivity under arid region. Commun Soil Sci Plant Anal. https://doi.org/10.1080/00103624.2020.1763385.

Hafiz MH, Farhat A, Ashfaq A, Hafiz FB, Wajid F, Carol Jo W, Fahad S, Gerrit H (2020a) Predicting kernel growth of maize under controlled water and nitrogen applications. Int J Plant Prod. https://doi.org/10.1007/s42106-020-00110-8.

Hafiz MH, Farhat A, Shafqat S, Fahad S, Artemi C, Wajid F, Chaves CB, Wajid N, Muhammad M, Hafiz FB (2018) Offsetting land degradation through nitrogen and water management during maize cultivation under arid conditions. Land Degrad Dev:1–10. https://doi.org/10.1002/ldr.2933.

Hafiz MH, Muhammad A, Farhat A, Hafiz FB, Saeed AQ, Muhammad M, Fahad S, Muhammad A (2019) Environmental factors affecting the frequency of road traffic accidents: A case study of sub-urban area of Pakistan. Environ Sci Pollut Res. https://doi.org/10. 1007/s11356-019-04752-8.

Hafiz MH, Wajid F, Farhat A, Fahad S, Shafqat S, Wajid N, Hafiz FB (2016) Maize plant nitrogen uptake dynamics at limited irrigation water and nitrogen. Environ Sci Pollut Res 24(3):2549–2557. https://doi.org/10. 1007/s11356-016-8031-0.

Haider SA, Lalarukh I, Amjad SF, Mansoora N, Naz M, Naeem M, Bukhari SA, Shahbaz M, Ali SA, Marfo TD, Subhan D, Rahul D, Fahad S (2021) Drought stress alleviation by potassium-nitrate-containing chitosan/montmorillonite microparticles confers changes in Spinacia oleracea L. Sustainability 13:9903. https://doi.org/10.3390/su13179903.

Hamza SM, Xiukang W, Sajjad A, Sadia Z, Muhammad N, Adnan M, Fahad S, et al. (2021) Interactive effects of gibberellic acid and NPK on morpho-physio-biochemical traits and organic acid exudation pattern in coriander (Coriandrum sativum L.) grown in soil artificially spiked with boron. Plant Physiol Biochem 167:884–900.

Hesham FA, Fahad S (2020) Melatonin application enhances biochar efficiency for drought tolerance in maize varieties:Modifications in physio-biochemical machinery. Agron J 112(4):1–22.

Huang Li Y, Li Xiao X, Zhang Yun B, Fahad S, Wang F (2021) dep1 improves rice grain yield and nitrogen use efficiency simultaneously by enhancing nitrogen and dry matter translocation. J Integrative Agric. https://doi.org/10.1016/S2095-3119(21)63795-4.

Hussain MA, Fahad S, Rahat S, Muhammad FJ, Muhammad M, Qasid A, Ali A, Husain A, Nooral A, Babatope SA, Changbao S, Liya G, Ibrar A, Zhanmei J, Juncai H (2020) Multifunctional role of brassinosteroid and its analogues in plants. Plant Growth Regul. https://doi.org/10.1007/s10725-020-00647-8.

Ibrar K, Aneela R, Khola Z, Urooba N, Sana B, Rabia S, Ishtiaq H, Mujaddad Ur Rehman, Salvatore M (2020) Microbes and environment: Global warming reverting the frozen zombies. In: Fahad S, Hasanuzzaman M, Alam M, Ullah H, Saeed M, Khan AK, Adnan M (eds) Environment, climate, plant and vegetation growth. Springer Publ Ltd, Springer Nature Switzerland AG. Part of Springer Nature. pp 607–634. https://doi.org/10.1007/978-3-030-49732-3.

Ikram U, Khadim D, Muhammad T, Muhammad S, Fahad S (2021) Gibberellic acid and urease inhibitor optimize nitrogen uptake and yield of maize at varying nitrogen levels under changing climate. Environ Sci Pollut Res. https://doi.org/10.1007/s11356-021-16049-w.

Ilyas M, Mohammad N, Nadeem K, Ali H, Aamir HK, Kashif H, Fahad S, Aziz K, Abid U (2020) Drought tolerance strategies in plants: A mechanistic approach. J Plant Growth Regul. https://doi.org/10.1007/s00344-020-10174-5.

Imran, Asad AK (2015) Phenological characteristics of Brassica Napus L. as influenced by biochar application and shoot cutting duration (days). Civi Environ Res. ISSN 2224-5790 (Paper) ISSN 2225-0514 (Online).

Imran, Asad AK, Fayaz A (2015a) Phenology of various rice genotypes as affected by different transplanting dates under cold climatic region of Khyber Pakhtunkhwa-Pakistan. J Environ Earth Sci. ISSN 2224-3216 (Paper) ISSN 2225-0948.

Imran, Asad AK, Inamullah, Luqman (2015b) Weeding stages and their effect on yield and yield components of rice in upper Swat, Pakistan. Pak J Weed Sci Res 21(4):555–563.

Imran, Inayat K, Izaz H, Rehman A, Anwar S, Ahmad F, Khan AA, Zada H (2015c) Growth and yield of maize hybrids as effected by different sowing dates in Swat Pakistan. J Pure Appl Biol 4(4).

Imran, Izaz H, Inayat K, Rehman A, Ahmad F, Zada H, Shahida N (2015d) Roots nodulation, yield and yield contributing parameters of mungbean cultivars as influenced by different Phosphorous level in Swat-Pakistan. Pure Appl Biol 4(4):557–567.

Imran, Khan AA, Ahmad F, Ullah I (2015e) Influence of hydrated calcium sulphate ($CaSO_4.2H_2O$) and nitrogen levels on water infiltration rate and maize varieties productivity in rainfed area of swat, Pakistan. J Chem Res 7(3):15–20.

Imran, Khan AA, Inam I, Ahmad F (2016a) Yield, and yield attributes of Mungbean (*Vigna radiata* L) cultivars as affected by phosphorous levels under different tillage systems. Cogent Food Agric 2:1151982.

Imran, Khan AA, Khan IU, Naveed S (2016b) Weeds density and late sown maize productivity influenced by compost application and seed rates under temperate environment. Pak J Weed Sci Res 22(1):169–181.

Iqbal, Bashir A, Inam U, Imran, Asad AK, Shazma A, Asad A, Khurrum S, Shehryar K (2016) Effect of phosphorus, sulphur and different irrigation levels on phenological traits of Triticale. Pure Appl Biol 5(2):303–310. http://dx.doi.org/10.19045/bspab.2016.50039.

Iqra M, Amna B, Shakeel I, Fatima K, Sehrish L, Hamza A, Fahad S (2020) Carbon cycle in response to global warming. In: Fahad S, Hasanuzzaman M, Alam M, Ullah H, Saeed M, Khan AK, Adnan M (eds) Environment, climate, plant and vegetation growth. Springer Publ Ltd, Springer Nature Switzerland AG. Part of Springer Nature. pp 1–16. https://doi.org/10.1007/978-3-030-49732-3.

Irfan M, Muhammad M, Muhammad JK, Khadim MD, Dost M, Ishaq AM, Waqas A, Fahad S, Saud S, et al (2021) Heavy metals immobilization and improvement in maize (Zea mays L.) growth amended with biochar and compost. Sci Rep 11:18416.

Jabborova D, Sulaymanov K, Sayyed RZ, Alotaibi SH, Enakiev Y, Azimov A, Jabbarov Z, Ansari MJ, Fahad S, Danish S, et al (2021) Mineral fertilizers improves the quality of turmeric and soil. Sustainability 13:9437. https://doi.org/10.3390/su13169437.

Jan M, Anwar-ul-Haq M, Adnan NS, Muhammad Y, Javaid I, Xiuling L, Depeng W, Fahad S (2019) Modulation in growth, gas exchange, and antioxidant activities of salt-stressed rice (Oryza sativa L.) genotypes by zinc fertilization. Arab J Geosci 12:775. https://doi.org/10.1007/s12517-019-4939-2.

Kamran M, Wenwen C, Irshad A, Xiangping M, Xudong Z, Wennan S, Junzhi C, Shakeel A, Fahad S, Qingfang H, Tiening L (2017) Effect of paclobutrazol, a potential growth regulator on stalk mechanical strength, lignin accumulation and its relation with lodging resistance of maize. Plant Growth Regul 84:317–332. https://doi.org/10.1007/s10725-017-0342-8.

Khadim D, Fahad S, Jahangir MMR, Iqbal M, Syed SA, Shah AK, Ishaq AM, Rahul D, et al (2021a) Biochar and urease inhibitor mitigate $NH_3$ and $N_2O$ emissions and improve wheat yield in a urea fertilized alkaline soil. Sci Rep 11:17413.

Khadim D, Saif-ur-R, Fahad S, Syed SA, Shah AK, et al (2021b) Influence of variable biochar concentration on yield-scaled nitrous oxide emissions, Wheat yield and nitrogen use efficiency. Sci Rep 11:16774.

Khan AZ, Imran, Asim M, Aiman K, Hasina G, Habib A, Said W (2016a) Impact of fertilizer priming on seed germination behavior and vigor of maize. Pure Appl Biol 5(4):744–751. http://dx.doi.org/10.19045/bspab.2016.50093.

Khan IA, Shah B, Khan A, Zaman M, Din MMU, Shah SRA, Junaid K, Adnan M, Ahmad N, Akbar R, Fayaz W, Rahman IU (2016b) Screening of different maize Cultivars against maize shootfly and red pumpkin beetle at Peshawar. J Ento Zool Stud 4(1):324–327.

Khan MMH, Niaz A, Umber G, Muqarrab A, Muhammad AA, Muhammad I, Shabir H, Shah F, Vibhor A, Shams H. A-H, Reham A, Syed MBA, Nadiyah MA, Ali TKZ, Subhan D, Rahul D (2021) Synchronization of Boron application methods and rates is environmentally friendly approach to improve quality attributes of Mangifera indica L. On sustainable basis. Saudi J Bio Sci. https://doi.org/10.1016/j.sjbs.2021.10.036.

Mahar A, Amjad A, Altaf HL, Fazli W, Ronghua L, Muhammad A, Fahad S, Muhammad A, Rafiullah, Imtiaz AK, Zengqiang Z (2020) Promising technologies for Cd-contaminated soils: Drawbacks and possibilities. In: Fahad S, Hasanuzzaman M, Alam M, Ullah H, Saeed M, Khan AK, Adnan M (eds) Environment, climate, plant and vegetation growth. Springer Publ Ltd, Springer Nature Switzerland AG. Part of Springer Nature. pp 63–92. https://doi.org/10.1007/978-3-030-49732-3.

Mahmood Ul H, Tassaduq R, Chandni I, Adnan A, Muhammad A, Muhammad MA, Muhammad H-ur-R, Mehmood AN, Alam S, Fahad S (2021) Linking plants functioning to adaptive responses under heat stress conditions: A mechanistic review. J Plant Growth Regul. https://doi.org/10.1007/s00344-021-10493-1.

Manzer HS, Saud A, Soumya M, Abdullah A. Al-A, Qasi DA, Bander MA. Al-M, Hayssam MA, Hazem MK, Fahad S, Vishnu DR, Om PN (2021) Molybdenum and hydrogen sulfide synergistically mitigate arsenic toxicity by modulating defense system, nitrogen and cysteine assimilation in faba bean (*Vicia faba* L.) seedlings. Environ Pollut 290:117953.

Md Jakir H, Allah B (2020) Development and applications of transplastomic plants; a way towards eco-friendly agriculture. In: Fahad S, Hasanuzzaman M, Alam M, Ullah H, Saeed M, Khan AK, Adnan M (eds) Environment, climate, plant and vegetation growth. Springer Publ Ltd, Springer Nature Switzerland AG. Part of Springer Nature. pp 285–322. https://doi.org/10.1007/978-3-030-49732-3.

Md. Enamul H, Shoeb AZM, Mallik AH, Fahad S, Kamruzzaman MM, Akib J, Nayyer S, K. M. Mehedi A, Swati AS, Md Yeamin A, Most SS (2020) Measuring vulnerability to environmental hazards: Qualitative to quantitative. In: Fahad S, Hasanuzzaman M, Alam M, Ullah H, Saeed M, Khan AK, Adnan M (eds) Environment, climate, plant and vegetation growth. Springer Publ Ltd, Springer Nature Switzerland AG. Part of Springer Nature. pp 421–452. https://doi.org/10.1007/978-3-030-49732-3.

Mubeen M, Ashfaq A, Hafiz MH, Muhammad A, Hafiz UF, Mazhar S, Muhammad Sami ul Din, Asad A, Amjed A, Fahad S, Wajid N (2020) Evaluating the climate change impact on water use efficiency of cotton-wheat in semi-arid conditions using DSSAT model. J Water Climate Change. https://doi.org/10.2166/wcc.2019.179/622035/jwc2019179.pdf.

Muhammad N, Muqarrab A, Khurram S, Fiaz A, Fahim N, Muhammad A, Shazia A, Omaima N, Sulaiman AA, Fahad S, Subhan D, Rahul D (2021) Kaolin and Jasmonic acid improved cotton productivity under water stress conditions. J Saudi Soc Agric Sci 28:6606–6614. https://doi.org/10.1016/j.sjbs.2021.07.043.

Muhammad Z, Abdul MK, Abdul MS, Kenneth BM, Muhammad S, Shahen S, Ibadullah J, Fahad S (2019) Performance of Aeluropus lagopoides (mangrove grass) ecotypes, a potential turfgrass, under high saline conditions. Environ Sci Pollut Res. https://doi.org/10.1007/s11356-019-04838-3.

Muhammad Tahir ul Q, Amna F, Amna B, Barira Z, Xitong Z, Ling-Ling C (2020) Effectiveness of conventional crop improvement strategies vs. omics. In: Fahad S, Hasanuzzaman M, Alam M, Ullah H, Saeed M, Khan AK, Adnan M (eds) Environment, climate, plant and vegetation growth. Springer Publ Ltd, Springer Nature Switzerland AG. Part of Springer Nature. pp 253–284. https://doi.org/10.1007/978-3-030-49732-3.

Muzammal R, Fahad S, Guanghui D, Xia C, Yang Y, Kailei T, Lijun L, Fei-Hu L, Gang D (2021) Evaluation of hemp (Cannabis sativa L.) as an industrial crop: A review. Environ Sci Pollut Res. https://doi.org/10.1007/s11356-021-16264-5.

Naveed, Muhammad I, Inayat K, Imtiaz K, Imran, Muhammad IK, Haroon K (2016) Anthelmintic, anti-lice, insecticidal, cytotoxic and phytotoxic potential of ethanolic extracts of two wild medicinal plants *Iphiona Grantioides* and *Plucheaarguta*. J Woulfenia 23(11):13–25.

Noor M, Naveed ur R, Ajmal J, Fahad S, Muhammad A, Fazli W, Saud S, Hassan S (2020) Climate change and costal plant lives. In: Fahad S, Hasanuzzaman M, Alam M, Ullah H, Saeed M, Khan AK, Adnan M (eds) Environment, climate, plant and vegetation growth. Springer Publ Ltd, Springer Nature Switzerland AG. Part of Springer Nature. pp 93–108. https://doi.org/10.1007/978-3-030-49732-3.

Qamar uz Z, Zubair A, Muhammad Y, Muhammad ZI, Abdul K, Fahad S, Safder B, Ramzani PMA, Muhammad N (2017) Zinc biofortification in rice: Leveraging agriculture to moderate hidden hunger in developing countries. Arch Agron Soil Sci 64:147–161. https://doi.org/10.1080/03650340.2017.1338343.

Rashid M, Qaiser H, Khalid SK, Al-Wabel MI, Zhang A, Muhammad A, Shahzada SI, Rukhsanda A, Ghulam AS, Shahzada MM, Sarosh A, Muhammad FQ (2020) Prospects of biochar in alkaline soils to mitigate climate change. In: Fahad S, Hasanuzzaman M, Alam M, Ullah H, Saeed M, Khan AK, Adnan M (eds) Environment, climate, plant and vegetation growth. Springer Publ Ltd, Springer Nature Switzerland AG. Part of Springer Nature. pp 133–150. https://doi.org/10.1007/978-3-030-49732-3.

Rehman M, Fahad S, Saleem MH, Hafeez M, Muhammad Habib ur R, Liu F, Deng G (2020) Red light optimized physiological traits and enhanced the growth of ramie (Boehmeria nivea L.). Photosynthetica 58(4):922–931.

Sadam M, Muhammad Tahir ul Q, Ghulam M, Muhammad SK, Faiz AJ (2020) Role of biotechnology in climate resilient agriculture. In: Fahad S, Hasanuzzaman M, Alam M, Ullah H, Saeed M, Khan AK, Adnan M (eds) Environment, climate, plant and vegetation growth. Springer Publ Ltd, Springer Nature Switzerland AG. Part of Springer Nature. pp 339–366. https://doi.org/10.1007/978-3-030-49732-3.

Safi UK, Ullah F, Mehmood S, Fahad S, Ahmad Rahi A, Althobaiti F, et al (2021) Antimicrobial, antioxidant and cytotoxic properties of Chenopodium glaucum L.. PLoS One 16(10):e0255502. https://doi.org/10.1371/journal.pone.0255502.

Sahrish N, Shakeel A, Ghulam A, Zartash F, Sajjad H, Mukhtar A, Muhammad AK, Ahmad K, Fahad S, Wajid N, Sezai E, Carol Jo W, Gerrit H (2022) Modeling the impact of climate warming on potato phenology. Eur J Agron 132:126404.

Sajid H, Jie H, Jing H, Shakeel A, Satyabrata N, Sumera A, Awais S, Chunquan Z, Lianfeng Z, Xiaochuang C, Qianyu J, Junhua Z (2020) Rice production under climate change: Adaptations and mitigating strategies. In: Fahad S, Hasanuzzaman M, Alam M, Ullah H, Saeed M, Khan AK, Adnan M (eds) Environment, Climate, Plant and Vegetation Growth. Springer Publ Ltd, Springer Nature Switzerland AG. Part of Springer Nature. pp 659–686. https://doi.org/10.1007/978-3-030-49732-3.

Sajjad H, Muhammad M, Ashfaq A, Waseem A, Hafiz MH, Mazhar A, Nasir M, Asad A, Hafiz UF, Syeda RS, Fahad S, Depeng W, Wajid N (2019) Using GIS tools to detect the land use/land cover changes during forty years in Lodhran district of Pakistan. Environ Sci Pollut Res. https://doi.org/10.1007/s11356-019-06072-3.

Saleem MH, Fahad S, Adnan M, Mohsin A, Muhammad SR, Muhammad K, Qurban A, Inas AH, Parashuram B, Mubassir A, Reem MH (2020a) Foliar application of gibberellic acid endorsed phytoextraction of copper and alleviates oxidative stress in jute (Corchorus capsularis L.) plant grown in highly copper-contaminated soil of China. Environ Sci Pollut Res. https://doi.org/10.1007/s11356-020-09764-3.

Saleem MH, Fahad S, Shahid UK, Mairaj D, Abid U, Ayman ELS, Akbar H, Analía L, Lijun L (2020c) Copper-induced oxidative stress, initiation of antioxidants and phytoremediation potential of flax (Linum usitatissimum L.) seedlings grown under the mixing of two different soils of China. Environ Sci Poll Res. https://doi.org/10.1007/s11356-019-07264-7.

Saleem MH, Rehman M, Fahad S, Tung SA, Iqbal N, Hassan A, Ayub A, Wahid MA, Shaukat S, Liu L, Deng G (2020b) Leaf gas exchange, oxidative stress, and physiological attributes of rapeseed (Brassica napus L.) grown under different light-emitting diodes. Photosynthetica 58(3):836–845.

Saman S, Amna B, Bani A, Muhammad Tahir ul Qamar, Rana MA, Muhammad SK (2020) QTL mapping for abiotic stresses in cereals. In: Fahad S, Hasanuzzaman M, Alam M, Ullah H, Saeed M, Khan AK, Adnan M (eds) Environment, climate, plant and vegetation growth. Springer Publ Ltd, Springer Nature Switzerland AG. Part of Springer Nature. pp 229–252. https://doi.org/10.1007/978-3-030-49732-3.

Samreen, Ibrar M, Lalbadshah, Shahida N, Imran, Inayat K (2016) Ethnobotanical study of subtropical hills of Darazinda, Takht-e-Suleman range F.R D.I. Khan, Pakistan. Pure Appl Biol 5(1):149–164. http://dx.doi.org/10.19045/bspab.2016.50020.

Saud S, Chen Y, Fahad S, Hussain S, Na L, Xin L, Alhussien SA (2016) Silicate application increases the photosynthesis and its associated metabolic activities in Kentucky bluegrass under drought stress and post-drought recovery. Environ Sci Pollut Res 23(17):17647–17655. https://doi.org/10.1007/s11356-016-6957-x.

Saud S, Chen Y, Long B, Fahad S, Sadiq A (2013) The different impact on the growth of cool season turf grass under the various conditions on salinity and drought stress. Int J Agric Sci Res 3:77–84.

Saud S, Fahad S, Cui G, Chen Y, Anwar S (2020) Determining nitrogen isotopes discrimination under drought stress on enzymatic activities, nitrogen isotope abundance and water contents of Kentucky bluegrass. Sci Rep 10:6415. https://doi.org/10.1038/s41598-020-63548-w.

Saud S, Fahad S, Yajun C, Ihsan MZ, Hammad HM, Nasim W, Amanullah Jr, Arif M, Alharby H (2017) Effects of nitrogen supply on water stress and recovery mechanisms in Kentucky bluegrass plants. Front Plant Sci 8:983. https://doi.org/10.3389/fpls.2017.00983.

Saud S, Li X, Chen Y, Zhang L, Fahad S, Hussain S, Sadiq A, Chen Y (2014) Silicon application increases drought tolerance of Kentucky bluegrass by improving plant water relations and morph physiological functions. SciWorld J 2014:1–10. https://doi.org/10.1155/2014/368694.

Senol C (2020) The effects of climate change on human behaviors. In: Fahad S, Hasanuzzaman M, Alam M, Ullah H, Saeed M, Khan AK, Adnan M (eds) Environment, climate, plant and vegetation growth. Springer Publ Ltd, Springer Nature Switzerland AG. Part of Springer Nature. pp 577–590. https://doi.org/10.1007/978-3-030-49732-3

Shafi MI, Adnan M, Fahad S, Fazli W, Ahsan K, Zhen Y, Subhan D, Zafar-ul-Hye M, Martin B, Rahul D (2020) Application of single superphosphate with humic acid improves the growth, yield and phosphorus uptake of wheat (Triticum aestivum L.) in calcareous soil. Agron 10:1224. https://doi.org/10.3390/agronomy10091224.

Shah F, Lixiao N, Kehui C, Tariq S, Wei W, Chang C, Liyang Z, Farhan A, Fahad S, Huang J (2013) Rice grain yield and component responses to near 2°C of warming. Field Crop Res 157:98–110.

Sidra K, Javed I, Subhan D, Allah B, Syed IUSB, Fatma B, Khaled DA, Fahad S, Omaima N, Ali TKZ, Rahul D (2021) Physio-chemical characterization of indigenous agricultural waste materials for the development of potting media. J Saudi Soc Agric Sci. https://doi.org/10.1016/j.sjbs.2021.08.058.

Subhan D, Zafar-ul-Hye M, Fahad S, Saud S, Martin B, Tereza H, Rahul D (2020) Drought stress alleviation by ACC deaminase producing *Achromobacter xylosoxidans* and *Enterobacter cloacae*, with and without timber waste biochar in maize. Sustainability 12(6286). https://doi.org/10.3390/su12156286.

Tariq M, Ahmad S, Fahad S, Abbas G, Hussain S, Fatima Z, Nasim W, Mubeen M, ur Rehman MH, Khan MA, Adnan M. (2018). The impact of climate warming and crop management on phenology of sunflower-based cropping systems in Punjab, Pakistan. Agri Forest Met 256:270–282.

Unsar NU, Muhammad R, Syed HMB, Asad S, Mirza AQ, Naeem I, Muhammad H ur R, Fahad S, Shafqat S (2020) Insect pests of cotton crop and management under climate change scenarios. In: Fahad S, Hasanuzzaman M, Alam M, Ullah H, Saeed M, Khan AK, Adnan M (eds) Environment, climate, plant and vegetation growth. Springer Publ Ltd, Springer Nature Switzerland AG. Part of Springer Nature. pp 367–396. https://doi.org/10.1007/978-3-030-49732-3.

Visualising carbon pathways: http://serc.carleton.edu/eet/carbon/index.html Interesting resource which allows students to create animations of parts of the carbon cycle.

Wahid F, Fahad S, Subhan D, Adnan M, Zhen Y, Saud S, Manzer HS, Martin B, Tereza H, Rahul D (2020) Sustainable management with mycorrhizae and phosphate solubilizing bacteria for enhanced phosphorus uptake in calcareous soils. Agriculture 10(334). https://doi.org/10.3390/agriculture10080334.

Wajid N, Ashfaq A, Asad A, Muhammad T, Muhammad A, Muhammad S, Khawar J, Ghulam MS, Syeda RS, Hafiz MH, Muhammad IAR, Muhammad ZH, Muhammad Habib ur R, Veysel T, Fahad S, Suad S, Aziz K, Shahzad A (2017) Radiation efficiency and nitrogen fertilizer impacts on sunflower crop in contrasting environments of Punjab. Pak Environ Sci Pollut Res 25:1822–1836. https://doi.org/10.1007/s11356-017-0592-z.

Wu C, Kehui C, She T, Ganghua L, Shaohua W, Fahad S, Lixiao N, Jianliang H, Shaobing P, Yanfeng D (2020) Intensified pollination and fertilization ameliorate heat injury in rice (Oryza sativa L.) during the flowering stage. Field Crops Res 252:107795.

Wu C, Tang S, Li G, Wang S, Fahad S, Ding Y (2019) Roles of phytohormone changes in the grain yield of rice plants exposed to heat: A review. PeerJ 7:e7792. https://doi.org/10.7717/peerj.7792.

Yang Z, Zhang Z, Zhang T, Fahad S, Cui K, Nie L, Peng S, Huang J (2017) The effect of season-long temperature increases on rice cultivars grown in the central and southern regions of China. Front Plant Sci 8:1908. https://doi.org/10.3389/fpls.2017.01908.

Zafar-ul-Hye M, Akbar MN, Iftikhar Y, Abbas M, Zahid A, Fahad S, Datta R, Ali M, Elgorban AM, Ansari MJ et al. (2021) Rhizobacteria inoculation and Caffeic acid alleviated drought stress in lentil plants. Sustainability 13:9603. https://doi.org/10.3390/su13179603.

Zafar-ul-Hye M, Muhammad N, Subhan D, Fahad S, Rahul D, Mazhar A, Ashfaq AR, Martin B, Jiˇrí H, Zahid HT, Muhammad N (2020a) Alleviation of cadmium adverse effects by improving nutrients uptake in bitter gourd through cadmium tolerant rhizobacteria. Environments 7(54). https://doi.org/10.3390/environments7080054.

Zafar-ul-Hye M, Tahzeeb-ul-Hassan M, Muhammad A, Fahad S, Martin B, Tereza D, Rahul D, Subhan D (2020b) Potential role of compost mixed biochar with rhizobacteria in mitigating lead toxicity in spinach. Sci Rep 10:12159. https://doi.org/10.1038/s41598-020-69183-9.

Zahida Z, Hafiz FB, Zulfiqar AS, Ghulam MS, Fahad S, Muhammad RA, Hafiz MH, Wajid N, Muhammad S (2017) Effect of water management and silicon on germination, growth, phosphorus and arsenic uptake in rice. Ecotoxicol Environ Saf 144:11–18.

Zahir SM, Zheng HG, Ala Ud D, Amjad A, Ata Ur R, Kashif J, Shah F, Saud S, Adnan M, Fazli W, Saud A, Manzer HS, Shamsher A, Wajid N, Hafiz MH, Fahad S (2021) Synthesis of silver nanoparticles using Plantago lanceolata extract and assessing their antibacterial and antioxidant activities. Sci Rep 11:20754.

Zia-ur-Rehman M (2020) Environment, climate change and biodiversity. In: Fahad S, Hasanuzzaman M, Alam M, Ullah H, Saeed M, Khan AK, Adnan M (eds) Environment, climate, plant and vegetation growth. Springer Publ Ltd, Springer Nature Switzerland AG. Part of Springer Nature. pp 473–502. https://doi.org/10.1007/978-3-030-49732-3.

# 16

# *Pollinators Ecology and Management*

**Abdel Rahman Mohammad Tawaha**
*Al Hussein Bin Talal University*

**Ekaterina Kozuharova and Hristo Valchev**
*Medical University of Sofia*

**Marian Hýbl and Denisa Kovářová**
*Mendel University in Brno*

**Petr Mráz**
*University of South Bohemia in České Budějovice*

**Mohamed Shebl Abd Elfattah**
*Suez Canal University*

**Laila Trioui**
*University Hassan II*

**Abdel Razzaq Al-Tawaha**
*Universiti Putra Malaysia*

**Ali M. Qaisi**
*University of Jordan Amman*

**Amanullah, Imran, and Shah Khalid**
*The University of Agriculture*

**Abdur Rauf**
*University of Swabi*

**Shah Fahad**
*Hainan University*
*The University of Haripur*

DOI: 10.1201/9781003286400-16

**CONTENTS**



## 16.1 Introduction

Pollination is the transfer of pollen from the anther of a flower to the stigma of the same or another flower which is known as self- or cross-pollination (Abrol 2015). Pollination of plant species that contribute to soil conservation and control erosion and those plant species that provide valuable resources for wildlife can be of considerable importance for both the present and the future (Hoopingarner and Waller 1992). Because of this crucial ecological service, the pollination and plant pollinators have been subjected in thousands of publications dealing with several issues in the last decades and recently with the pollination shortage and pollinators decline (Ollerton 2017; Ahmad et al. 2016). This service is provided by several factors, which could be done by wind, animals, or mechanical pollination. Pollinators are essential for fruit trees, oilseed, forage and fiber crops, vegetables, nuts, and seed production. Pollinators are affecting 35% of the world's crop production, increasing the production of 87 of the major food crops worldwide, in addition to many plant-derived medicines in the world's pharmacies (FAO 2006). The animal pollination is considered to be one of the most important types which are provided by vertebrates (mammals, bats, birds, and rodents) and invertebrates (mostly insects) (Osman and Shebl 2019). Around 87.5% of angiosperms are pollinated by animals (Ollerton et al. 2011), and there are approximately 350,000 known species of pollinators servicing the approximately 352,000 species of flowering plants (Paton et al. 2008). Nearly one-third of the pollination services come from pollen transfer mediated by insects (Klein et al. 2007). The total economic contribution of animal pollination to agriculture is €153 billion (~$200 billion) annually, which is considered a great contribution to our food security (Gallai et al. 2009). However, animal pollination usually plays an important role in promoting and maintaining genetic variation in plant populations, which in turn supports the species for more adaptability to new and changing environments. Because of this, animal or insect pollinators are considered to be of fundamental importance for plant diversity conservation and ecosystem functioning (Wilcock and Neiland 2002). A strong and clear mutually beneficial relationship exists between the flower visitors and their associated flowers, and this mutualism is a key interaction in terrestrial ecosystems (Procter et al. 1996). Studies on plant–pollinator interactions, pollination ecology, and pollination biology have started since the eighteen century, and till now, there are still more information to be discovered in terms of pollination and pollinator diversity (Waser 2006; Ollerton 2017). There is a gap in plant–pollinator interactions studies due to lack of sampling interval and lack of sampling across different habitats and geographical regions. On the other hand, studies were concentrated in some regions, and very little was known in some other parts of the world such as parts of the Mediterranean, African, and Asian regions (Eradley et al. 2006).

Honeybee is the most widespread and known plant pollinator and was attracting more attention than other pollinators (Ollerton et al. 2012; Smith and Saunders 2016). Despite this fact, other insect pollinators (butterflies, wasps, native bees, hoverflies, and beetles) or other vertebrates have a great value in terms of pollination services of some plants. The effectiveness is varied due to their abundance, propensity for touching the anthers, carrying pollen and contacting stigma, and moving to other plants for doing pollination (Herrera 1987; Rodriguez-Rodriguez et al. 2013).

Hymenoptera, particularly bees (Apoidea which are highly diverse biologically and ecologically) and sphecid wasps, are the major and most important plant pollinators in addition to some dipterans (Kevan and Baker 1983). Notably, among more than 20,000 known bee species worldwide, less than 5% are social and even fewer (1%) are domesticated and used in managed pollination services (Michener 2007). Bees and their floral plant hosts are of great importance in both natural and agricultural ecosystems, and it is the predominant and most important plant pollinators (Nabhan and Buchmann 1997; Dafni and O'Toole 1994; Daily 1997; Kremen and Miles 2012). Given their ecological services, the wild bees are complement honey bees and other managed pollinators in their pollination services (Freitas and Paxton 1998; Garibaldi et al. 2013). The presence of native bees, in addition to honeybees, can enhance and increase the efficiency of the pollination services (Greenleaf and Kremen 2006). They are dependent on several products produced by angiosperm flowers such as pollen, nectar, and oils. This results in special morphological adaptations for both bees and plants. The needs of some plants for pollination by only one single species of bees are absolute. Monolectic bees are known in some bee species of several genera such as *MacroapisMelitta* and *Systropha* (James and Pitts-Singer 2008). Some other groups of bees visit plants of one single plant family (oligolectic) and others visit a wide range of plants (polylectic) (Shebl et al. 2013; Shebl et al. 2015).

There is a long debate regarding the pollinator decline and pollination crisis recently which will affect our food security due to lack of pollination services. This decline is due to several environmental threats and factors including epidemic of some diseases and parasites, agricultural intensification, urbanization, climate change, habitat loss and fragmentation, availability of food, pollution, invasive alien species, and pesticides (LeBuhn et al. 2012). As the global pollinators decline, concerns about the shortage of pollination services have raised in several countries, where extensive farming practices reduce pollinator diversity in agricultural ecosystems as well as pollination services delivered to crops (Nieto 2014). Still much knowledge is needed to evaluate the status of pollinator populations' decline and inconsequence the impact on pollination services provided by them.

## 16.2 Plant Dependence on Animal Pollinators for Seed Production

The term "breeding systems" refers to both the sexual and asexual propagation of plants. Each plant taxon has its own peculiarities with regard to pollination and fertilization. The various reproductive systems in plants can be summarized in three main groups: xenogamy (cross-fertilization), autogamy (self-fertilization), and asexual propagation (apomixis and vegetative propagation). These breeding mechanisms find their place in the adaptation of plants to the environment. The evolutionary processes and breeding systems are correlated. The pollen/ovule ratio (P/O) is a good indicator of the angiosperms' breeding systems. The reproductive structure of a plant population is determined by the distribution of pollen (e.g. the chance for pollination between individuals with close genotypes), the distribution of seeds, and the ability for apomixis and vegetative propagation. A general explanation for diversity in plant breeding systems is offered by sex-allocation theory, which postulates a trade-off between the allocations of resources to the two sexual functions. Pollinators collect their food in the most energy-efficient way according to the "optimal foraging theory". The varying nature of the flower resources (nectar and pollen) requires a flexible system of "learning" in the pollinators. Flower constancy (pollinators following particular flowers for foraging) is related to the gene flow of the pollinated plants.

### 16.2.1 Biology of Reproduction, Breeding Systems and Evolutionary Trends in Plants

The term "reproduction systems (breeding systems)" includes all systems, sexual and asexual for the propagation of plants (Richards 1996; Dafni 1992). Animals generally have a generation that is formed by the combination of the gametes of two different parents (hermaphrodites are a rarity). Most often male and female functions are divided into different individuals with particular sexual dimorphism. Choice of a breeding partner and the models of the mating behavior can be very complex, involving territoriality and fights between males for a chance to leave a generation. All of this as well as migrations affects the hereditary features of the generation, determining the genetic structure of populations, their variability, and ultimately it influences the evolutionary patterns and evolutionary guidelines. Unlike animals, plants are immobile within a generation, have no behavior, do not choose a mating partner, and yet have complex and effective mechanisms to regulate genetic variability. Reproductive systems in plants are significantly more diverse. The hereditary features of the offspring, the genetic structure of the populations, the nature of the variability, and the evolutionary trends in plants are dependent on how pollination and fertilization processes occur. Each plant taxon has peculiarities with regard to pollination and fertilization (Richards 1990; Richards 1996; Richards 1997; Richards 2001; Ellstrand 1992; Ellstrand et al. 1999; Barrett 1992; Barrett 2002; Körner 2003; Wedderburn and Richards 2006; Charlesworth 2006; Glover 2007).

The various reproductive systems in plants can be summarized in three main groups although there is a transition between them: xenogamy (cross-fertilization), autogamy (self-fertilization), and asexual propagation (apomixis and vegetative propagation). In many plant populations, the three or at least two of the reproduction types are present as steady evolutionary strategies. Xenogamy is a fusion of male and female gametes from genetically diverse individuals of the same species. It increases the recombination of genetically variable material so that a population has greater variability than an autogamous population and has a higher adaptive potential. Autogamy is a process that merges male and female gametes derived from one individual (the same flower, the same inflorescence, another inflorescence but on the same individual). This mechanism limits the genetic variability of the population but can stabilize some features and provide isolated individuals with an opportunity for reproduction. Self-fertilization is less common in perennial species but is common in annuals. Often, self-fertilization is optional and may be a successful strategy in unfavorable climatic conditions (e.g. rainy and cold periods) or in the absence of pollinators (Stebbins 1974; Tootill 1983; Richards 1990; Richards 1996; Eckert and Schaefer 1998; Duan et al. 2009; Briggs and Walters 2016). Asexual reproduction is the formation of new individuals from a parent without gamete merging – vegetative or apomixis. Some authors include vegetative propagation within the apomixis (Terziiski and Yurukova-Grancharova 2011). Apomixis senso stricto is defined as seed formation without the sexual process – the formation of a diploid germ or somatic sac from a somatic cell or directly from a megaspore maternal cell. These breeding mechanisms find their place in the adaptation of plants to the environment and the evolutionary processes. It is a well-known fact that in some areas of the world where the conditions are extreme the individuals of a plant taxon are propagated vegetatively. Apomixis is a pathway for speciation that finds its place in cases of hybridization and hetero-polyploidization (Richards 1990; Richards 1996; Hörandl and Paun 2007; Terziiski and Yurukova-Grancharova 2011).

When autogamy is possible but not obligatory, plants "prefer" to receive pollen from individuals of the same species but with a different genotype. This favors greater variability and adaptability (de Nettancourt 1977). The prevention of autogamy and the promotion of xenogamy can take place at different stages of reproduction – from the opening of the flowers to the first division of the zygote – and by various mechanisms (Faegri and van der Pijl 1971; Handel 1983; Richards 1990). Separation of the male and female function in space (hercogamy, e.g. stigma above the anthers) and time (dichogamy, anthers, and stigma mature at different times). Many examples can be listed and *Haberlea rhodopensis* Friv. (Generiaceae) is one of them (Figure 16.1). Self-incompatibility occurs when pollination and fertilization are impossible between male and female parts of the same genet – flower, inflorescence, or branch. Self-incompatibility is a genetically controlled system which causes a plant to reject its own pollen from the same genet – the flower or inflorescence of the same plant individual (de Nettancourt 1977; de Nettancourt 1984; Harris et al. 1989; Richards 1990; Charlesworth et al. 2005). Rejection can take place at different stages of pre-fertilization or maturation processes – sporophytic or gametophytic self-incompatibility (Richards 1990). The transition from xenogamy (cross-fertilization) to autogamy (self-fertilization) is accompanied by some changes in flower phenology and pattern, as well as a reduction in P/O. These changes reduce the energy expenditure for the flower and favor self-pollination (Plitmann Levin 1989). P/O is a better indicator of the angiosperms' breeding systems than the size or morphology of the flower (Cruden 1976). Studies by Lloyd (1965), Baker (1967), Arroyo (1973), and Cruden (1973, 1976) show that the flowers of self-incompatible and other xenogamy species produce more pollen grains than close (related) self-compatible and/or autogamous species. Compared to other flower

**FIGURE 16.1** *Haberlea rhodopensis* Friv. (Generiaceae) – an example of self-pollination prevention by space separation of male and female function (herkogamy) and anthers maturing before stigma (proterandry).

characteristics, the number of ovules in the ovary does not decrease. This ratio reflects the potential for a sufficient number of pollen grains to reach each stigma so that the result is maximum seed production at minimal energy consumption. Thus P/O is associated with breeding systems. Autogamous flowers have lower P/O than xenogamous ones. Cleistogamous flowers usually have the lowest P/O (Cruden 1976). In his studies, Cruden (1976) shows that: (1) P/O can be placed on an equal principle with other parts of the plant breeding system; (2) there is a strong link between the breeding system, P/O, and the habitat that the plant inhabits; and (3) some plants tend to reduce pollen production and this indicates that P/O is a good indicator of breeding systems from morphological characteristics. Breeding systems are divided into the following five classes: bonded xenogamy (cross-fertilization), facultative xenogamy, facultative autogamy, bonded autogamy (self-fertilization), and cleistogamy. The evolutionary change from class to class is reflected by a significant reduction in P/O. Studies show that the gradient in P/O values is associated with combined, transient propagation systems. This trend is found in both anemophilous and zoophilous plants (Lloyd 1965, 1987, 1983; Baker 1967; Cruden 1973; Cruden 1976; Gibbs et al. 1975; Queller 1984; Schoen 1977; Charnov 1979; Charnov 1982; Charlesworth and Charlesworth 1981; Campbell et al 1986; Philbrick and Anderson 1987; Ritland and Ritland 1989; Plitmann and Levin 1989; Mione and Anderson 1992; Lloyd and Webb 1992; Gallardo et al 1994; Wang et al 2004; Porcher and Lande 2005). Taking into account that successful pollination is not only dependent on the limitations of the quantity and quality of pollen carried but is also dependent on the distribution of the reproductive resources (the sex-allocation theory, Charnov 1982), it can be assumed that the increased number of ovules is not just an indicator of autogamy, but a strategy for maximizing the number of seeds (Gangoiti et al. 2008; Etcheverry et al. 2012). Michalski and Durka (2010) compare P/O and autogamy levels in 107 angiospermous and interpret the results in the context of pollination, the life form, and phylogenetic relationships between species. There is a significant dependence between P/O and the crossing rate, but taking into account other factors this relationship becomes ambiguous. In insect-pollinated species, this dependence is approximate. At the same time, the large number of ovules is not necessarily proportional to high reproductive success and a high number of seeds. This is represented by a low percentage of ripened seeds relative to the ovules in the ovary. The dependence is influenced by a large number of factors, including lack of pollen and/or resource scarcity (Willson and Price 1980; Bierzychudek 1981; Montalvo and Ackerman 1987; Burd 1994; Juenger and Bergelson 1997; Medrano et al. 2000; Ashman et al. 2004; Hill et al. 2008; Jakobsson et al 2009; Duffy and Johnson 2011; Fernández et al. 2012; Reich et al. 2014). Determining the limiting factor is often a difficult task with ambiguous results (Medrano et al. 2000; Kudo et al. 2001; Wesselingh 2007). Nevertheless, testing the P/O ratio is a convenient approach to evaluate the breeding system, particularly when other tests are not possible (Dafni 1992; Kozuharova et al. 2015; Adnan et al. 2016). The simultaneous application of a direct method of the examination of breeding systems with the isolation of flowers from pollinator access described by Dafni (1992) and an indirect method by estimating the P/O ratio with subsequent result and analysis comparison makes it possible to refine the specifics. This is a field where more research is needed.

### 16.2.2 Theory in Angiosperms, Reproductive Structure of Their Populations and Pollinators' Behavior

The reproductive structure of a species and the individual population is determined by the distribution of pollen, the distribution of seeds, and the chance and nature of pollination between individuals with close genotypes and/or phenotypes. A general explanation for diversity in plant breeding systems is offered by sex-allocation theory, which postulates a balance in the allocation of resources of the two sexual functions. It explains the high rate of hermaphroditism in the angiospermous with the more efficient resource investment in this function compared to the less efficient resource investment in the same-sex function. This theory has been proven by many experimental tests, including testing for the character and synchronization of nectar secretion in male and female functions at the flower and inflorescence levels. Resource allocation is in line with pollinator behavior. It is flexible and complex because an efficient collection of pollen and/or nectar is a response to a complex system of factors. The varying nature of flower resources (nectar and pollen) requires a flexible learning system for pollinators (this is particularly expressed in bees) and the following of a particular food source, which is referred to as "flower constancy" which is in direct relation with the gene flow of pollinated plant organisms (Levin and Kerster 1974; Pyke 1978a,b; Levin 1978; Waddington et al. 1981; Waddington 1981; Zimmerman 1983; Lloyd 1983; Ritland and Ritland 1989; Richards 1990; Grant 1995; Goulson 1999; Osborne et al. 1999; Campbell 2000; Barrett 2002; Carlson 2007). Knowing the processes and dependencies that lead to the release of nectar has not only a purely practical meaning for the production of honey but above all these are fundamental aspects that clarify certain features related to breeding systems and mechanisms, such as geitonogamy and its avoidance. The calorific value of nectar, concentration, viscosity, and the composition of sugars are variables reflecting on the behavior of pollinators, the direction of their movement, and, hence, the genetic exchange in plant populations (Corbet 1978; Corbet et al. 1979; Waddington et al. 1981; Waddington 1983; Kay 1984; Burquez and Corbet 1991). The dynamics of nectar secretion as a final result reflects on the gene exchange characteristic (Waddington et al. 1981; Waddington 1981; Waddington 1983; Richards 1990; Richards 1996; Richards 2001; Goulson 1999; Stout and Goulson 2002; Carlson 2007; Pacini and Nepi 2007).

Ehrlich and Raven (1969) point out that for the precise clarification of the problem about what degree of the gene flow is limited, there is a need to conduct a large number of experimental studies over a long period of time. Subsequently, a number of experiments and field studies have been conducted on pollination behavior (the list is huge and is not presented here). Of particular interest are surveys of bumblebees bees from this point of view, which concern the range of food collection, the direction of flights, including marking by microscopic tagging, radar tracking, genetic markers, etc. respectively they are aiming at clarifying the problems of the gene flow within the plants (Osborne et al. 1996; Osborne et al. 1999; Cresswell et al. 2002; Goverde et al. 2002; Chapman et al. 2003; Darvill et al. 2004; Pasquet et al. 2008; Knight et al. 2005; Lepais et al. 2010).

According to the “optimal foraging theory”, pollinators collect their food in the most energy-efficient way – visiting adjacent flowers and inflorescences (Levin 1978; Levin 1979; Waddington 1983; Richards 1990; Stout et al. 1998). The reproductive structure of a population is a function of its spatial structure. It also depends on the flower constancy of the pollinators, which affects inter-species crossing and the exchange of genetic information (Levin 1978; Levin 1979). The tendency for pollinators to limit their visits to a particular species or certain flower morph is defined as “flower constancy” (Levin et al. 1971; Heinrich 1975; Dafni 1992). It is an important prerequisite for effective pollination in the plant community, which is made up of many different species. Species with a similar flower structure and a similar flowering period potentially compete for the same pollinators. This may be a selective force in terms of flowering time or flower morphology (Levin et al. 1971; Heinrich 1975; Waser 1983; Ranta et al. 1981; Campbell 1985; Arroyo and Dafni 1993; Chittka et al. 1999; Gegear and Laverty 2005; Raine and Chittka 2007). An important factor determining the pollinator activity, pollinator success, and seed production is the size of the population or its fragments – the larger size is generally associated with higher activity (Heinrich 1979a,b; Levin 1978; Handel 1983; Pleasants 1980; Petanidou et al. 1995). Ground bees are polylectic (Brian 1951; Free 1970; Macior 1994). Their flower constancy reflects the level of their individual experience (Heinrich 1976; Heinrich 1979a,b). Their choice depends on many factors such as the depth of the corolla, the quantities of nectar and pollen, and the relationship between the species (Brian 1957). The high energy efficiency they achieve is the result of their behavioral specialization – they do not revisit the flowers from which they have collected nectar (Goulson et al. 1998). The biological significance of this behavior is multifunctional. On one hand, the likelihood of self-pollination is reduced, but above all it stimulates pollinators to produce more effective pollinating actions with minimal energy consumption.

The main disadvantage of inbreeding compared to free crossbreeding is the reduction of genetic diversity. One of the negative consequences is a reduction in the probability of successful adaptation when environmental conditions are changing. Angiosperms in the evolution process have developed a number of mechanisms for avoiding inbreeding such as dichogamy and hercogamy (see above). Moreover, often the structure of the inflorescence and its development are coordinated with the pollinator behavior so that the flowers in the female phase are visited before those in the male phase (Faegri and van der Pijl 1971; Waddington et al. 1981; Waddington 1983; Navarro-Pérez et al. 2017). Self-incompatibility is another mechanism to avoid inbreeding (Richards 1990; Handel 1983). On the other hand, this biological specificity has some limitations and adverse effects. For example, *Filipendula rubra* (Rosaceae) is a rare, self-incompatible plant species with a significant possibility of vegetative propagation by branch populations. The larger the branch, the lower the seed production is. This paradox is a result of the specific behavior of pollinators to feed with minimal energy consumption – collecting pollen and nectar from the contiguous flowers. When these flowers are from the same branch, they do not form seeds because they are self-incompatible, a phenomenon called an “inbreeding depression” (Aspinwall and Cristian 1992). Examples of plants with self-incompatible flowers and inbreeding depression are many (Willis 1993; Husband and Schemske 1995). Inbreeding depression in a broader sense is expressed in decreased fertility and decreased viability of the offspring as a result of inbreeding. Classical genetic research and modern molecular approaches have shown that the inbreeding depression is mainly caused by the presence and manifestation of recessive deleterious mutations in the plant populations (Charlesworth and Willis 2009). Inbreeding depression is the main conservation issue for the small populations of rare plant species (Hedrick and Kalinowski 2000; Wright et al. 2008). The inbreeding depression increases significantly in unfavorable environmental conditions (Armbruster and Reed 2005).

At the plant community level, the overlap in resource exploitation is avoided due to the varying length of the proboscis of the bees. In this way, the flowers with deep corolla tubes are visited by long-proboscis ground bees and the flowers with a shallow tube from short proboscis ones. This leads to a reduction in the competition for resources among pollinators and optimizes the effectiveness of pollination (Heinrich 1976). The co-adaptive mechanisms operate not only at the level of separate plant species and pollinators but also at the ecosystem level (Heinrich 1976; Totland 1993). Even in parasitic plants and their hosts, which grow sympatric and bloom at the same time, the overlapping of the niches (overlapping of resources used by different species) is only partial (Ollerton et al. 2007; Arafat et al. 2016). The large variety of pollinators in the community is a factor that greatly contributes to avoiding overlapping niches and lowering pollinator competition, resulting in higher reproductive success in plant species (Klein 2009).

The lack of suitable pollinators in extreme and pioneer habitats is considered as one of the main reasons for transition from outcrossing to inbreeding (autogamy) even in plants with a high degree of flowers co-adaptation in both annual and perennial taxa, which affects the genetic variability and genotype fixing (Grant and Grant 1965; Stebbins 1974; Richards 1982; Richards 1990; Herlihy and Eckert 2002). It is necessary to take into account the specificity of the breeding mechanism in each particular plant species in order to assess its breeding structure and the potential genetic consequences of negative repercussions.

## 16.3 Ecological Reliance of Pollinators on Plant Food Resources

Many animal species visit flowers to collect pollen as their primary food source and also to obtain other rewards, like nectar, oil, resin, or esters. While doing so, they pollinate the flowers (Johnson and Steiner 2000). Although pollinators and flowering plants pollinated by animals depend on each other, and at first glance they seem to act as pure altruists, both in fact operate selfishly. A certain cost–benefit ratio applies to all participants who, of course, aim to balance it for their own benefit. Because the process of rewards production is quite costly to a plant, it has to be balanced up by sufficient reproduction opportunities obtained in exchange for the rewards. On the other hand, the energy spent by pollinators on flying and foraging activity must be adequate to the energy obtained from flower rewards (Southwick et al. 1981). This competition for benefits gave rise to many plant species and pollinators that remain a part of pollination systems. In addition to the continual coevolutionary competition for benefits, pollinator communities are now exposed to additional selection pressures affecting all ecosystems globally and resulting not in an unprecedented rate of changes in the environment as well as in the resources on which pollinators, plants, and their pollination systems depend. The main causes are the intensification of industry, agriculture, the growing human population, and the changes that accompany this transformation. These are mainly landscape fragmentation, environmental contamination by agrochemicals, urbanization, and change of plant communities toward alien/invasive species (Khan et al. 2017; Steffan-Dewenter 2003; Harris and Johnson 2004; Benton et al. 2002; Williams et al. 2011; Vanbergen 2013). As a result, one can observe more extinction of individual pollinator species, plants, and in some cases whole pollination systems (Weller 1994; Ladley and Kelly 1995; Mawdsley et al. 1998; Robertson et al. 1999; Elmqvist 2000).

## 16.4 Impact of Landscape Fragmentation on Pollinators

Rapid environmental degradation has been observed worldwide for a long time. It is caused by expanding human population and associated vast agricultural and industrial development as well as by increasing share of built-up areas (Harris and Johnson 2004; Saunders et al. 1991; Steffan-Dewenter 2003). Consequently, the spatial patterns of habitat layout and the structure of the entire landscape are also changing dramatically. This is characterized by further destruction and fragmentation of natural and semi-natural habitats resulting in isolated remnants of natural vegetation surrounded by a transformed matrix with typically predominant alien vegetation and anthropogenically used areas (agricultural, industrial or urban). It is the landscape fragmentation that is considered to be one of the main reasons for the loss of biodiversity and mass extinction of biota in general (Rathcke and Jules 1993; Turner 1996).

## 16.5 Changes in Abundance and Diversity of Pollinators

Compared to undisturbed habitats, most fragmented habitats provide lower resource availability. A reduction in sizes of plant populations situated within fragmented habitats can result in a fall of pollinator populations due to lower availability of food resources (Rathcke and Jules 1993; Kremen and Ricketts 2000) as well as of non-food ones (e.g. nesting places) (Kremen and Ricketts 2000; Cane 2001). The fragment's trophic capacity is largely influenced by its size and structure. There is a general assumption that with the fragment size the diversity and abundance of resources will increase, which will positively affect the diversity and abundance of pollinators who are able to persist within the fragment (Saunders et al. 1991; Baz and Garcia-Boyero 1995). However, even in large fragments, poor pollen flow (i.e. transfer) can occur if the habitats are too far apart. And also, the distance preventing the pollen flow is very individual. While large pollinators, such as robust birds, moths, or bats can cover distances of up to 8–25 km, smallest pollinators, such as bugs, small moths, and butterflies, can encounter difficulties when dealing with distances of several hundred meters or, in extreme cases, even only 50 m (Feinsinger et al. 1987; Rathcke and Jules 1993; Debinski and Holt 2000). Thus, the nature of the fragment is also influenced by the degree of isolation which can significantly affect not only the plant population and its density but also species and genetic diversity of pollinators.

## 16.6 Changes in Pollinator Behavior

Habitat fragmentation can result in behavioral changes of residual pollinator species. Because pollinators usually behave to maximize food intake per unit of time and energy spent on feeding, their behavior can change in relation to changes in food availability (Pyke et al. 1977; Southwick et al. 1981). They can adjust the distance of flights between flowers to the amount of nectar obtained from one flower (Harris and Johnson 2004). Bumblebees feeding on highly rewarding flowers, for instance, fly only short distances. Lower attendance of plants located within the fragments may, therefore, lead to accumulation of nectar in flowers, increased likelihood of visits paid to flowers of the same plant, and thus to higher levels of geitonogamy (self-pollination) (Ågren 1996; Bosch and Waser 1999; Cane and Tepedino 2001).

## 16.7 Pollinator Sensitivity to Fragmentation

Pollinators differ in their sensitivity to fragmentation and changes in resource availability in relation to their physiological demands, flight capacity, level of specialization, and ecological demands. One of the main factors seems to be the degree of food specialization of individual species (Bond 1994; Johnson and Steiner 2000). Food specialists are generally regarded as a very vulnerable category of pollinators because of their dependency on resources provided by a specific plant species (or a narrow group of species). Due to their limited flexibility, such pollinator species are unable to respond to rapidly changing environments (Kremen and Ricketts 2000; Baz and Garcia-Boyero 1995). In contrast, a larger degree of tolerance to fragmentation has been observed in less specialized species (Ghazoul 2005). Another vulnerable category is endemic species due to their low plasticity (Noss and Csuti 1997; Summerville and Crist 2001). Also, species with low dispersal ability are threatened by fragmentation as a result of poor genetic diversity within small populations (Noss and Csuti 1997; Schultz and Crone 2001). Pollinators able to cover larger distances are again less threatened not only because of their genetic diversity but also due to the possibility of a food-resource accumulation from various areas of one or more fragments (Rathcke and Jules 1993). Furthermore, pollinator species with long lifespans may be particularly vulnerable because their feeding period usually exceeds the flowering period of any host plant, and thus such species require a wide range of flowering plant species over time (Cane 2001). In some cases, however, fragmented habitats get colonized by new, less sensitive pollinator species (Aizen and Feinsinger 1994; Dick 2001) that can be even favored over native species in a disturbed landscape (Ghazoul 2005).

# REFERENCES

Abrol DP (2015) Pollination biology Vol.1 Pests and pollinators of fruit crops. Springer International Publishing, Switzerland, pp 1–452.

Adnan M, Shah Z, Arif M, Khan MJ, Mian IA, Sharif M, Alam M, Basir A, Ullah H, Rahman I, Saleem N (2016) Impact of rhizobial inoculum and inorganic fertilizers on nutrients (NPK) availability and uptake in wheat crop. Can J Soil Sci 96:169–176.

Ågren J (1996) Population size pollinator limitation and seed set in the self-incompatible herb Lythrum salicaria. Ecology 77:1779–1790.

Ahmad W, Hassan G, Ali M, Khan N, Ishaq M, Afridi K, Shah IA, Adnan M, Tajudin R, Ullah L, Rehman AU, Hussain B (2016) Estimates of heritability, genetic advance and correlation in F3 populations of Wheat. Pure Appl Biol 5(4):1142–1150.

Aizen MA, Feinsinger P (1994) Habitat fragmentation, native insect pollinators and feral honey bees in Argentine Chaco Serrano. Ecol Appl 4:378–392.

Arafat Y, Shafi M, Khan MA, Adnan M, Basir A, Rahman IU, Arshad M, Khan A, Saleem N, Romman M, Rahman Z, Shah JA (2016) Yield response of Wheat cultivars to zinc application rates and methods. Pure Appl Biol 5(4):1260–1270.

Armbruster P, Reed DH (2005) Inbreeding depression in benign and stressful environments. Heredity 95(3):235–242.

Arroyo J, Dafni A (1993) Interspecific pollen transfer among co-occuring heteromorphic and homomorphic species. Isr J Bot 41:225–232.

Arroyo MTK (1973) Chiasma frequency evidence on the evolution of autogamy in Limnanthes floccosa (Limnanthaceae). Evolution 27:679–688.

Ashman TL, Knight TM, Steets JA, et al. (2004) Pollen limitation of plant reproduction: ecological and evolutionary causes and consequences. Ecology 85(9):2408–2421.

Aspinwall N, Cristian T (1992) Pollination biology seed production and pollination structure in Queen-of-the prairy Filipendula rubra (Rosaceae) at Botkin Fen Missouri. Am J Bot 79(5):488–494.

Baker HG (1967) The evolution of weedy taxa in the Eupatorium microstemon species aggregate. Taxonomy 16:293–300

Barrett SCH (1992) Heterostylous genetic polymorphisms: model systems for evolutionary analysis. Evol Funct Heterostyly 1:29.

Barrett SCH (2002) The evolution of plant sexual diversity. Nat Rev Genet 3:274–284.

Baz A, Garcia-Boyero A (1995) The effects of forest fragmentation on butterfly communities in central Spain. J Biogeogr 22:129–140.

Benton TG, Bryant DM, Cole L, et al (2002) British Ecological Society2Blackwell Science Ltd Linking agricultural practice to insect and bird populations: a historical study over three decades. J Appl Ecol 39:673–687.

Bierzychudek P (1981) Pollinator limitation of plant reproductive effort. Am Nat:838–840.

Bohan DA, Boffey CW, Brooks DR, et al (2005) Effects on weed and invertebrate abundance and diversity of herbicide management in genetically modified herbicide-tolerant winter-sown oilseed rape. Proc Royal Soc B 272:463–474.

Bond WJ (1994) Do mutualisms matter? Assessing the impact of pollinator and disperser disruption on plant extinction. Philos Trans R Soc London Ser B 344:83–90.

Bosch M, Waser NM (1999) Effects of local density on pollination and reproduction in Delphinium nuttallianum and Aconitum columbianum (Ranunculaceae). Am J Bot 86:871–879.

Brian AD (1951) The pollen collected by bumblebees. J Anim Ecol 20:191–194.

Brian AD (1957) Differences in the flowers visited by four species of bumblebees and their causes. J Anim Ecol 26:71–96.

Briggs D, Walters SM (2016) Plant variation and evolution. Cambridge University Press.

Burd M (1994) Bateman's principle and plant reproduction: the role of pollen limitation in fruit and seed set. Bot Rev 60(1) 83–139.

Burquez A, Corbet SA (1991) "Do flowers reabsorb nectar?" Funct Ecol:369–379.

Campbell CS, Famous NC, Zuck MG (1986) Pollination bilogy of Primula laurentiana (Primulaceae) in Maine. Rhodora 88:253–260.

Campbell DR (1985) Poliinator sharing and seed set of Stellaria pubera: competition for pollination. Ecology 66(2):544–553.

Campbell DR (2000) Experimental tests of sex-allocation theory in plants. Trends Ecol Evol 15(6):227–232.

Cane JH (2001) Habitat fragmentation and native bees: a premature verdict? Conserv Ecol 5(3):3.

Cane JH, Tepedino VJ (2001) Causes and extent of declines among native North American invertebrate pollinators: detection evidence and consequences. Conserv Ecol 5(1):1

Carlson JE (2007) Male-biased nectar production in a protandrous herb matches predictions of sexual selection theory in plants. Am J Bot 94(4):674–682.

Chapman RE, Wang J, Bourke AFG (2003) Genetic analysis of spatial foraging patterns and resource sharing in bumble bee pollinators. Mol Ecol 12(10):2801–2808.

Charlesworth D (2006) Evolution of plant breeding systems. Curr Biol 16(17):726–735.

Charlesworth D, Charlesworth B (1981) Allocation of resources to male and female functions in hermaphrodites. Biol J Linn Soc 15:57–74.

Charlesworth D, Vekemans X, Castric V, et al (2005) Plant self-incompatibility systems: a molecular evolutionary perspective. New Phytol 168(1):61–69.

Charlesworth D, Willis JH (2009) The genetics of inbreeding depression. Nat Rev Genet 10(11):783–796.

Charnov EL (1979) Simultaneous hermaphroditism and sexual selection. Proc Nat Acad Sci USA 76:2480–2484.

Charnov EL (1982) The theory of sex allocation (Vol. 355). Princeton University Press. Princeton, NJ.

Chittka L, Thomson JD, Waser NM (1999) Flower constancy insect psychology and plant evolution. Naturwissenschaften 86(8):361–377.

Corbet SA (1978) Bee visits and the nectar of Echium vulgare L and Sinapis alba L. Ecol Entomol 31:25–37.

Corbet SA, Unwin DM, Prŷs-Jones OE (1979) Humidity nectar and insect visits to flowers with special reference to Crataegus Tilia and Echium. Ecol Entomol 41:9–22.

Cresswell JE, Osborne JL, Bell SA (2002) A model of pollinator mediated gene flow between plant populations with numerical solutions for bumblebees pollinating oilseed rape. Oikos 98(3):375–384.

Cruden RW (1973) Reproductive biology of weedy and cultivated Mirabilis (Nyctaginaceae). Am J Bot 60:802–809.

Cruden RW (1976) Pollen-ovule rations: a conservative indicator of breeding systems in flowering plants. Evolution 31:32–46.

Dafni A (1992) Pollination ecology. IRL Press. 270 pp.

Dafni A, O'Toole C (1994) Pollination syndromes in the Mediterranean: generalizations and peculiarities. In: Arianoutsou M, Groves R (eds) Plant-animal interactions in Mediterranean-type ecosystems. [Tasks Veg. Sci., 34.]-Dordrecht, Boston and London, pp 125–135.

Daily GC (1997) Nature's services Societal dependence on natural ecosystems. Island Press, Washington DC, 392 pp.

Darvill B, Knight ME, Goulson D (2004) Use of genetic markers to quantify bumblebee foraging range and nest density. Oikos 107(3):471–478.

De Nettancourt D (1977) Incompatibility in angiosperms. Springer-Verlag, Berlin.

De Nettancourt D (1984) Incompatibility. In: Linskens HF, Heslop-Harrison J (eds) Encyclopedia of plant physiology; New Series Vol 17 Cellular Interactions. Springer-Verlag, Berlin, pp 624–639.

Debinski DM, Holt RD (2000) A survey and overview of habitat fragmentation experiments. Conserv Biol 14:342–355.

Dick CW (2001) Genetic rescue of remnant tropical trees by an alien pollinator. Proc Royal Soc B 268:2391–2396.

Duan YW, Zhang TF, He YP, et al (2009) Insect and wind pollination of an alpine biennial Aconitum gymnandrum (*Ranunculaceae*). Plant Biol Stuttgart 11(6):796–802.

Duffy KJ, Johnson SD (2011) Effects of pollen reward removal on fecundity in a self-incompatible hermaphrodite plant. Plant Biol 13(3):556–560.

Eckert CG, Schaefer A (1998) Does self-pollination provide reproductive assurance in Aquilegia canadensis (Ranunculaceae). Am J Bot 85(7):919–924.

Ehrlich PR, Raven PH (1969) Differentiation of populations. Science 165(3899):1228–1232.

Ellstrand NC (1992) Gene flow by pollen: implications for plant conservation genetics. Oikos:77–86.

Ellstrand NC, Prentice HC and Hancock JF (1999) Gene flow and introgression from domesticated plants into their wild relatives. Ann Rev Ecol Systemat 539–563.

Elmqvist T (2000) Pollinator extinction in the Pacific Islands. Conserv Biol 14:1237–1239.

Eradley C, Roth D, Clarke J, et al (2006) Pollinators and pollination: a resource book for policy and practice. Published by the African Pollinator Initiative (API), pp 1–77.

Etcheverry AV, Alemán MM, Figueroa Fleming T, et al (2012) Pollen: ovule ratio and its relationship with other floral traits in Papilionoideae (Leguminosae): an evaluation with Argentine species. Plant Biol 14(1):171–178.

Faegri K, van der Pijl L (1971) The principles of pollination ecology, 2nd edn. Pergamon Press, New York, 291p.

FAO (2006) Protocol to Detect and Monitor Pollinator Communities Guidance for Practitioners.

Feinsinger P, Linhart YB, Busby WH et al (1987) Disturbance, pollinator predictability, and pollination success among Costa Rican cloud forest plants. Ecology 68:1294–1305.

Fernández JD, Bosch J, Nieto-Ariza B, et al (2012) Pollen limitation in a narrow endemic plant: geographical variation and driving factors. Oecologia 170(2):421–431.

Free JB (1970) The flower constancy of bumblebees. J Anim Ecol 39:395–402.

Freitas BM, Paxton RJ (1998) A comparison of two pollinators: the introduced honey bee *Apis mellifera* and an indigenous bee *Centris tarsata* on cashew *Anacardium occidentale* in its native range of NE Brazil. J Appl Ecol 35(1):109–121.

Gallai N, Salles JM, Settele J, et al (2009) Economic valuation of the vulnerability of world agriculture confronted with pollinator decline. Ecol Econ 68:810–821.

Gallardo R, Dominguetz E, Munoz J (1994) Pollen-ovule ratio pollen size and breeding system in Astragalus (Fabaceae) subgenus Epiglottis: a pollen and seed allocation approach. Am J Bot 81(12):1611–1619.

Gangoiti MV, Cortizo AM, Arnol V, et al (2008) Opposing effects of bisphosphonates and advanced glycation end-products on osteoblastic cells. Eur J Pharmacol 600(1):140–147.

Garibaldi LA, Dewenter R, Winfree R, et al (2013) Wild pollinators enhance fruit set of crops regardless of honey bee abundance. Science 339:1608–1611.

Gegear RJ, Laverty TM (2005) Flower constancy in bumblebees a test of the trait variability hypothesis. Anim Behav 69(4):939–949.

Ghazoul J (2005) Buzziness as usual? Questioning the global pollination crisis. Trends Ecol Evol 20:367–373.

Gibbs PE, Milne C, Carrillo M (1975) Correlation between the breeding system and recombination index in five species of Senecio. New Phytol 75:619–626.

Glover BJ (2007) Understanding flowers and flowering: an integrated approach. Oxford University Press, New York, 452 pp.

Goulson D (1999) Foraging strategies of insects for gathering nectar and pollen, and implications for plant ecology and evolution. Perspectives in Plant. Ecol Evol Systemat 2(2):185–209.

Goulson D, Hawson SA, Stout JC (1998) Foraging bumblebees avoid flowers already visited by conspecifics or by other bumblebee species. Anim Behav 55(1):199–206.

Goverde M, Schweizer K, Baur B, et al (2002) Small-scale habitat fragmentation effects on pollinator behaviour: experimental evidence from the bumblebee Bombus veteranus on calcareous grasslands. Biol Conserv 104(3):293–299.

Grant V (1995) Sexual selection in plants: pros and cons. Proc Nat Acad Sci 92(5):1247–1250.

Grant V, Grant K (1965) Flower pollination in the Phlox family. Columbia University Press, 180 pp.

Greenleaf S, Kremen C (2006) Wild bees enhance honey bees pollination of hybrid sunflower. Proc Natl Acad Sci USA, 103(37):13890–13895.

Handel SN (1983) Pollination ecology plant population structure and gene flow in in the evolution of the Angiosperms in "Pollination biology" ed Leslie Real. Academic Press, pp 163–202.

Harris LF, Johnson SD (2004) The consequences of habitat fragmentation for plant–pollinator mutualisms. Int J Trop Insect Sci 24:29–43.

Harris PJ, Weinhandi JA, Clarke AE (1989) Effect on in Vitro pollen growth of an isolated style glycoprotein associated with self-incompatibility in Nicotiana alata. Plant Physiol 89:360–367.

Hedrick PW, Kalinowski ST (2000) Inbreeding depression in conservation biology. Ann Rev Ecol Systemat 31(1):139–162.

Heinrich B (1975) Bee flowers: a hypothesis on flower variety and blooming times. Evolution 29:325–334

Heinrich B (1976) The foraging specialization of individual bumblebees. Ecol Monogr 46:105–128.

Heinrich B (1979a) "Majoring" and "minoring" by foraging bumblebees Bombus vagus: an experimental analysis. Ecology 60:245–255.

Heinrich B (1979b) Bumblebee economics. Harvard University Press, Harvard, 245 pp.

Herlihy CR, Eckert CG (2002) Genetic cost of reproductive assurance in a self-fertilizing plant. Nature 416 (6878):320–323.

Herrera CM (1987) Components of pollinator "quality": comparative analysis of a diverse insect assemblage. Oikos 50:79–90.

Hill LM, Brody AK, Tedesco CL (2008) Mating strategies and pollen limitation in a globally threatened perennial Polemonium vanbruntiae. Acta Oecologica 33(3):314–323.

Hoopingarner RA, Waller GD (1992) Crop pollination. In: Graham J (ed) The hive and the honey bee. Dadant and Sons, Hamilton, IL, pp 1043–1082 (Revised Edition).

Hörandl E, Paun O (2007) Patterns and sources of genetic diversity in apomictic plants: implications for evolutionary potentials. Apomixis: Evolution, mechanisms and perspectives. ARG Gantner Verlag KG, Lichtenstein, pp 169–194.

Husband BC, Schemske DW (1995) Magnitude and timing of inbreeding depression in a diploid population of Epilobium angustifolium (Onagraceae). Heredity 75(2):206–215.

Jakobsson A, Lázaro A, Totland Ø (2009) Relationships between the floral neighborhood and individual pollen limitation in two self-incompatible herbs. Oecologia 160(4):707–719.

James RR, Pitts-Singer TL (2008) Bee pollination in agricultural ecosystems. Oxford University Press, New York, 232 pp.

Johnson SD, Steiner KE (2000) Generalization versus specialization in plant pollination systems. Trends Ecol Evol 15:140–143.

Jones EL, Leather SR (2012) Invertebrates in urban areas: a review. Eur J Entomol 109:463–478.

Juenger T, Bergelson J (1997) Pollen and resource limitation of compensation to herbivory in scarlet gilia Ipomopsis aggregate. Ecology 78(6):1684–1695.

Kay QON (1984) Differences between sexes in floral morphology nectar production and insect visits in dioecious species Silene dioica. New Phytol 98:515–529.

Kevan PG, Baker HG (1983) Insects as flower visitors and pollinators. Ann Rev Entomol 28:407–453.

Khan A, Gul F, Jamal Y, Ali J, Adnan M, Khan MA, Naeem I, Bacha Z (2017) Weeds management in maize hybrids with night plowing. Pak J Weed Sci Res 22(2):179–191.

Klein AM (2009) Nearby rain forest promotes coffee pollination by increasing spatio-temporal stability in bee species richness. For Ecol Manage 258:1838–1845.

Klein AM, Vaissière BE, Cane JH, Steffan-Dewenter I, et al (2007) Importance of Pollinators in Changing Landscapes for World Crops. Proc Biol Sci / Royal Soc 274(1608):303–313.

Knight ME, Martin AP, Bishop S, et al (2005) An interspecific comparison of foraging range and nest density of four bumblebee (Bombus) species. Mol Ecol 14(6):1811–1820.

Knight TM, Steets JA, Vamosi JC, et al (2005) Pollen limitation of plant reproduction: pattern and process. Ann Rev Ecol Evol Systemat:467–497.

Körner C (2003) Alpine plant life: functional plant ecology of high mountain ecosystems. Springer-Verlag.

Kozuharova E, Richards AJ, Hale M, et al (2015) Another brick in the wall of the Oxytropis Campestris complex with an emphasis of three members of this group from Pirin Mts, The Balkans. Fabaceae: Classification Nutrient Composition and Health Benefits.

Kremen C, Miles A (2012) Ecosystem services in biologically diversified versus conventional farming systems: benefits externalities and trade-offs. Ecol Soc 17:40.

Kremen C, Ricketts T (2000) Global perspectives on pollination disruptions. Conserv Biol 14:1226–1228.

Kudo G, Maeda T, Narita K (2001) Variation in floral sex allocation and reproductive success within inflorescences of Corydalis ambigua (Fumariaceae): pollination efficiency or resource limitation? J Ecol 89(1):48–56.

Ladley JJ, Kelly D (1995) Explosive New Zealand mistletoe. Nature 378:766.

LeBuhn G, Droege S, Connor EF, et al (2012) Detecting insect pollinator declines on regional and global scales. Conserv Biol: J Soc Conserv Biol 27(1):113–120.

Lepais O, Darvill B, O'connor S, et al (2010) Estimation of bumblebee queen dispersal distances using sibship reconstruction method. Mol Ecol 19(4):819–831.

Levin DA (1978) Pollinator behavior and breeding structure of plant populations. In: Richards AJ (ed) The pollination of flowers by insects, pp 135–150.

Levin DA (1979) Pollinator foraging behavior: genetic implications for plants. In: Solbrig O Jain S Johnson G Raven P (eds) Topics of plant population biology, Columbia University Press, pp 131–153.

Levin DA, Kerster HW (1974) Gene flow in seed plants. In: Dobzhansky T, et al (eds) Evolutionary biology. © Plenum Press, New York.

Levin DA, Kerster HW, Niedzlek M (1971) Competition for pollinators between simultaneously flowering species. Am Nat 104(939):455–467.

Lloyd DG (1965) Evolution of self-compatibility and racial differentiation in Leavenworthia (Cruciferae). Contr Gray Herb 195:3–134.

Lloyd DG (1983) Evolutionarily stable sex ratios and sex allocations. J Theor Biol 105:525–539.

Lloyd DG (1987) Allocasions to pollen seeds and pollination mechanisms in self-fertilizing plants. Funct Ecol 1:83–89.

Lloyd DG, Webb CJ (1992) The evolution of heterostyly. In: Barrett SCH (ed) Evolution and function of heterostyly. Springer-Verlag, Berlin, pp 151–178.

Macior LW (1994) Pollen foraging Dynamics of subalpine bumblebees (Bombus Latr.) Plant Species Biol 9:99–106.

Mawdsley NA, Compton SG, Whittaker RJ (1998) Population persistence pollination mechanisms and figs in fragmented tropical landscapes. Conserv Biol 12:1416–1420.

Medrano M, Guitian P, Guitián J (2000) Patterns of fruit and seed set within inflorescences of Pancratium maritimum (Amaryllidaceae): nonuniform pollination, resource limitation, or architectural effects. Am J Bot 87(4) 493–501.

Michalski SG, Durka W (2010) Pollen and ovule production in wind-pollinated species with special reference to Juncus. Plant Systemat Evol 286(3–4):191–197.

Michener CD (2007) The world of the bees, 2nd edn. Johns Hopkins University Press, Baltimore, pp. 1–953.

Mione T, Anderson GJ (1992) Pollen- ovule rations and breeding system evolution in Solanum section Basarthrum (Solanaceae). Am J Bot 79:279–287.

Montalvo AM, Ackerman JD (1987) Limitations to fruit production in Ionopsis utricularioides (Orchidaceae). Biotropica:24–31.

Morales CL, Traveset A (2009) A meta-analysis of impacts of alien vs. Native plants on pollinator visitation and reproductive success of co-flowering native plants. Ecol Lett 12:716–728.

Nabhan GP, Buchmann SL (1997) Services provided by pollinators. In: Daily GC (ed) Nature's services: societal dependence on natural ecosystems. Island Press, Washington DC, pp 133–150.

Navarro-Pérez ML, López J, Rodríguez-Riaño T, et al (2017) Confirmed mixed bird–insect pollination system of Scrophularia trifoliata L a Tyrrhenian species with corolla spots. Plant Biol. Online Version of Record published before inclusion in an issue. https://doi.org/10.1111/plb.12548.

Nieto A (2014) European red list of bees. Publication Office of the European Union, Luxembourg, 98 pp. ISBN: 978-92-79-44512-5.

Noss RF, Csuti B (1997) Habitat fragmentation. Principles Conserv Biol 2:269–304.

Ollerton J (2017) Pollinator diversity: distribution ecological function and conservation. Ann Rev Ecol Evol Systemat 48:353–376.

Ollerton J, Price V, Armbruster WS, et al (2012) Overplaying the role of honey bees as pollinators: a comment on Aebi and Neumann 2011. Trends Ecol Evol 27:141–42.

Ollerton J, Stott A, Allnutt E, et al (2007) Pollination niche overlap between a parasitic plant and its host. Oecologia 151(3) 473–485.

Ollerton J, Winfree R, Tarrant S (2011) How many flowering plants are pollinated by animals? Oikos 120:321–326.

Osborne JL, Clark SJ, Morris RJ, et al (1999) A landscape-scale study of bumble bee foraging range and constancy using harmonic radar. J Appl Ecol 36(4):519–533.

Osborne JL, Williams IH, Carreck NL, et al (1996) June Harmonic radar: a new technique for investigating bumblebee and honey bee foraging flight. In: VII International Symposium on Pollination 43, pp 159–164.

Osman, M.A.M., Shebl, M.A. Vulnerability of Crop Pollination Ecosystem Services to Climate Change. In Climate Change Impacts on Agriculture and Food Security in Egypt; Ewis Omran, E.S., Negm, A., Eds.; Springer: Cham, Switzerland, 2020.

Pacini E, Nepi M (2007) Nectar production and presentation. In: Nicolson SW, Nepi M, Pacini E (eds) Nectaries and nectar. Springer, Netherlands, pp 167–214.

Pasquet RS, Peltier A, Hufford MB, et al (2008) Long-distance pollen flow assessment through evaluation of pollinator foraging range suggests transgene escape distances. Proc Nat Acad Sci 105(36):13456–13461.

Paton AJ, Brummitt N, Govaerts R, et al (2008) Towards target 1 of the global strategy for plant conservation: a working list of all known plant species—progress and prospects. Taxon 57:602–611.

Petanidou T, den Nijs JCM, Ostermeijer JGB, Ellis-Adam AC (1995) Pollination ecology and patch dependent reproductive success of the rare peraennial Gentiana pneumonanthe L. New Phytol 129:155–163.

Philbrick CT, Anderson GJ (1987) Implications of pollen- ovule rations and pollen size for the reproductive biology of Potamogeton and autogamy in aquatic angiosperms. Systemat Bot 12:98–105.

Pleasants IM (1980) Competition for bumblebee pollinators in Rocky mountain communities. Ecology 61(6):1446–1459.

Plitmann U, Levin DA (1989) Breeding systems in the Polemoniaceae. Plant Systemat Evol 170:205–214.

Porcher E, Lande R 2005. The evolution of self-fertilization and inbreeding depression under pollen discounting and pollen limitation. J Evol Biol 18(3):497–508.

Procter M, Yeo P, Lack A (1996) The natural history of pollination, 2nd edn. Harper and Collins Publications, London.
Pyke GH (1978a) Optimal foraging in bumblebees and coevolution with their plants. Oecologia 36:281–293.
Pyke GH (1978b) Optimal foraging: movement patterns of bumblebees between inflorescences. Theor Popul Biol 13:72–98.
Pyke GH, Pulliam HR, Charnov EL (1977) Optimal foraging: a selective review of theory and test. Q Rev Biol 52:137–154.
Queller DC (1984) Pollen- ovule rations and hermafrodite sexual allocation strategies. Evolution 38:1148–1151.
Raine NE, Chittka L (2007) Flower constancy and memory dynamics in bumblebees (Hymenoptera: Apidae: *Bombus*). Entomologia Generalis 29(2/4):179.
Ranta E, Teras I, Lungberg H (1981) Phenological spread in flowering of bumblebee pollinated plants. Acta Botanica Fennica 18:229–236.
Rathcke BJ, Jules ES (1993) Habitat fragmentation and plant-pollinator interactions. Curr Sci 65:273–277.
Reich PB, Hobbie SE, Lee TD (2014) Plant growth enhancement by elevated $CO_2$ eliminated by joint water and nitrogen limitation. Nat Geosci 7(12):920–924.
Richards AJ (1982) The influece of minor structural changes in the flower of breeding system and speciation in Epipactis Zinn. (Orchidaceae). In: Armstrong JA, Powell JM, Richards AJ (eds) Pollination and evolution. Royal Botanic Gardens, Sydney, pp 9–28.
Richards AJ (1990) Plant breeding systems. Unwin Human, London, University Press, Cambridge, 529pp.
Richards AJ (1997) Plant breeding systems, 2nd edn. Garland Science.
Richards AJ (1996) Breeding systems in flowering plants and the control of variability. Folia geobotanica et phytotaxonomica 31:283–293.
Richards AJ (2001) Does low biodiversity resulting from modern agricultural practice affect crop pollination and yield. Ann Bot 88(2):165–172.
Ritland C, Ritland K (1989) Variation of sex allocation among eight taxa of the Mimulus guttatus species complex (Scrophulariaceae). Am J Bot 76:1731–1739.
Robertson AW, Kelly D, Ladley JJ, et al (1999) Effects of pollinator loss on endemic New Zealand mistletoes (Loranthaceae). Conserv Biol 13:499–508.
Rodriguez-Rodriguez MC, Jordano P, Valido A (2013) Quantity and quality components of effectiveness in insular pollinator assemblages. Oecologia 173:179–190.
Saunders DA, Hobbs RJ, Margules CR (1991) Biological consequences of ecosystem fragmentation: a review. Conserv Biol 5:18–32.
Shebl M, Kamel S, Mahfouz H (2013) Bee fauna (Apoidea: Hymenoptera) of the Suez Canal Region Egypt. J Apic Sci 57(1):33–44.
Shebl MA, Patiny S, Michez D (2015) Supplementary note on the solitary bee fauna from the Suez Canal region of Egypt (Hymenoptera: Apoidea). J Melittology 47:1–5.
Schoen DJ (1977) Morphological phenological and pollen-distribution evidence of autogamy and xenogamy in Gilia achilleifolia (Polemoniaceae). Systemat Bot 2:280–286.
Schultz CB, Crone EE (2001) Edge-mediated dispersal behaviour in a prérie butterfly. Ecology 82:1879–1892.
Sih A, Baltus M (1987) Patch size pollinator behavior and pollinator limitation in catnip (*Nepeta cataria*). Ecology 68(6):1679–1690.
Smith TJ, Saunders ME (2016) Honey bees: the queens of mass media despite minority rule among insect pollinators insect. Conserv Diver 9:384–390.
Southwick EE, Loper GM, Sadwick SE (1981) Nectar production composition energetics and pollinator attractivness in spring flowers of western New York. Am J Bot 68:994–1002.
Stebbins GL (1974) Flowering plants: evolution above species level. Edward Arnold Ltd, London, 399p.
Steffan-Dewenter I (2003) Importance of habitat area and landscape kontext for species richness of bees and wasps in fragmented orchard meadows. Conserv Biol 17:1036–1044.
Stout JC, Goulson D (2002) The influence of nectar secretion rates on the responses of bumblebees (Bombus spp.) to previously visited flowers. Behav Ecol Sociobiol 52(3):239–246.
Stout JC, Goulson D, Allen JA (1998) Repellent scent-marking of flowers by a guild of foraging bumblebees (Bombus spp.) Behav Ecol Sociobiol 43(4–5):317–326.
Summerville KS, Crist TO (2001) Effects of experimental habitat fragmentation on patch use by butterflies and skippers (Lepidoptera). Ecology 82:1360–1370.
Terziiski D, Yurukova-Grancharova P (2011) Phenomena apomixis in plants Ecoprogress Sofia.

Tootill E (1983) The Penguin Dictionary of Botany. Penguin Books, London, 326 pp.
Totland O (1993) Pollination in alpine Norway: flowering phenology insect visitors and visitation rates in two plant communities. Can J Bot 71:1072–1079.
Turner IM (1996) Species loss in fragments of tropical rain forest: a review of the evidence. J Appl Ecol 33:200–209.
Vanbergen AJ (2013) Threats to an ecosystem service: pressures on pollinators. Front Ecol Environ 11:251–259.
Waddington KD (1981) Factors influencing pollen flow in bumblebee-pollinated Delphinium virescens. Oikos:153–159.
Waddington KD (1983) Pollen flow and optimal outcrossing distance. Am Nat:147–151.
Waddington KD, Allen T, Heinrich B (1981) Floral preferences of bumblebees (Bombus edwardsii) in relation to intermittent versus continuous rewards. Anim Behav 29:779–784.
Wang YQ, Zhang DX, Chen ZY (2004) Pollen histochemistry and pollen: ovule ratios in Zingiberaceae. Ann Bot 94(4):583–591.
Waser NM (1983) The adaptive nature of floral traits: ideas and evidence. Pollination Biol:241–285.
Waser NM (1998) Pollination angiosperm speciation and the nature of species boundaries. Oikos 81:198–201.
Wedderburn F, Richards AJ (2006) Variation in within morph incompatibility inhibition sites in heteromorphic Primula L. New Phytol 116(1):149–162.
Weller SG (1994) The relationship of rarity to plant reproductive biology In: Bowles ML, Whelan CJ (eds) Restoration of endangered species. Cambridge University Press, pp 90–170.
Wesselingh RA (2007) Pollen limitation meets resource allocation: towards a comprehensive methodology. New Phytol 174(1):26–37.
Wilcock C, Neiland R (2002) Pollination failure in plants: why it happens and when it matters. Trends Plant Sci 7:270–277.
Williams NM, Cariveaub D, Winfreeb R, et al (2011) Bees in disturbed habitats use but do not prefer alien plants. Basic Appl Ecol 12:332–341.
Williams NM, Kremen C (2007) Resource distributions among habitats determine solitary bee off spring production in a mosaic landscape. Ecol Appl 17:910–921.
Willis JH (1993) Partial self-fertilization and inbreeding depression in two populations of Mimulus guttatus. Heredity 71(2) 145–154.
Willson MF, Price PW (1980) Resource limitation of fruit and seed production in some Asclepias species. Can J Bot 5820:2229–2233.
Wright LI, Tregenza T, Hosken DJ (2008) Inbreeding in breeding depression and extinction. Conserv Genet 9(4):833.
Zimmerman M (1983) Plant reproduction and optimal foraging experimental nectar manipulations in Delphinium nelsonii. Oikos 57–63.

# Index